Dahlem Workshop Reports
Life Sciences Research Report 44
Productivity of the Ocean: Present and Past

The goal of this Dahlem Workshop:
to explore the sensitivity of ocean
productivity to global environmental
changes over different time scales

Life Sciences Research Reports

Series Editor: Silke Bernhard

Held and published on behalf of the

Stifterverband für die Deutsche Wissenschaft

Sponsored by:

Senat der Stadt Berlin
Wilhelm Woort-Stiftung im
Stifterverband für die Deutsche Wissenschaft

Productivity of the Ocean: Present and Past

W.H. Berger, V.S. Smetacek and G. Wefer, Editors

Report of the Dahlem Workshop on
Productivity of the Ocean: Present and Past
Berlin 1988, April 24–29

Rapporteurs:
K.W. Bruland, W.B. Curry, T.D. Herbert,
P.A. Jumars, B. von Bodungen, P.J. leB. Williams

Program Advisory Committee:
W.H. Berger, V.S. Smetacek, and G. Wefer, Chairpersons
L.A. Codispoti, M. Sarnthein, E. Suess,
H.R. Thierstein, J.R. Toggweiler

A Wiley–Interscience Publication

John Wiley & Sons 1989
Chichester · New York · Brisbane · Toronto · Singapore

Copy Editors: J. Lupp, K. Klotzle

Photographs: E.P. Thonke

With 4 photographs, 87 figures, 19 tables, 4 color plates

Library of Congress Cataloging-in-Publication Data

Dahlem Workshop on Productivity of the Ocean : Present and Past (1988 : Berlin, Germany)
Productivity of the ocean : present and past : report of the Dahlem Workshop on Productivity of the Ocean, Present and Past, Berlin 1988, April 24–29 / W.H. Berger, V.S. Smetacek, and G. Wefer, editors ; rapporteurs, K.W. Bruland . . . [et al.].
p. cm.—(Dahlem workshop reports) (Life sciences research report ; 44)
"Held and published on behalf of the Stifterverband für die Deutsche Wissenschaft; sponsored by Senat der Stadt Berlin, Wilhelm Woort-Stiftung im Stifterverband für Deutsche Wissenschaft"—P. facing t.p.
"A Wiley-Interscience publication."
Includes bibliographies and indexes.
ISBN 0 471 92246 3
1. Marine productivity—Congresses. I. Berger, Wolfgang H. II. Smetacek, V. S. III. Wefer, G. (Gerold) IV. Stifterverband für Deutsche Wissenschaft. V. Berlin (Germany : West). Senat. VI. Wilhelm Woort-Stiftung. VII. Title. VIII. Series. IX. Series: Life sciences research report ; 44.
QH1.8. M34D34 1988
574.5'2636—dc19

89–5339
CIP

British Library Cataloguing in Publication Data

Dahlem Workshop on Productivity of the Ocean ; Present and Past. (1988 : Berlin Germany)
Productivity of the Ocean.—(Dahlem workshop reports. Life sciences research report ; 44)
1. Oceans. Bed. Micropalaeontology
I. Title II. Berger, W. H. (Wolfgang Helmut), 1937–
III. Smetacek, V. S. IV. Wefer, G. V. Series
560.92

ISBN 0 471 92246 3

Typeset by Photo·graphics, Honiton, Devon
Printed and bound in Great Britain
by The Bath Press Ltd, Bath, Avon

This Dahlem Workshop Report is dedicated to the late A.P. Crary, visionary and dedicated Chief Scientist at the U.S. National Science Foundation (NSF), who encouraged and supported the early continuous deep core drilling efforts, and B. Lyle Hansen, inventive and industrious drilling expert who, while at USA CRREL, succeeded in accomplishing the first core drilling through a polar ice sheet at Camp Century, Greenland, in 1966.

Table of Contents

The Dahlem Konferenzen

Founders

Recognizing the need for more effective communication between scientists, the Stifterverband für die Deutsche Wissenschaft*, in cooperation with the Deutsche Forschungsgemeinschaft**, founded Dahlem Konferenzen in 1974. The project is financed by the founders and the Senate of the City of Berlin.

Name

Dahlem Konferenzen was named after the district of Berlin called *Dahlem*, which has a long-standing tradition and reputation in the sciences and arts.

Aim

The task of Dahlem Konferenzen is to promote international, interdisciplinary exchange of scientific information and ideas, to stimulate international cooperation in research, and to develop and test new models conducive to more effective communication between scientists.

The Concept

The increasing orientation towards interdisciplinary approaches in scientific research demands that specialists in one field understand the needs and problems of related fields. Therefore, Dahlem Konferenzen has organized workshops, mainly in the Life Sciences and the fields of Physical, Chemical and Earth Sciences, of an interdisciplinary nature.

Dahlem Workshops provide a unique opportunity for posing the right questions to colleagues from different disciplines who are encouraged to state what they do not know rather than what they do know. The aim is

---

* *The Donors Association for the Promotion of Sciences and Humanities, a foundation created in 1921 in Berlin and supported by German trade and industry to fund basic research in the sciences.*

** *German Science Foundation.*

not to solve problems or to reach a consensus of opinion, the aim is to define and discuss priorities for further research.

Topics

The topics are of contemporary international interest, timely, interdisciplinary in nature, and problem oriented. Dahlem Konferenzen approaches internationally recognized scientists to suggest topics fulfilling these criteria. Once a year, the topic suggestions are submitted to a scientific board for approval.

Program Advisory Committee

A special Program Advisory Committee is formed for each workshop. It is composed of 6–7 scientists representing the various scientific disciplines involved. They meet approximately one year before the workshop to decide on the scientific program and define the workshop goal, select topics for the discussion groups, formulate titles for background papers, select participants, and assign them their specific tasks. Participants are invited according to international scientific reputation alone. Exception is made for younger German scientists. Invitations are not transferable.

Dahlem Workshop Model

Since no type of scientific meeting proved effective enough, Dahlem Konferenzen had to create its own concept. This concept has been tested and varied over the years. It is internationally recognized as the *Dahlem Workshop Model*. Four workshops per year are organized according to this model. It provides the framework for the utmost possible interdisciplinary communication and cooperation between scientists in a period of 4½ days.

At Dahlem Workshops 48 participants work in four interdisciplinary discussion groups. Lectures are not given. Instead, selected participants write background papers providing a review of the field rather than a report on individual work. These papers, reviewed by selected participants, serve as the basis for discussion and are circulated to all participants before the meeting with the request to formulate written questions and comments to them. During the workshop, each of the four groups prepares reports reflecting their insights gained through the discussion. They also provide suggestions for future research needs.

Publication

The group reports written during the workshop together with the revised background papers are published in book form as the Dahlem Workshop

Reports. They are edited by the editor(s) and the Dahlem Konferenzen staff. The reports are multidisciplinary surveys by the most internationally distinguished scientists and are based on discussions of advanced new concepts, techniques, and models. Each report also reviews areas of priority interest and indicates directions for future research on a given topic.

The Dahlem Workshop Reports are published in two series:

1) Life Sciences Research Reports (LS), and
2) Physical, Chemical, and Earth Sciences Research Reports (PC).

Director

Silke Bernhard, Dr med., Dr phil. h.c.

Address

Dahlem Konferenzen
Tiergartenstr. 24–27
D-1000 Berlin (West) 30

Tel.: (030) 262 50 41

THE DAHLEM WORKSHOP MODEL

MONDAY	TUESDAY	WEDNESDAY	THURSDAY	FRIDAY
A. Opening (P) B. Introduction (P) C. Selection of Problems for the Group Agendas (S) 1 2 3 4	1 2 3 4	1 3	F. Report Session 1 2 3 4	G. Distribution of the Reports H. Reading Time I. Discussion of the Group Reports (P)
D. Presentation of Group Agendas (P) E. Group Discussions (S) 1 2		2 4		J. Groups Meet to Revise their Reports (S) 1 2 3 4

KEY: (P) = Plenary Session;
(S) = Simultaneous Sessions;
○ = one discussion group

Explanation of the Dahlem Workshop Model

A. Opening
Background information is given about Dahlem Konferenzen and the Dahlem Workshop Model.

B. Introduction
The goal and the scientific aspects of the workshop are explained.

C. Selection of Problems for the Group Agenda
Each participant is requested to define priority problems of his choice to be discussed within the framework of the workshop goal and his discussion group topic. Each group discusses these suggestions and compiles an agenda of these problems for their discussions.

D. Presentation of the Group Agenda
The agenda for each group is presented by the moderator. A plenary discussion follows to finalize these agendas.

E. Group Discussions
Two groups start their discussions simultaneously. Participants not assigned to either of these two groups attend discussions on topics of their choice.
The groups then change roles as indicated on the chart.

F. Report Session
The rapporteurs discuss the contents of their reports with their group members and write their reports, which are then typed and duplicated.

G. Distribution of Group Reports
The four group reports are distributed to all participants.

H. Reading Time
Participants read these group reports and formulate written questions/ comments.

I. Discussion of Group Reports
Each rapporteur summarizes the highlights, controversies, and open problems of his group. A plenary discussion follows.

J. Groups Meet to Revise their Reports
The groups meet to decide which of the comments and issues raised during the plenary discussion should be included in the final report.

Preface

In the last few years it has become apparent that the content of carbon dioxide in the atmosphere has changed considerably on time scales of thousands of years. The mechanisms producing these changes must largely be sought in the ocean, and especially in the fluctuating productivity of the ocean.

The dynamics of ocean productivity and its role in the marine carbon cycle are poorly understood despite (or perhaps because of) a flood of new results emerging over the last few years. Many of these results come from new approaches, such as satellite imaging (see Plates 1 to 3), time series sediment trapping, *in situ* seafloor experiments, high-resolution seismic profiling (Plate 4), and recovery of undisturbed sediments through special coring devices, including hydraulic piston coring in deep ocean drilling (Plate 4).

This workshop aimed to bring together marine biologists, geochemists, and geologists to assess the state of knowledge and ignorance concerning the processes which lead to the export of organic matter from the photic zone, its transit to the seafloor, and its burial within the sedimentary record. Our goal was to clarify how productivity changes through time and how it can be reconstructed from the biogenous signals in the sediments. Such reconstruction is necessary if we are to model the CO_2 fluctuations on medium to long time scales.

The central themes which emerged during the workshop were (*a*) continuous versus episodic production and associated export from the photic zone, (*b*) carbon flux from transfer of particulate carbon versus transfer of dissolved organic matter, (*c*) role of seafloor processes in remobilization of carbon and nutrients, and (*d*) long-term reconstruction of productivity and its relationships to the fluctuations of carbon dioxide through geologic time. These topics are discussed in the background papers. In addition, four group reports summarize the results of discussions and outline areas of consensus and controversy.

The Editors
Bremen, July 1988

Productivity of the Ocean: Present and Past
eds. W.H. Berger, V.S. Smetacek and G. Wefer, pp. 1–34
John Wiley & Sons Limited

Ocean Productivity and Paleoproductivity—An Overview

W.H. Berger[1], V.S. Smetacek[2], and G. Wefer[3]

[1] *Scripps Institution of Oceanography*
University of California, San Diego
La Jolla, CA 92093, U.S.A.

[2] *Alfred-Wegener-Institut für Polar- und Meeresforschung*
2850 Bremerhaven, F.R. Germany

[3] *Geowissenschaften*
Universität Bremen
2800 Bremen 33, F.R. Germany

Abstract. Ocean productivity helps control the partitioning of carbon between the large ocean reservoir and the relatively small atmospheric reservoir. Productivity fluctuations, therefore, are important in providing feedback to climatic changes. From this viewpoint, a central problem is to understand the workings of the "biological pump," which pulls carbon out of surface waters and sequesters it at depth and within sediments. The efficiency of the pump depends on the leakage of carbon out of the pelagic foodweb and the associated downward transport of organic matter, both in particulate (POC) and in dissolved (DOC) form. These processes are strongly influenced by seasonality and by episodic events. Bacterial accretion of DOC on POC below the photic zone in productive regions may be important in linking the two pumps.

Favorable sites for burial of organic carbon are on the upper continental slope, for several reasons: (*a*) the coastal setting results in high, pulsating productivity, (*b*) the shelf supplies additional carbon, and (*c*) settling and decomposition in the water column are of short duration. The Pleistocene record in sediments of the continental slope shows large fluctuations in the burial rates of organic carbon, which are interpreted as productivity fluctuations. Such fluctuations appear to be in phase with changes in atmospheric carbon dioxide content.

In ancient warmwater oceans the rules governing productivity variations may have been entirely different from those applicable today and in the late Tertiary. Thus, empirical correlations between productivity and proxy signals become increasingly suspect. Instead, the basic principles of productivity control in entirely different geochemical and geographic settings have to be identified and understood, to allow reconstruction of paleoproductivity.

THE PROBLEM OF CHANGING OCEAN PRODUCTIVITY

Perhaps the most exciting discovery of this decade, regarding the global carbon cycle, was the demonstration that the carbon dioxide content of the glacial-age atmosphere was about one-third lower than typical Holocene values (200 ppm vs. 300 ppm). The evidence is contained in the air trapped within polar ice (Berner et al. 1980; Delmas et al. 1980; Barnola et al. 1987). The amplitude of the CO_2 changes, accompanying the glacial–interglacial variations in (deuterium-derived) temperature, and the rapidity of change make it extremely likely that the chief factors responsible are to be found in the carbon chemistry of the ocean. The ocean holds about 60 times more carbon than the atmosphere; a small change in its chemistry can produce a marked change in the atmospheric reservoir with which it communicates. Evidence for this link between atmospheric CO_2 and ocean carbon chemistry can be found, for example, in the $\delta^{13}C$ record of deep-sea foraminifera (see Fig. 1).

There are a number of ways in which the ocean can change its ability to hold CO_2. The most important one, in all likelihood, is a change in its productivity (Broecker 1982; Broecker and Peng 1986). Photosynthesis, and the export of organic carbon from the sunlit zone as particulate organic carbon (POC) and as dissolved organic carbon (DOC), acts as a "biological pump" which lowers the partial pressure of CO_2 in surface waters. The atmospheric CO_2 content orients itself mainly on the pCO_2 of these (carbon-impoverished) warm surface waters of the global ocean. The high CO_2 values (in excess of 600 ppm in places; see Broecker and Peng 1982) which build up below the photic zone in many parts of the ocean can only be seen by the atmosphere in a few restricted areas of upwelling and deep mixing.

THE BASIC TASK

The need to understand the intriguing interrelationships between productivity, ocean chemistry, atmospheric carbon dioxide, and climate has taken on a new urgency as we enter an era of climatic warming from industrial CO_2 (Jones et al. 1987). Thus, these matters have now captured the interest and imagination of biological, chemical, and geological oceanographers, resulting in a rapidly expanding effort on the subject (see, e.g., Bruland et al. 1984; Sundquist and Broecker 1985; and recent releases of the U.S. Global Ocean Flux Study (GOFS) Planning Office, Woods Hole Oceanographic Institution).

A basic task, common to all approaches, is to reconstruct in some detail the workings of the biological pump and to discover how its efficiency

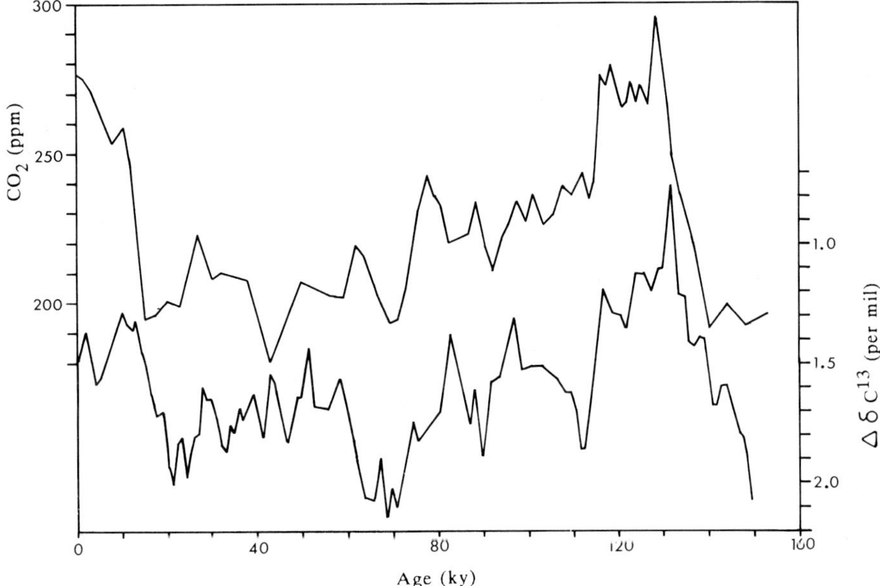

Fig. 1—Carbon dioxide concentrations in the Vostok ice core from Antarctica (Barnola et al. 1987), compared with a productivity-related carbon isotope signal from the eastern tropical Pacific (difference between the $\delta^{13}C$ values of planktonic and benthic foraminifera; Shackleton et al. 1983), show that ocean productivity and atmospheric CO_2 tend to vary together. Time scale of Barnola et al. is adjusted here to the one of Shackleton et al. by correlation of the deuterium signal in the ice with the oxygen isotope signal in the sediment.

changes through time, on various time scales. Unfortunately, how the pump works is not understood even on a quite elementary level. Instead, biologists argue among themselves about the most fundamental issues: what nutrients control primary production, where in the water column does photosynthesis take place, what is the ratio between internal cycling and throughput production in the productive zone, what is the relative importance of the major types of primary producers, and so on. Lately, the role of the dissolved organic matter (DOM) has become a burning question.

Bypassing the debate, geochemists put traps below the euphotic zone, catch the output, and describe it in terms of empirical equations, such as $J(z)_{,} = f\ (PP,\ z)$ (Suess 1980; Betzer et al. 1984). While the depth, z, is easily determined, doubts remain as to trapping efficiency and hence the accuracy of the flux, J(z). These doubts have been laid aside for the time being. For a value for primary production, PP, of course the biologists must still be consulted. Ecology cannot be avoided after all. This has become ever more apparent as the short-term variability of J(z), seen in multi-sample traps, has emerged as a major problem.

We believe that the integration of information from longer time scales has much to offer regarding the attempt to understand the biological pump. The long-range view emphasizes fundamentals; it is difficult to extrapolate into the past using empirical equations derived from present patterns alone. By focusing on the task of understanding changes in productivity on many different time scales, we shall greatly enhance our insight into the ocean carbon cycle and ultimately into how the ocean controls atmospheric CO_2, and hence interacts with climate.

In this hopeful spirit, we planned the Dahlem workshop on ocean productivity for the purpose of cross-disciplinary education. The idea was to get the biologists to tell the geoscientists what is going on within the ocean, and the geoscientists to tell the biologists which of these processes are important on the long time scales.

FRAMEWORK OF THE WORKSHOP

The concepts and information which are necessary to read the sedimentary record of the productivity of past oceans: this was the basic agenda of the conference.

One outstanding problem which arose during discussions was this: can the record be regarded as the remainder of export from the euphotic factory, diminished by road tolls on the way down and by import duties at the seafloor? Or should we adopt an entirely different analogy: that of trash left over from (seasonal) Oktoberfest events and from (sporadic) rock festivals? If the latter analogy is better, what chance is there to reconstruct the typical or average productivity (as seen on global maps) from the sediment which largely contains the memories of happenings?

In the overview which follows we have attempted to recognize the conflict in conceptualization—steady factory output versus trash from spasmodic happenings—without trying to resolve it. The factory analogy guided our planning while the event analogy emerged as a unifying theme of the conference (see Legendre and Le Fevre, Peinert et al., and Wefer, all this volume).

The topics of the working groups reflect the flow of organic carbon in the pelagic environment (cf. Romankevich 1984), covering primary production to export production, to transit flux through bathyal depths and to the seafloor, where the carbon rain feeds the benthos and where part of it is incorporated into sediments, there providing the memory which we need to tap in order to find the rules for secular change (Fig. 2). Each of the invitees was asked to address an important issue within this structure and focus on the controversies. The group reports, in contrast, give a summary of the

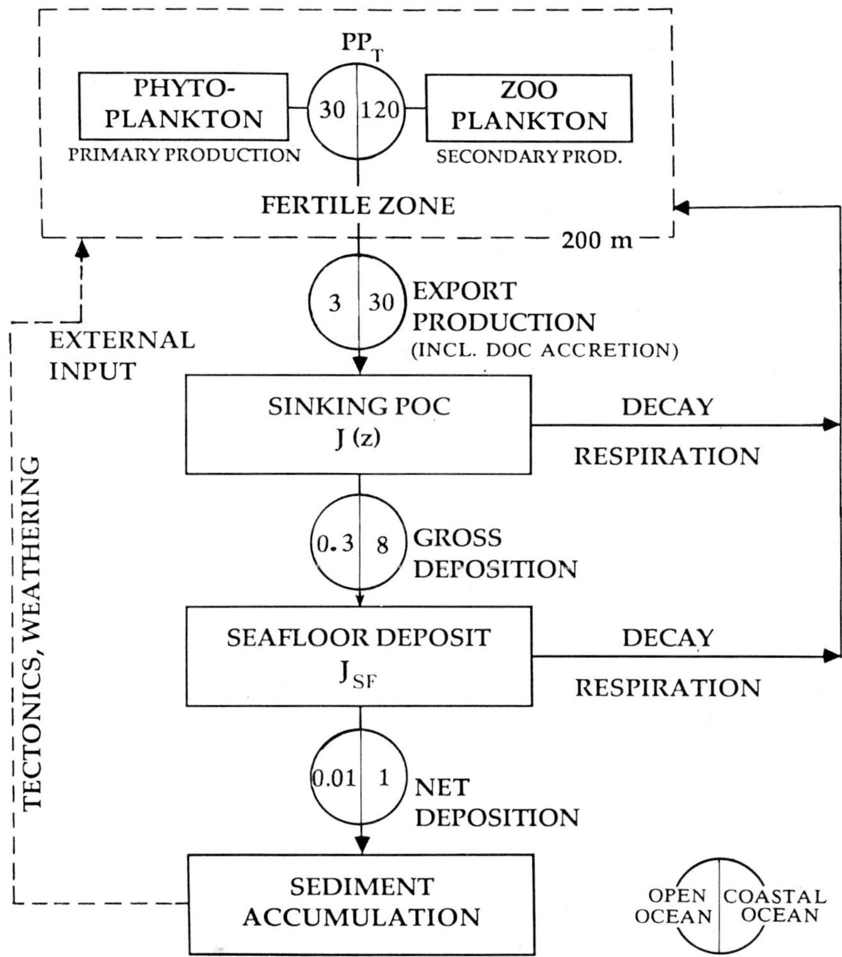

Fig. 2—Sketch of transfer of particulate organic carbon in the ocean, from primary production to burial in the sediment, showing the main elements of the "biological pump." Numbers are fluxes in g C $m^{-2}y^{-1}$; to the left in each circle a typical open ocean value; to the right a typical coastal ocean value.

discussions, showing where consensus prevails and where disagreement persists.

The flow chart for organic carbon (Fig. 2) represents two pathways: one for pelagic conditions (PP = 30 g C m^{-2} y^{-1}) and one for coastal oceans (PP = 120 g C m^{-2} y^{-1}). A factor of four in productivity somehow translates into a factor of 100 in burial. This blue ocean/green ocean dichotomy is fundamental to the workings of the pump. Of course, emphasis on the burial factor is primarily a geological consideration, while the biologists may well be visualizing differences in pelagic and benthic community structure instead. The dichotomy is pervasive, and episodicity probably plays a large role in it.

PRIMARY PRODUCTION

Primary production is, in principle, the rate of photosynthetic fixation of carbon by chlorophyll-containing organisms. "Gross" production refers to the total carbon fixed while "net" production is what is left after the carbon used for respiration has been subtracted. The broad patterns of primary production are reflected in many biologically mediated properties of the surface ocean: chlorophyll content (= color), phosphate content (= nutrients), and abundance of planktonic organisms (= standing stock). That these properties are highly correlated with each other has been appreciated for some time (see Fig. 3).

The overall pattern corresponds closely to satellite color images (Plate 1), as well as to maps of productivity (Steemann Nielsen and Aabye Jensen 1957; Koblents-Mishke et al. 1970; Berger, this volume). This latter correspondence may well reflect the fact that Steemann Nielsen used these earlier maps as guidance in contouring productivity, while calibrating the contours with the Galathea radiocarbon measurements.

The most commonly used operational definition for primary production is based on the amount of radioactive CO_2 taken up by growing organisms in seawater, in a given time, under suitable conditions including the correct level of irradiation (Steemann Nielsen and Aabye Jensen 1957). When doing the experiments, one assumes that uptake is photosynthetic, that none of the cells were eaten during the experiment, and that natural conditions are correctly simulated. Although these assumptions are open to question, the radiocarbon index, which may be called Steemann Nielsen productivity, is demonstrably useful as a mappable index of photosynthetic activity and is highly correlated with the other important properties of the productive system. We should expect good correspondence of this index with new production since its pattern closely follows the distribution and availability of nutrients, wherever light is not limiting.

The Steemann Nielsen productivity, then, is our preferred definition of primary production. It represents the level of activity of photosynthetic organisms in a column of water (as well as some of the associated heterotrophic activity) and is given as the amount of carbon fixed per square meter per unit time. Typical values for the open ocean range from 25 g C m^{-2} y^{-1} in the very centers of subtropical gyres to 250 g C $m^{-2}y^{-1}$ in coastal ocean regions with strong mixing processes—a range of a factor of ten (Fig. 4).

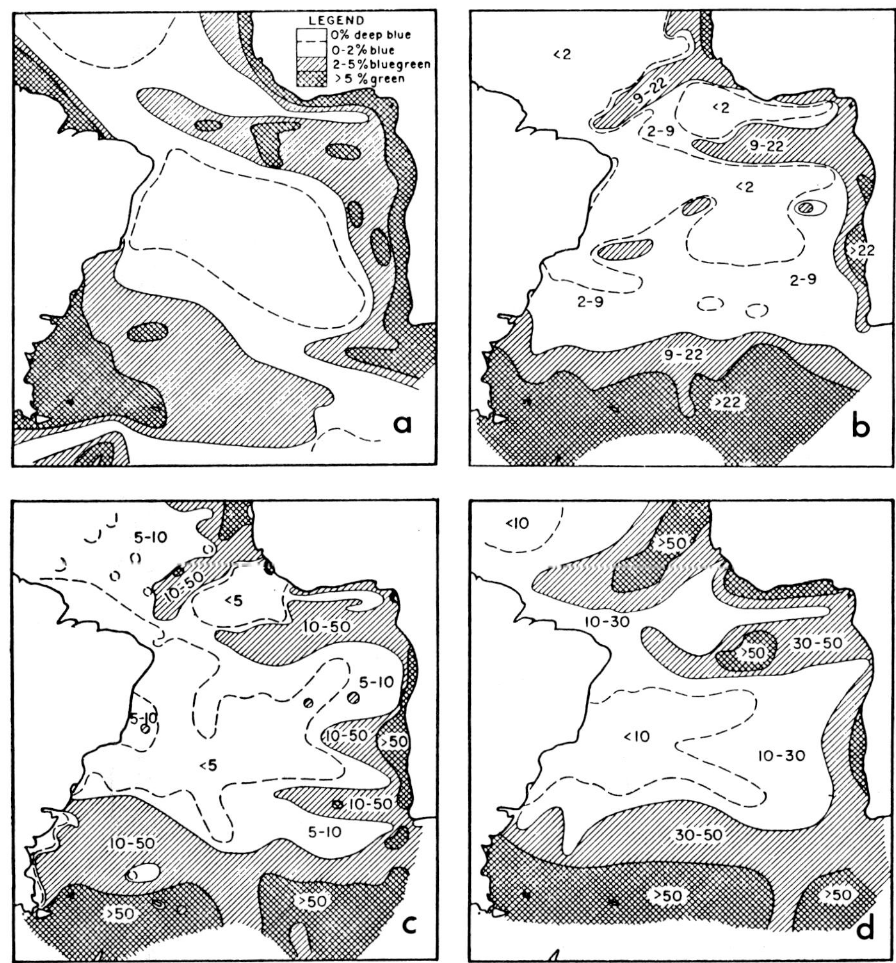

Fig. 3—Surface ocean properties reflecting productivity: (*a*) ocean color; (*b*) phosphate concentration in μg/l, upper 50 m; (*c*) abundance of planktonic organisms in thousands per liter, upper 50 m; (*d*) abundance of zooplankton in no./4 liter, upper 50 m. From Steemann Nielsen and Aabye Jensen (1957), summarized from earlier workers: (*a*) Schott, (*c*) Hentschel, (*d*) Hentschel and Wattenberg.

The total annual primary production of the world ocean, by the definition here adopted, is approximately 30 gigatons of carbon (Platt and Subba Rao 1975; Berger et al. 1987). One gigaton is 10^{15} grams. Other definitions, of course, yield different totals (see Sundquist, in Sundquist and Broecker 1985). One quarter of the total production is delivered by slightly less than 10% of the ocean area (Fig. 4), one half of the production by about 30%.

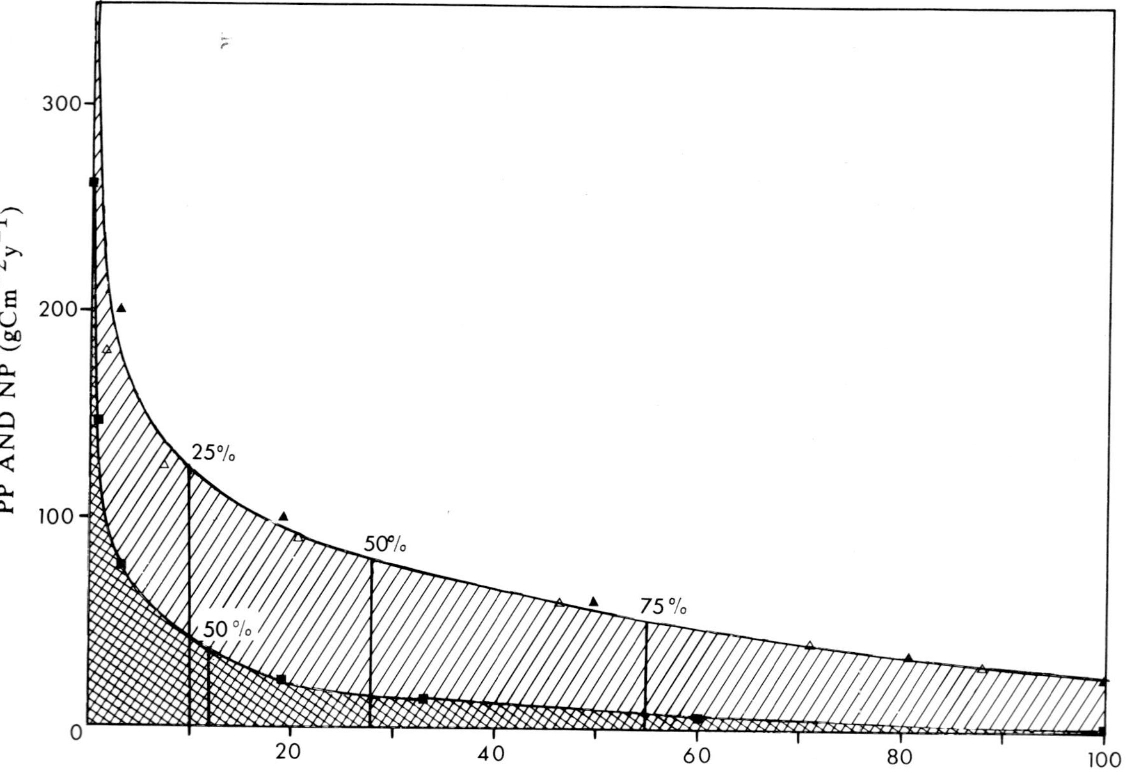

Fig. 4—Histogram of primary production levels in the world ocean based on the maps of Berger et al. (1987) (see also Berger, this volume). Points are taken from their table (filled symbols, compilation map; open symbols, synthetic map). Fraction of new production is calculated from $f=PP/400-PP^2/340{,}000$ (equation fitted to the data of Eppley and Peterson (1979); see text). Vertical lines: 25% sections for primary production and 50% sections for new production.

NEW PRODUCTION

The need to understand the processes shunting organic matter out of the pelagic food web (traditionally seen as a chain from algae to fish) led to the introduction and development of the concepts of "new" and "regenerated" production (Dugdale and Goering 1967; Eppley and Peterson 1979; see Eppley, this volume). New production is that portion of photosynthesis which depends on inorganic nitrate supply from below the euphotic zone, while regenerated production depends on the recycling of nutrients within that zone. The new production is the one of interest in the present context: it delivers the export to deeper waters and, ultimately, the supply to the seafloor and the sediment. Regenerated production is a function of the time scale considered, since regeneration rates differ greatly between the organisms involved, largely as a function of size (Peinert et al., this volume).

Naturally, the proportion of new production of primary production (NP/PP, or "f-ratio") will be lowest in the central gyre where the supply of new nutrients is minimal, and highest in the coastal ocean where mixing with deeper waters is vigorous. Eppley and Peterson (1979) propose values of 6%, 13%, 18%, and 30% for productivities of 26, 51, 73, and 124 g C m^{-2} y^{-1}, going from oligotrophic to inshore waters. From these values it is readily seen that the expected f-ratio may be expressed as

$$f = PP / 410, \tag{1}$$

where PP stands for primary production. Thus, for example, for a primary production of 100 g C m^{-2} y^{-1} one would expect a new production value of just under 25%. This implies that the new production NP is given as

$$NP = f \times PP = PP^2 / 410, \tag{2}$$

that is, NP is a function of the square of the productivity.

Equation 1 is valid for productivities up to, say, 150 g C m^{-2} y^{-1}. At higher productivities it must overestimate the f-ratio, if the limit of 50% set by Eppley and Peterson is accepted. To take account of such a limit, we suggest an equation of the form

$$f = PP/a - PP^2/b. \tag{3}$$

For a = 400 and b = 340 000, this equation provides an excellent fit to the data given by Eppley and Peterson, between productivity values from 0 to 500 g C m^{-2} y^{-1}. We have used it to estimate the portion of new production in the primary production histogram for the world ocean (Fig. 4). It is seen that under these assumptions 50% of all new production occurs in 12% of the ocean area (that is, mainly in coastal waters) and that the total for the ocean is approximately 6 Gt C per year. A multi-box model study by Postma (1971) gave the same value, and subsequent studies also are in general agreement.

EXPORT PRODUCTION

Just as the productivity of a forest might be defined by the amount of wood it produces, rather than by the amount of leaf litter, for example, so the ocean productivity can be gauged by the amount of fallout it generates from the photic zone: that is, by the amount of organic matter which will accumulate in a trap set below this zone. For traps immediately below the photic zone, on the whole, the export production is expected to be the same as the new production, as required for steady state. However, in any given particular situation, steady state cannot prevail entirely because of patchiness of productivity in space and time, and horizontal advection.

An inequality between new production and export production should occur especially in upwelling areas where much of the import and export of nutrients and organisms takes place across horizontal gradients, rather than vertically only. Other inequalities may arise from bacterial aufwuchs on settling particles, deriving sustenance from dissolved organic matter below the photic zone.

For deeper traps the export will quickly decrease. In the upper 1000 meters (where the export concept is useful) the amount trapped is a function of 1/depth, as becomes evident when plotting the amount of organic matter trapped against the logarithm of depth (Fig. 5). A value for export production, therefore, must be accompanied by the depth for which it applies. For this reason alone it is necessary to keep the two concepts "new production" and "export production" well separated.

Suess (1980) and (on the basis of the same data compilation) Betzer et al. (1984) have proposed equations to relate PP to the export as follows:

$$J(z) = PP \ / \ (0.0238 \times z + 0.212) \tag{4}$$

(Suess) which reduces to

$$J(z) = 40 \times PP \ / \ z \tag{5}$$

for depths below a few hundred meters, and

$$J(z) = 0.409 \times PP^{1.41} \ / \ z^{0.628} \tag{6}$$

(Betzer et al.) which asserts that there is a nonlinear relationship between PP and export, in the same sense as for Equation 2, except less powerful. For Equation 6, decay with depth is slowed compared to Equation 4.

On the average, the ratio of export production to the overlying primary production is 20% at 100 meters and 10% at 200 meters, based on a compilation of data up to and including 1985 (Berger et al. 1987). This proportion can vary greatly, however, depending on the circumstances. It is strongly suspected that low productivities do indeed yield a smaller proportion of export than do high ones, as proposed by Betzer et al. (1984), and that highly variable productivity yields more than productivity which is

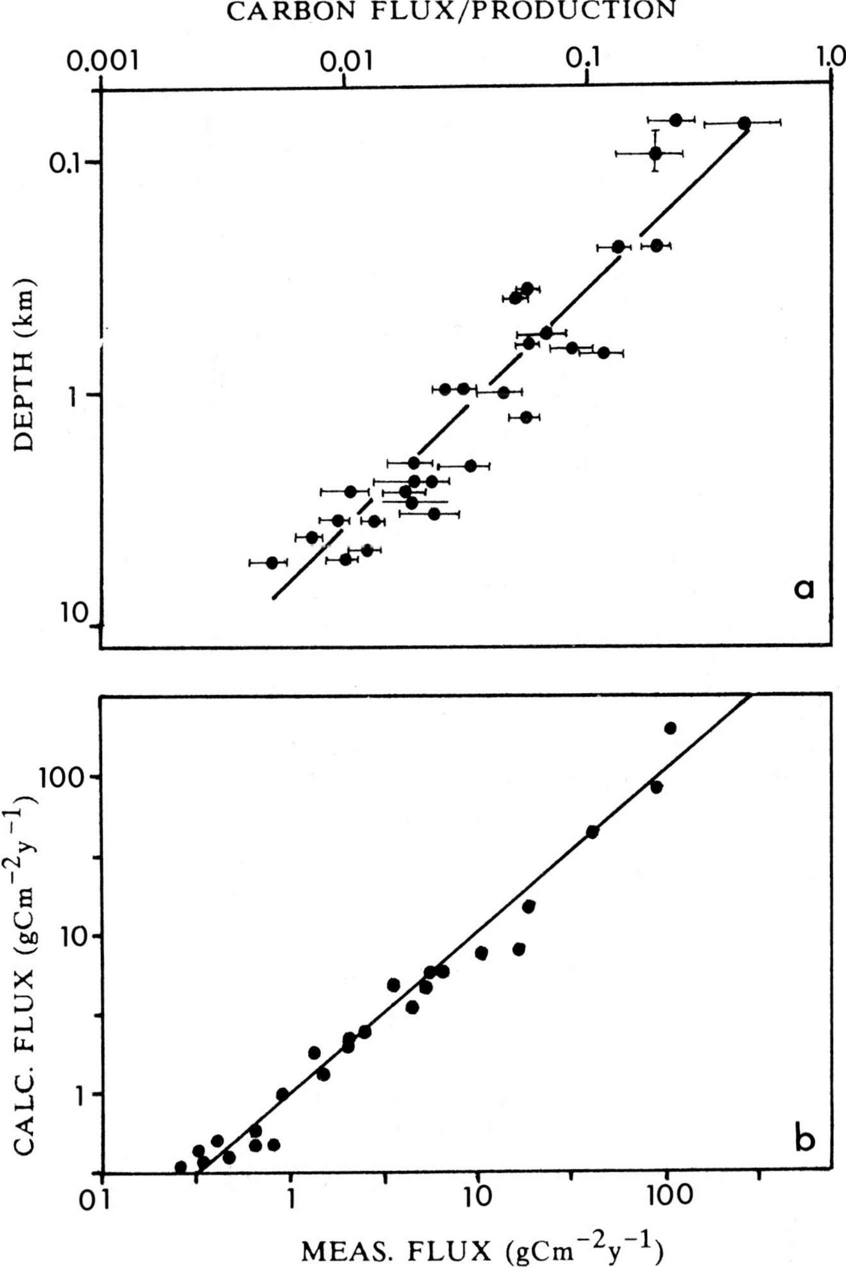

Fig. 5—Fluxes of particulate organic carbon, normalized to mean annual productivity: (*a*) according to Suess (1980); (*b*) according to Betzer et al. (1984).

less variable. Thus, the factor of six observed as a common range for primary production in the global ocean could easily be increased to a factor of between 12 and 36 for export production, between central gyres and upwelling areas.

The total annual export production in the world ocean at 100 meters depth is approximately 6 gigatons of carbon, at 200 meters depth it is 3 gigatons, and at 500 meters depth 1.2 gigatons (Berger et al. 1987). This estimate is probably on the conservative side, relying largely on open ocean results. Of the global export flux, a minimum of 20% pertains to coastal ocean areas (shelf seas and boundary zone), assuming the same proportional output relative to primary production. However, the coastal ocean export from the photic zone is near 50% of the total, if the power functions of Eppley and Peterson (1979) and of Betzer et al. (1984) apply.

EXPORT AND EPISODICITY

The (assumed) nonlinear relationship between productivity and fallout, as mentioned, derives from basic principles: the dependency of new production on nutrient supply from below and the episodic nature of nutrient injection into the photic zone. This is true for the open ocean with seasonally varying productivity (Platt and Harrison 1985), and especially in the coastal ocean setting. The coastal ocean injection scenario is backed by observations on diatom blooms (see, e.g., Summerhayes, in Thiede and Suess 1983). Impressive large-scale evidence is now available from satellite-derived snapshots of surface temperature and chlorophyll distributions, showing the familiar filaments and eddies off California and the U.S. East Coast (see, e.g., Woods Hole Oceanographic Institution 1987).

Why should the carbon export from variable systems be so much higher than from stable ones? The answer, most likely, lies within the workings of the food chain, that is, in the efficiency of transfer from one trophic level to the next (Fig. 6). Whenever autotrophs and heterotrophs are in dynamic balance this transfer is efficient. Recycling within the water column will dominate: the organic matter produced by photosynthesis is largely eaten by the abundant heterotrophs waiting to be fed, within and just below the euphotic zone. Much of this recycling is done within a community of very small organisms, producing particles which settle extremely slowly, if at all, and which are readily decomposed by bacteria. A "food-ring" results with relatively little leakage, that is, regenerated production dominates.

Variability, on the other hand, involves a change in conditions and hence a lack of fully co-adapted autotrophs, heterotrophs, and decomposers. Instead, sporadic nutrient injection (partly seasonal) stimulates rapid growth of autotrophs which will bloom, not being grazed sufficiently to prevent exponential increase. We are mainly dealing here with "new" production.

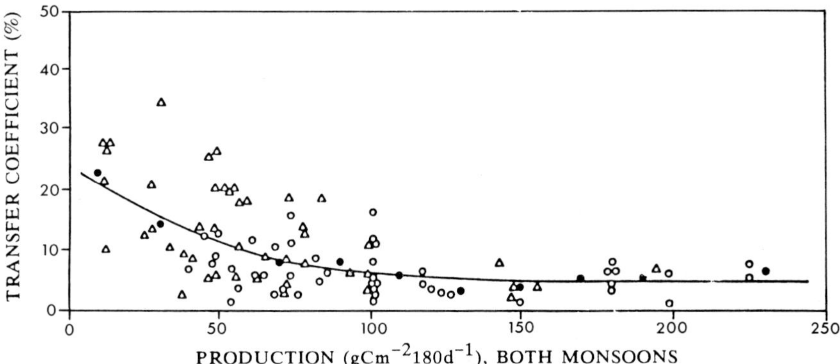

Fig. 6—Transfer coefficient from primary to secondary production as a function of primary production. Open circles, S.W. monsoon; triangles, N.E. monsoon; filled circles, average for both monsoons. Redrawn from Cushing (1973).

Aggregate formation follows blooming—apparently an effort to provide seeds for the next generation, at least in diatoms (Smetacek 1985). Much of the resulting output will settle quite readily, for example, as masses of coccolithophores or diatoms. In coastal regions, fecal matter output is important: loaded with silt, shells, or both the pellets sink out before being broken up and decomposed (Dunbar and Berger 1981).

In sum, export production implies new production and new production implies nutrient injection, which is a sporadic process largely determined by wind stress, either directly or indirectly. Wind stress, of course, is subject to seasonal fluctuations, interannual variations, and long-term climatic changes. As a result of episodicity favoring export, opportunistic species will dominate the memory bank on the seafloor. The notable exception to this pattern will be fossilizable heterotrophs which do without the oligotrophic food ring: the protists with symbiotic algae. The most abundant pelagic foraminifera belong here, as well as certain radiolaria.

THE DOM PROBLEM

Traditionally, the downward flux of organic matter is seen in terms of a rain of particles toward the seafloor, which is attenuated by breakup and oxidation. This concept has recently been challenged vigorously, raising the possibility of substantial downward flux of DOM (see Toggweiler, this volume). The proposed downward flux is based on DOM measurements of Sugimura and Suzuki (1988), which suggest that the amount of DOC has been severely underestimated in the past. These authors put forward the striking proposition that the downward flux of DOM accounts for much of the apparent oxygen utilization (AOU) at depth.

If the results of Sugimura and Suzuki should prove to be generally applicable, the export from the photic zone and its role in providing the transit flux through the water column has to be seen in a new light. Since oxygen minima are observed to form below the more productive regions, the DOC and POC dynamics must be closely coupled. While some DOC, presumably, would move through the ocean, as do other dissolved substances by advection and diffusion, a large portion must be tied to the formation, settling, and destruction of particles. Thus, downward DOC transfer should in fact be simulated by bacterial aufwuchs on particles settling out of the photic zone, with subsequent resolubilization. DOM destruction below the photic zone can be achieved by bacterial respiration on very small suspended particles and (to a lesser extent) by bacterial growth on settling POC. The bacterial aufwuchs—a by-pack on the POC rain—would then become part of the organic carbon flux seen below productive regions and would normally be counted as export from the photic zone.

The importance of bacteria in coupling the POM and DOM cycles is apparent from the recent results of Karl et al. (1988) and Cho and Azam (1988), among others (see also Angel; Bishop; and Bruland et al., all this volume).

TRANSIT FLUX

Regardless of whether it contains a substantial component of bacterial by-pack, once settling organic matter has passed through the upper 1000 meters of the water column it presumably has been stripped of the more nutritious components. Thus, it undergoes much less change during the rest of the voyage toward the seafloor. Exactly what happens in midwater and at bathyal depths is something of a mystery because life processes there are not well constrained by quantitative concepts and measurements. The abundance of deep-living organisms varies geographically, however, and one may reasonably assume that this is because the supply of food varies with the overlying productivity.

The equations of Suess (1980) and of Betzer et al. (1984) make no distinction between export and transit flux. The distinction between transit flux and export production is useful, however, not only to emphasize the contrast between the rapid decay within the export zone and the much slower decay in the transit zone, but also because of the potential problem of bacterial aufwuchs within the export zone accreting additional organic matter to settling particles.

If bacterial aufwuchs is to be considered, an equation describing the transit flux might usefully consist of three terms: fast decay, accretion, and slow decay. These three terms would have the same general form (consisting of a scaling factor, a productivity function, and a depth function) because all three depend on the concentration of particles at depth:

$$J(z) = k_1 \times PP^a/z^b + k_2 \times PP^c/z^d + k_3 \times PP^g/z^h, \quad (7)$$

where J(z) is the mass flux in transit at depth z, as before. The exponents on PP express the nonlinearity of the relationship between export and primary production mentioned above, and those on z set the power of particle breakup and decay with depth. The existing data do not allow a distinction between sinks and sources. Instead, they can be described very well by combining the first two terms and simplifying the equation as follows:

$$J(z) = 17 \times PP/z + PP/100. \quad (8)$$

It says that the flux at 100 meters is 18% of primary production most of which decays rapidly according to the 1/z rule, with about 6% of the export being resistant to decay and persisting during transit to the ocean floor (Eq. J4 of Berger et al. 1987).

A similar fit is achieved by

$$J(z) = 9 \times PP/z + 0.7 \times PP/z^{0.5} \quad (9)$$

in which case some decay is allowed in the resistant fraction. The equation of Betzer et al. (1984) (Equation 6, above) subsumes all processes into the first term, combining different decay rates into some kind of average. Their equation does, however, emphasize the nonlinearity of the effect of primary production on the export.

The treatment exemplified by these calculations implies a more or less steady rain of particles disintegrating on their way to the seafloor. However, there is no steady rain, and there is no regular disintegration with depth.

The background drizzle of particles is greatly modified by episodic events (Honjo 1982 and many others; see also Bishop and Wefer, both this volume). In the Sargasso Sea south of Bermuda, for example, continuous trapping over an eight year period showed marked seasonal and interannual variations for the total flux at 3200 m, which ranged from 16 to 120 mg $m^{-2} y^{-1}$ (Deuser 1987; see Fig. 7a). In high latitudes seasonality of flux can produce extreme fluctuations. For instance, time series trap experiments in the Bransfield Strait (Antarctica) show that particle flux is restricted to a surprisingly short period during the austral summer, and that total flux at a water depth of 1588 m varied between 1900 and 1 mg $m^{-2}y^{-1}$ (Fig. 7b). Clearly, the application of flux equations under such conditions will be difficult: over which time-span is the flux to be integrated, and which value of productivity is to be taken as appropriate for the phase-shifted flux at depth?

Very little is known about disintegration at depth. The consensus is that the bulk of transit flux is in large, fast-sinking vehicles that are generated from the suspended pool of small particles by aggregation or by filtering

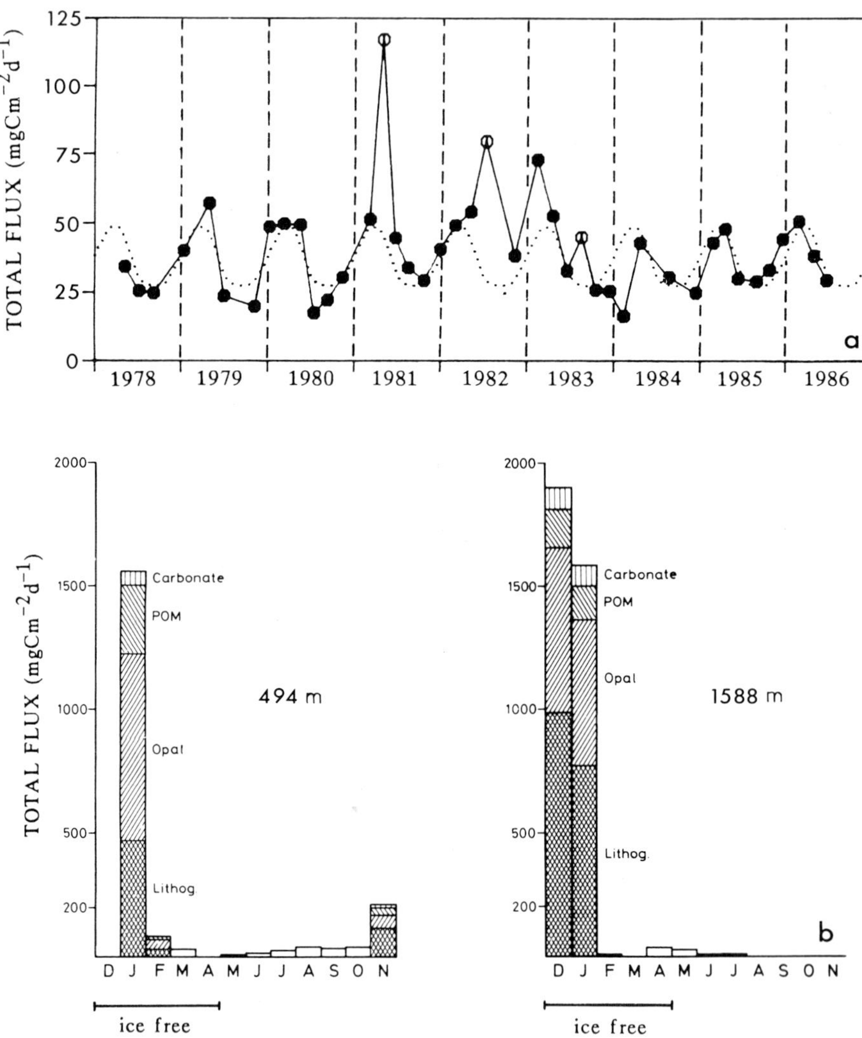

Fig. 7—Variability of particle flux into sediment traps. Upper: flux rates at 3200 m in the Sargasso Sea (total particle flux). Redrawn from Deuser (1987). Lower: flux rates at 494 m and 1588 m water depth in Bransfield Strait, from Dec. 1, 1983, to Nov. 25, 1984 (POM). Redrawn from Wefer et al. (1988).

organisms (feces, appendicularian houses, etc.). However, the processes responsible for the fate of these particles in the water column need yet to be identified and studied. The role of bacteria has to be clarified, in terms of organic by-packing and aggregation of particles. Also, the active downward transfer of organic matter by migrating midwater organisms has to be considered (Angel and Bruland et al., both this volume). The importance of such transfer is unknown, but it too could conceivably aid in producing an apparent increase in flux at midwater depth (e.g., Karl and Knauer 1984). Other factors, such as horizontal advection of organic matter at depth from the continental slope (see Walsh, this volume), might likewise disturb the simple scenario underlying the flux equations.

BENTHIC RESPIRATION

By far the greatest portion of the organic matter arriving at the seafloor is oxidized near its surface and fuels the activities of benthic organisms here (see Reimers, this volume). Thus, the abundance of benthic biomass reflects the supply of organic matter, as is evident from biomass distributions (e.g., Rowe 1983). In addition to mass, benthic communities differ greatly in their structure, depending on food supply. Both species composition and size distribution are affected (Thiel 1983; Jumars and Wheatcroft, this volume). Of special interest to paleoceanographers is the fact that the benthic foraminiferal assemblage responds to changes in supply of organic matter (Lutze and Coulbourn 1984; Gooday 1988; Altenbach and Sarnthein, this volume). Traditionally, much of this response has been interpreted in terms of depth distributions (e.g., Haake et al. 1982).

Oxygen consumption on the deep-sea floor is a substantial portion of the overall oxygen utilization. If we assume that 1–2% of the overlying primary production reaches the deep-sea floor, bottom respiration should take up oxygen corresponding to about 0.3 GtC/y. This is 0.02 μmol per liter per year for the entire ocean, or 3 μmol/l per hundred years in the deep ocean, below 1000 m. Apparent oxygen utilization in the deep ocean increases at roughly 10 μmol/1 per hundred years (see, e.g., Broecker and Peng 1982). Thus, about 30% of the AOU increase at depth is from bottom respiration. Along the continental margins the proportion must be considerably higher (cf. Jahnke and Jackson 1987), and presumably this contributes significantly to the strong oxygen minimum in many coastal regions. If this is true, and DOC correlates with AOU in the sense proposed by Sugimura and Suzuki (1988), DOC may be taken up at the sediment surface, which would provide the correspondence seen.

Respiration on the seafloor has been measured by a number of workers over the last 20 years or so. On the continental slopes off the U.S. coasts there is a marked difference in oxygen uptake between the Pacific and the

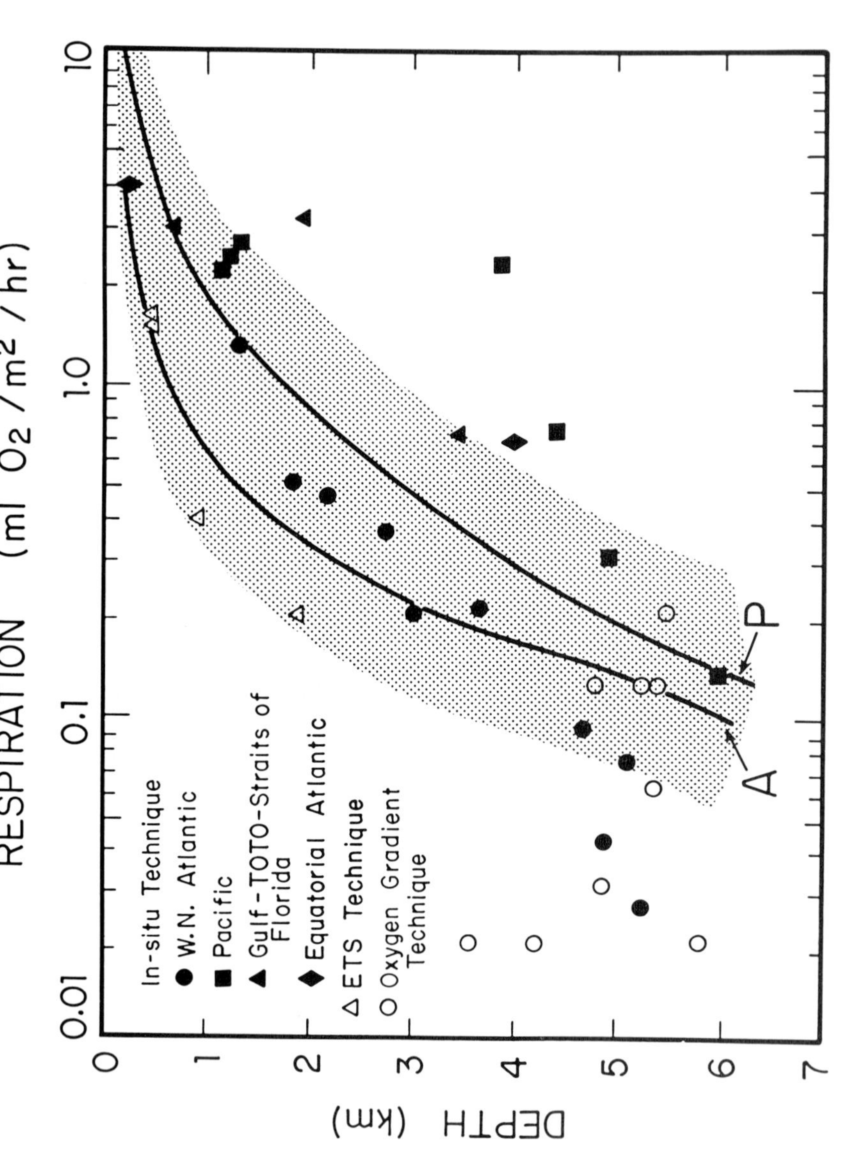

Fig. 8—Calculated oxygen uptake on the seafloor as a function of depth off the East Coast (line marked A) and off the Pacific Coast (line marked P), based on regular decrease of primary production away from the coast, on typical depth-to-distance relationship for the Atlantic and the Pacific margins and on J-flux equation (see text). Shaded zone denotes factor of two about the estimates. Scatter diagram is from Smith and Hinga (1983) with the fit (and shaded zone) added by Berger et al. (1987).

Atlantic: the Pacific having the higher rates for the same depths (Smith and Hinga 1983). This difference can be modeled to a first approximation if it is assumed that the supply of organic carbon follows a transit equation of the type given above. The primary productivity is set as PP=f(d), where d is the distance from the shore, and the depth is expressed as the typical bathymetric profile (z=f(d)). Results of such a fit (Fig. 8) suggest that the differences in hypsography of eastern and western margins are sufficient to explain the main features of the respiration rates observed (Berger et al. 1987).

At great depth the Atlantic values tend to be low, while the Pacific ones are high, with respect to the average fit (Fig. 8). Mechanisms which could be responsible for these patterns are differences in winnowing and shelf-bypassing (see Walsh, this volume), as well as differences in the efficiency of downslope transport, which depends on intensity of bioturbation and steepness of the slope. In view of this, it appears premature to generalize that productivity is higher in the Pacific than in the Atlantic (Hinga 1985) based on differences in respiration on the seafloor in offshore transects.

BURIAL AND PRESERVATION

The great difference in the supply of organic matter to the slope and to the abyssal environment which is evident in the respiration patterns (Fig. 8) is yet another manifestation of the blue-green contrast, which permeates all arguments about ocean carbon flux (or should). The higher supply to the upper slope, of course, results in greater burial of carbon in this setting. Here, where bioturbation is intense, organic matter is actively moved into the sediment, down from the sediment interface into an oxygen-poor subsurface layer. On the whole, sedimentation rates are highly correlated with depth of bioturbation. In addition, the sediment cover itself may have a protective function ("sealing factor"). Thus, the preserved fraction of the organic carbon arriving on the seafloor may be estimated from the sedimentation rate. The following equation is meant as a rule of thumb and is based on a data set given in Müller and Suess (1979), using Equation 9 to estimate supply to the seafloor:

$$F\text{ (sed) }/F\text{ (bot) (percent)} = 2.5\ S - S^2/50, \tag{10}$$

where F(bot) and F(sed) are flux arriving on the bottom and flux accumulating in the sediment, respectively, and S is in cm/ka. The equation is valid over the range of 2 to 20 cm/ka. At sedimentation rates below 1 cm/ka it is sufficient to multiply the rate S by two in order to get a rough estimate of the preserved fraction, in percent of supply to the seafloor. The suggestion that the preservation ratio is correlated to oxygen content of the bottom water overlying the sediment is controversial (Emerson, in Sundquist and

Broecker 1985). At very low oxygen values, however, when metazoans are excluded from the benthos, there does seem to be such an effect (see Jumars et al., this volume).

Oxidation of organic matter continues within the sediment at a lesser rate, mediated by different kinds of bacteria (see Bender and Heggie 1984; Emerson et al. 1985). When the free oxygen is used up, nitrate is reduced to deliver oxygen for combustion (as well as manganese and iron oxyhydroxides), and subsequently sulfate. Iron sulfide then precipitates. Because of the high concentration of sulfate in seawater it is potentially a large source for oxygen. Methane can develop within organic-rich sediments: a process whose effect on sound velocity can be seen in the seismic stratigraphy of upwelling margins, and which (presumably) increases the propensity for sliding and other disturbances of the record in this setting.

The more or less regular trend of decrease of organic matter abundance with age of sediment, in any one site, has been the subject of much discussion (e.g., Heath et al. 1977). A great number of different decay rates have been published for organic matter within the sediment column. These are likely to be a matter of the time scale considered. Over a time scale of 20,000 years, for example, a half-life of 10,000 years might be expected, while on a scale of millions of years the apparent half-life (now applicable to much more resistant types of organic matter) will be counted likewise in the millions of years.

The processes governing burial and preservation of biogenic materials other than organic matter is of great interest, of course, in the context of paleoproductivity. Substantial literature exists on the subject (see in Berger 1981; Emerson and Bender 1981; Leinen et al. 1986; Lyle et al. 1988). It is not possible here to review developments in this field. However, two important points need to be made. First, the output of carbonate and silica from the productive system cannot be expected to be related to primary production in a straightforward fashion, because the primary producers coccolithophorids and diatoms thrive under different conditions, which are not solely a function of productivity, and because heterotrophs (foraminifera, radiolarians) can contribute a large proportion of the fallout. Thus, the structure of the foodweb strongly enters the relationship. Second, carbonate and silica preservation patterns apparently are just as complicated as those for organic matter (Barron and Baldauf, this volume). Both carbonate and opal redissolve on the seafloor because of general undersaturation with their mineral phrases. This undersaturation is a result of global (rather than regional) excess production of the phases with respect to the overall geochemical input of their constituents to the ocean. In addition, undersaturation patterns strongly depend on deep circulation.

In the case of silica, high productivity has the effect of overwhelming the dissolution rate on the seafloor so that there can be good correspondence between silica accumulation and production above some threshold value. However, silica reacts with minerals within the sediment, so that it may disappear as recognizable productivity witness. In the case of carbonate, productivity has both a positive effect, through supply of carbonate, and a negative one, through supply of organic carbon which, due to CO_2 development, helps dissolve the carbonate soon after arrival (Emerson and Bender 1981).

PALEOPRODUCTIVITY FROM ORGANIC CARBON

The concept of changing ocean productivity stands at the beginning of paleoceanography: Arrhenius (1952) introduced it to explain the Pleistocene sedimentary record in the eastern equatorial Pacific, offering evidence from changes in the size distribution of *Coscinodiscus nodulifer*, a centric diatom. His arguments are now familiar and widely applied: an increase in the planetary temperature gradient during glacial periods brings increased wind velocities which stimulates upwelling of nutrient-rich thermocline water. The same principle applies in an eastern boundary setting, as shown in the CLIMAP studies (1976, 1981), whose temperature reconstructions suggest increased coastal upwelling.

Recent developments have focused on the quantitative reconstruction of productivity, using transfer equations on sediment proxies and carbon isotopes in foraminifera.

A quantitative regional reconstruction of Pleistocene productivity was attempted 10 years ago by Müller and Suess (1979), using organic matter abundance in sediments. Compiling data for the present ocean, they calibrated organic matter abundance and sedimentation rates against productivity and wrote an empirical transfer equation of the form:

$$PaP = C\,p\,(1 - \phi) \,/\, 0.0030\, S^{0.3}, \tag{11}$$

which says that the paleoproductivity (in g C $m^{-2}\,y^{-1}$) may be estimated as the product of organic carbon percentage C, dry sediment density p, and sediment volume (percent of total), divided by a function of sedimentation rate. In short, productivity is estimated as a function of organic matter weight in the sediment, with a moderate correction for sedimentation rate. For a range of sedimentation rates of a factor of 3 in a given core, for example, the correction is maximally 40%.

Müller and Suess applied their equation to a core off N.W. Africa and found that oceanic productivity was approximately the same as today during past interglacial stages but was higher by a factor of 2 to 3 during glacial stages (see Fig. 9). The higher productivity of the glacial ocean was attributed to increased coastal and equatorial upwelling, as well as to generally increased mixing due to stronger temperature gradients, winds, and currents.

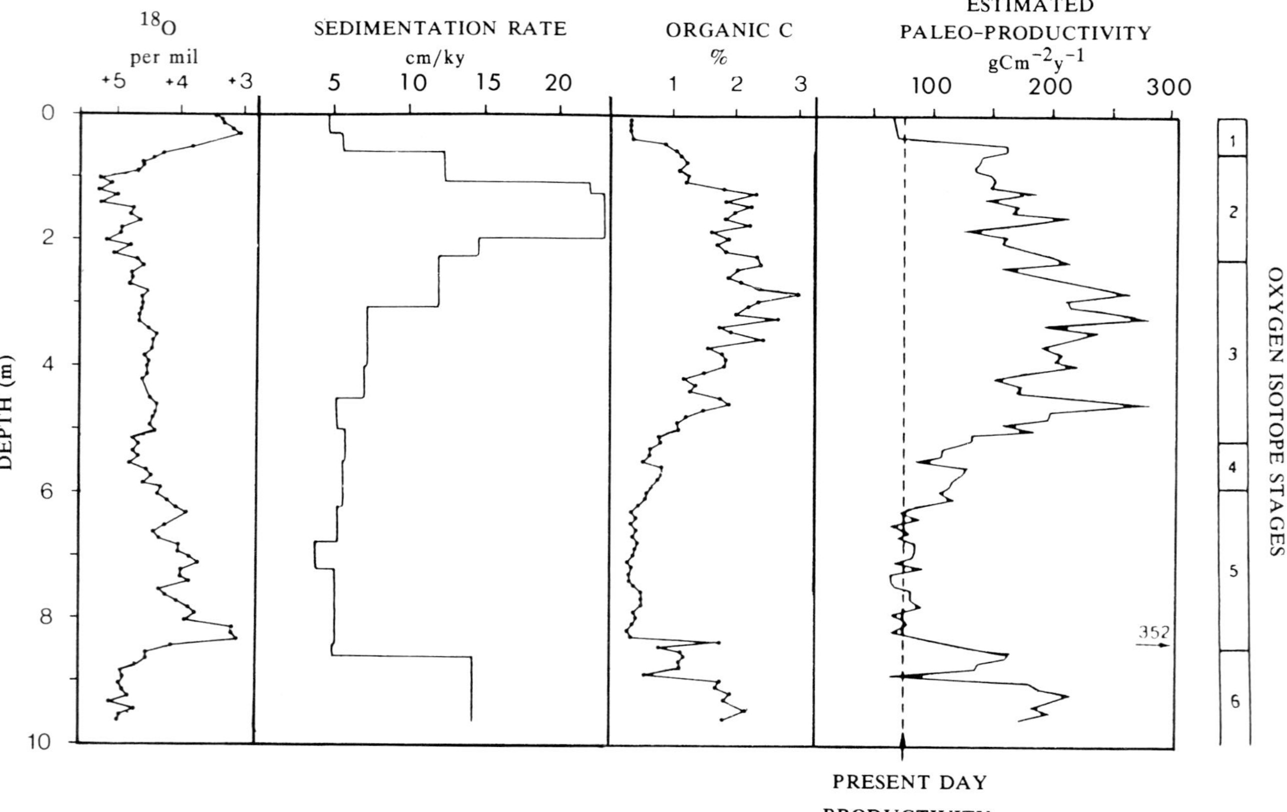

Fig. 9—Paleoproductivity estimates for Meteor Core 12392–1 on the continental rise off the Spanish Sahara, N.W. Africa. Sedimentation rate and organic carbon content are input variables from which productivity is calculated. From Müller and Suess (1979) and Müller et al. (1983).

It is likely that increased low latitude upwelling, which appears to have been a general phenomenon, contributed to the reduction of atmospheric pCO_2 during glacial time. Boyle (1986) suggests an effect of 25 ppm for a doubling of the rate of upwelling.

Additional applications of the method are in Müller et al. (1983) and Sarnthein et al. (1987). The latter pointed out that in calibrating carbon flux to carbon accumulation, the flux to the seafloor must be used rather than the overlying productivity. They give the following transfer equation (here slightly simplified):

$$PaP = 15.9\ C^{0.66}\ S^{-0.05}\ (p(1 - \phi))^{0.66}\ z^{0.32}. \qquad (12)$$

The simplification is that the bulk sedimentation rate and the organic-free bulk sedimentation rate, which they distinguish, are combined. The error is negligible.

Equations 11 and 12 reduce to much the same rule-of-thumb if some simplifying assumptions are made. Considering that there is an overall correspondence of organic carbon percentage and sedimentation rate (Heath et al. 1977; Müller and Suess 1979), we can set approximately

$$C = a\ S^{0.6}, \qquad (13)$$

and we note that the Müller and Suess equation reduces to

$$PaP = k\ C^{0.5} \qquad (14)$$

for any one core (z=constant), if bulk density and porosity do not vary very much.

Similarly, Equation 12 becomes

$$PaP = k\ C^{0.6} \qquad (15)$$

where k must be set for the Holocene productivity (k=PP (Recent)/C (Recent)$^{0.6}$). Equations 14 and 15 indicate that the range of fluctuations in productivity, at any one site, is roughly the same as the range in the square root of the ratio of maximum and minimum organic carbon content. Refinements beyond such a guess may be difficult to achieve, given the various sources of error (admixture of terrigenous carbon, uncertainties in instantaneous sedimentation rate, uncertainty in f-ratio variation). The C_{org} signal as such seems to be quite reliable: considering that such a small fraction of the organic carbon delivered is finally buried, it is remarkable that percentages correlate as well as they do between cores, even in areas of very low sedimentation rate (see Finney et al. 1988).

Applying the rule-of-thumb represented by Equation 15 to a series of organic carbon records from Angola Basin and the Walvis Ridge (Fig. 10), it appears that productivity in these regions increased by about a factor of two in the last 5 million years, with the most notable increase just after 3

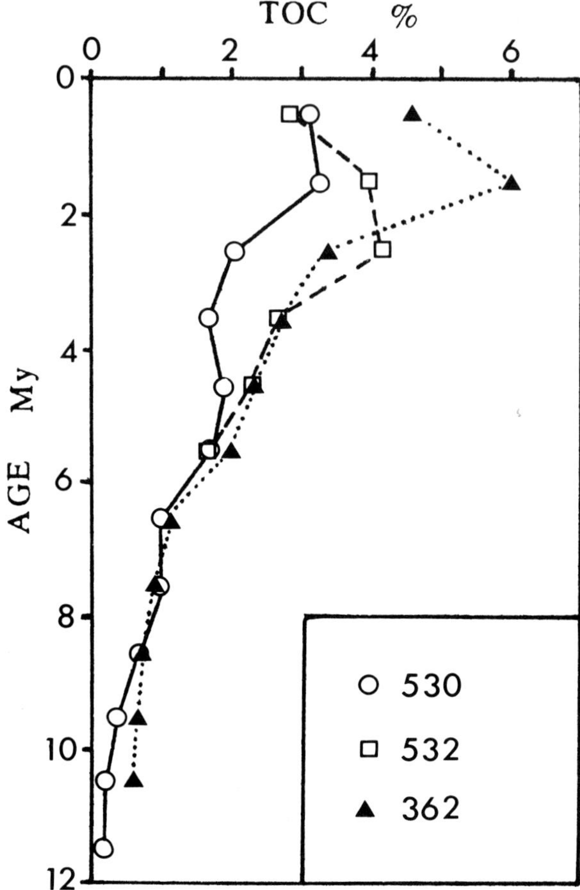

Fig. 10—Organic carbon content in deep-sea drilling sites in Angola Basin (DSDP 530) and on Walvis Ridge (DSDP 562 and DSDP 362). From P.A Meyers et al. in Thiede and Suess (1983: p. 459).

million years ago (cf. the results of Sarnthein et al. 1984, off N.W. Africa). Allowance has to be made, of course, for the general increase in sedimentation rate during this time span and also for the diagenetic destruction of organic matter. Nevertheless, the trend of increasing productivity is most likely real and reflects increased mixing and upwelling in the southeastern Atlantic over this period.

The interpretation of the fluctuations of organic carbon content in terms of paleoproductivity is complicated by the possibility of varying supply of terrigenous organic carbon and shelf-produced (i.e., redeposited) carbon, especially in the type of situation studied by Müller and Suess, where the glacially exposed shelf is not that far away. In addition, changes in oxygen concentration at the site of deposition may conceivably influence the burial fraction (see Emerson, in Sundquist and Broecker 1985). However, Pedersen (1983) believed that enhanced preservation during glacials could be ruled out as a cause of the organic carbon fluctuations he observed in the eastern equatorial Pacific.

PALEOPRODUCTIVITY FROM OTHER METHODS

A quantitative map for glacial productivity in the Atlantic is provided by Mix (this volume). He applies the transfer method of Imbrie and Kipp (1971), which was employed so successfully in the reconstruction of temperature fields (CLIMAP group 1976), to reanalyze the CLIMAP data in terms of productivity. Interestingly, when applied to a core in the eastern Atlantic, his results differ from earlier ones based on organic matter. Earlier studies used the abundance of certain planktonic species as an indicator of productivity variation (e.g., Prell and Curry 1981; Pisias and Rea 1988).

The reconstruction of paleoproductivity from carbon isotopes in foraminifera likewise has great potential and should complement the transfer equations much as oxygen isotope studies complement temperature transfer methods (Berger and Vincent 1986; Altenbach and Sarnthein, this volume). A link between ocean productivity and carbon isotopes was first postulated by Tappan (1968), who equated organic carbon deposition (in the margins) with ocean fertility and read increased deposition of organic carbon from a change in the $\delta^{13}C$ of carbonates toward heavier values.

A similar approach was taken by Vincent and Berger (in Sundquist and Broecker 1985) in their reconstruction of the timing and magnitude of the Monterey upwelling episode, which coincides with high-latitude cooling and (most likely) with the buildup of Antarctic ice 15 million years ago. They suggested that the sequestration of organic carbon in the upwelling-dominated sediments enhanced the polar cooling process by drawdown of the total carbon content in the ocean, and hence of atmospheric carbon dioxide. The same argument concerning upwelling and pCO_2 drawdown applies to the glacial periods within the Pleistocene, of course. If we accept the arguments of Shackleton and Pisias (in Sundquist and Broecker 1985), that a change in the carbon system preceded a change in sea level, we could then envisage increased upwelling preceding ice buildup and decreased upwelling preceding melting. A study of carbonate preservation fluctuations (e.g., Berger and Keir 1984; Crowell, in Sundquist and Broecker 1985; Peterson and Prell, in Sundquist and Broecker 1985) with such a scenario in mind could be instructive; the ocean's saturation should respond to sequestration and release of organic carbon on this time scale.

The difference between planktonic and benthic $\delta^{13}C$ compositions has emerged as one of the most fruitful tools in the estimation of paleoproductivity (Shackleton et al. 1983; Curry and Crowley 1987). Broecker (1982) introduced the concept that this difference is controlled by nutrient concentration in deep waters and used it to back his explanation for the changes in atmospheric CO_2 discovered in ice cores (Berner et al. 1980; Delmas et al. 1980). Broecker's model was boldly extrapolated into the late Pleistocene by Shackleton et al. (1983), who inferred worldwide changes in productivity from the carbon isotope chemistry of foraminifera in a core taken in the eastern equatorial Pacific, and converted the inferred productivity fluctuations to atmospheric changes in pCO_2. The delta $\delta^{13}C$ signal of their core, in fact, bears a strong resemblance to subsequent measurements in ice cores (see Fig. 1) and, if properly scaled, can be made to hindcast the pCO_2 fluctuations reasonably well (Shackleton et al. 1983; Shackleton and Pisias, in Sundquist and Broecker 1985).

This success of paleoceanographic methods is gratifying; it clearly shows that the carbon system of the ocean is in close correspondence with the pCO_2 changes in the atmosphere through geologic time. However, a word of caution is perhaps appropriate. To obtain pCO_2 from proxies, a model must be adopted and its parameters fitted to the glacial-postglacial pCO_2 change. Thus, an observed correspondence further back in time will then show that the model gives consistent results, but it does not prove that the model correctly describes the mechanisms controlling atmospheric pCO_2. Once Broecker's model is accepted, for example, the oxygen isotope record of sea level change can be used for similar backward extrapolation of pCO_2 (Keir and Berger 1983), but with the model-dependency rather more obvious than for the $\delta^{13}C$ signal.

The various quantitative studies on paleoproductivity, and newly emerging ones based on biomarkers (Prahl and Muehlhausen, this volume), provide the foundation for long-range geochemical ocean modeling (Keir, this volume). These efforts herald the beginning of a new phase in paleoceanography: an integration of a detailed reconstruction of physical, chemical, and biological history of the ocean, especially of the Pleistocene and Neogene ocean.

PRODUCTIVITY OF ANCIENT OCEANS

For ancient oceans, i.e., oceans which existed before there was any or much ice in polar regions, the patterns of circulation and of productivity control cannot be assumed to have been the same as now. Instead, the basic principles of how the ocean works in such a completely different setting have to be discovered and applied, and very large changes in the geographic and geochemical framework have to be admitted in order to reconstruct reasonable scenarios (Arthur and Schlanger 1979; Thierstein, this volume).

Taking this long-range view, Bramlette (1965) argued that the productivity of the ocean must ultimately be controlled by the rate of nutrient input from land. To explain the plankton extinctions at the end of the Cretaceous, he suggested that widespread transgression (as occurred in the late Cretaceous) would eventually turn off the supply of nutrients and starve the ocean's photosynthetic machinery, causing extinction from food chain collapse. Bramlette's postulate of nutrient modulation by sea level change reappears in Broecker's (1982) phosphate extraction model, in a moderated form, for the purpose of controlling pCO_2.

The question of which process ultimately controls the productivity of the ocean is a fascinating one and probably cannot be attacked without taking into account both the controls on phosphate and on nitrate concentrations in the deep ocean (see Codispoti, this volume). Quite possibly, iron has also to be considered (Martin and Gordon 1988). The nature of these controls is complex and must be different for the different nutrients. Supply from outside sources is only one factor; internal cycling is just as important. As Bogoyavlenskiy (1967) pointed out, the rate at which silica can be precipitated is limited by the rate at which it is dissolved in deep water and at the seafloor, because most of it is recycled., Similarly, the concentration of nutrients in the ocean is largely a function of the rate of remineralization of organic matter, be it particulate or dissolved.

To model paleoproductivity correctly, then, at least three rates have to be constrained: the supply from outside the ocean, the rate of remineralization, and the rate of delivery of deep water to the photic zone. The first two control burial rate; the second two control concentration and supply rate of nutrients to surface waters. Such data are difficult to come by. The marine phosphorus cycle, for example, is understood but in rough outline even for the present ocean (Froelich et al. 1982). For the past we are dealing largely with guesswork, as is illustrated in the various attempts to unravel the origin of phosphorites (e.g., Arthur and Jenkyns 1981).

STARVED OCEANS?

In a warm ocean, as in the early Tertiary and in the Cretaceous, a low supply of oxygen should result in decreased remineralization of all nutrients: the mass of carbon and associated nutrients which can be readily recycled is tied to the amount of oxygen available to do the job. Applying the concept of remineralization, it may be argued that the most slowly remineralizing nutrient will be limiting. An increased loss of nitrate through denitrification, in a warm, oxygen-poor ocean, would simulate slowed remineralization and produce a nitrate-starved, low fertility system (Berger and Roth 1975; Codispoti and Piper 1975).

A positive feedback system could be set up by the denitrification mechanism, whereby short-term warming (and decreased oxygen supply at depth) results in a loss of fertility, and hence in a change in the partitioning of CO_2, such that the atmosphere obtains a larger share, favoring further warming. Also, such an ocean should become enriched with other nutrient materials (silicate, phosphate, trace elements), which would fail to precipitate at low concentrations. Such positive feedback mechanisms, and associated changes in biogenous facies, could translate subtle changes in Milankovitch radiation input into substantial lithologic variations, producing rhythmic bedding in pelagic sediments (see Herbert et al., this volume).

Biogenic sediments in a nutrient-starved ocean would not be well fractionated into calcareous facies and nutrient-rich facies but would simulate inorganic precipitation, with a high abundance of oolites, and phosphatized and silicified shallow water carbonates. In the present ocean, of course, shallow water carbonates precipitate in essentially phosphate-free and silica-free environments, except in certain upwelling areas.

Nitrate destruction is not the only conceivable mechanism of warm ocean starvation, of course. Trace elements may be in short supply, such as iron or molybdenum. For the present, it has been proposed (Martin and Gordon 1988) that most of the iron necessary for photosynthesis is delivered by the atmosphere. In times of widespread transgression and weak zonal winds, supply of aeolian sediment would be greatly reduced. On the other hand, trace elements may be available as chelates in a greatly expanded DOC pool, in an oxygen-poor ocean, so that a wind source would be unimportant. Even today and in the Pleistocene chelates in the DOC pool may play a greater role than is allowed for in the hypothesis of Martin and Gordon. Incidentally, an expanded DOC reservoir in warm oceans could, upon proper stimulation through temperature and oxygen variation, change atmospheric pCO_2 on short notice by release or uptake of carbon.

In addition to input reduction, destruction, or slowed remineralization, nutrients could be trapped semipermanently in deep saline bottom water, in an ocean dominated by haline circulation and with episodic replacement of abyssal waters. Again, such a system would be quite sensitive to disturbance, through alternating sequestration and injection of nutrients. Sporadic nutrient supply from a deep reservoir would be stressful to pelagic organisms adapted to low fertility (producing blooms of opportunist algae such as *Braarudosphaera*) and would also be inimical to the growth of coral reefs (as is seen when contaminating tropical lagoons with sewage). Nutrient excess has been invoked in explaining the drowning of reefs during mid-Cretaceous time, e.g., as a consequence of this type of nutrient-generated stress (Hallock and Schlager 1986).

The safe distance of Cretaceous oceans provides opportunities for speculation which might provide insights into the workings of systems that are closer by. For example, it is quite possible that an analog to a quasi-stagnant bottom water mass existed not too long ago, at the end of the last glacial (Worthington 1968). In such a cold (1 degree below present) and saline (1 permil above present) water, body nutrients could have been trapped and destroyed. By reducing the productivity of the ocean in the earliest Holocene such a process could have contributed to the rapid rise of pCO_2 at the beginning of the Altithermal.

OCEAN ASYMMETRY AND PALEOPRODUCTIVITY

The overall pattern of deep circulation, which is so important in the remineralization of nutrients and their resupply to the waters above, is reflected in a number of proxy signals, including carbonate and silica preservation and $\delta^{13}C$ distributions in benthic foraminifera. One important index is the asymmetry between the Atlantic and the Pacific, regarding these proxies. An asymmetric ocean, such as we have now thanks to the vigorous production of North Atlantic Deep Water, is on the whole better ventilated than a more symmetric one and presumably mixes more rapidly. Thus, one would expect a more vigorous recycling of nutrients. However, asymmetry is likely to be correlated with other important phenomena, such as range of oxygen content and size of the DOC pool and coastal and equatorial upwelling. Thus, the resulting controls on productivity are by no means obvious. A careful study of the effects on productivity of changing asymmetry, using evidence from silica deposition (Barron and Baldauf, this volume) and from carbon isotopes (Miller and Fairbanks, in Sundquist and Broecker 1985), should bear rich fruit in understanding long-term nutrient dynamics.

The phenomenon most readily studied as a function of geologic time, and one of the most promising, is that of the history of coastal upwelling. More generally, what is sought is the development of the blue ocean/green ocean dichotomy. With but little exaggeration, one might say that the biogeochemistry of the ocean is governed by coastal processes during those times when upwelling processes are strong. During such times, nutrients and carbon are pulled out of the system in the margins. Thus, the record of coastal upwelling (Suess and Thiede 1983; Thiede and Suess 1983) provides a memory bank of an important controlling function. In contrast, during such times the pelagic deep-sea sediments act more as monitors of the system than as a control, providing a comparably reduced sink for carbon and nutrients. The memory of pelagic sediments, of course, tends to be less influenced by the variabilities and vicissitudes of regional noise which overprints the signals in the margins.

The task is to tie the signals from the pelagic deep-sea record to the record in the margins. The marginal biogenic sediments contain the clues to the trapping efficiency of shelf seas, which varied with sea level, oxygen content, nutrient concentration, and physical mixing along the continental edge. This trapping efficiency, which determines the overall partitioning of biogenous sediments between margin and deep-sea floor in terms of basin-shelf fractionation, may be taken as one measure of paleoproductivity. Fluctuations in margin dominance, in regard to organic carbon deposition, should produce corresponding fluctuations in the pCO_2 of the atmosphere over short and intermediate time scales (10^2 to 10^5 years).

Acknowledgements. We are most grateful to the many colleagues who came to Berlin to exchange ideas on the topics touched on here (and more). Without their contributions to our education in matters of productivity and paleoproductivity this overview could not have been written. We hope they will find it rewarding, as we did, to view the various findings (including their own) in the broad context of the long-range dynamics of global ocean productivity. We owe special thanks to Dr. Silke Bernhard and her helpful and efficient team, who made our week-long seminar a pleasurable experience as well as a productive one, and who took pains to diminish ours, as editors. Heike Schmidt assisted in preparing the figures. The preparatory work for the conference, at the University of Bremen in 1986, was made possible by the Humboldt Foundation.

REFERENCES

Arrhenius, G.O.S. 1952. Sediment cores from the east Pacific. Rep. Swed. Deep Sea Exped. 1947–1948, vol. 5, 288pp.

Arthur, M.A., and Jenkyns, H.C. 1981. Phosphorites and paleoceanography. In: Proc. 26th Internat. Geol. Congress, Geology of Oceans, pp. 83–96. *Oceanol. Acta, Spec. Issue.*

Arthur, M.A., and Schlanger, S.O. 1979. Cretaceous "oceanic anoxic events" as causal factors in development of reef-reservoired giant oil fields. *Am. Assn. Petrol. Geol. Bull.* **63**: 870–885.

Barnola, J.M.; Raynaud, D.; Korotkevich, Y.S.; and Lorius, C. 1987. Vostok ice core provides 160,000-year record of atmospheric CO_2. *Nature* **329**; 408–414.

Bender, M.L., and Heggie, D.T. 1984. Fate of organic carbon reaching the sea floor: a status report. *Geochim. Cosmochim. Acta* **48**: 977–986.

Berger, W.H. 1981. Paleoceanography: the deep-sea record. In: The Sea, ed. C. Emiliani, vol. 7, pp. 1437–1519. New York: Wiley.

Berger, W.H.; Fischer, K.; Lai, C.; and Wu, G. 1987. Ocean productivity and organic carbon flux. I. Overview and maps of primary production and export production. Univ. California, San Diego, SIO Reference 87–30.

Berger, W.H., and Keir, R.S. 1984. Glacial-Holocene changes in atmospheric CO_2 and the deep-sea record. In: Climate Processes and Climate Sensitivity, eds. J.E. Hansen and T. Takahashi. *Geophys. Monog.* **29**: 337–351. Washington, D.C.: Amer. Geophys. Union.

Berger, W.H., and Roth, P.H. 1975. Oceanic micropaleontology: progress and prospects. *Rev. Geophys. Space Phys.* **13 (3)**: 561–585, 624–635.

Berger, W.H., and Vincent, E. 1986. Deep-sea carbonates: reading the carbon-isotope signal. *Geol. Rundschau* **75**: 249–269.

Berner, W.; Oeschger, H.; and Stauffer, B. 1980. Information on the CO_2 cycle from ice core studies. *Radiocarbon* **22**: 227–235.

Betzer, P.R.; Showers, W.J.; Laws, E.A.; Winn, C.D.; DiTullio, G.R.; and Kroopnick, P.M. 1984. Primary productivity and particle fluxes on a transect of the equator at 153°W in the Pacific Ocean. *Deep-Sea Res.* **31**: 1–11.

Bogoyavlenskiy, A.N. 1967. Distribution and migration of dissolved silica in oceans. *Int. Geol. Rev.* **9 (2)**: 133–153.

Boyle, E.A. 1986. Deep ocean circulation, preformed nutrients, and atmospheric carbon dioxide: theories and evidence from oceanic sediments. In: Mesozoic and Cenozoic Oceans, ed. K.J. Hsü. *AGU Geodynam. Ser.* **15**: 49–59.

Bramlette, M.N. 1965. Massive extinctions in biota at the end of Mesozoic time. *Science* **148**: 1696–1699.

Broecker, W.S. 1982. Ocean chemistry during glacial time. *Geochim. Cosmochim. Acta* **46**: 1689–1705.

Broecker, W.S., and Peng, T.-H. 1982. Tracers in the Sea. Palisades, NY: Eldigio Press.

Broecker, W.S., and Peng, T.-H. 1986. Carbon cycle: 1985. Glacial to interglacial changes in the operation of the global carbon cycle.*Radiocarbon* **28 (2A)**: 309–327.

Bruland, K.W., ed. 1984. Global Ocean Flux Study – Proceedings of a Workshop. Washington, D.C.: National Academy Press.

Cho, B.C., and Azam, F. 1988. Major role of bacteria in biogeochemical fluxes in the ocean's interior. *Nature* **332**: 441–443.

CLIMAP. 1976. The surface of the ice-age Earth. *Science* **191**: 1131–1137.

CLIMAP. 1981. Seasonal reconstructions of the Earth's surface at the last glacial maximum. *Geol. Soc. Amer. Chart Ser.* **MC–36**.

Codispoti, L.A., and Piper, D.Z. 1975. Marine phosphorite deposits and the nitrogen cycle. *Science* **188**: 15–18.

Curry, W.B., and Crowley, T.J. 1987. The $\delta^{13}C$ of equatorial Atlantic surface waters: implications for ice age pCO_2 levels. *Paleocean.* **2**: 489–517.

Cushing, D.H. 1973. Production in the Indian Ocean and the transfer from the primary to the secondary level. In: The Biology of the Indian Ocean, ed. B. Zeitzschel, pp. 475–486. New York: Springer.

Delmas, R.J.; Ascencio, J.-M.; and Legrand, M. 1980. Polar ice evidence that atmospheric CO_2 20,000 years BP was 50% of present. *Nature* **284**: 155–157.

Deuser, W.G. 1987. Variability and hydrography and particle flux: transient and long-term relationships. In: Particle Flux in the Ocean, eds. E.T. Degens, E. Izdale, and S. Honjo. *Mitt. Geol. Palaeont. Inst. Univ. Hamburg* **62**: 179–193.

Dugdale, R.C., and Goering, J.J. 1967. Uptake of new and regenerated forms of nitrogen in primary productivity. *Limnol. Ocean.* **12**: 196–206.

Dunbar, R.B., and Berger, W.H. 1981. Fecal pellet flux to modern bottom sediment of Santa Barbara Basin (California) based on sediment trapping. *Geol. Soc. Amer. Bull.* **92**: 212–218.

Emerson, S., and Bender, M. 1981. Carbon fluxes at the sediment-water interface of the deep-sea: calcium carbonate preservation. *J. Marine Res.* **39**: 139–161.

Emerson, S.; Fischer, K.; Reimers, C.; and Heggie, D. 1985. Organic carbon dynamics and preservation in deep-sea sediments. *Deep-Sea Res.* **32**: 1–21.

Eppley, R.W., and Peterson, B.J. 1979. Particulate organic matter flux and planktonic new production in the deep ocean. *Nature* **282**: 677–680.

Finney, B.P.; Lyle, M.W.; and Heath, G.R. 1988. Sedimentation at MANOP Site H (eastern equatorial Pacific) over the past 400,000 years: climatically induced redox variations and their effects on transition metal cycling. *Paleocean.* **3**: 169–189.

Froelich, P.N.; Bender, M.L.; Luedtke, N.A.; Heath, G.R.; and DeVries, T. 1982. The marine phosphorus cycle. *Am. J. Sci.* **282**: 474–511.

Gooday, A.J. 1988. A response by benthic foraminifera to the deposition of phytodetritus in the deep sea. *Nature* **332**: 70–73.

Haake, F.-W.; Coulbourn, W.T.; and Berger, W.H. 1982. Benthic foraminifera: depth distribution and redeposition. In: Geology of the Northwest African Continental Margin, eds. U. von Rad, K. Hinz, M. Sarnthein, and E. Seibold, pp. 632–657. New York: Springer.

Hallock, P., and Schlager, W. 1986. Nutrient excess and the demise of coral reefs and carbonate platforms. *Palaios* **1**, 389–398.

Heath, G.R.; Moore, T.C.; and Dauphin, J.P. 1977. Organic carbon in deep-sea sediments. In: The Fate of Fossil Fuel CO_2 in the Oceans, eds. N.R. Andersen and A. Malahoff, pp. 605–625. New York: Plenum.

Hinga, K.R. 1985. Evidence for a higher average primary productivity in the Pacific than in the Atlantic Ocean. *Deep-Sea Res.* **32**: 117–126.

Honjo, S. 1982. Seasonality and interaction of biogenic and lithogenic particulate flux at the Panama Basin. *Science* **218**: 883–884.

Imbrie, J., and Kipp, N.G. 1971. A new micropaleontological method for quantitative paleoclimatology: application to a Late Pleistocene Caribbean core. In: The Late Cenozoic Glacial Ages, ed. K.K. Turekian, pp. 71–181. New Haven: Yale Univ. Press.

Jahnke, R.A., and Jackson, G.A. 1987. Role of sea floor organisms in oxygen consumption in the deep North Pacific Ocean. *Nature* **329**: 621–623.

Jones, P.D.; Wigley, T.M.L.; and Raper, S.C.B. 1987. The rapidity of CO_2-induced climatic change: observations, model results and palaeoclimatic implications. In: Abrupt Climatic Change, eds. W.H. Berger and L.D. Labeyrie, pp. 47–55. Dordrecht: Reidel.

Karl, D.M., and Knauer, G.A. 1984. Vertical distribution, transport, and exchange of carbon in the northeast Pacific Ocean: evidence of multiple zones of biological activity. *Deep-Sea Res.* **31**: 221–243.

Karl, D.M.; Knauer, G.A.; and Martin, J.H. 1988. Downward flux of particulate organic matter in the ocean: a particle decomposition paradox. *Nature* **332**: 438–441.

Keir, R.S., and Berger, W.H. 1983. Atmospheric CO_2 content in the last 120,000 years: the phosphate-extraction model. *J. Geophys. Res.* **88 (C10)**: 6027–6038.

Koblents-Mishke, O.J.; Volkovinsky, V.V.; and Kabanova, J.G. 1970. Plankton primary production of the World Ocean. In: Scientific Exploration of the South Pacific, ed. W.S. Wooster, pp. 183–193. Washington, D.C.: U.S. Nat. Acad. Sci.

Leinen, M.; Cwienk, D.; Heath, G.R.; Biscaye, P.E.; Kolla, V.; Thiede, J.; and Dauphin, J.P. 1986. Distribution of biogenic silica and quartz in recent deep-sea sediments. *Geology* **14**: 199–203.

Lutze, G.-F., and Coulbourn, W.T. 1984. Recent benthic foraminifera from the continental margin of Northwest Africa: community structure and distribution. *Mar. Micropal.* **8**: 361–401.

Lyle, M.; Murray, D.W.; Finney, B.P.; Dymond, J.; Robbins, J.M.; and Brooksforce,

K. 1988. The record of late Pleistocene biogenic sedimentation in the eastern tropical Pacific Ocean. *Paleocean.* **3**: 39–59.

Martin, J.H., and Gordon, R.M. 1988. Northeast Pacific iron distributions in relation to phytoplankton productivity. *Deep-Sea Res.* **35**: 177–196.

Müller, P.J.; Erlenkeuser, H.; and von Grafenstein, R. 1983. Glacial-interglacial cycles in oceanic productivity inferred from organic carbon contents in eastern North Atlantic sediment cores. In: Coastal Upwelling, Part B, eds. J. Thiede and E. Suess, pp. 365–398. New York: Plenum.

Müller, P.J., and Suess, E. 1979. Productivity, sedimentation rate, and sedimentary organic matter in the oceans. I. Organic carbon preservation. *Deep-Sea Res.* **26A**: 1347–1362.

Pedersen, T.F. 1983. Increased productivity in the eastern equatorial Pacific during the last glacial maximum (19,000 to 14,000 B.P.). *Geology* **11**: 16–19.

Pisias, N.G., and Rea, D.K. 1988. Late Pleistocene paleoclimatology of the central equatorial Pacific: sea surface response to the southeast trade winds. *Paleocean.* **3**: 21–37.

Platt, T., and Harrison, W.G. 1985. Biogenic fluxes of carbon and oxygen in the ocean. *Nature* **318**: 55–58.

Platt, T., and Subba Rao, D.V. 1975. Primary production of marine microphytes. Photosynthesis and productivity in different environments. In: International Biological Programme, vol. 3, pp. 249–279. Cambridge: Cambridge Univ. Press.

Postma, H. 1971. Distribution of nutrients in the sea and the oceanic nutrient cycle. In: Fertility of the sea, ed. J.D. Costflow, pp. 337–349. New York: Gordon and Breach.

Prell, W.L., and Curry, W.B. 1981. Faunal and isotopic indices of monsoonal upwelling: western Arabian Sea. *Oceanol. Acta* **4**: 91–98.

Romankevich, E.A. 1984. Geochemistry of Organic Matter in the Ocean. Heidelberg: Springer.

Rowe, G.T. 1983. Biomass and production of the deep-sea macrobenthos. In: Deep-Sea Biology. The Sea, vol. 8, ed. G.T. Rowe, pp. 97–121. New York: Wiley.

Sarnthein, M.; Thiede, J.; Pflaumann, U.; Erlenkeuser, H.; Fütterer, D.; Koopmann, B.; Lange, H.; and Seibold, E. 1984. Atmospheric and oceanic circulation patterns off Northwest Africa during the past 25 million years. In: Geology of the Northwest African Continental Margin, eds. U. von Rad et al., pp. 545–604. Berlin: Springer.

Sarnthein, M.; Winn, K.; and Zahn, R. 1987. Paleoproductivity of oceanic upwelling and the effect on atmospheric CO_2 and climatic change during deglaciation times. In: Abrupt Climatic Change, eds. W.H. Berger and L.D. Labeyrie, pp. 311–337. Dordrecht: D. Reidel.

Shackleton, N.J.; Hall, M.A.; and Shuxi, C. 1983. Carbon isotope data in core V19–30 confirm reduced carbon dioxide concentration in the ice age atmosphere. *Nature* **306**: 319–322.

Smetacek, V. 1985. Role of sinking in diatom life-history cycles: ecological, evolutionary and geological significance. *Marine Biol.* **84**: 239–251.

Smith, K.L., and Hinga, K.R. 1983. Sediment community respiration in the deep sea. In: Deep-Sea Biology. The Sea, vol. 8, ed. G.T. Rowe, pp. 331–371. New York: Wiley.

Steemann Nielsen, E., and Aabye Jensen, E. 1957. Primary oceanic production, the autotrophic production of organic matter in the oceans. *Galathea Rep.* **1**: 49–136.

Suess, E. 1980. Particulate organic carbon flux in the oceans: surface productivity and oxygen utilization. *Nature* **288**: 260–263.

Suess, E., and Thiede, J., eds. 1983. Coastal Upwelling: Its Sedimentary Record. Part A: Responses of the Sedimentary Regime to Present Coastal Upwelling. New York: Plenum.

Sugimura, Y., and Suzuki, Y. 1988. A high temperature catalytic oxidation method of non-volatile dissolved organic carbon in seawater by direct injection of liquid sample. *Marine Chem.* **24**: 105–131.

Sundquist, E.T., and Broecker, W.S., eds. 1985. The Carbon Cycle and Atmospheric CO_2: Natural Variations Archean to Present. *Geophys. Monog.* **32**. Washington, D.C. Amer.Geophys. Union.

Tappan, H. 1968. Primary production, isotopes, extinctions and the atmosphere. *Paleogeog. Pal. Pal.* **4**: 187–210.

Thiede, J., and Suess, E. 1983. Coastal Upwelling: Its Sedimentary Record. Part B: Sedimentary Records of Ancient Coastal Upwellings. New York: Plenum.

Thiel, H. 1983. Meiobenthos and nanobenthos of the deep sea. In: Deep-Sea Biology. The Sea, vol. 8, ed. G.T. Rowe, pp. 167–230. New York: Wiley.

Wefer, G.; Fischer, G.; Fütterer, D.; and Gersonde, R. 1988. Seasonal particle flux in the Bransfield Strait (Antarctica). *Deep-Sea Res.*, in press

Woods Hole Oceanographic Institution. 1987. Map of relative intensity of spectral lines related to chlorophyll concentrations, based on data collected by NASA Nimbus-7, Dec. 1981. U.S. Global Flux Study Planning Office.

Worthington, L.V. 1968. Genesis and evolution of water masses.*Meteor. Monog.* **8**: 63–67.

Productivity of the Ocean: Present and Past
eds. W.H. Berger, V.S. Smetacek and G. Wefer, pp. 35–48
John Wiley & Sons Limited

Food Web Structure and Loss Rate

R. Peinert[1], B. von Bodungen[1], and V.S. Smetacek[2]

[1] *Institut für Meereskunde*
2300 Kiel, F.R. Germany

[2] *Alfred Wegener Institut für Polar- und Meeresforschung*
2850 Bremerhaven, F.R. Germany

Abstract. Seasonal and regional patterns of sedimentation are related to processes in the uppermost water layers of the oceans, that is, new and regenerated production and processes which accelerate the sinking particles or help to retain matter in suspension. During spring blooms, new production results in high losses of phytoplankton cells after nutrient depletion. Grazing by herbivores determines quantity and composition of sinking particles. Whereas swarm grazers promote high loss rates by producing fast-sinking fecal matter, copepod grazing favors retention of matter in productive surface layers. The gearing of herbivorous copepod grazing to spring phytoplankton growth is decisive for the ratio of sedimented to retained essential elements. Retention of matter is most efficient in nutrient-regenerating systems. Retention efficiency will decline with deviations from steady state conditions, brought about by nutrient injections, life cycle-related behavior of dominant grazers, and succession patterns within the heterotrophs. Quantitative budgets of pelagic processes and sedimentation are as yet uncertain, especially with respect to the role played by dissolved organic matter.

INTRODUCTION

In general, productivity of ocean surface water is reflected in the organic carbon content of the underlying sediments. The relationship, however, is not simple. The wide range of regional and seasonal variation in quantity and composition of sedimenting matter, as measured with sediment traps, is a function of the overlying pelagic system. In general, high sedimentation rates characterize systems dominated by new production whereas lower vertical fluxes indicate regenerating type systems (Eppley and Petersen 1979). More specifically, sedimentation is tied to a seasonal succession of stages of plankton development, as shown by Smetacek et al. (1984) in a long-term study of the Kiel Bight shallow water ecosystem (Baltic Sea).

The seasonal succession of new and regenerated production in temperate latitudes (spring blooms followed by summer steady state regenerating systems) is reflected, to some extent, in the annual change in spatial patterns on a global scale: on a longitudinal transect, winter conditions in the high northern latitudes would coincide with different stages of spring development at temperate latitudes and post-bloom steady state conditions at lower latitudes. The same applies for upwelling systems and their transition into regenerating systems of the central oceanic gyres along latitudinal transects.

The great variation of vertical fluxes on a time scale of weeks to months (Wefer, this volume) indicates that sinking particles are formed by many different types of biological processes in the overlying water columns. They can be divided into chains of processes that retain biogenic matter in suspension ("retention chains") and others that accelerate losses via sedimentation ("export chains"). Rapid sinking of matter produced in the euphotic zone indicates that large particles contribute a major portion of sedimenting material, that is, packaging of small particles into larger transport vehicles takes place. Examining the origin and the seasonal and regional conditions of the production and fate of such particles is of particular interest.

NEW PRODUCTION: DOMINANCE OF AUTOTROPHS

Seasonal spring blooms in temperate shallow waters, that are dominated by autotrophs and exhibit simple food web interactions, are the best investigated recurrent new production systems. They are physically controlled and triggered by a threshold ratio between incident radiation and turbulent mixing that often favors diatoms as primary producers at high initial nutrient concentrations. Blooms are inherently transient and are terminated by exhaustion of winter-accumulated new nutrients. A chain of processes leading to the sedimentation of phytoplankton and freshly formed phytodetritus follows when grazing pressure is low. The export potential of the bloom is determined by initial nutrient availability, which is a function of concentration and the depth of mixing. Hence, physical conditions have a marked impact on the magnitude of flux. Vertical flux can be calculated according to the budget:

$$N_{flux} = N - (PON + DON + DIN), \quad (1)$$

where N_{flux} is the vertical flux of nitrogen, be it passive sinking of particles or active downward transport by vertically migrating organisms; N is the initial reactive nitrogen pool; and PON, DON, and DIN represent particulate, dissolved organic and inorganic nitrogen pools, respectively, that are retained in the mixed layer.

The spring bloom signal is not only recorded at the shallow sea floor but also regularly transmitted to great depths in the open ocean, with only a small time lag to processes in the euphotic zone (Deuser 1986). In new production regimes with a small grazer population, the transport vehicles with high sinking velocities originate from the autotrophs themselves.

Diatoms are most common in new production systems and certain features of their biology are important for mediating sedimentation, as discussed by Smetacek (1985). Under favorable growth conditions their ability to decrease sinking velocity by means of cell-sap density regulation minimizes vertical losses against the drag of gravity. Even positive buoyancy has been recorded (Villareal 1988). When environmental conditions deteriorate, especially after nutrient depletion, sinking velocities of the individual cells increase (Bienfang et al. 1982). Mucous production by senescent cells and their entanglement by spines promotes aggregate formation; scavenged mineral particles, together with heavy silica frustules, additionally increase sinking velocities. Within the small-sized coccolithophorids, the production of palmelloid stages as part of their life cycle provides large vehicles for rapid downward transport from surface layers. Cadee (1985) collected large numbers of coccolithophorid palmelloid stages in sediment traps in the North Sea. Earlier, Honjo (1982) observed a sedimentation event of these autotrophs in the Panama Basin that included mineral particles. The direct sedimentation of diatoms and coccolithophorids to the seafloor can be expected to provide geological signals in terms of intact opal or carbonate skeletons. Phytoplankters without mineral skeletons, which are common in new production regimes such as the gelatinous colony forming *Phaeocystis* sp., have recently been found to also sediment in large quantities during spring in the Barents Sea (Wassmann, personal communication).

Diatom species composition in the water column can differ from trap collections (Peinert 1986), and these again can deviate from recordings on the deep-sea floor (Takahashi 1986). This differential sedimentation may possibly be due to breakup of aggregates, varying frustule dissolution rates, and resting spore formation. Different behavior of diatom species may reflect differences in the life cycle strategies of neritic species (benthic resting stage) and truly oceanic ones (pelagic resting stage), as discussed by Smetacek (1985). While both types of diatoms would sink out of near-surface layers, differential sedimentation to greater depths would reduce the amount of material reaching deeper water layers or the seafloor and thus have a direct impact on the geological record.

Apart from spring blooms, *upwelling areas* are important new production sites on a global scale, with high exports via sedimentation, as reflected by high carbon content of the underlying sediments. Phytoplankton development during upwelling and horizontal water transport exhibits a pattern in space that is comparable to the seasonal spring development. The cold, nutrient-rich upwelled water carries a seeding population, and the phase of phytoplankton biomass buildup is followed by sedimentation later and further offshore. Upwelling systems can be intensively exploited by herbivores. The direct sinking of autotrophs, however, contributes significantly to the vertical flux.

Whereas spring blooms and upwelling systems are the classical new production regimes, *new production at the event scale in space and time*, that is superimposed on regenerating conditions in stratified waters, is not well documented. In a Gulf Stream warm-core ring, phytoplankton stocks increased by a factor of 4 and primary production by a factor of 20 at the time scale of days after a physical mixing event. Concomitantly, the size spectrum of the autotrophs significantly shifted towards larger species (Hitchcock et al. 1987).

The relationship between such new production events and their contribution to the short-term export of matter by sedimentation is very difficult to assess. Such events change the ratio between new and total production (f-ratio) within the entire productive layer. If this deviation is episodic, as is most likely in the above named examples, the additional autotrophic biomass may again sediment shortly thereafter if not utilized by heterotrophs. Even if taken up by heterotrophs, however, it will sediment eventually, albeit at a different time scale depending on food web efficiency and modified by food web interactions.

In contrast to episodic events, continuous new production proceeds below the seasonal thermocline in stratified waters, driven by upward diffusion or entrainment of new nutrients from below the euphotic zone. An increasing f-ratio at depth could be a widespread phenomenon. Although only a minor portion of the total primary production of the water column would be new production occurring at the lower reaches of the euphotic zone, it could account for the bulk of exported particulate matter. The pelagic system in general would be of a regenerating type, and overall losses by sedimentation would be low. The upper part of the euphotic zone, however, would represent the truly regenerating system.

Under these conditions, it is unlikely that direct autotrophic contributions to organic particle fluxes are important. Phytoplankton, however, could favor the aggregation of various particulates by mucous production, as heterotrophic mucus feeders and jelly plankters do (Alldredge and Newell 1986). Smetacek and Pollehne's (1986) discussion of excess carbohydrate (mucous) production by phytoplankton at high ambient light levels and low nutrient concentrations implies that this would be most important in the very top surface layers. Heterotrophic mucous production as a source for marine snow, however, can take place throughout the water column. Both origins can provide rapid transport vehicles for removing the smallest particles.

HETEROTROPHIC INFLUENCE ON VERTICAL FLUXES

There is no simple relationship between herbivore grazing and sedimentation. Fecal production by metazoans generally increases particle size of suspended matter and thus increases sinking velocity. However, the fate and degree of modification undergone by food particles eventually egested as feces differs widely and ranges from hailstorm-like sedimentation to retention in the layer of their production.

Grazing by Swarm Feeders

Swarm organisms, such as herbivorous fish, krill and salps, are patchily distributed in relation to their food; they commonly produce large feces that tend to sink out rapidly before they can be recycled. Because of the short food chain, these organisms accelerate export of essential elements from surface layers.

Large fecal matter from planktivorous fish can temporarily dominate sedimentation, as reported from the Peru upwelling system (Staresinic et al. 1984) and the Mediterranean Sea (Heussner et al. 1987). In the latter case, however, much larger amounts were collected by a trap at 50 m depth in contrast to one at 100 m depth. This implies rapid sinking of matter from the uppermost layer and subsequent deceleration beneath the euphotic zone, presumably due to disintegration and utilization by other organisms. Whereas the anchoveta fecal strings from the Peru upwelling contained large amounts of diatom cells and broken frustules, those from the Mediterranean Sea, originating from horse mackerel and sardines, were filled with fragments of planktonic crustaceans.

Salps can exhibit extremely high water clearance rates. During a mass occurrence in waters off Ireland, salps were reported to have terminated a diatom bloom prior to nutrient depletion. Numerous intact diatom cells or unbroken frustules were found in their fecal matter that dominated sedimentation during that time (Peinert 1985; Bathmann 1988). The hail of sedimenting salp fecal matter is inherently episodic and can well account for large, short-term variations in the flux of particulates at a given locality. Salps could contribute to irregular variability between years since no time patterns have yet been found for extreme differences in abundance.

Small-scale variation in sedimentation is also reported for euphausiids. In environments as different as Antarctic waters and the Norwegian Coastal Current, their densely packed, rapidly sinking fecal strings can account for the bulk of material leaving productive surface layers (Peinert 1986; von Bodungen et al. 1987). The impact of krill feeding, however, differs from that of salps with respect to the magnitude and composition of export. Grazing by krill does not contribute intact diatom frustules to vertical flux, since the food is thoroughly shredded in the stomach. Rapid disintegration of fecal strings in midwater depths (von Bodungen et al. 1987) prolongs the residence time in the water column and promotes further repackaging and degradation. Krill fecal strings are thus of minor importance for the geological record in terms of visible markers.

It can be concluded that short-term, locally intensive grazing impoverishes the pelagic system with respect to essential elements. It results in small-scale variation in sedimentation patterns in space and time that depend on the grazers' residence time in the respective water bodies. High sinking velocities of large fecal matter, however, do not necessarily imply a rapid and quantitative transfer to greater depths or to the seafloor. Grazing by krill and salps during spring blooms promotes losses prior to nutrient depletion. In areas of nutrient supply by upwelling or nonlimiting nutrient reserves (Antarctic waters), large stocks of herbivorous fishes in the former and krill in the latter region can be maintained, respectively. In regenerating systems, locally impoverished areas will arise from such kind of grazing. Essential elements will not be replenished until a breakdown of seasonal stratification. Consequently, food resources of less motile, smaller-sized organisms, such as protozoans and small copepods, will decline as well.

Impact of Herbivorous Copepod Grazing

Copepods also produce fecal pellets that are large in relation to the size of their food, and increased sinking rates should follow from Stoke's law. Copepods are unable to undertake significant horizontal migrations and their patchiness is much less marked as compared to swarm organisms. Vertical migration transports them into new water bodies if current shear prevails. The more evenly copepods are horizontally distributed, however, the more this resembles a simple exchange of individuals and not a new colonization of a water body. Because of their widespread abundance and importance as grazers, the impact of copepods on the vertical flux of particles deserves special attention. There is now evidence from field experiments that sinking copepod fecal pellets are not the transport vehicles that account for *major portions* of sedimenting particles (Cadee 1985; Peinert et al. 1987; Pilskaln and Honjo 1987). The mechanisms preventing mass sinking of copepod fecal matter are likely a combination of coprophagy, mechanical pellet disintegraton, and bacterial breakdown. The more intensive the interactions of different organisms within a complex food web are, the more efficient the retention mechanisms should be. Small et al. (1987) deployed sediment traps both in the thermocline and beneath the euphotic zone and found that only the latter trap collected copepod fecal pellets. Fecal matter was obviously recycled efficiently above the thermocline. Vertical migration of copepods and other heterotrophs interlink the two layers of the euphotic zone with greater depths.

In this context the biology of copepods is of considerable interest. Important features are factors determining size of overwintering stocks, mechanisms that trigger their ontogenetical vertical migration, feeding behavior, and reproductive cycle in relation to food supply.

Hypothetical spring phytoplankton developments under increasing grazing pressure of copepods and the resultant impact on sedimentation patterns are schematically depicted in Fig. 1: in temperate shallow water systems,

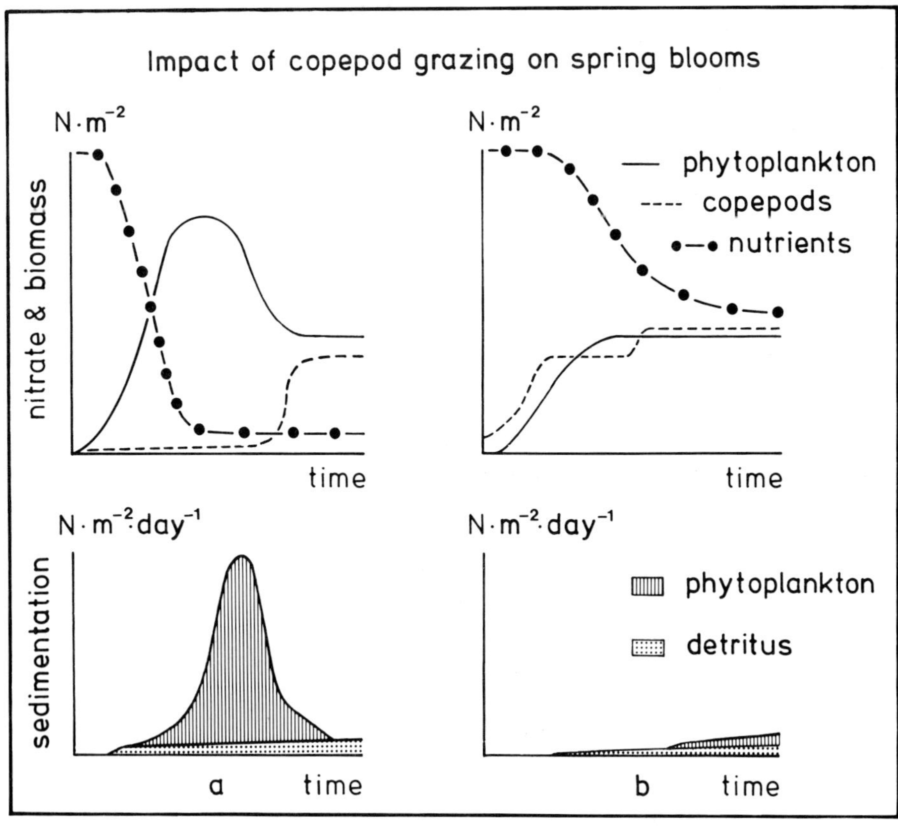

Fig. 1—Hypothetical influence of low and heavy copepod grazing on spring bloom development and pelagic sedimentation in terms of reactive nitrogen in the mixed layer. (*a*) Phytoplankton biomass accumulates until nutrient exhaustion and sediments result, as copepod grazing starts late. Loss rates are major and dominated by phytoplankton and fresh phytodetritus. (*b*) Early onset of copepod grazing prevents phytoplankton accumulation, nutrient exhaustion, and resultant sedimentation of autotrophs. Loss rates are minor and dominated by detritus.

the onset of breeding by the few overwintering copepods is coupled to spring phytoplankton growth. Generation times at low spring temperatures are at a scale of weeks. Both factors contribute to decoupling phytoplankton growth from herbivore grazing. Hence, mass sedimentation of algae takes place after nutrient depletion (Fig. 1a). In contrast, phytoplankton growth and copepod grazing are tightly coupled in the subarctic Pacific gyre. Here, breeding of the dominant herbivorous copepods is reported to begin prior to the onset of phytoplankton growth (Frost et al. 1983). Heavy grazing pressure exerted on the early stages of bloom growth prolongs the period during which reduced phytoplankton stocks grow on new nutrients. The greater the proportion of essential elements retained in surface waters by copepod grazing, the lower the spring sedimentation rates are (Fig. 1b). All observed spring developments should, in principle, be located somewhere on a continuum between these extreme examples.

An example for close coupling of phytoplankton and herbivores during spring (resembling Fig. 1b) was found in the Norwegian Current at 67°N (Peinert et al. 1987). Here, the early ascent of a large number of copepods to surface waters and their heavy grazing prevented rapid increase of phytoplankton biomass, and new nutrients remained available till long after spring, despite favorable hydrographic conditions for a biomass buildup. Sedimentation rates, recorded with sediment traps in different depths, were low. Detritus, including fecal pellets, comprised the bulk of settled matter. Phytoplankton contribution to trap collections was insignificant.

This example indicates that grazer control on spring phytoplankton accumulation can also be found in certain areas of the North Atlantic, despite differences in the overwintering strategy of the major resident herbivorous copepods. We do not know whether the above situation recurs annually in the Norwegian Current: the timing of ontogenetical vertical migration of *Calanus finmarchicus* to surface layers is known to vary over a range of weeks (Lie 1968). However, the timing mechanism (internal clock, some environmental trigger) is not known. The question of why the dominant copepods in the two northern oceans have differing life cycles can only be speculated upon.

In the Norwegian Sea, the sedimentation of copepod fecal pellets appears to be related to the size and age structure of the dominant *C. finmarchicus* population. In May/June a portion of the copepod population already commenced descent from surface waters for hibernation. In surface layers reproduction went on, shifting the composition of the population towards younger stages (unpublished data). This coincided with a change in numbers and size of suspended and sedimented copepod fecal pellets (Bathmann et al. 1987). Suspended pellet numbers were high in early May and dominated by large ones, originating from adult copepods. In late June, however, suspended numbers had decreased and smaller ones dominated (presumably from juvenile *C. finmarchicus* stages and other smaller species). Sediment trap collections, however, revealed that only very low numbers of mostly large pellets sedimented during the early period, whereas a pulse (comprised of smaller ones) was recorded in late June. High numbers of suspended large-sized pellets thus coincided with low loss rates by sedimentation and vice versa. These findings support the hypothesis that matter which is incorporated in copepod fecal pellets can indeed be retained in suspension. The efficiency, however, can vary and may depend on mechanisms closely related to population structure and abundance of the copepods themselves. If each copepod generation contributed to the overwintering stock, such pulses should be observed when the respective stages descend to great depths. Such behavior would have considerable bearing on the vertical flux.

Another factor mediating sedimentation rates at different time scales is the predation by carnivores on hibernating herbivorous copepods. Active *Euchaeta* sp. were collected together with dormant *C. finmarchicus* in deep water layers of the Norwegian Current in winter (unpublished data). Predation by carnivores decreases the size of the overwintering stock and thus the number of herbivorous copepods that ascend to surface waters the following spring. This influences the spring herbivorous grazing potential and, hence, the retention capability for essential elements during this period. Furthermore, any contribution of fecal matter from carnivorous grazing at great depth to the particle flux during autumn and winter would be sustained by organic substance produced months before. Hibernating herbivores that are stationary in midwater layers at a time scale of months may be exploited more efficiently by a deep living community of carnivores and raptorial detritus feeders than passively sinking particles of the same size spectrum ever could be.

Impact of Other Organisms

Aside from variability in vertical fluxes of detritus, changes in the contribution of different organisms themselves to sedimenting matter are indicative of changes in the pelagic environment. Whereas mass-sinking events of autotrophs can be related to a general decoupling of primary producers and heterotrophs, far less information is available to explain the sedimentation of heterotrophic organisms. Very little is known about the effects of changes in the physical environment, alterations of food supply and life cycles on sedimentation patterns, and the respective impacts are not easily separated.

Figure 2 gives an example of sequential sedimentation of organisms from sediment traps deployed in the Norwegian Current at 1000 m depth. Pteropod shells sediment during autumn only, coinciding with the annual maximum of the total flux (Wefer, this volume). Nothing is known, however, about the mechanisms governing their direct sedimentation, whether induced by starvation, old-age mortality, or predation. In the water column, copepod biomass declines during autumn (Lie 1968), when pteropods (*Limacina* sp.) are still present (Wiborg 1955). This annual succession from herbivores of different feeding type (filterers and mucus feeders) might well generally increase sedimentation rates. This would be the case if, e.g., *Limacina* sp. do not retain matter in the pelagic food web as efficiently as the copepods do.

Factors governing the sedimentation patterns of the protozoans (foraminifera, radiolarians, tintinnid ciliates) reside, in all likelihood, in their life cycles. Thus, foraminifera descend to depth for reproduction, discarding the adult shell in the process (Berger 1976). It is possible for other protozoans with mineral skeletons to exhibit similar behavior, if not for reproduction, then for dormancy. Thunell and Honjo (1987) linked foraminifera flux to temperature changes, presumably as a life cycle triggering factor.

The contribution of minipellets to mass fluxes is considered to be of minor quantitative importance. According to Stoke's law they exhibit slow sinking velocities because of their small size (10–50 μm diameter). The hitherto unknown minipellet producers (possibly large radiolarians) could serve to reprocess matter that otherwise is rapidly exported to great depths (Gowing and Silver 1985). The collection of increased minipellet numbers at times in sediment traps could indicate the uncoupling of food web interactions that deteriorate their degradation in the overlaying water column.

The production of dissolved organic matter (DOM) and its fate in changing seasonal food web interactions can be of quantitative importance for the vertical flux of particulates. In general, the DOM load of a water column follows the seasonal plankton production cycle, with an increase at the onset and a decrease at the termination of phytoplankton growth and high values in between. During spring growth a portion of the new nutrients will be channelled into the DOM pool, resulting in their retention within the surface layer. The nutrients bound in the DOM pool will eventually become available to the summer phytoplankton population. However, the rate of remineralization via the microbial loop is an unsolved question. The size of the DOM pool in surface layers might well be larger than assumed until recently (Susuki et al. 1985). If the higher values are proven correct, attention will have to be refocused on fundamental questions pertaining to sources and sinks of the DOM pool and to factors regulating its turnover rate. The answers to these questions should have far-reaching implications for our present conceptual understanding of the functioning of pelagic systems.

CONCLUDING REMARKS

We have attempted to show that magnitude and seasonality of vertical flux can be a simple function of physical (mixed layer depth) and chemical (new

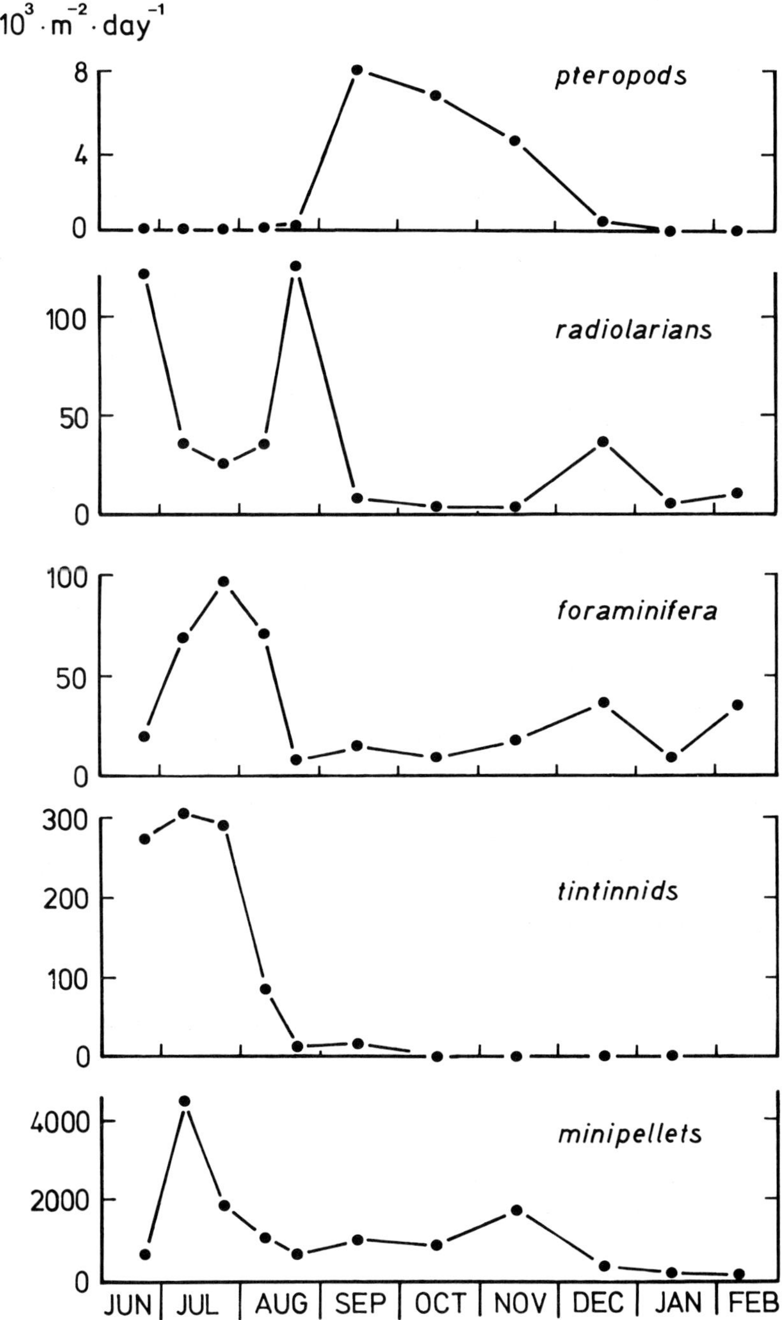

Fig. 2—Different organisms and minipellets collected by a moored automatic sediment trap in the Norwegian Current at 1000 m depth (67°N, 06°E) between June 1986 and February 1987.

nutrient concentration) factors in the case of spring blooms terminated by nutrient exhaustion. Input to the DOM pool is one major unknown here. Herbivore grazing complicates this relationship, as biological factors acting at various time scales can significantly modify annual export patterns. The factors regulating transition from new to regenerating systems are of particular interest in this context. Presumably, they regulate the amount of biogenic elements carried over from one system type to the next, and hence the magnitude and timing of the vertical export. A general increase in oceanic production, for instance, would not necessarily result in higher sedimentation rates within extant seasonal patterns. It could lead to more fundamental changes in the pattern of oceanic vertical flux by favoring a particular type of "export process chain" over another (e.g., greater increase in euphausiids or salp biomass relative to that of copepods). More knowledge of the biology and population regulatory mechanisms of the major herbivores in relation to the environment will be necessary before we can begin to consider the possible quantitative impact of changing oceanic productivity on the sedimentary record.

Acknowledgements. Thanks to M. Angel, W. Berger, P. Bienfang, and C. Reynolds for helpful criticism and valuable suggestions. Captains and crews of research vessels Littorina, Meteor, and Poseidon are thanked for their collaboration at sea. Space and numbers of citations in this paper have been limited, and papers by Alldredge, Berger, Billett, Eppley, Flynn, Gowing, Glover, Margalef, Pitcher, Reid, and Smayda were deleted from an earlier draft. The work was funded by the Deutsche Forschungsgemeinschaft via Sonderforschungsbereich 313. This is publ. nr. 50 of the Sonderforschungsbereich 313 at University of Kiel and publ. nr. 110 of the Alfred Wegener Institut für Polar- und Meeresforschung, Bremerhaven.

REFERENCES

Alldredge, A.L., and Hartwig, E.O., eds. 1986. Aggregate Dynamics in the Sea. Workshop report, Office of Naval Research, American Institute of Biological Sciences.

Bathmann, U. 1988. Mass occurrence of *Salpa fusiformis* in the spring of 1984 off Ireland: implications for sedimentation processes. *Marine Biol.* **97**: 127–135.

Bathmann, U.; Noji, T.T.; Voss, M.; and Peinert, R. 1987. Copepod fecal pellets: abundance, sedimentation and content at a permanent station in the Norwegian Sea in May/June 1986. *Mar. Ecol. Prog. Ser.* **38**: 45–51.

Berger, W.H. 1976. Biogenous deep sea sediments: production and preservation. In: Chemical Oceanography, vol. 5, eds. J.P. Riley and R. Chester, pp. 265–387. London: Academic.

Bienfang, P.K.; Harrison, P.J.; and Quarmby, L.M. 1982. Sinking rate response to depletion of nitrate, phosphate and silicate in four marine diatoms. *Marine Biol.* **67**: 295–302.

Cadee, G.C. 1985. Macroaggregates of *Emiliana huxleyi* in sediment traps. *Mar. Ecol. Prog. Ser.* **42**: 193–196.
Deuser, W.G. 1986. Seasonal and interannual variations in deep-water particle fluxes in the Sargasso Sea and their relation to surface hydrography. *Deep-Sea Res.* **33**: 225–246.
Eppley, R.W., and Petersen, B.J. 1979. Particulate organic matter flux and planktonic new production in the deep ocean. *Nature* **282**: 677–680.
Frost, B.W.; Landry, M.R.; and Hasset, P.R. 1983. Feeding behaviour of large calanoid copepods *Neocalanus cristatus* and *N. plumchrus* from the subarctic Pacific. *Deep-Sea Res.* **30**: 1–13.
Gowing, M.M., and Silver, M.W. 1985. Minipellets: a new and abundant size class of marine fecal pellets. *J. Marine Res.* **43**: 395–418.
Heussner, S.; Monaco, A.; and Fowler, S.W. 1987. Characterisation and vertical transport of settling biogenic particles in the northwestern Mediterranean. In: Particle Flux in the Ocean, eds. E.T. Degens, E.I. Izdar, and S. Honjo. Mitt. Geol.-Paläont. Inst., Univ. Hamburg. *SCOPE/UNEP Sonderband* **62**: 127–147.
Hitchcock, G.L.; Langdon, C.; and Smayda, T.J. 1987. Short-term changes in the biology of a Gulf Stream warm-core ring: phytoplankton biomass and productivity. *Limnol. Ocean.* **32**: 919–928.
Honjo, S. 1982. Seasonality and interaction of biogenic and lithogenic particulate flux at the Panama Basin. *Science* **218**: 883–884.
Lie, U. 1968. Variation in the quantity of zooplankton and propagation of *C. finmarchicus* at station M in the Norwegian Sea 1959–66. *Fisk. Dir. Skr. Ser. Havsunders.* **14**: 121–128.
Peinert, R. 1985. Saisonale und regionale Aspekte der Produktion und Sedimentation von Partikeln im Meer. Univ. Kiel, *Ber. Sonderforschungsbereich 313*.
Peinert, R. 1986. Production, grazing and sedimentation in the Norwegian Coastal Current. In: The Role of Freshwater Outflow in Coastal Marine Ecosystems, ed. S. Skreslet, pp. 361–374. Berlin: Springer–Verlag.
Peinert, R.; Bathmann, U.; von Bodungen, B.; and Noji, T. 1987. The impact of grazing on spring phytoplankton growth and sedimentation in the Norwegian Current. In: Particle Flux in the Ocean, eds. E.T. Degens, E.I. Izdar, and S. Honjo. Mitt. Geol.-Paläont. Inst., Univ. Hamburg. *SCOPE/UNEP Sonderband* **62**: 149–164.
Pilskaln, C.H., and Honjo, S. 1987. The fecal pellet fraction of biogeochemical particle fluxes to the deep sea. *Glob. Biogeochem. Cyc.* **1**: 31–48.
Small, F.L.; Knauer, G.A.; and Tuel, M.D. 1987. The role of sinking fecal pellets in stratified euphotic zones. *Deep-Sea Res.* **34**: 1705–1712.
Smetacek, V. 1985. Role of sinking in diatom life-history cycles: ecological, evolutionary and geological significance. *Marine Biol.* **84**: 239–251.
Smetacek, V., and Pollehne, F. 1986. Nutrient cycling in pelagic systems: a reappraisal of the conceptual framework. *Ophelia* **26**: 401–428.
Smetacek, V.; von Bodungen, B.; Knoppers, B.; Peinert, R.; Pollehne, F.; Stegmann, P.; and Zeitzschel, B. 1984. Seasonal stages characterizing the annual cycle of an inshore pelagic system. *Rapp. P.-v. Cons. Int. Explor. Mer.* **183**: 126–135.
Staresinic, N.; Hovey Clifford, C.; and Hulbert, E.M. 1984. Role of the southern Anchovy. *Engraulis ringens*, in the downward transport of particulate matter in the Peru coastal upwelling. In: Coastal Upwelling: Its Sedimentary Record, eds. E. Suess and J. Thiede. New York: Plenum.
Susuki, Y.; Sugimura, Y.; and Itoh, T. 1985. A catalytical oxidation method for the determination of total nitrogen dissolved in seawater. *Marine Chem.* **16**: 83–97.

Takahashi, K. 1986. Seasonal fluxes of pelagic diatoms in the subarctic Pacific 1982–1983. *Deep-Sea Res.* **33**: 1225–1251.

Thunell, R.C., and Honjo, S. 1987. Seasonal and interannual changes in planktonic foraminiferal production in the North Pacific. *Nature* **328**: 335–337.

Villareal, T.A. 1988. Positive buoyancy in the oceanic diatom *Rhizosolenia debyana* H. PERAGALLO. *Deep-Sea Res.*, in press.

von Bodungen, B.; Fischer, G.; Nöthig, E.-A.; and Wefer, G. 1987. Sedimentation of krill faeces during spring development in Bransfield Stait, Antarctica. In: Particle Flux in the Ocean, eds. E.T. Degens, E.I. Izdar, and S. Honjo. Mitt. Geol.-Paläont. Inst., Univ. Hamburg. *SCOPE/UNEP Sonderband* **62**, 243–257.

Wiborg, K.F. 1955. Zooplankton in relation to hydrography in the Norwegian Sea. *Fisk. Dir. Skr. Ser. Havsunders.*, vol. 11, No. 4.

Productivity of the Ocean: Present and Past
eds. W.H. Berger, V.S. Smetacek and G. Wefer, pp. 49–63
John Wiley & Sons Limited

Hydrodynamical Singularities as Controls of Recycled versus Export Production in Oceans

L. Legendre[1] and J. Le Fèvre[2]

[1]*Département de biologie, Université Laval*
Québec, Québec, Canada, G1K 7P4

[2]*Laboratoire d'Océanographie Biologique*
Université de Bretagne Occidentale
29287 Brest Cedex, France

Abstract. Phytoplankton production in oceans is either recycled in the euphotic zone or exported to the remainder of the ecosystem. The fate of marine primary production is discussed within the framework of a conceptual model, which specifies five major bifurcations where part of the production may be channeled into export pathways: *(a)* production of large versus small cells; *(b)* sinking or not of large cells; *(c)* grazing versus accumulation (and microphagy) of large cells; *(d)* recycling of small cells in the microbial food loop or aggregation; *(e)* sinking versus accumulation (and microphagy) of aggregates. It is shown that hydrodynamical singularities play a major role at each bifurcation in favoring production export over *in situ* recycling, which influences both present marine ecosystems and geological records.

INTRODUCTION

The oceanic environment is characterized by the presence of hydrodynamical singularities that occur over a wide range of spatio-temporal scales. Examples of such singularities are: *(a)* on the vertical axis, the pycnocline, and the ice-water interface; *(b)* on the horizontal plane, eddies, fronts, upwelling areas, and coastal zones; *(c)* along the time axis, temporal transitions in vertical stability of the water column with periodicities ranging from annual to semidiurnal (tidal) and sometimes smaller (e.g., Langmuir circulation). Before addressing the question as to whether such physical singularities play a role in the global production of oceans, some important terms must be defined.

In the pelagic environment, primary production is mainly effected by microscopic algae (phytoplankton), ranging in size from <1 μm to 2000 μm. Biological oceanographers often distinguish between two types of phytoplankton production: *(a)* new production (P_N), which is derived from nutrients supplied by hydrodynamical processes, and *(b)* regenerated production (P_R), which is fueled by nutrients regenerated locally by heterotrophs (see Eppley, this volume). The importance of P_N versus P_R is usually estimated from the relative uptake by phytoplankton of NO_3 (replenished by physical processes) versus NH_4 and urea (regenerated *in situ*). Total primary production is $P_T = P_N + P_R$, and the ratio of new to total production ("f-ratio") is $f = P_N/P_T$. The concepts of new and regenerated production refer to the origin of nutrients used by the primary producers, not directly to the fate of primary production in marine ecosystems. In this chapter, "recycled production" and "export production" will be used instead to identify pathways in the food web. It will be shown that part of primary production derived from "new" nutrients is generally recycled in the upper water column, while part of production fueled by "regenerated" nutrients may be exported. Within the context of this chapter, export production includes fluxes of biogenic material from surface waters both as sinking and as swimming particles, up to fish and whales (see Bishop, this volume). New and export production become mathematically equivalent only on global scales and when all forms of inputs and outputs are taken into account. For nitrogen, this includes nitrogen fixation and denitrification (Codispoti, this volume) as well as fluxes of dissolved organic matter (Toggweiler, this volume).

Goldman (1988) proposed a model of phytoplankton production in open oceanic waters. This model distinguishes between two production systems in the euphotic zone that are separated both spatially and temporally. The first system (the microbial food loop) consists of very small phytoplankters (<5 μm, ultraplankton), heterotrophic bacteria, and protozoa; its primary production would be the one measured by standard incubations at sea. Only a small part of this production would be exported, since most of the energy stored by ultraplankton is dissipated within the loop (e.g., Smith et al. 1986). The second system is based on larger cells (e.g., diatoms), growing in response to short-lived and localized mixing events, which would allow rapid bursts in production that remain mostly undetected by bottle incubations. These large cells would account for much of the export production from the euphotic zone.

According to Goldman (1988), growth of large cells in the open ocean would reflect biological events that are patchy and on temporal and spatial scales not easily amenable to quantitative measurement. It will be argued here that this is the case because hydrodynamical forcing in open oceanic waters is generally weak. Biological events that are sporadic in oceanic waters may become systematic in other environments. It will be shown that this occurs in relation with hydrodynamical singularities which, in our view, play a major role in favoring production export over *in situ* recycling.

BIFURCATION MODEL

The role of hydrodynamical singularities in the global production of oceans will be discussed within the framework of a conceptual model where five major bifurcations are specified (Fig. 1). The first bifurcation sets the conditions for the production of relatively large phytoplankters versus ultraplankters (<5 μm). The large cells are expected to sink rapidly out of the euphotic zone; the second bifurcation sets conditions for large cells to remain instead in the euphotic zone. At the third bifurcation, algae that remain in the euphotic zone may either be grazed by herbivores or accumulate in hydrodynamical traps, where they are degraded into smaller particles and may be grazed by microphagous plankters. Contrary to large cells, ultraplankton are expected to remain in the euphotic zone and become part of the microbial food loop; the fourth bifurcation gives conditions under which very small cells become aggregated into larger particles. At the fifth bifurcation, these aggregates may either sink to depth or accumulate in hydrodynamical traps where microphagy can take place.

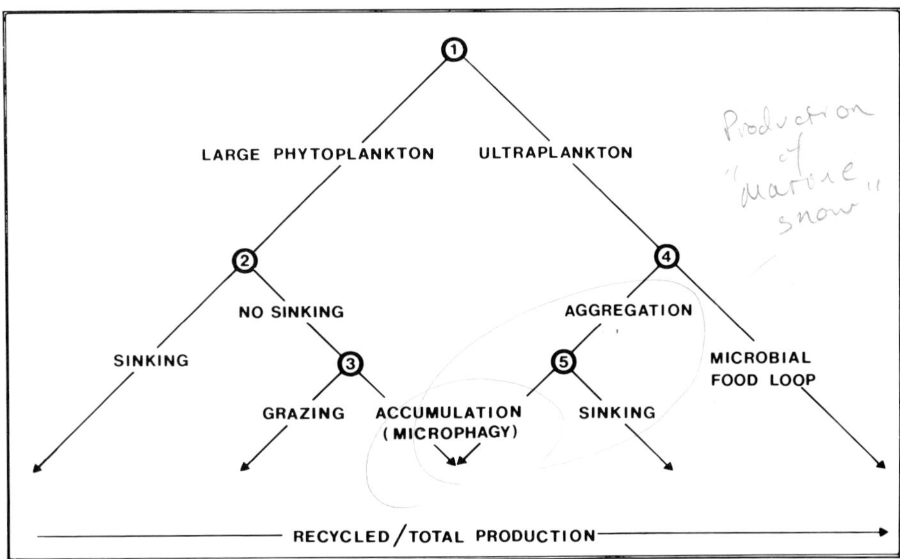

Fig. 1—Bifurcation model of export production (downward arrows) in oceans. The model (see text) does not imply hierarchy among physical factors causing the bifurcations since several bifurcations may take place simultaneously.

At each bifurcation, part of the primary production may be channeled into export pathways, which does not preclude coexistence with recycling pathways as recognized by Goldman (1988) for open oceanic waters. It will be shown that hydrodynamical singularities are involved in the five bifurcations. Given the format of Dahlem Konferenzen papers, no overall review of biological production at hydrodynamical singularities will be attempted here; instead, selected examples of key mechanisms will be presented with preference given to papers containing comprehensive reference lists.

Bifurcation 1: Small versus Large Phytoplankton

It is generally accepted (refs. in Goldman 1988) that sustained export production requires short food chains based on large primary producers. If hydrodynamical singularities do play a role in favoring production export, they must positively influence the growth of relatively large cells (>5 μm). In the oceans, large forms (including chains of small cells) generally belong to diatoms and dinoflagellates, but not all diatoms nor dinoflagellates are large. To the contrary, some species are only a few micrometers in diameter. Species >5 μm also include most coccolithophores.

There are several cases in which large cells are associated with hydrodynamical singularities. In the annual cycle, blooms of diatoms occur when "new" nutrients are abundant and vertical stratification confines the cells above a "critical depth" (Sverdrup 1953). These conditions trigger blooms at various times of the year, most often spring and autumn. In areas where vertical mixing is such as to maintain high concentrations of "new" nutrients throughout the summer and simultaneously keep the surface mixed layer shallower than the critical depth, large diatoms continuously dominate phytoplankton biomass (e.g., Levasseur et al. 1984). On time scales that extend from a few days to several months, hydrodynamical conditions can therefore channel primary production into large diatoms.

On the vertical axis, biomass at the ice-water interface in polar regions is dominated by diatoms in the spring and at the beginning of the summer (refs. in Horner 1985). Diatoms and/or dinoflagellates can also dominate subsurface chlorophyll maxima during the summer, as in the western Mediterranean (Estrada et al. 1985) or on the northwest European shelf (refs. in Le Fèvre 1986). This would also occur in oceanic waters (e.g., Gulf Stream warm core rings; Hitchcock et al. 1987), where outgrowths of large phytoplankton have been observed in response to sporadic mixing events. On the horizontal plane, diatoms are also associated with various structures such as upwelling areas (refs. in Wangersky 1977) and nutrient-rich coastal mixed layer, while both diatoms and dinoflagellates may dominate the biomass in frontal areas (refs. in Le Fèvre 1986).

The reasons why hydrodynamical singularities favor large cells were summarized by Margalef (1978). Rather than discussing detailed physiological responses of species to light and nutrients, Margalef associated life forms of phytoplankton to the amount of "auxiliary" (i.e., mechanical, nonphotosynthetic) energy supplied to the system and degraded therein. Large cells generally have lower surface/volume ratios than small cells so that they require richer nutrient environments. In addition, nonmotile cells such as diatoms would rapidly sink out of the euphotic zone in the absence of relatively strong vertical mixing; however, strong mixing may deepen the mixed layer below the critical depth so that there is a narrow window open for large diatoms in the spectrum of auxiliary energy. A number of hydrodynamical singularities offer the high nutrient and turbulence conditions required by diatoms. Singularities with lower levels of auxiliary energy would rather support large dinoflagellates since these motile organisms do not require the same degree of turbulence as diatoms.

Bifurcation 2: Fate of Large Phytoplankton

According to Stokes' model, the speed of a spherical particle falling through water is a direct function of both the density of the particle and the square of its radius, and is an inverse function of water viscosity (see Smayda 1970; Margalef 1978). For example, a particle 100 μm in diameter sinks 10^4 times more rapidly than a 1 μm particle. Despite large departures from Stokes' model, due to nonspherical shapes and surface properties of the cells, large cells have a tendency to sink and be exported while most small cells remain in the euphotic zone where they are recycled. In addition, sinking rates of several groups, including diatoms and dinoflagellates, are influenced by the physiological state of the organisms (see Smayda 1970; Peinert et al., this volume).

Large cells can sink very fast, at rates which may exceed 100 m d^{-1} (refs. in Goldman 1988). Aggregation of cells in the water column accelerates sinking, including that of coccolithophores. The question at hand is therefore not "why do large cells sink?" but rather "why do some of the large cells not sink out of the euphotic zone before being grazed by herbivores or accumulated in hydrodynamical traps?" Examples of such traps include fronts and Langmuir cells because of the associated convergent circulation, certain types of eddies, and the pycnocline, where sinking rates often decrease sharply.

In order to answer the above question, it is useful to first consider an example in which large phytoplankton are grazed. A simple case of herbivorous grazing can be found in Vives (1971), for an upwelling area of the Mediterranean. Figure 2 shows the relationship between upward water velocity (an indication of nutrient flux) and the total numbers of pelagic

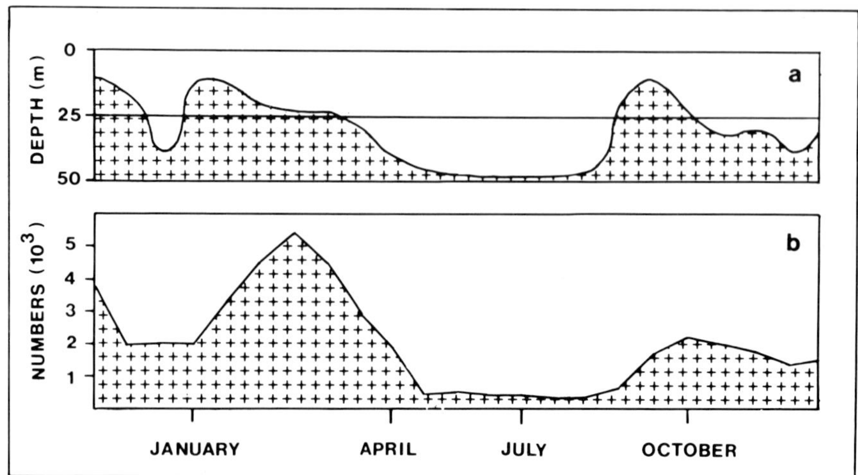

Fig. 2—Time series of (*a*) upward water velocities (shaded: > 2 m d^{-1}) from 50 to 25 m and 25 to 1 m and (*b*) numbers of pelagic copepods, in an upwelling area off the Spanish coast in the Mediterranean (after Vives 1971).

copepods (an approximate indication of herbivores). It can be seen that while decreases in copepod numbers closely match decreases in upwelling, copepods do peak about 3 weeks to 1 month after increases in upward water velocity, which is consistent with the generation time of most herbivorous species. In other words, while the herbivore community may collapse almost as soon as its food source is severed, it takes about one generation time to build up in response to favorable conditions. The latter must therefore last long enough if the response is to take place at all.

In the previous section (bifurcation 1), it has been explained why the growth of large cells occurs at those hydrodynamical singularities where vertical mixing does match the time characteristics of phytoplankton. Furthermore, the dynamics of herbivorous grazing are such that the time characteristics of hydrodynamical singularities must also match the pace of zooplankton life history in order for phytoplankton not to be lost to sinking. In the next section it will be shown that different hydrodynamical time scales may lead to very different heterotrophic pathways; in few cases, however, is the transfer from large primary producers to heterotrophs possible in the euphotic zone without minimum persistence of hydrodynamical singularities.

Bifurcation 3: Grazing versus Accumulation

The third bifurcation concerns the fate of large phytoplankton remaining in the euphotic zone. It can be viewed as an alternative between an herbivore pathway, through which algal production is exported to higher and higher trophic levels such as zooplankton and ultimately fish, and a decomposer pathway through which phytoplankton biomass goes to microheterotrophs, with bacteria as the first link. Consistent with previous attempts at conceptualizing this critical step (Legendre et al. 1986; Le Fèvre and Frontier 1988), the view advocated here is that the switching between the two pathways is governed by the degree of matching between the temporal characteristics of physical phenomena (i.e., hydrodynamical singularities) and those of biotic responses, which usually involve a time lag between the driving physical events and the resulting changes in pelagic food chains.

The simplest case would be that of a permanent or quasi-permanent physical process, such as the Mediterranean upwelling described above or the ice-water interface (refs. in Horner 1985), where part of the production is channeled in the herbivore pathway. More complex physical singularities are the frontal systems, ranging in size from small river plumes to major oceanic structures and whose dynamics span over a comparable range of temporal rates. Frontal systems have important biological effects and, as exemplified by Fig. 3, often govern the taxonomic composition of

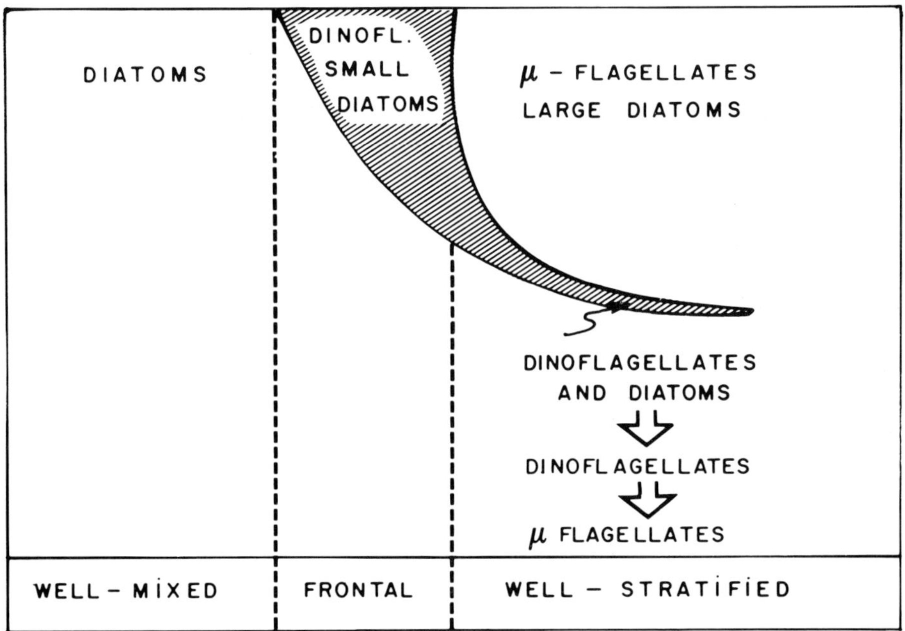

Fig. 3—Generalized depth distribution of phytoplankton in some frontal areas with surface and subsurface chlorophyll maxima (shaded); the open arrows refer to a seasonal succession within the subsurface chlorophyll maximum. From Demers et al. (1986) after data from several sources, by permission of Springer-Verlag.

phytoplankton. The effects on subsequent links in food chains, and more generally the switching between the major trophic pathways, can be shown to depend on the temporal characteristics of frontal dynamics, and especially on the periodicity and duration of nutrient input through mixing processes. The Ligurian Front in the Mediterranean is an oceanic thermohaline structure where high phytoplankton standing stocks are found near the surface in the frontal region together with high-density patches of the copepods *Calanus helgolandicus, Centropages typicus*, and *Euchirella rostrata* (Boucher 1984). The best-defined frontal maxima are those of late copepodites (which may be the most actively feeding stages), at least in the case of *C. helgolandicus* (herbivore) and *E. rostrata* (carnivore). This suggests that phytoplankton production at the front is channeled into a fully fledged pelagic food chain; the latter would be given sufficient time to build up because changes in the major features of this front take months.

Short periodicities in nutrient enrichment may result in building up an herbivore food chain as well, as exemplified by the ecosystem at the Celtic Sea shelf-break. When the seasonal thermocline has been established, internal tides are generated there through the interaction between the barotropic tidal wave and bottom topography, producing pycnocline oscillations on a semidiurnal frequency, with a maximum range (ca. 60 m) over the shelf-break. As a result, wind-induced vertical mixing is periodically enhanced (Mazé et al. 1986) and nutrients are pulsed into the euphotic zone with a semidiurnal periodicity. For the plankton, these pulses are equivalent to a continuous flux and the outcome is similar to that found in an upwelling environment: increased standing stocks of both phytoplankton and zooplankton are found in coincidence with a band of cool water over the shelf-break (refs. in Le Fèvre 1986) and clearcut maxima of herbivorous copepods have been observed at spots where the energy of internal tides is maximal (Le Fèvre and Frontier 1988). Shelf-break areas, including that of the Celtic Sea, often support large commercial fisheries, these singularities being thus cases where temporal characteristics of physical processes channel plankton production into the herbivore export pathway.

On the northwest European shelf, major thermal fronts are found in summer at the boundaries between tidally well-mixed and seasonally stratified waters. High phytoplankton standing stocks are often found along the fronts on their stratified side (e.g., Pingree et al. 1975). However, despite a number of speculations that these frontal systems should play a major role in the overall pelagic production in the area, no relationship has been observed between them and the abundance of herbivorous zooplankton or fisheries activity (refs. in Le Fèvre 1986; see also Fogg et al. 1985). The dynamics of nutrient input at these tidal fronts were described by Morin et al. (1985), who showed that the nutrient reserve is found in the bottom water on the stratified side of the front and that some of it is made available through intensified mixing from neap to spring tides; nutrient-rich water is then found along the front on the well-mixed side and nutrient input to the surface water in the stratified area takes place through turbulent cross-frontal exchanges. With some time lag, with respect to the variations in tidal range, phytoplankton growth is stimulated for a few days every second week; this is consistent with the actual observations of high chlorophyll values in the area. A number of convergent observations suggest that in the intervals between the episodes of active growth, high phytoplankton biomasses are often maintained by mechanical accumulation through frontal convergence, with little growth or no growth at all taking place (refs. in Le Fèvre 1986). Herbivorous zooplankton cannot keep pace with such sporadic phytoplankton outgrowths. They will not readily feed upon senescent cells either, so that the accumulated biomass tends to be recycled instead by a microheterotrophic community of decomposers with bacteria as the first link. The process may result in the buildup of an accumulation biotope (see Le Fèvre 1986, pp. 225–232) where bacteria, mostly associated with particles, are grazed upon by microflagellates and ciliates; microheterotrophs in the accumulation biotope may, in turn, be grazed upon by microphagous metazoans such as appendicularians, thaliaceans, or the pteropod genus *Limacina*. Together with observations of high bacterial activity on European tidal fronts, especially in coincidence with nongrowing patches of the dinoflagellate *Gyrodinium aureolum*, indirect evidence of this trophic pathway has been recorded in the form of maxima in the abundance of appendicularians, doliolids, and *Limacina* (see Le Fèvre 1986; Le Fèvre and Frontier 1988) associated with either the tidal fronts or the pycnocline. The latter is another hydrodynamical singularity where both increased production of large cells (bifurcation 1) and accumulation of particulate material can occur.

Bifurcation 4: Fate of Small Cells — Microbial Food Loop versus Aggregation

As explained above, ultraplankton are expected to remain in the euphotic zone where, according to Goldman (1988), their normal fate is to be recycled within a microbial community ("microbial food loop"), which is mainly composed of bacteria and protozoans and in which most of the energy is dissipated. Escape from this loop (rightmost arrow in Fig. 1) could, in principle, be mediated by metazoans grazing directly on ultraplankton. However, Fenchel (1984) points out that the only demonstrated case of what he calls "baleen whale" feeding by large organisms on bacterial-size plankton is found in appendicularians, while the authors of the demonstration (King et al. 1980) conclude that this appendicularian shunt is not a major process either with respect to appendicularian food requirements or with respect to bacterial grazing by very small flagellates. More generally, possible metazoan consumers of very small particles include filter feeders (mainly tunicates) and a few other organisms with a ciliary mucous collecting mechanism such as *Limacina*. With the exception of appendicularians, most of these are unable to significantly collect particles <2 μm, as pointed out by Fenchel (1984) who concludes that "in the pelagic food chains there are limits to size ratios between prey and predator species." Aggregation into larger particles will thus be the major way through which escape from the microbial loop will take place, through either grazing or sedimentation.

Although a variety of physical, chemical, and biological processes may be involved (e.g., the formation of "marine snow" mentioned below), at least some hydrodynamical singularities are known to be effective in aggregating very small particles into larger ones. A complex case is the one of the air-water interface, where particle generation takes place through various processes involving micro-bubble physics, the chemistry of the surface film, and bacterial activity (refs. in Le Fèvre 1986). More generally, any structure that can act as a hydrodynamical trap (see bifurcation 2) will concentrate particles of various sizes and be instrumental in the buildup of aggregates.

Bifurcation 5: Sinking versus Accumulation

Once aggregated into larger particles, ultraplankton cells may either sink to deeper waters or be accumulated, at least temporarily, in hydrodynamical traps. According to Silver et al. (1986), sedimentation of intact cells occurs mainly through their incorporation into "marine snow," i.e., aggregates of organic material derived from organic substrates such as appendicularian mucous houses. The sources reviewed by these authors show, however, that this mode of export, while possibly important as a source of organic material for deep waters, always involves less than 1% of the ultraphytoplankton in the euphotic zone.

Several types of hydrodynamical singularities can be instrumental in accumulating microbial aggregates and detrital particles. Among them are the fronts, because of their associated convergent circulation, which actually play a role in both the formation of aggregates from smaller particles (bifurcation 4) and the degradation of unconsumed large phytoplankters (bifurcation 3). As in the above mentioned microbial loop, microheterotrophic activity in the accumulation biotope will, in part, result in mineralization of organic matter, i.e., recycling. Biomass export will also occur if only because bacteria will tend to be associated with particles, a condition under which they are more readily consumed by zooplankton (King et al. 1980; see also Angel, this volume). More generally, because of the presence of debris and protozoan grazers of bacteria, particle size will tend to be larger than in the midwater microbial loop, i.e., more likely to be within the reach of metazoan microphagous zooplankters.

Microphagy leads to production export in at least two different ways, one of which is the sinking of fecal pellets to deeper waters. Alldredge (1984) mentions that fecal pellets produced by thaliaceans and pteropods sink much faster than those of euphausids or copepods and she quotes several sources indicating that the flux of fecal material from thaliaceans to the benthos may be quite significant in some situations. The second way is predation. It is widely recognized, for instance, that some fish larvae, especially flatfish, feed on appendicularians. Thaliaceans also have predators, e.g., hyperiid amphipods such as *Phronima* that carve a shelter out of the body of a salp and feed at least in part on the material of their house. Such crustaceans, with a size in the centimetric range, are in turn a perfectly suitable food for large pelagic fish (e.g., tuna).

Little is known, however, of the relative importance of the microphagous pathway in production export to higher-order consumers. Given estimates found in the literature of conversion efficiencies, an educated guess would be that no more than 15% of the bacterial biomass produced per unit time is passed on to metazoan grazers. It is worth noting that most of the latter are gelatinous plankters, whose body contains at least 95% water (see Alldredge 1984) and which therefore, despite their relatively large sizes, do not represent a large carbon biomass. It may also be worth noting that the success in redirecting microbial production toward metazoans, however inefficient, is achieved through a dramatic shortening of the food chain ("baleen whale feeding"). At least in the case of thaliaceans, the transfer is facilitated by their ability to increase their numbers in an explosive way through asexual multiplication, which might enable their populations to cope with the pace of the microbial community.

Recycled versus Export Production

The above discussion mainly concerned export production. Recycled production, however, takes place at each bifurcation in Fig. 1, where the ratio of recycled to total production increases from left to right in the diagram. Production and sinking of large cells (bifurcation 1) are accompanied by exudation of organic matter in the euphotic zone, at a rate considered rather small by some authors while others quote values up to 50% of the net annual production (refs. in Prieur and Legendre 1988). Bifurcation 2 leads to grazing by herbivores and therefore to excretion of dissolved organic matter in the euphotic zone. Bifurcations 3 and 5 converge toward the accumulation biotope, where probably more than 85% of the biomass is recycled. Export of ultraplankton through sinking is only possible if the cells are incorporated into large organic aggregates, in which recycling by bacterial activity is very high. Finally, small cells in the microbial food loop (bifurcation 4) are almost entirely recycled. In Fig. 1 there is also a general decrease in the size of photosynthetically derived particles from left to right, which suggests an allometric control of the ratio of recycled to total production.

HYDRODYNAMICAL SINGULARITIES AND GLOBAL PRODUCTION

One way to assess the importance of hydrodynamical singularities in global marine production would be to estimate the proportion of total primary production that actually takes place at these singularities. This approach would not distinguish, however, between recycled and export production, which would obscure the significance of hydrodynamical singularities since the fates of recycled and export production are not the same either ecologically or geologically.

It has been shown above that hydrodynamical singularities deeply influence the fate of primary production in oceans. This is of major significance for present marine ecosystems, and also for the interpretation of geological records as these only reflect export production. Walsh et al. (1981) report the effects of hydrodynamical singularities (upwelling, river runoff, tidal mixing, and so on) on bottom sedimentation. Nutrient enrichment is often associated with such hydrodynamical singularities on the continental shelf; actual losses of photosynthetic carbon to the bottom depend on the activity of the pelagic food web, those areas with less active food webs (e.g., Southwest African shelf) losing more than areas with active ones (e.g., Peru). A fraction of production exports in the form of sinking cells and fecal pellets is lost from the food web to bottom sediments, since 1 to 25% of the primary production on the shelf is eventually buried on the continental slope. Export production may therefore be reflected in geological records, and sinking versus grazing may determine the contribution of pelagic ecosystems to bottom sediments.

The bifurcation model took advantage of Goldman's (1988) idea, that both new and regenerated production could operate side by side in open oceanic waters, given their different temporal and spatial scales. Putting this idea into a more general hydrodynamical framework made it possible to understand under which conditions primary production is channeled into export pathways. The recognition that hydrodynamical singularities play a major role in channeling phytoplankton production is consistent with the hypothesis of Legendre et al. (1986), as reformulated by Prieur and Legendre (1988), that "enhanced biological production occurs at ergoclines as the consequence of the matching or resonance of physical rates with biological rates," hydrodynamical singularities being here a more general term for "ergoclines." In addition, the bifurcation model recognizes the fundamental role of size rather than trophic levels in marine ecosystems.

In reviewing the literature for this chapter, we found very few studies in which simultaneous information was available on hydrodynamical processes, phytoplankton biomass, and/or production by size classes, sinking, and grazing by zooplankton. Without such data it is not possible to understand and assess the fate of primary production in pelagic ecosystems, and thus to model fish production. In addition, without convincing estimates of the relative importance of recycled versus export production in oceans, as well as grazing versus sinking, geological records cannot be interpreted in terms of past marine ecosystems. Assessing the oceanographic conditions controlling the relative importance of export versus recycled pathways, therefore, remains a major task in oceanography. This task becomes much easier once it is understood that hydrodynamical singularities are a key factor controlling bifurcations that lead to either recycling or export of pelagic primary production.

Acknowledgements. We thank Drs. M.V. Angel, W.H. Berger, P.K. Bienfang, L.A. Codispoti, D.H. Cushing, R.W. Eppley, L. Fortier, S. Frontier, R. Gersonde, B.T. Hargrave, P.M. Holligan, A.R. Longhurst, R. Margalef, J.F. Minster, and B.M. Mycke for their comments and suggestions. Financial support from the French Ministère des Affaires Étrangères to J.L.F. made possible a trip to Canada to write this paper, and a research grant from the Natural Sciences and Engineering Research Council of Canada to L.L. as well as a contract from the Direction des Recherches et Études Techniques, Ministère de la Défense, Paris, to the Université de Bretagne Occidentale, were instrumental in the completion of the work.

REFERENCES

Alldredge, A.L. 1984. The quantitative significance of gelatinous zooplankton as pelagic consumers. In: Flows of Energy and Materials in Marine Ecosystems: Theory and Practice, ed. M.J.R. Fasham, pp. 407–433. New York: Plenum.

Boucher, J. 1984. Localization of zooplankton populations in the Ligurian marine front: role of ontogenic migration. *Deep-Sea Res.* **31**: 469–484.

Demers, S.; Legendre, L.; and Therriault, J.C. 1986. Phytoplankton responses to vertical tidal mixing. In: Tidal Mixing and Plankton Dynamics, eds. J. Bowman, C.M. Yentsch, and W.T. Peterson, pp. 1–40. Berlin: Springer.

Estrada, M.; Vives, F.; and Alcaraz, M. 1985. Life and the productivity of the open sea. In: Key Environments: Western Mediterranean, ed. R. Margalef, pp. 148–197. Oxford: Pergamon.

Fenchel, T. 1984. Suspended marine bacteria as a food source. In: Flows of Energy and Materials in Marine Ecosystems: Theory and Practice, ed. M.J.R. Fasham, pp. 301–315. New York: Plenum.

Fogg, G.E.; Egan, B.; Floodgate, G.D.; Jones, D.A.; Kassab, J.Y.; Lochte, K.; Rees,

E.I.S.; and Turley, C.M. 1985. Biological studies in the vicinity of a shallow-sea tidal mixing front. VII. The frontal ecosystems. *Phil. Trans. R. Soc. Lon. B* **310**: 555–571.

Goldman, J.C. 1988. Spatial and temporal discontinuities of biological processes in pelagic surface waters. In: Toward a Theory of Biological–Physical Interactions in the World Ocean, ed. B.J. Rothschild, pp. 273–296. Dordrecht: Kluwer.

Hitchcock, G.L.; Langdon, C.; and Smayda, T.J. 1987. Short-term changes in the biology of a Gulf Stream warm-core ring: phytoplankton biomass and productivity. *Limnol. Ocean.* **32**: 919–928.

Horner, R.A., ed. 1985. Sea Ice Biota. Boca Raton: CRC Press.

King, K.R.; Hollibaugh, J.T.; and Azam, F. 1980. Predator-prey interactions between the larvacean *Oikopleura dioica* and bacterioplankton in enclosed water columns. *Marine Biol.* **56**: 49–57.

Le Fèvre, J. 1986. Aspects of the biology of frontal systems. *Adv. Mar. Biol.* **23**: 163–299.

Le Fèvre, J., and Frontier, S. 1988. Influence of temporal characteristics of physical phenomena on plankton dynamics, as shown by north-west European marine ecosystems. In: Toward a Theory on Biological–Physical Interactions in the World Ocean, ed. B.J. Rothschild, pp. 245–272. Dordrecht: Kluwer.

Legendre, L.; Demers, S.; and Lefaivre, D. 1986. Biological production at marine ergoclines. In: Marine Interfaces Ecohydrodynamics, ed. J.C.J. Nihoul, pp. 1–29. Amsterdam: Elsevier.

Levasseur, M.; Therriault, J.C.; and Legendre, L. 1984. Hierarchical control of phytoplankton succession by physical factors. *Mar. Ecol. Prog. Ser.* **19**: 211–222.

Margalef, R. 1978. Life-forms of phytoplankton as survival alternatives in an unstable environment. *Oceanol. Acta* **1**: 493–509.

Mazé, R.; Camus, Y.; and Le Tareau, J.Y. 1986. Formation de gradients thermiques à la surface de l'océan, au-dessus d'un talus, par interaction entre les ondes internes et le mélange dû au vent. *J. Cons. Int. Explor. Mer* **42**: 221–240.

Morin, P.; Le Corre, P.; and Le Fèvre, J. 1985. Assimilation and regeneration of nutrients off the west coast of Brittany. *J. Mar. Biol. Assn. UK* **65**: 677–695.

Pingree, R.D.; Pugh, P.R.; Holligan, P.M.; and Forster, G.R. 1975. Summer plankton blooms and red tides along tidal fronts in the approaches to the English Channel. *Nature* **258**: 672–677.

Prieur L., and Legendre, L. 1988. Oceanographic criteria for new phytoplankton production. In: Toward a Theory on Biological-Physical Interactions in the World Ocean, ed. B.J. Rothschild, pp.71–112. Dordrecht: Kluwer.

Silver, M.W.; Gowing, M.M.; and Davoll, P.J. 1986. The association of photosynthetic picoplankton and ultraplankton with pelagic detritus through the water column (0–2000m). In: Photosynthetic Picoplankton, eds. T. Platt and W.K.W. Li. *Can. Bull. Fish. Aquat. Sci.* **214**: 311–341.

Smayda, T.J. 1970. The suspension and sinking of phytoplankton in the sea. *Ocean. Mar. Biol. Ann. Rev.* **8**: 353–414.

Smith, R.E.H.; Harrison, W.G.; Irwin, B.; and Platt, T. 1986. Metabolism and carbon exchange in microplankton of the Grand Banks (Newfoundland). *Mar. Ecol. Prog. Ser.* **34**: 171–183.

Sverdrup, H.U. 1953. On conditions for the vernal blooming of phytoplankton. *J. Cons. Int. Explor. Mer* **18**: 287–295.

Vives, F. 1971. L'affleurement d'eau sur la côte catalane et les indicateurs biologiques (copépodes). *Inves. Pesq.* **35**: 161–169.

Walsh, J.J.; Premuzic, E.T.; and Whitledge, T.E. 1981. Fate of nutrient enrichment on continental shelves as indicated by the C/N content of bottom sediments. In: Ecohydrodynamics, ed. J.C.J. Nihoul, pp. 13–49. Amsterdam: Elsevier.

Wangersky, P.J. 1977. The role of particulate matter in the productivity of surface waters. *Helgoländer Wiss. Meer.* **30**: 546–564.

Productivity of the Ocean: Present and Past
eds. W.H. Berger, V.S. Smetacek and G. Wefer, pp. 65–83
John Wiley & Sons Limited

Is the Downward Dissolved Organic Matter (DOM) Flux Important in Carbon Transport?

J.R. Toggweiler

Geophysical Fluid Dynamics Laboratory/NOAA
Princeton University
Princeton, NJ 08542, U.S.A.

Abstract. A new method for measuring the concentration of dissolved organic carbon (DOC) and dissolved organic nitrogen (DON) in seawater has recently been applied to the study of the material balance in the oceanic water column. These measurements suggest that the downward transport of organic carbon and nitrogen in the dissolved organic phase is every bit as important as the downward transport in sinking particles. It appears that DOC and DON are the most important organic substrates supporting the consumption of oxygen and the remineralization of nitrate below the thermocline. Although still controversial, these findings are supported by a model study which shows that the vertical transport of organic matter cannot be attributed solely to the fast-sinking particles caught in sediment traps. A characterization of the vertical flux as such produces a model nutrient distribution which bears little resemblance to observed distributions.

INTRODUCTION

Sediment traps and large volume filtration systems have now been in use for more than a decade. These technologies give oceanographers a means of directly sampling the vertical flux of particulate material from the upper ocean. In the pre-trap era oceanographers could only infer what the characteristics of the sinking particles were from the distributions of biologically altered chemicals in the ocean, such as the nutrients O_2 and CO_2, and the composition of sediments on the seafloor. The distributions of these elements have suggested that most of the organic material which falls from the productive layers of the upper ocean is consumed within the upper kilometer or two of the ocean (Wyrtki 1962; Fiadeiro and Craig 1978).

Before the advent of direct sampling it was widely assumed that the sinking material represented some kind of degraded residue from the biological production in the upper ocean. Broecker and Peng (1982, p.1 of their text on chemical oceanography) characterize the sinking material as follows: "While much of the plant material is consumed by animals living in the surface ocean, some insolubles and indigestibles (i.e., fecal matter) move into the deep sea under the influence of gravity."

After a decade of direct sampling, it is quite clear that this characterization is incorrect. Upper ocean sediment trap results, summarized by Martin et al. (1987), show that most of the sinking organic matter flux in the open ocean is consumed in the upper 200–300 m. Most of the material which escapes consumption in the upper kilometer sinks straight to the bottom. Bishop et al. (1987) show that the decrease in the flux of organic carbon with depth in the upper few hundred meters is highly correlated with animal biomass. Subsurface animal populations apparently find and consume the sinking material very shortly after it falls out of the productive upper layers, suggesting that this material is hardly "indigestible." In the North Atlantic, bottom cameras have recorded the arrival of intact phytoplankton aggregations, which fall to the bottom in conjunction with the collapse of the spring bloom at the surface (Billet et al. 1983). The blanket of fresh material over the sediment surface disappears within a few weeks as the benthos feast on this manna from the upper ocean.

If sedimenting particulate matter is in fact readily digestible, and is consumed primarily near the surface and at the bottom, what accounts for the remineralization of nitrate and phosphate, and the consumption of oxygen in the interior of the ocean? Two recent papers by Suzuki et al. (1985) and Sugimura and Suzuki (1988) have unveiled a new method for measuring the total concentrations of DON and DOC in seawater. These papers report that the concentrations of DOC and DON are about four times higher in surface water and two times higher in deep water than previously thought. These papers also show that DOC and DON concentrations decrease strongly with depth, mirroring the increases in dissolved inorganic carbon and nitrate. In this review I will try to show that the new DOC and DON results appear to fill this void in our characterizations of the downward organic matter fluxes in the ocean. Downward advection and mixing of DOC and DON may well be more important in balancing the upward fluxes of CO_2 and nitrate from the deep sea than downward fluxes of sinking particles. While advection and mixing are inherently less efficient modes of vertical transport than particle sinking, these processes may simply move much bigger pools of organic matter.

Phytoplankton and other organisms in the upper ocean food webs are known to excrete dissolved organic compounds. We know that dissolved organic matter represents the primary substrate supporting bacteria in the ocean. How does the new DOC and DON data fit our conceptions of oceanic ecosystems? The C:N ratio in Suzuki and Sugimura's DOM is relatively low, about 7–8. It does not make much sense, ecologically, that the phytoplankton would excrete so much nitrogen bound up in a form that will be passively mixed out of the upper ocean. I would like to conclude by reviewing some ideas and speculations regarding the origin and destruction of oceanic DOM.

BACKGROUND

The measurement of dissolved organic carbon in seawater is, theoretically, a straightforward process. One first filters the water to remove the particulate carbon and then drives off inorganic CO_2 by acidification. The remaining carbon is oxidized using a strong chemical oxidant, UV radiation, or some combustion process, and the CO_2 yield from the oxidation is determined by IR spectroscopy. To measure DON one oxidizes a water sample and measures total nitrate. The difference between the nitrate concentrations in oxidized and unoxidized samples represents the DON.

Historically, the measurement of DOC in seawater has been fraught with disagreement and contention. Estimates of the DOC content of seawater have covered a great range (for a review see Sugimura and Suzuki 1988). In recent years, chemical oceanographers in the West have generally settled on wet chemistry oxidation techniques and have favored a lower range of DOC concentrations. These "standard" techniques yield deep water concentrations which fall in the 30–50 μmol/l range, while surface water runs as high as 100 μmol/l. Oceanographers in the Soviet Union, using a dry combustion technique, have maintained that DOC levels might be twice this high.

Total organic carbon (TOC = DOC + POC on unfiltered samples) was measured during some of the GEOSECS cruises using a standard wet oxidation technique. (The particulate organic carbon (POC) represents at most 10% of the TOC in surface water, and a few percent in subsurface water.) This semi-global data set shows the highest TOC values in mid and low latitude surface waters. TOC decreases strongly in the upper few hundred meters and remains largely invariant in deeper waters. High latitude surface waters have TOC contents intermediate between mid latitude surface water and deep water.

The technique used by Suzuki et al. (1985) and Sugimura and Suzuki (1988) for measuring DOC and DON is a high temperature combustion process which utilizes a special platinum catalyst. Various investigators have tried to use high temperature combustion methods to measure DOC in the past (R. Barber, personal communication) but only Suzuki and Sugimura seem to have made them work reliably. Other investigators are now

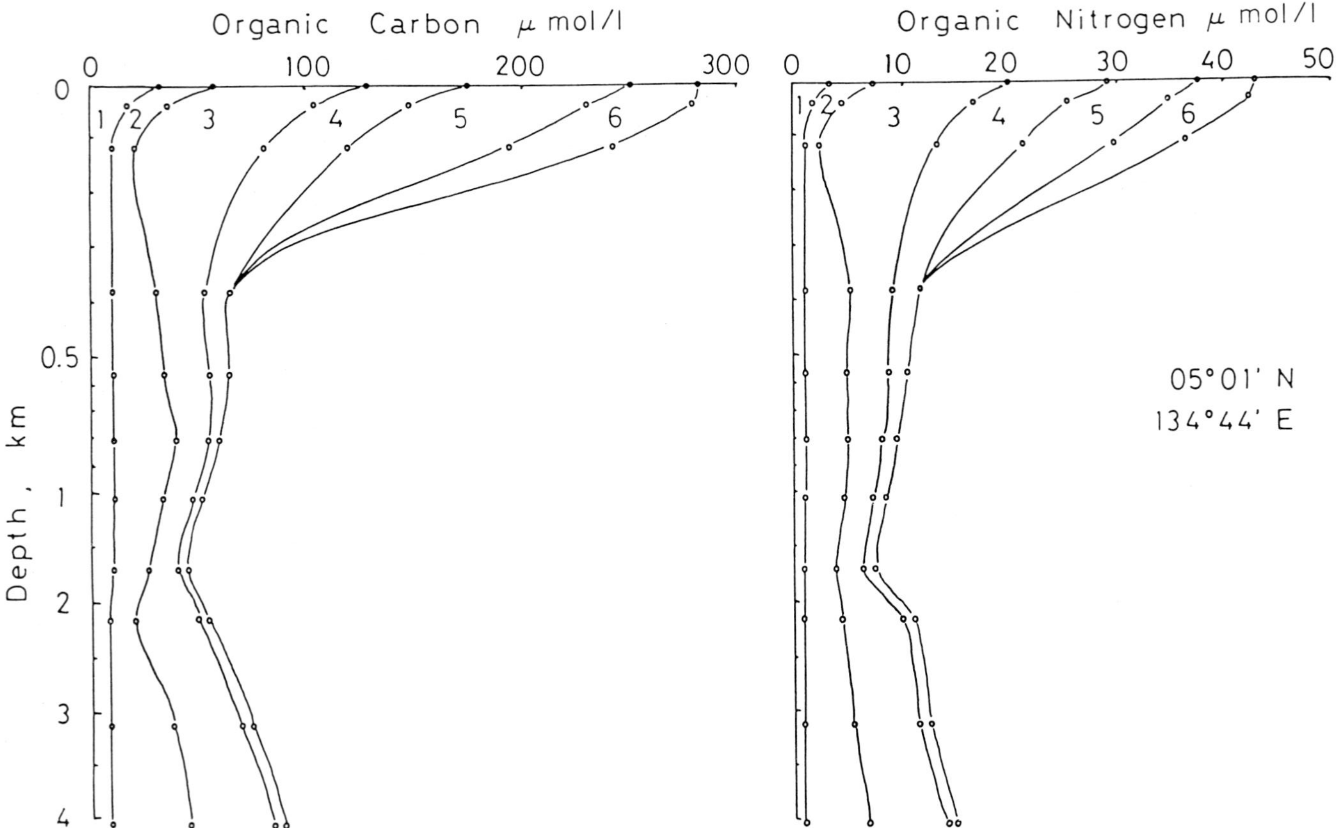

Fig. 1—Vertical profiles of DOC and DON at 5°N, 135°E, in the equatorial North Pacific. DOC and DON are subdivided into molecular size pools as follows (units – daltons): 1: < 1800; 2: 1800–4000; 3: 4000–20,000; 4: 20,000–60,000; 5: 60,000–100,000; 6: > 100,000. From Sugimura and Suzuki (1988).

attempting to replicate these results using similar equipment. Because no one else has yet done so, one must be cautious in fully accepting these results.

CHARACTERISTICS OF SUZUKI AND SUGIMURA'S DOM POOL

Figures 1 and 2 show vertical profiles of DOC and DON at two stations in the western North Pacific in which the DOM is separated by molecular size. Molecular size ranges are given in the caption to figure 1. The total DOC and total DON concentrations are given by the envelope of size fraction curves. Surface water concentrations are about 300 μmol/l for DOC and 40 μmol/l for DON. The surface DON values are particularly noteworthy because they are almost as large as nitrate values in the subsurface nitrate maximum. Surface water DOC and DON measurements made using the standard techniques yield only 60–90 μmol/l and 4–10 μmol/l, respectively. At each station in figures 2 and 3 the DOC and DON concentrations decline through the upper kilometer, reaching minimum values between 1000 and 1500 m. Concentrations increase below these depths. This sort of structure in the lower parts of the thermocline and deep water is not seen in DOM profiles made using the standard techniques.

Note that most of the water column structure below a few hundred meters can be attributed to size fraction 3. Taking into account the relative scales used in the DOC and DON plots, one sees that the largest amounts of DON relative to DOC are in fraction 3. Component 6, the largest size fraction, would appear to be the most labile; it totally disappears in the upper 400 m at the equatorial station and is hardly present in the 20°N station. Component 1, the smallest, appears to be the most refractory, since its top to bottom concentrations are the least variable. Component 2 has its highest concentrations in the deepest water.

Williams and Druffel (1987) have reported that the ^{14}C age of the deep water DOC extracted by the standard techniques is about 6000 years. Williams and Druffel speculate that about half of the deep water DOC is composed of extremely refractory material which has cycled through the ocean many times. It would have lost most of its ^{14}C to decay and would have to be almost completely resistant to microbial oxidation. A small subcomponent may have a terrestrial origin (Mantoura and Woodward 1983). Jackson and Williams (1985), again using the standard techniques, have found that a large and ubiquitous component of the marine DOM has a very high C:N ratio of about 25. There is probably considerable overlap between the refractory component responsible for the old radiocarbon ages and the component with high C:N ratios. Degens (1970) reports that most of the marine DOM compounds oxidized by the standard techniques have molecular weights less than 5000 Daltons.

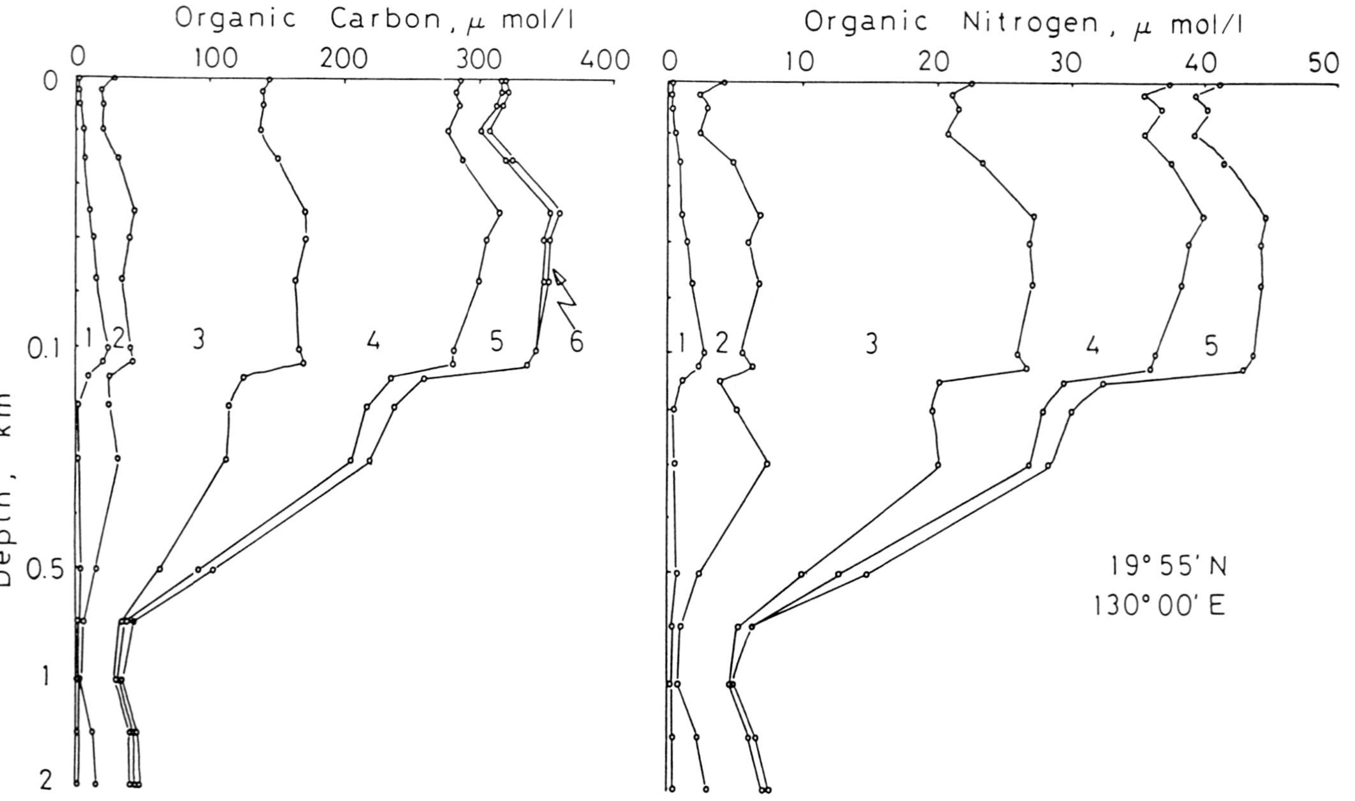

Fig. 2—Vertical profiles of DOC and DON at 20°N, 130°E, in the western North Pacific. DOC and DON are subdivided into the molecular size pools identified in the caption to Fig. 1. From Sugimura and Suzuki (1988).

Given what we know about oceanic turnover rates, the vertical distributions of Suzuki and Sugimura's DOM size fractions suggest that the bulk of their material persists in the ocean for some limited amount of time, perhaps several hundred years. In contrast to the material described by Williams and Druffel, this material appears to be susceptible to microbial breakdown. The largest size fraction has a fairly low C:N ratio of about 6. Most of the DOM compounds analyzed by Suzuki and Sugimura have molecular weights well in excess of 5000 Daltons.

If we assume that the high temperature combustion method extracts the refractory DOC characterized by Williams and Druffel (1987) and Jackson and Williams (1985), it would appear from the limited water column characterizations above that this very old material would have to be concentrated in Suzuki and Sugimura's smaller size fractions 1 and 2, the fractions with the smallest vertical gradients. Fractions 1 and 2 have molecular sizes up to 4000 Daltons and when combined account for about 20 to 50 μmols of DOC per liter of seawater. The C:N ratios of the DOM in fractions 1 and 2 are somewhat higher than in the larger fractions, but not nearly as high as in the material identified by Jackson and Williams. On the basis of water column characterizations, concentrations, molecular size, and C:N ratios, the size fractions 1 and 2 appear to have much in common with the refractory DOM extracted using the standard techniques. The remaining DOM extracted by Suzuki and Sugimura, the mystery substance, consists of larger molecules with lower C:N ratios that are produced and broken down in the ocean over relatively short time scales.

PROPERTY/PROPERTY ANALYSIS

Figure 3 reproduces a plot of DOC vs. AOU (apparent oxygen utilization) from Sugimura and Suzuki (1988) for three western N. Pacific stations. The slope of a line drawn through the data points is close to -1. Sugimura and Suzuki cite this correlation as evidence that the disappearance of DOC "sufficiently explains the amount of AOU in the water." The correlation between AOU and DOC, taken at face value, is inconsistent with the sediment trap results; surely the sinking POC is responsible for some part of the oxygen consumption. Below I will show that a simple correlation between AOU and DOC is misleading because it does not take into account the "preformed" concentrations of these substances.

A preformed concentration represents the concentration of a chemical substance which is present in a given water mass when the water mass was last at the surface. The preformed AOU is, by definition, zero because one usually assumes that surface waters have oxygen contents which are in equilibrium with the atmosphere. The preformed DOC is an unknown because we do not have any DOC measurements from high latitudes which

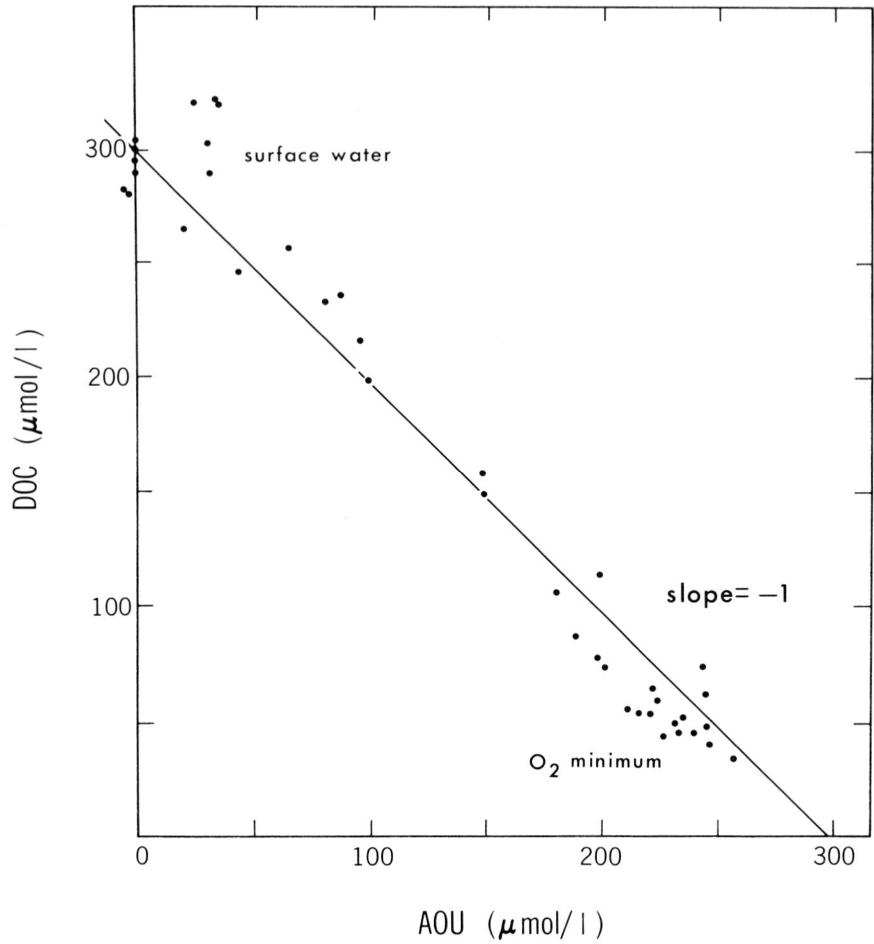

Fig. 3—Plot of DOC vs. AOU from three stations in the western North Pacific. From Sugimura and Suzuki (1988).

were made using Suzuki and Sugimura's technique. For this reason property/property diagrams constructed using oxygen or AOU as one of the properties (e.g., figure 3) are especially vulnerable to misinterpretation. Figure 3 gives the impression that the properties of the subsurface water masses evolved from the local, zero AOU, 300 μmol/l DOC mid latitude surface water. All that one can reasonably conclude from figure 3 is that the subsurface water masses evolved from one, or several, preformed compositions which lie on the left-hand axis of the plot.

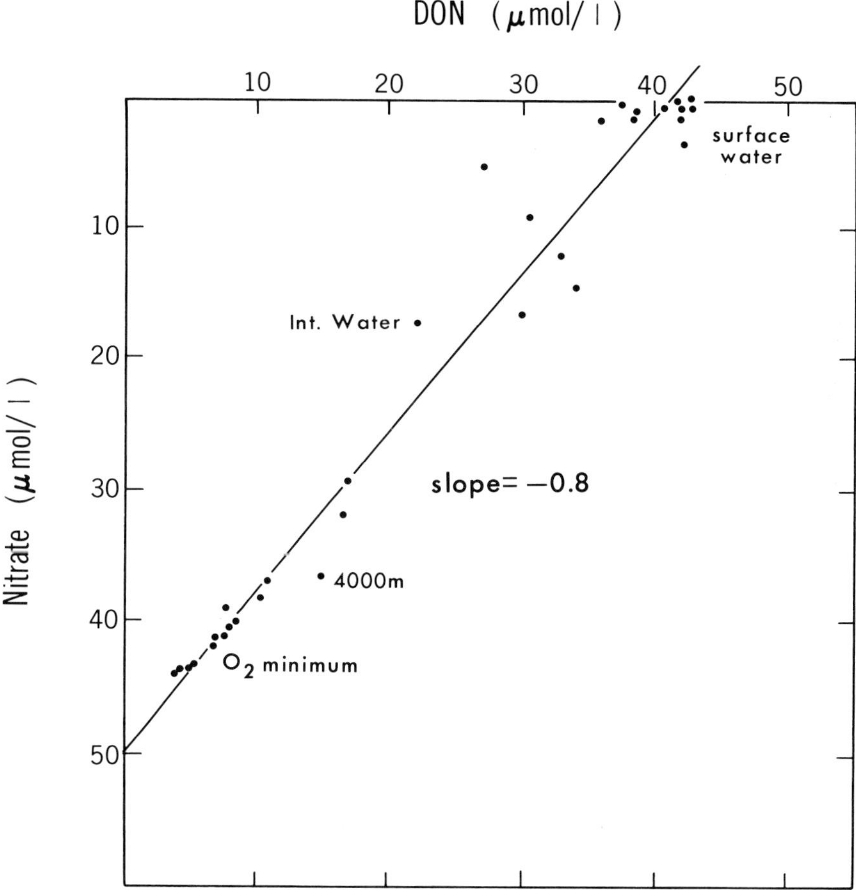

Fig. 4—Plot of DON vs. nitrate from data tabulated in Suzuki et al. (1985). The data include two vertical profiles at 20° and 40°N along approximately the 140°E meridian, and several surface water samples.

Figure 4 shows a plot of DON vs. nitrate made up from data tabulated in Suzuki et al. (1985). Surface water and O_2 minimum water define the end points in figure 4. Water from the thermocline, intermediate water, and deep water fill in the middle regions of the trend. The slope of the line drawn through these points is −0.8. This is a more meaningful correlation than the AOU vs. DOC correlation in figure 3 because there is enough data available for DON and nitrate to suggest what the preformed components might be.

I can roughly identify two preformed components from the data set. One is an actual surface water sample collected to the northeast of Japan. It has a temperature of 2.6°, a salinity of 33.2 per mil, a nitrate content of 17 μmol/l, and a DON content of 22 μmol/l. I have labeled it "Int. Water," because there is a reasonable chance that it represents the preformed nitrate and DON compositions of the low salinity North Pacific Intermediate Water. The deepest sample collected in either of the papers is from 4000 m at 5°N and 134°E. I have interpolated the DON value for this sample from figure 13c in Sugimura and Suzuki (1988) and have taken a NO_3 value from the same depth at the nearest GEOSECS station (241, almost 5,000 km due east). I label this point "4000 m" and assume that it retains some of the preformed composition of Antarctic Bottom Water.

Because the North Pacific O_2 minimum lies below the low-salinity intermediate water and above the deep water, the O_2 minimum end member composition for DON and NO_3 evolves from these two preformed components, not the local surface water. The observations that the two preformed components lie close to the trend line joining the local surface water and O_2 minimum water reflects, I think, two facts: *(a)* low and mid latitude surface waters are the source of most of the long-lived DON, and *(b)* the DON and nitrate compositions of the outcropping deeper water masses are only weakly altered at the surface. The DOM which breaks down in the O_2 minimum may be produced in the local surface water but the actual pathway by which it reaches the ocean's interior probably includes a detour through higher latitudes.

The slope of −0.8, relating DON and nitrate, reflects the fact that 80% of the nitrate remineralization below the thermocline is due to the oxidation of DON. The remineralization of PON in sinking particles contributes only 20%. This same 80/20 partitioning must hold for the consumption of oxygen and remineralization of CO_2 below the thermocline as well.

We can now go back to figure 3 and ask how preformed DOC affects one's interpretation of this diagram. From the characteristics of the DON I infer that the preformed DOC in North Pacific Intermediate Water and Antarctic Bottom Water is lower than the DOC in mid latitude surface water. Therefore, I expect that the DOC and AOU composition of O_2 minimum water evolves from preformed components which lie well off the trend in figure 3. Based on the DON vs. NO_3 slope (−0.8), the C:N ratio in the DOM (7), and the stoichiometry of oxygen consumption relative to nitrate release (1/9), I would say that the true $\Delta AOU/\Delta DOC$ slope is about −1.6, not −1.0. One can extrapolate a line with this slope from the cluster of O_2 minimum points to the zero AOU axis. From the intercept I predict an average preformed DOC value of roughly 175 μmol/l for the O_2 minimum water.

One cannot partition nitrate remineralization and oxygen consumption in the upper thermocline between DOM and sinking particles using this data. The zone in which most of the particulate organic matter is remineralized is shallower than 300 m (Martin et al. 1987). Water in this depth range has nitrate concentrations less than 7 or 8 μmol/l. The DON vs. NO_3 plot in figure 4 has no well-defined trend in this part of the plot. For this reason, the property/property plots in figure 4 do not preclude an important role for particles in the upper thermocline.

It seems to me that a large part of Suzuki and Sugimura's DOM is produced, in the Pacific at least, within warm surface waters and is broken down over several hundred years in subsurface waters. Long-term ocean mixing processes, which exchange properties between low and mid latitude surface water and deep water, dominate the DON vs. nitrate plot. The true measure of the role of DOM vs. particles in the vertical exchange is obscured in the AOU vs. DOC plot by the resetting of the AOU component to zero and the lack of DOC data from high latitudes.

MODELING STUDIES

Modeling studies have just begun to address the question of how the characterization of the vertical flux of organic matter affects the distribution of nutrients and CO_2 in the ocean and the ocean's chemical cycles. Initial results indicate that it does indeed make a great deal of difference how the flux is characterized. If the upward flux of nutrients into the surface layers of the ocean solely gave rise to a sinking flux of organic particles, which were remineralized according to the vertical scaling demonstrated by Martin et al.'s sediment traps, the ocean's nutrient distribution would be radically altered and the ocean's particle production would be higher than is currently measured.

This conclusion was reached in a 3-D model study by R. Najjar at Princeton University, as reported in Toggweiler et al. (1987) and Sarmiento et al. (1988). The physical model used in these experiments is an idealized, two-hemisphere, flat-bottomed sector model of a single ocean basin. It is driven by zonally and meridionally averaged winds, and zonally averaged surface temperatures and salinities.

Chemically, the model contains only phosphorus and has no external sources or sinks. If one ignores the effects of denitrification and nitrogen fixation, nitrate could be substituted for PO_4 in this model with a suitable stoichiometric correction. In order to avoid having to predict the surface nutrient field, the model restores the predicted phosphate concentration at each surface layer grid point toward the observed zonally averaged surface concentration in each latitude band as compiled from the GEOSECS data set. The same fraction of phosphate removed from the model surface box per time step is removed from all model boxes in the upper 100 m. The total amount of phosphate removed from the upper 100 m at a given location represents the model's predicted new production for the euphotic zone at that point. The flux of phosphate from the upper 100 m is then numerically partitioned into all subsurface layers immediately below according to the sediment trap-derived scaling function in Martin et al. (1987). The model is initialized with a uniform phosphate field in which the concentration in every grid box is equal to the global average.

Figure 5 shows a zonally averaged north–south section of phosphate predicted by the model. Figure 6 shows the phosphate distributions along

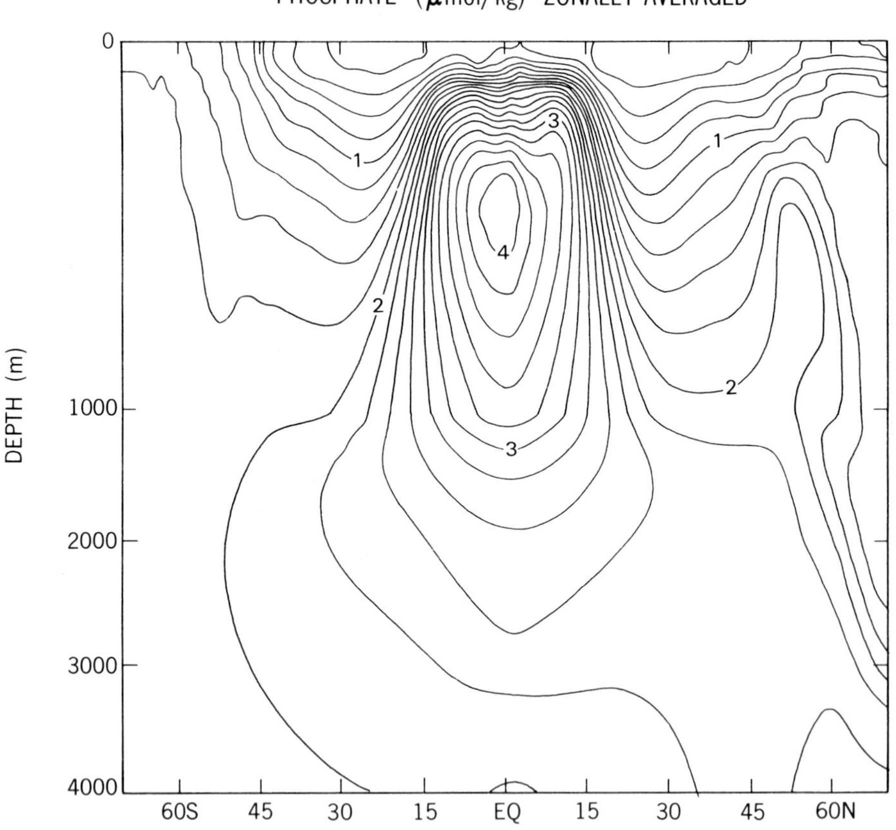

Fig. 5—North-south section of phosphate predicted by a 3-D nutrient cycle model run in an idealized sector ocean basin. See text for description. Results shown are zonally averaged (units: μmols/kg). From Toggweiler et al. (1987).

north–south GEOSECS tracks in the western Pacific and western Atlantic for comparison. Figure 7 shows a map of the predicted phosphate concentration at 180 m and a map of the model's new production, where the phosphorus flux units have been scaled up by a factor of 130 to be equivalent to a flux of organic carbon. The model develops an intense, shallow nutrient maximum beneath the tropics, in which phosphate concentrations exceed 8 μmols/kg adjacent to the eastern boundary. In the ocean, observed phosphate maxima, in contrast, are much deeper, have concentrations which rarely exceed 3 μmols/kg, and are not so focused within the tropics.

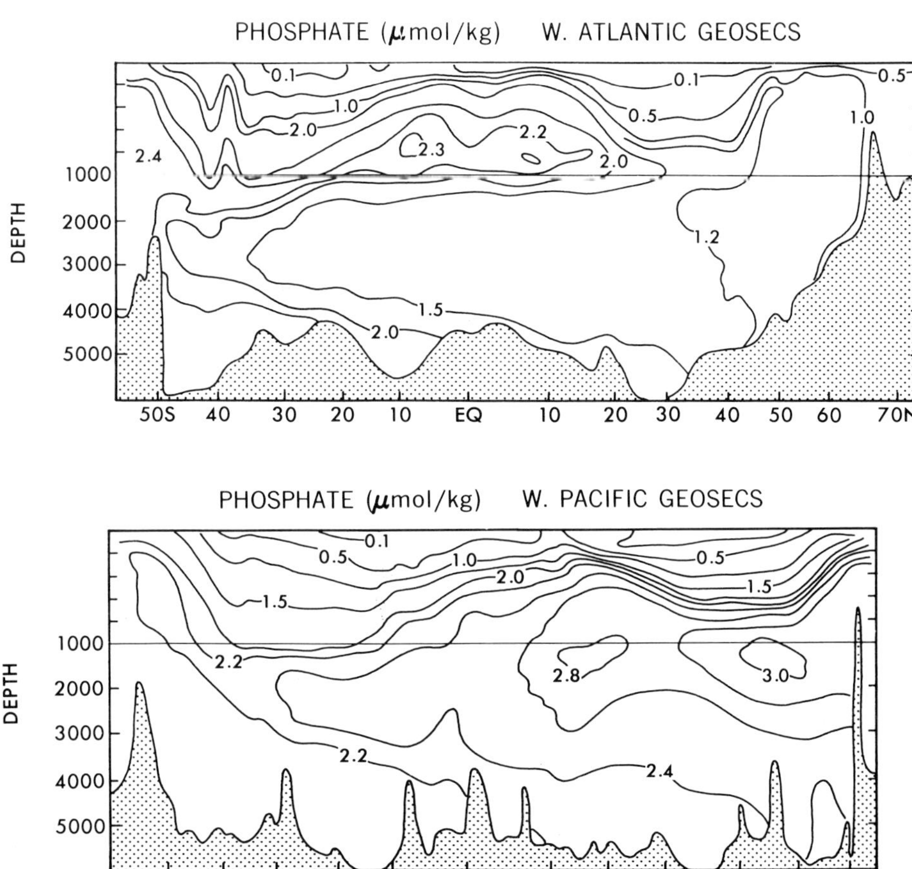

Fig. 6—North-south sections of phosphate along the W. Atlantic (upper) and W. Pacific (lower) GEOSECS tracks (units: μmols/kg).

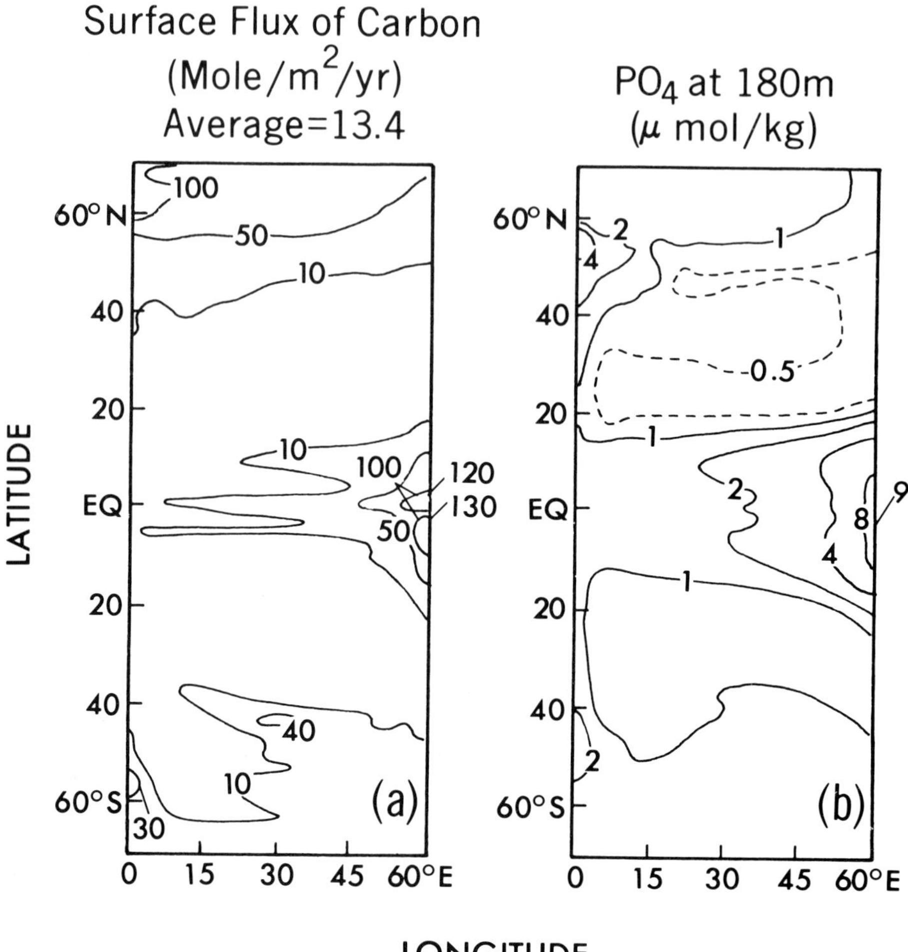

Fig. 7—(a) Map of equivalent particulate organic carbon flux (new production) at 100 m predicted by the model (units: mols C/m^2/yr). *(b)* Map of phosphate concentrations at 180 m predicted by the sector model (units: μmols/kg). From Sarmiento et al. (1988).

The intense phosphate maximum develops in the model because phosphate is trapped within converging subsurface layers associated with the wind-driven equatorial upwelling. When this high phosphate water upwells, it produces particles which are remineralized immediately below, leading to still more particle production, and so on. The model predicts new production well in excess of 50 mols C/m^2/yr within the eastern equatorial zone. Outside of the tropics, the shallow nutrient remineralization leads to high upper thermocline phosphate concentrations generally, and a basin-average new production of 13 mols C/m^2/yr, almost 10 times higher than the organic carbon fluxes actually caught in sediment traps (Martin et al. 1987).

In hindsight, the result is not entirely unexpected, given the intensity with which observed vertical fluxes of carbon decrease with depth immediately below the photic zone. Nutrients taken up in the euphotic zone are simply not moved very far by sinking particles. Furthermore, lengthening the scale over which the flux is remineralized in the model does not make the unrealistic features of the phosphate distribution disappear. It is not simply the vertical scale which produces the unrealistic features but also the vertical trajectory assumed in having the new production remineralized immediately below where it is produced.

The most realistic simulations have been derived when half of the new production is put into a long-lived dissolved organic phosphorus pool (breakdown rate = 1/200 yr^{-1}), which is allowed to advect with the water. This mechanism allows much of the organic production and nutrients from the equatorial upwelling zone to be transported out of the tropics without being trapped. It also allows some of the DOM to be mixed into the deeper strata of the ocean before being remineralized. It deepens and disperses the phosphate maximum and lowers the basin-average particulate carbon flux into good agreement with sediment trap results.

We chose the parameters for the DOM experiment to yield a DOM pool consistent in size with that observed by Suzuki and Sugimura, assuming a C–P ratio of ~100 in the DOM. It is worth pointing out that this same result cannot be obtained by putting half the new production into a pool of small particles which can advect with the water. The particulate organic matter pool in the ocean is 10–100 times smaller than the DOM pool. If a significant fraction of the ocean's new production ended up in nonsinking particles, the turnover time for the pool would be measured in days, instead of years. Material this ephemeral could not advect or mix far enough to make any difference.

We readily acknowledge that the circulation field produced by the sector model in these experiments is not entirely realistic. We are currently repeating this set of experiments in a global general circulation model which includes real continental boundaries, bottom topography, and observed surface wind stresses. The global model has been thoroughly evaluated by running predictive simulations of the distributions of natural and bomb-produced radiocarbon (Toggweiler et al. 1989). An effect which has not been included in the model, and which might be very important, is downward organic matter transport along continental margins. Organic matter production along continental margins is much higher per unit area than in the open ocean. If there is a direct flux of organic matter from the surface to the deep sea, along continental margins over the globe which exceeds the open ocean flux (Walsh 1983), downward particulate transport along the ocean's margins may circumvent the problem raised by the sediment trap observations.

SOURCES AND SINKS OF SUZUKI AND SUGIMURA'S DOM

Why should marine organisms produce DOM which ties up scarce nutrients in forms which are so difficult to break down? DOM is produced by most elements of marine food webs. Lancelot and Billen (1985) provide an excellent review of the literature on the production of DOM associated with primary production. They report that the release of photosynthetically produced DOC by healthy, growing phytoplankton ranges from 0 to 10% of the primary production. Several studies report higher percentages of release by organisms under low nutrient conditions, particularly when the phytoplankton are dominated by flagellates. Populations dominated by diatoms characteristically release low percentages of their primary production as DOC. Caron et al. (1985) report that heterotrophic microflagellates release 10% of the carbon they ingest as DOC.

Microbiologists maintain that DOC provides the main substrate for bacterial growth in the ocean (Ducklow 1983). It is usually assumed that the bacteria derive most of their energy by breaking down the primary exudates of the phytoplankton and animal populations. Microbiologists view the DOC on which the bacteria feed as a small, labile pool which turns over on time scales of hours or days. Estimates of bacterial biomass and production suggest that the fraction of the primary phytoplankton production which passes through the bacteria may be 20–50% (Ducklow 1983), somewhat greater than the DOM fraction detected by ^{14}C incubations.

One of the most thorny issues concerning the origin of marine DOM involves the transformation of the relatively simple compounds produced and utilized by marine organisms into the complex, long-lived, and uncharacterizable compounds which form the bulk of the ocean's DOM. Only 10–20% of the DOM can be characterized as identifiable compounds or classes of compounds (Williams 1975).

The primary exudates of the phytoplankton are mainly low molecular weight metabolites and polymeric carbohydrates (Lancelot and Billen 1985). The carbohydrates are quantitatively more important, giving the pool of excreted DOC a distinctly high C:N ratio. The metabolites are readily usable by bacteria whereas the high molecular weight carbohydrates must be broken down. In order to use the high molecular weight substrates, the bacteria produce extracellular enzymes to break these compounds down into small molecules which can be absorbed through their cell walls. The enzyme reactions are not 100% efficient, leading to condensation reactions between the carbon-rich primary exudates with the nitrogen-rich enzymes. This sort of condensation reaction has been separately proposed by Duursma (1965) and Sieburth and Jensen (1969) as a possible source of the long-lived, low C:N, DOM in seawater.

The same enzymes which break down the primary exudates are probably capable of breaking down the condensation products as well. When the primary exudates are abundant, the production of condensation products may exceed the breakdown of these compounds, with the net result being an increase in the condensed organic matter pool. When the primary exudates are not available, the condensation products may be consumed, albeit slowly.

There are sedimentary and terrestrial analogues for the formation and breakdown of condensed DOM (Rice 1982; Melillo et al. 1984). Melillo et al. describe the nitrogen immobilization by decaying forest litter as a batch process. The decay of fresh, nitrogen-poor substrates results in nitrogen uptake from the surroundings, as bacterial enzymes are consumed in the decay process. Initially, the nitrogen concentration in the organic matter goes up, but nitrogen uptake eventually gives way to nitrogen release as the entire pool of organic material is consumed. The hypothesis outlined above for oceanic DOM represents the same type of process operating in a continuous mode. The seasonal batch processing of forest litter is replaced in the ocean by a physical separation between the zone of active condensed-DOM formation, where phytoplankton and animal exudates are being produced, and a zone of net destruction, where the exudates are not available.

This hypothesis could probably be put to the test with a well-resolved survey of an ocean basin which included some seasonal time series. A key prediction is a systematic variation in the C:N ratio. If the terrestrial analogy holds, the highest C:N ratios in the bulk DOM should be found in the surface waters, with the lowest C:N in the most condensed fractions. Suzuki and Sugimura's size fraction work suggests that there are distinct pools within the DOM which should be characterizable with new approaches. There is also a great need to know whether or to what degree dissolved organic phosphorus is included in the additional DOM extracted by Suzuki and Sugimura's technique.

CONCLUSIONS

The new higher DOC and DON concentrations reported by Suzuki and Sugimura, although not yet reproduced by other investigators, may revolutionize our models of how the vertical material balance is maintained in the ocean. This work is revolutionary, not simply because it shows the pool of oceanic DOM is bigger, but also because it shows that this pool is much more dynamic and more integral to the ocean's chemical cycles than previously thought. The DOM oxidized by Suzuki and Sugimura's technique appears to consist of larger molecules which contain more nitrogen than the DOM oxidized by standard methods. Oceanic distributions suggest that Suzuki and Sugimura's DOM is relatively long-lived in the ocean, but is considerably more degradable than the bulk DOM extracted by the standard methods. Relationships between DOM concentrations and concentrations of other geochemical properties suggest that 80% of the nitrate remineralization, organic matter respiration, and oxygen consumption below the thermocline is due to DOM which has been mixed downward from the upper ocean or advected laterally from high altitudes.

Given the potential impact of these new results, we know very little about mechanisms which govern the size of the ocean's DOM pools. Sugimura and Suzuki's DOC profiles, if extrapolated to the whole ocean, suggest that the ocean's DOC pool is at least twice as big as the atmospheric CO_2 reservoir. There are no obvious feedbacks controlling the size of this reservoir like those controlling the inorganic CO_2 pool (e.g., the effects of pH on $CaCO_3$ solubility). The ocean's DOM, therefore, represents a present-day wild card in our efforts to understand the ocean's chemical cycles.

Acknowledgements. I would like to acknowledge the helpful comments of J. Sarmiento, R. Najjar, L. Legendre, W. Berger, J. Bishop, and R. Keir in preparing this review. I would also like to thank Wendy Marshall and Phil Tunison at GFDL for their help in preparing the manuscript and figures.

REFERENCES

Billet, D.S.M.; Lampitt, R.S.; Rice, A.L.; and Mantoura, R.F.C. 1983. Seasonal sedimentation of phytoplankton to the deep sea benthos. *Nature* **302**: 520–522.

Bishop, J.K.B.; Stepien, J.C.; and Wiebe, P.H. 1987. Particulate matter distributions, chemistry, and flux in the Panama Basin: response to environmental forcing. *Prog. Ocean.* **17**: 1–59.

Broecker, W.S., and Peng, T.-H. 1982. Tracers in the Sea. Palisades, NY: Eldigio Press.

Caron, D.A.; Goldman, J.C.; Anderson, O.K.; and Dennett, M.R. 1985. Nutrient cycling in a microflagellate food chain. II. Population dynamics and carbon cycling. *Mar. Ecol. Prog. Ser.* **24**: 243–254.

Degens, E.T. 1970. Molecular nature of nitrogenous compounds in seawater and recent sediments. In: Organic Matter in Natural Waters, ed. D.W. Hood. *Inst. Mar. Sci. Alaska.* **1**: 77–106.

Ducklow, H. 1983. Production and fate of bacteria in the oceans. *Bioscience* **33**: 494–501.

Duursma, E.K. 1965. The dissolved organic constituents of seawater. In: Chemical Oceanography, eds. J.P. Riley and G. Skirrow, vol. 1, 1st ed. pp. 433–475. London: Academic Press.

Fiadeiro, M.E., and Craig, H. 1978. Three-dimensional modeling of tracers in the deep Pacific Ocean. I. Salinity and oxygen. *J. Marine Res.* **36**: 323–355.

Jackson, A.J., and Williams, P.M. 1985. Importance of dissolved organic nitrogen and phosphorus to biological nutrient cycling. *Deep-Sea Res.* **32**: 223–235.

Lancelot, C., and Billen, G. 1985. Carbon-nitrogen relationships in nutrient metabolism of coastal marine ecosystems. In: Advances in Aquatic Microbiology, eds. H.W. Jannasch and P.J. leB. Williams, vol. 3, pp. 263–321. London: Academic Press.

Mantoura, R.F.C., and Woodward, E.M.S. 1983. Conservative behavior of riverine dissolved organic carbon in the Severn Estuary: chemical and geochemical implications. *Geochim. Cosmochim. Acta* **47**: 1293–1309.

Martin, J.H.; Knauer, G.A.; Karl, D.M.; and Broenkow, W.W. 1987. VERTEX: carbon cycling in the northeast Pacific. *Deep-Sea Res.* **34**: 267–285.

Melillo, J.M.; Naiman, R.J.; Aber, J.D.; and Linkins, A.E. 1984. Factors controlling mass loss and nitrogen dynamics of plant litter decaying in northern streams. *Bull. Amer. Sci.* **35**: 341–356.

Rice, D.L. 1982. Detritus nitrogen problem: new observations and perspectives from organic geochemistry. *Mar. Ecol. Prog. Ser.* **9**: 153–162.

Sarmiento, J.L.; Toggweiler, J.R.; and Najjar, R. 1988. Ocean carbon cycle dynamics and atmospheric pCO_2. *Phil. Trans. R. Soc. Lond. A* **325**: 3–21.

Sieburth, J.M., and Jensen, A. 1969. Studies on algal substances in the sea. II. The formation of gelbstoff (humic material) by oxidates of Phaeophyta. *J. Exp. Mar. Biol. Ecol.* **3**: 275–289.

Sugimura, Y., and Suzuki, Y. 1988. A high temperature catalytic oxidation method of non-volatile dissolved organic carbon in seawater by direct injection of liquid sample. *Marine Chem.* **24**: 105–131.

Suzuki, Y.; Sugimura, Y.; and Itoh, T. 1985. A catalytic oxidation method for the determination of total nitrogen dissolved in seawater. *Marine Chem.* **16**: 83–97.

Toggweiler, J.R.; Dixon, K.; and Bryan, K. 1989. Simulations of radiocarbon in a coarse resolution world ocean model. *J. Geophys. Res.*, in press.

Toggweiler, J.R.; Sarmiento, J.L.; Najjar, R.; and Papademetriou, D. 1987. Models of chemical cycling in the oceans: a progress report. Ocean Tracers Lab. Tech. Rep. #4, Atmospheric and Oceanic Sciences Program, Princeton Univ., 61 pp.

Walsh, J.J. 1983. Death in the sea: enigmatic phytoplankton losses. *Prog. Ocean.* **12**: 1–86.

Williams, P.J. leB. 1975. Biological and chemical aspects of dissolved organic material in seawater. In: Chemical Oceanography, eds. J.P. Riley and G. Skirrow, vol.2, 2nd ed., pp. 301–363. London: Academic Press.

Williams, P.M., and Druffel, E.R.M. 1987. Radiocarbon in dissolved organic matter in the central North Pacific Ocean. *Nature* **330**: 246–248.

Wyrtki, K. 1962. The oxygen minima in relation to ocean circulation. *Deep-Sea Res.* **9**: 11–23.

Productivity of the Ocean: Present and Past
eds. W.H. Berger, V.S. Smetacek and G. Wefer, pp. 85–97
John Wiley & Sons Limited

New Production: History, Methods, Problems

R.W. Eppley

Institute of Marine Resources,
Scripps Institution of Oceanography
University of California, San Diego
La Jolla, CA 92093-0218, U.S.A.

Abstract. There are at least four general methods of measuring new production: (*a*) sediment traps at the base of the euphotic zone, (*b*) bottle experiments for assessing nitrate and carbon dioxide assimilation rate, (*c*) the disequilibrium of dissolved ^{234}Th with respect to its parent ^{238}U, and (*d*) nutrient, oxygen, carbon dioxide, and tracer gas distributions in the context of circulation models. The latter method has the greater history. The circulation models in the historical work were implicit.

Images of ocean color provide a new way of viewing phytoplankton standing stocks on a global scale. This approach, along with modeling, may lead to an understanding of export production from the euphotic zone at the important time and space scales.

Estimates of ocean annual primary production of carbon have oscillated over the years between about 20 and 60 Gt. Current estimates are in the upper part of this range, with new production of carbon 5–10 Gt.

HISTORICAL NOTE

Eric Mills (1982) has written a fascinating history of the development of ideas in plankton dynamics. Those ideas most pertinent to the new production concept, from the perspective of planktology, include Brandt's concepts at the turn of the century, on the significance of nutrients for phytoplankton, and Nathansohn's idea that "nutrient salts must accumulate in deep water as organisms sink and decay" (Mills 1982). Given these insights, the notion of export production from the euphotic zone was implicit in the classic plankton production studies in the 1920s and 1930s at Plymouth and Oslo although not explicitly stated. For example, Harvey (1928) included a diagram of food web interactions in which the cycle is closed by means of "corpses and excreta" as substrate for the regeneration of phosphates and nitrates. He came close to the new production idea when he stated (p. 169) that the fertility of the ocean depends primarily on two factors, "namely the length of time taken by the corpses of marine organisms and excreta to decay, and the length of time taken by the phosphates and nitrates so formed to come again within range of algal growth."

Cooper (1933), and Atkins before him, estimated primary production from changes in nutrient concentrations over the seasons in the English Channel, recognizing that it was net production (photosynthesis less the respiration of both plants and animals) that was measured. "All the figures are minima and would be increased by the substances taking part in the life cycle more than once" (Cooper 1933). Redfield, Smith, and Ketchum (1937) used a 4-layer model to calculate phosphorus dynamics in the Gulf of Maine. Their assumptions included explicit statements, such as (*a*) all synthesis of organic phosphorus from inorganic P takes place in the upper layer, (*b*) all downward movement of P is due to sinking, and (*c*) all upward movement of P is due to eddy processes. Riley's models clearly embody the idea of export production as well, especially where he estimated primary production (and much more) from oxygen consumption below the euphotic zone (Riley 1951).

METHODS OF MEASURING NEW PRODUCTION

Definitions

New production has been defined by Dugdale and Goering (1967) as the primary production in the euphotic zone resulting from nutrient inputs from outside the euphotic zone, such as from deep water, the atmosphere, or from land. Export production is used in different ways by different authors and should be defined for each situation or context. In this paper, export production refers to the flux of organic matter from the euphotic zone; in this definition it is equivalent to new production.

Incubation Experiments

The idea of measuring a quantity equivalent to new production in a simple incubation experiment came from Dugdale and Goering (1967). Their success in pioneering the use of ^{15}N tracers to measure rate processes contributed in other ways as well, but the new production paradigm has proved especially important. With ^{15}N-labelled ammonium, nitrate, urea, amino acids, and other N-compounds one can assess utilization in short-term experiments carried out aboard ship. Away from the influence of land, phytoplankton assimilation of ammonium is a measure of production based upon recycling within the euphotic zone, while assimilation of nitrate represents production based upon nutrients regenerated at depth and brought into the euphotic zone by the circulation. Within a few years a global estimate of new production could be attempted, i.e., the photosynthetic assimilation of carbon by phytoplankton that results from the assimilation of nitrate. The new production was interpreted as the export flux out of the euphotic zone. That first result was consistent with Broecker's geochemical model of nutrient input to the surface layer (Eppley and Peterson 1979).

Sorokin (1985) has used short-term measurements of phosphate uptake by phytoplankton and bacteria, along with indirect estimates of phosphate regeneration rate, to estimate regenerated production. Since ^{14}C primary production measures the sum of new and regenerated production, *sensu* Dugdale and Goering (1967), new production can be estimated by difference.

These essentially instantaneous rate measurements of new production should be equivalent to export production from the euphotic zone if measurements are time averaged over an appropriate period. Some of the new production may support a buildup of plankton biomass (Peinert et al., this volume) or of dissolved organic carbon (Toggweiler, this volume) that is not immediately exported (Sarmiento et al. 1987). In the central oceanic gyres, where plankton concentrations vary little over time, short-term experiments may represent export production. Where seasonal variation is significant, biomass changes average to zero and the seasonal average export is equal to new production (Sarmiento et al. 1987).

Particle Flux as a Measure of Export Production

Summaries of the flux of particulate carbon measured using sediment traps show a strong dependence of the flux on sampling depth and the rate of primary production in the overlying surface waters (reviewed by Hargrave 1985). The overall relation was nonlinear although some data showed a linear relation. The nonlinear relationship is consistent with results from incubation experiments (Fig. 1). The relation between particle flux and primary production is one factor in bringing together geochemists, paleoceanographers, and biologists in common cause in international programs such as the Joint Global Ocean Flux Study (JGOFS).

Seasonal Changes in Oxygen, Nutrients, and Carbon Dioxide as Measures of New Production

As noted earlier, this approach has historical significance; it was used by Atkins and Cooper in the 1920s and 1930s to assess the magnitude of phytoplankton production in the English Channel. It was recognized early that this method would underestimate primary production since regeneration

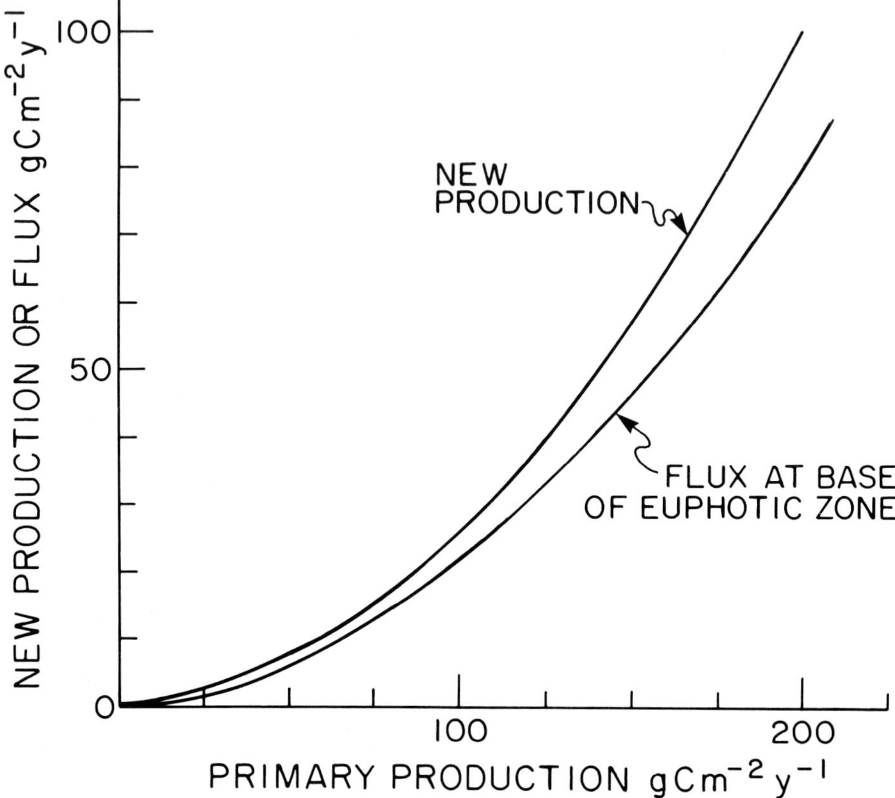

Fig. 1—Empirical relations between new production (measured in bottle experiments), particle flux (measured by sediment traps), and primary production (measured with ^{14}C). The new production relation is from Eppley and Peterson (1979): $P_{new}/P_{total} = 0.0025P_{total}$. Production units are gC $m^{-2}y^{-1}$.

The particle flux relation is that of Betzer et al. (reviewed and supported by Hargrave 1985): $F = AD^aP^b$, where F is the carbon flux (gC $m^{-2}y^{-1}$) at depth D (meters); P units are as before; and A, a, and b are constants.

The new production relation applies only to the euphotic zone. The depth of the euphotic zone was therefore used as D in the Betzer et al. equation. Values of the depth of the euphotic zone as a function of P were taken from Table 21, p. 88, of Parsons and Takahashi (1973).

of the substances measured clearly took place via the respiration of the plankton (Cooper 1933; Harvey 1928). Cooper called it "net production." More recently, it has been called "net community production" (Codispoti et al. 1986; Minas et al. 1986). Net production nowadays refers to autotrophic production less the respiration of the autotrophs themselves. Net community production measurements include losses due to all the organisms present, not only the autotrophs.

It was also noted early that the seasonal depletion of nutrients and carbon dioxide or the buildup of oxygen (even after correction for exchange with the atmosphere) exceeded the buildup of the biomass of plankton. It was recognized that the grazing and sinking losses were responsible. The sinking losses were quantified by Ketchum and Corwin (1965) in their drogue-following study of phosphorus, chlorophyll, and oxygen in the Gulf of Maine by constructing budgets over discrete depth intervals.

Effects of mixing and advection must be considered if the time rates of change of nutrients and metabolic gases are to be used to estimate rates of biological processes. Recent studies have attempted to do this. An example is the seasonal study of inorganic carbon on the Bering Sea shelf (Codispoti et al. 1986), where net community production was very close to the primary production estimated using ^{14}C incubations. The companion studies of nitrate-based new production in the same volume enhance the significance of the work.

French workers have described export production as "production du base" (Minas et al. 1986, and references therein). Seasonal changes in phosphate and oxygen concentrations were used for this purpose in the Mediterranean and the results compared with ^{14}C measurements of primary production (Minas et al. 1988). "Diagram analysis" and a mixing model are used, such that departures of concentration from those expected on the basis of mixing between end members indicate changes due to biological processes and gas exchange with the atmosphere. The technique has been applied also to upwelling areas off N.W. Africa, Peru, and S.W. Africa (Minas et al. 1986). The rate of heat gain of upwelled water brought to the surface was used to establish a time frame for converting the observed changes in concentration to rates.

Among the most sophisticated approaches for determining oxygen consumption rates, and hence export production, are those of Jenkins for the Beta Triangle region of the North Atlantic and for the time series station off Bermuda. Tritium and ^{3}He tracer distributions were used to estimate the velocity field. In the Beta Triangle, extensive hydrographic observations provided an independent view of the circulation. The oxygen distribution below the euphotic zone to a depth of 1 km could then be interpreted as a rate of oxygen utilization (Jenkins 1982). The large time and space integration accomplished in this approach appears to be greatly superior to the snapshot view provided by results of incubation experiments carried out with water bottle samples. The results indicated rates that were higher than expected rates of ^{14}C primary production, suggesting to Jenkins that incubation methods greatly underestimated primary production or that new production was a larger-than-expected fraction of primary production. Water bottle sampling from ships undersamples biological activity and this could contribute to the view, developed from ^{14}C production estimates, that the central oceanic gyres have relatively low biological production.

The time series data off Bermuda show a seasonally developed subsurface oxygen maximum (Jenkins and Goldman 1985), analogous to that in the central gyre of the North Pacific (Shulenberger and Reid 1981). Most of this signal is biological, rather than due to physical processes. Converted to a rate of primary production, the oxygen data imply a rate of new production exceeding the primary production rates measured at these sites with the ^{14}C method. The Bermuda time series allowed Jenkins and Goldman (1985) to extract a mean seasonal, depth-integrated oxygen anomaly in the upper waters, 1961–70, equivalent to about 60 gC m^{-2} y^{-1}, and similar to the Beta Triangle estimate.

The nitracline often breaks the surface in winter off Bermuda, and the consumption of this new nitrogen, resulting from deep winter mixing, would support considerable new production. Jenkins and Goldman (1985) also suggested that episodic wind mixing might bring sufficient nutrients into the upper waters to support a new production of the magnitude their calculations require.

Thorium/Uranium Isotope Disequilibria and New Production

Thorium is particle-reactive and has isotopes with a variety of half-lives. It is an excellent tracer for studies of particle dynamics in the ocean. Concentrations of dissolved ^{234}Th in the euphotic zone are often less than that expected from the decay of the parent ^{238}U and the disequilibrium can be used to calculate a residence time for the dissolved thorium. This residence time is inversely proportional to primary production (Coale and Bruland 1985), ranging from 2–3 days to >100 days. If all the particles to which thorium sorbs are biogenic, organic, and produced by local processes, then the residence time can be used, with measurements of particulate organic carbon, to give rates of new production in carbon units. This has great promise for assessing new production since measures of concentration also provide rates of the process, a property unique to radiotracers.

The dissolved ^{234}Th distribution in the North Pacific central gyre shows a depletion peak at about 70 m in summer (Fig. 2), coincident with the seasonal, subsurface oxygen maximum (Coale and Bruland 1987). These features are tens of meters above the nitracline in summer, suggesting they may not be due to utilization of nitrate. Recall that use of molecular nitrogen, by phytoplankton with the appropriate genes, results also in new production in the nitrogen fixation process (Dugdale and Goering 1967).

New Production Calculated Indirectly from an "f-ratio"

The ratio of new production to ^{14}C primary production has been called the f-ratio as shorthand for writing out the full name of the ratio each time it

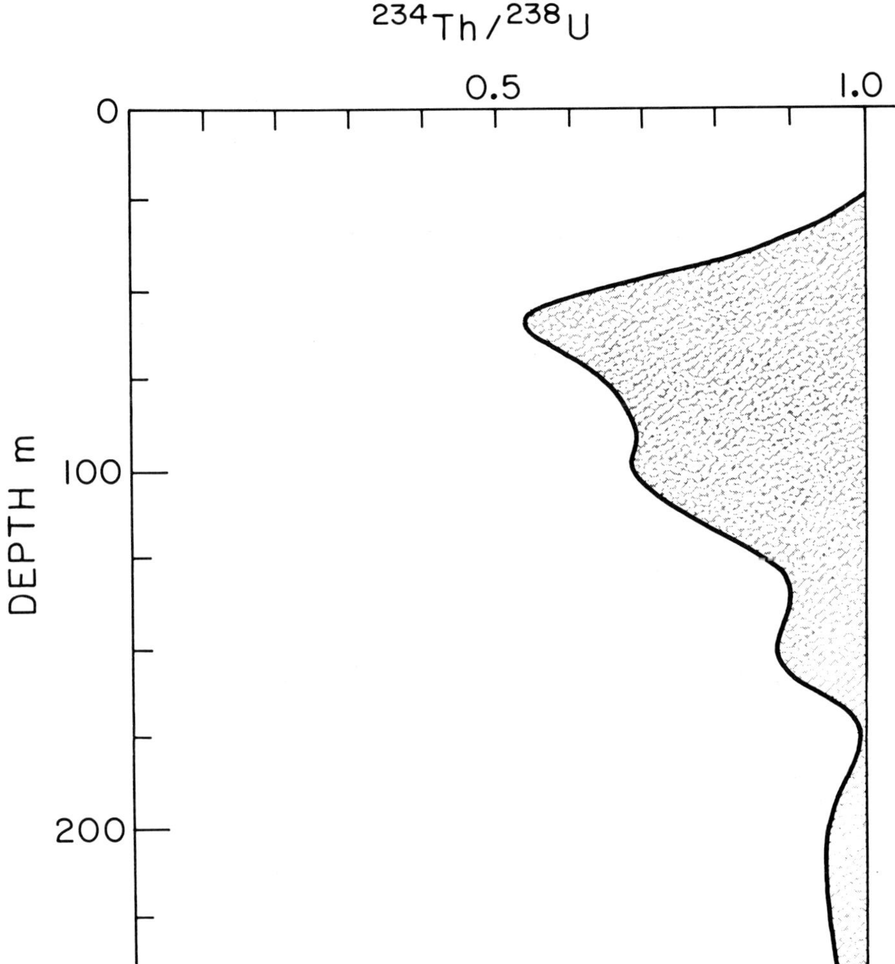

Fig. 2—Vertical profile of dissolved $^{234}Th/^{238}U$ ratio at VERTEX station 4 in the North Pacific central gyre. The vertical line at Th/U = 1 represents radioactive equilibrium. The filled area represents a deficiency of dissolved ^{234}Th with respect to its parent ^{238}U, probably because biogenic particles scavenge the thorium and then sink out. Redrawn from Coale and Bruland (1987).

is used (Eppley and Peterson 1979). Harrison et al. (1987) showed that the ratio varies with ambient nitrate concentration. This is supported by other results (Fig. 3). Thus ambient nitrate and ^{14}C primary production may provide an indirect estimate of nitrate-based new production, providing the variation of the f-ratio with nitrate concentration and other variables is sufficiently general and that measurements are made at the appropriate time

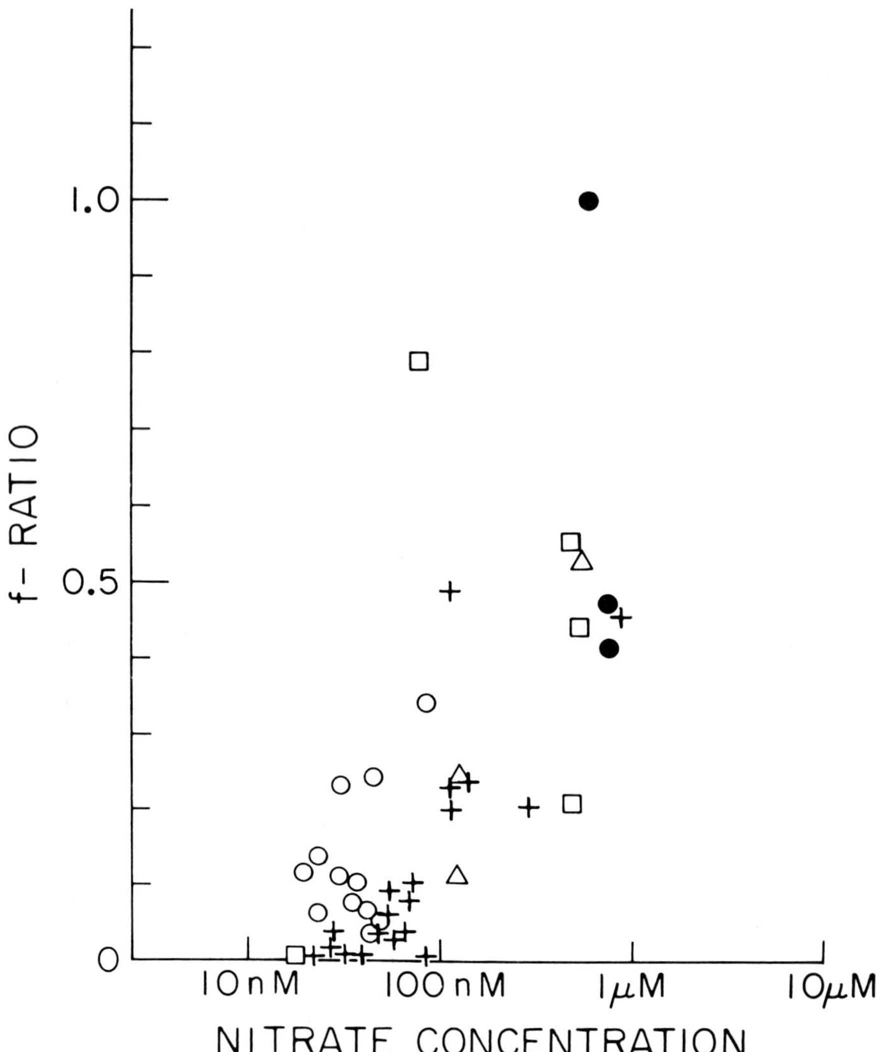

Fig. 3—Variation in f-ratio (the ratio of new production to total production) with the ambient concentration of nitrate. Data points from studies off southern California. Different symbols represent different cruises. Each point represents a sample from a single depth in the euphotic zone. Nitrate was measured using a chemiluminescent analyzer. Harrison et al. (1987) have discussed this relation in detail; the f-ratio took on maximum values of order 0.59–0.86 in their study. They also noted that ambient ammonium concentration influences the f-ratio by interfering with nitrate assimilation. Ammonium was not measured in the southern California study; ammonium concentration could be a major source of the experimental variability.

and space scales. Ammonium often inhibits use of nitrate by phytoplankton and its concentration must be known in order to choose an appropriate f-ratio (Harrison et al. 1987). Maximum values of the f-ratio appear to vary among regions.

The Promise of Remote Sensing

Chlorophyll and its related pigments can be measured with acceptable accuracy in the upper part of the euphotic zone by remote sensing. The near-surface chlorophyll concentration in many regions of the ocean is related to the depth-integrated pigment content of the euphotic zone. Models, both empirical and analytical, have been prepared as tools for estimating primary production from remotely sensed chlorophyll. These efforts are in a rapid state of development and a listing of recent work is beyond the present scope of this chapter. The models often include light, temperature, and mixed layer depth as well as chlorophyll concentrations. At least two of these, temperature and irradiance at the sea surface, can also be measured by satellite remote sensing. It may become possible to estimate water column primary production using only satellite information for important regions of the ocean.

The large-scale views of photosynthetic pigment distribution afforded by ocean color images are ideal for visual assessment of relative productivity and its spatial and temporal variability. For example, Feldman (1986) showed that the productive habitat (the area of high chlorophyll) of the eastern equatorial Pacific varied more than 10-fold between years. Monthly composite images have revealed the timing and spatial extent of the spring bloom in the North Atlantic (Esaias et al. 1986). For all their promise, satellite images of ocean color provide biased information; cloud cover restricts spatial coverage. Moreover, Abbott and Zion (1987) found that images useful for chlorophyll analysis coincided with windy days off northern California. Thus images were available only during certain meteorological conditions that varied on a time scale of days. Sporadic pulses of production would be observed only if they occurred during windy days.

Estimating new production using satellite data may be feasible. Possible approaches are to use temperature-nitrate relationships (Kamykowski and Zentara 1986) to identify water containing nitrate from a temperature image. The presence of nitrate determines the f-ratio and thus new production as a fraction of the primary production. A regional climatology of the nitracline depth might also be useful in this context, especially if nitracline depth covaried with sea surface temperature.

MODELING NEW PRODUCTION

Models of nitrate input to the euphotic zone provide estimates of new production. In the North Atlantic, the nutrient input from deep winter

TABLE 1 Selected recent estimates of new production.

	Primary Production		New Production		f-ratio	Reference
	gC $m^{-2}y^{-1}$	Gt*	gC $m^{-2}y^{-1}$	Gt*		
Central Ocean						
(Global based on N. Pacific)	130	42	18	5.9	0.14	2
	26	3.8	3.3	0.24	0.06	6
Sargasso Sea			50			1
Coastal Regions						
N.E. Pacific	250	9.0	42	1.5	0.17	2
E. Tropical Pacific	834	0.15	417	0.08–0.2	0.50	3
Bering Sea Shelf			~100**		~1**	5
Upwelling						
E. Pacific	420	0.15	85	0.03	0.20	2
Equatorial Pacific	176	1.9	77	0.85–1.9	0.44	3
Peru	2.8+		0.59+		0.21	4
N.W. Africa	3.5+		2.3+		0.64	4
S.W. Africa			1.1+			4
Global		19–24		3.4–4.7	0.18–0.20	6
		51		7.4	0.145	2

* Gt = 10^9 tons C y^{-1}
** Spring bloom only
\+ Daily rates

Reference Key: 1) Jenkins and Goldman 1985; 2) Martin et al. 1987; 3) Chavez and Barber 1987; 4) Minas et al. 1986; 5) Codispoti et al. 1986; 6) Eppley and Peterson 1979.

mixing drives a spring-summer new production, with a maximum near latitude 50°N (Yentsch 1986). Chavez and Barber (1987) used a circulation model of Wyrtki for the equatorial Pacific to estimate vertical flow and nitrate input. The corresponding new production estimate is the higher of their two values in Table 1.

Modeling the ocean carbon cycle so as to include new production is a major theme in the Global Ocean Flux Study (GOFS) (Sarmiento et al. 1987). Many of the limitations and possibilities, beyond the scope of this review, are discussed in the GOFS report (Sarmiento et al. 1987). There is promise that regional meso-scale and even basin-scale circulation models will be able to assimilate remotely sensed data, ultimately on scales fine enough to include processes important for new production (i.e., the synoptic or "event" scale), in addition to variability associated with meso- and basin-scale processes. Many processes will probably have to be approximated from climatologies and indirect parameterization. Not only the physics and chemistry, but also the biological processes must be handled correctly since the f-ratio, the time scale of coupling between new production and export, and the character of the sinking material are properties of the local food web.

Most of the information needed for global-scale synthesis of primary and new production comes from incubation experiments with ^{14}C and ^{15}N done prior to the development of trace metal clean methods, about 1980 (cf. Martin et al. 1987). The validity of the historical data remains controversial and perhaps must be evaluated for each data set. If older rates are correct, then they imply long-term secular increase in ocean productivity.

The validity of flux measurements with shallow sediment traps is also uncertain, as are the assumptions relating to ventilation, vertical and horizontal circulation, and sometimes time scale estimates in the many model-based estimates of export production. Space does not permit a case by case discussion. Suffice it to say that all the published rates of new production have been criticized in print or in public discussion.

RECENT ESTIMATES OF NEW PRODUCTION

Global estimates of primary production have increased since about 1980, and new production estimates even more (Table 1). For example, several authors have suggested that the annual average f-ratio for oligotrophic central oceans may be higher than the 0.06 value for oligotrophic central gyres proposed earlier (Eppley and Peterson 1979). New production based upon the sediment trap studies in the VERTEX program in the Pacific, for example Martin et al. (1987), is higher than the earlier estimates and the f-ratio was 0.14 (Table 1). The VERTEX ^{14}C primary production results are also higher. Similarly, higher primary production values in the recent work give higher rates of new production for the equatorial Pacific (Chavez and Barber 1987). Estimates of new production based upon the composite ocean color images of the N. Atlantic suggest rates higher than expected (W. Esaias, personal communication). Study on the time and space scales appropriate to the process, and extending our view of the oceans with satellite remote sensing is promoting a revolution in oceanography. These are exciting times for the participants.

Acknowledgements. Much of the inspiration for understanding new production derives from fellow participants in meetings of the Global Ocean Flux Study. I am also especially grateful to Carl Lorenzen, George Knauer, and Bernt Zeitzschel for learning opportunities. E.H. Renger carried out the chemiluminescent nitrate analyses, aided by Elizabeth Orellana. Space and number of citations for this paper have been limited. Papers by V. Asper, W. Berger, B. von Bodungen, Y. Collos, C. Oudot, R. Peinert, T. Probyn, R. Sambrotto, G. Slawyk, V. Smetacek, R.C. Smith, S.V. Smith, J. Walsh, N. Welschmeyer, and others were deleted from an earlier draft. This work was supported by the U.S. National Science Foundation grant OCE 86-13685 and by a grant from the U.S. Department of Energy.

REFERENCES

Abbott, M.R., and Zion, P.M. 1987. Spatial and temporal variability of phytoplankton pigments off northern California during Coastal Ocean Dynamics Experiment I. *J. Geophys. Res.* **92**: 1745–1755.

Chavez, F.P., and Barber, R.T. 1987. An estimate of new production in the equatorial Pacific. *Deep-Sea Res.* **34**: 1229–1243.

Coale, K.H., and Bruland, K.W. 1985. ^{234}Th:^{238}U disequilibria within the California Current. *Limnol. Ocean.* **30**: 22–33.

Coale, K.H., and Bruland, K.W. 1987. Oceanic stratified euphotic zone as elucidated by ^{234}Th:^{238}U disequilibria. *Limnol. Ocean.* **32**: 189–200.

Codispoti, L.A.; Friedrich, G.E.; and Hood, D.W. 1986. Variability in the inorganic carbon system over the southeastern Bering Sea shelf during spring 1980 and spring–summer 1981. *Cont. Shelf Res.* **5**: 133–160.

Cooper, L.H.N. 1933. Chemical constituents of biological importance in the English Channel, November, 1930 to January, 1982, Part I. Phosphate, silicate, nitrate, nitrite, ammonia. *J. Mar. Biol. Assn. UK* **18**: 677–728.

Dugdale, R.C., and Goering, J.J. 1967. Uptake of new and regenerated forms of nitrogen in primary productivity. *Limnol. Ocean.* **23**: 196–206.

Eppley, R.W., and Peterson, B.J. 1979. Particulate organic matter flux and planktonic new production in the deep ocean. *Nature* **282**: 677–680.

Esaias, W.E.; Feldman, G.C.; McClain, C.R.; and Elrod, J.A. 1986. Monthly satellite-derived phytoplankton pigment distribution for the North Atlantic Ocean Basin. *EOS, Trans. Amer. Geophys. Union* **67**: 835–837.

Feldman, G.C. 1986. Variability in the productive habitat in the eastern equatorial Pacific. *EOS, Trans. Amer. Geophys. Union* **67**: 106–108.

Hargrave, B.T. 1985. Particle sedimentation in the ocean. *Ecol. Model.* **30**: 229–246.

Harrison, W.G.; Platt, T.; and Lewis, M.R. 1987. f-Ratio and its relationship to ambient nitrate concentration in coastal waters. *J. Plankton Res.* **9**: 235–248.

Harvey, H.W. 1928. Biological Chemistry and Physics of Sea Water. Cambridge: Cambridge Univ. Press.

Jenkins, W.J. 1982. Oxygen utilization rates in North Atlantic subtropical gyre and primary production in oligotrophic systems. *Nature* **300**: 246–248.

Jenkins, W.J., and Goldman, J.C. 1985. Seasonal oxygen cycling and primary production in the Sargasso Sea. *J. Marine Res.* **43**: 465–491.

Kamykowski, D., and Zentara, S.-J. 1986. Predicting plant nutrient concentrations from temperature and sigma-t in the upper kilometer of the world ocean. *Deep-Sea Res.* **33**: 89–105.

Ketchum, B.H., and Corwin, N. 1965. The cycle of phosphorus in a plankton bloom in the Gulf of Maine. *Limnol. Ocean.* **10** (supp.): R148–R161.

Martin, J.H.; Knauer, G.A.; Karl, D.M.; and Broenkow, W.W. 1987. VERTEX: carbon cycling in the northeast Pacific. *Deep-Sea Res.* **34**: 267–285.

Mills, E.L. 1982. Saint-Simon and the oceanographers: patterns of change in the study of plankton dynamics 1897–1946. In: Selected Works of Gordon A. Riley, ed. J.S. Wroblewski, pp. 3–18. Halifax, Nova Scotia: Dalhousie Univ. Press.

Minas, H.J.; Minas, M.; Coste, B.; Gostan. J.; Nival, P.; and Bonin, M.-C. 1988. Production de base et de recyclage; une revue de la problématique en Méditerranée nord-occidentale. In: Océanographie Pélagique Méditerranéane, eds. H.J. Minas and P. Nival. *Oceanol. Acta*, in press.

Minas, H.J.; Minas, M.; and Packard, T.T. 1986. Productivity in upwelling areas deduced from hydrographic and chemical fields. *Limnol. Ocean.* **31**: 1180–1204.

Parsons, T.R., and Takahashi, M. 1973. Biological Oceanographic Processes. Oxford: Pergamon.

Redfield, A.C.; Smith, H.P.; and Ketchum, B.H. 1937. The cycle of organic phosphorus in the Gulf of Maine. *Biol. Bull.* **73**: 421–443.

Riley, G.A. 1951. Oxygen, phosphate and nitrate in the Atlantic Ocean. *Bull. Bingham Ocean. Coll.* **13**: 1–126.

Sarmiento, J.; Frost, B.; and Wroblewski, J., eds., 1987. Workshop on Modeling in GOFS. US GOFS Rep. No. 4. Woods Hole, MA: US GOFS Planning Office, Woods Hole Oceanographic Institution.

Shulenberger, E., and Reid, J.L. 1981. The Pacific shallow oxygen maximum, deep chlorophyll maximum, and primary productivity, reconsidered. *Deep-Sea Res.* **28**: 901–919.

Sorokin, Y.I. 1985. Phosphorus metabolism in planktonic communities of the eastern tropical Pacific Ocean. *Mar. Ecol. Prog. Ser.* **24**: 87–97.

Yentsch, C.S. 1986. Estimates of "new production" in the mid North Atlantic. In: Workshop on Upper Ocean Processes, eds. R. Eppley and H. Ducklow, US GOFS Rep. No. 3, pp. 31–33. Woods Hole, MA: US GOFS Planning Office, Woods Hole Oceanographic Institution.

Standing, left to right:
Louis Legendre, Richard Eppley, Ulrich Bathmann, Victor Smetacek, Robbie Toggweiler, Colin Reynolds, Wolf Berger

Seated, left to right:
Jean-François Minster, Bodo von Bodungen, Gene Feldman, Peter leB. Williams, Gerhard Fischer

Productivity of the Ocean: Present and Past
eds. W.H. Berger, V.S. Smetacek and G. Wefer, pp. 99–115
John Wiley & Sons Limited

Group Report
Export Productivity from the Photic Zone

P.J. leB. Williams and B. von Bodungen, Rapporteurs

U. Bathmann	L. Legendre
W.H. Berger	J.-F. Minster
R.W. Eppley	C.S. Reynolds
G.C. Feldman	V.S. Smetacek
G. Fischer	J.R. Toggweiler

GOAL: Modes of Export Production; Responses to Climatic Variation

INTRODUCTION

We met and discussed, as a very diverse group, the factors controlling the export of material from the trophogenic zone of the oceans. We found very quickly, although not surprisingly, that plankton biologists were concerned with what remained in the upper parts of the ocean rather than what was lost. It is thus to be expected that we are singularly unsure of the details of the very first stage, i.e., the export stage, of the process that eventually leads to the formation of the sediments. In preparing this report we endeavored to establish what we know, that is, the prevailing hypotheses. In doing so the gaps in our knowledge soon became evident and we quickly found ourselves producing a long list of unanswered but key questions.

The terms below will be used frequently in the report and without spending time on their exact definitions it is important to consider their general meaning. For the purpose of our discussion we will take them to mean:

Primary Production (partial synonyms: "productivity," "gross production," etc.): This is taken to be the photosynthetic rate of the phytoplanktonic community.

New Production: Given a confined system, such as the mixed layer of an oceanic area, new production is that part of primary production which has its nitrogen requirement satisfied by the input of exogenous combined nitrogen. The source of nitrogen could be deep water nitrate (or perhaps organic nitrogen), nitrogen fixed by nitrogen-fixing organisms or atmospheric input of ammonia or nitrate. In coastal and shelf regions it would also include nutrient input from rivers and sediments. It seems to be equivalent, at least in many instances, to the older term *net (community) production*, although the latter is somewhat more operational, derived from measurements of net oxygen or biomass change over a period of time such as a day or season. There are advantages in retaining these two terms with their somewhat separate meanings.

Export Production: In a general way this refers to the transfer of biogenic material (e.g., by particle settling, dissolved organic matter (DOM) advection, or carried by vertical zooplankton migrations) from one zone in the ocean system to another. As it turns out, this term will be used by different people to refer to essentially a hierarchy of rates and processes because of the various depth horizons and procedures used to determine the exported production. Export production from the primary system, the photosynthetic zone, or on time scales longer than a year will be equal to new production, although they may, or in most cases will, be separated both in space and time.

Limiting Nutrient: This term is used by aquatic ecologists in two ways (see Blackman 1905). One use, originating from the concept of Liebig, presumes that a single chemical factor will eventually arrest the extent growth of an individual or community: it may be a major cellular constituent (nitrogen, phosphorus) or a trace metal (iron, molybdenum, etc.). If the environment is supplemented with this chemical then the population will grow, and if the supplementation is continued then control will eventually shift to another factor. In this context the empirical relationship between the major plankton constituents is important (the rationalization of Redfield (1958)) for it enables us, from an inspection of the initial composition of an environment, to anticipate what might eventually limit the extent of growth of the planktonic community. This does not concern itself with the rate growth of the community. The other use of the term "limiting nutrient" is in what may be referred to as the "Monod" sense, i.e., the factor limits the *rate* rather than the *extent* of growth. In this case, strictly, the factor does not need to be chemical; it could be physical (e.g., light). In principle, in the latter case, more than one factor can limit at one time, and during growth the control may shift from one nutrient to another.

What do we know, or believe we know, about the temporal and spatial relationships between export production and magnitude of new and primary production, pelagic community structure and the ways that they are controlled by the physics and chemistry of the environment? To start off, we attempted to identify current hypotheses dealing with the export of material from the trophogenic zone:

1. Particulate flux is dominated by large particles; their export shows strong seasonality (e.g., Bishop et al. 1977, 1987; Smetacek and Pollehne 1986).
2. Large particles are formed both actively by zooplankton and passively by aggregation of small particles (e.g., Smayda 1970, 1971).
3. The proportion of annual primary production exported is inversely related to depth of the mixed layer (e.g., Hargrave 1975).
4. The higher the rate of new production, the greater the loss via sinking (e.g., Eppley and Peterson 1979).
5. Nitrogen is the short-term limiting element for planktonic production (e.g., Dugdale and Goering 1967).
6. Sinking organic particles rather than water-transported DOM contribute the bulk of downward carbon flux.

These hypotheses have been with us for varying lengths of time and naturally are being continuously qualified and refined. Recently, additional hypotheses have been put forward which challenge the universality of some of the above:

1. DOM concentrations, 3- to 4-fold higher than previously believed, have been reported for the upper few hundred meters. These results balance nitrate and dissolved organic nitrogen profiles through the water column, implying that the vertical transfer of organic matter can occur in the dissolved state as well as in the particulate (see Toggweiler, this volume).
2. Great improvements in our procedures for the measurement of iron in the marine environment have revealed, in the N.E. North Pacific and perhaps also the Southern Ocean, that iron levels are sufficiently low to control the extent of phytoplankton growth. This questions the universal role of nitrogen, not only in the present oceans but also in the past, as a limiting factor in the "Liebig" sense (Martin and Fitzwater 1988).
3. Although sediment traps collect fecal pellets of larger (e.g., salps and euphausiids) and smaller (e.g., protozoans) zooplankton, the fecal pellets of the intermediate-sized and abundant zooplankton (the copepods) are rare in trap collections. This points to some selective reworking of copepod pellets.

In order to structure the discussion of the factors regulating the magnitude and composition of exported production, we explored the various interactions between the physicochemical environment and community structure. We used this as a framework to explore what we think we understand about the pelagic ecosystem and exported production, and perhaps more importantly, to discover what questions we were unable to answer.

PHYSICAL FORCING FUNCTIONS

Legendre and Le Fèvre (this volume) have proposed a conceptual model in which phytoplankton production may be channeled into export through various pathways. They have argued that hydrodynamical singularities play a major role at each bifurcation in their model, thus influencing present marine ecosystems and also potentially the geological records, as these are both governed by export production. This role can probably be extended to instabilities of physical forcing parameters.

As far as geological records are concerned, there is a need to shift attention from primary production towards export production. This is because *there does not seem to be a simple direct relationship between phytoplankton production measured in surface waters and the amount of all forms of biogenic material exported to depth*. For example, one mole of nitrate imported in surface waters may fuel the export of very different quantities of biogenic carbon, according to the types of pelagic ecosystems which are themselves a reflection of hydrodynamical forcing. This is of very great significance for global flux studies *since there is probably no stoichiometric relationship between the upward flux of nitrate and the downward flux of biogenic carbon*. This suggests that sedimentologists should try to relate their records to export production rather than primary production.

Another point of significance when comparing biological, chemical, and sedimentological data is the fact that *export production does not refer to the same quantities*. For biological oceanographers, export production most often refers to all biogenic material leaving the productive layer as particles, from minute bacteria to whales. Chemical oceanographers might include dissolved material as well. On the other hand, sedimentologists would normally consider only those particles collected in sediment traps, or just those potentially preserved in the geological records. There are advantages in referring to this loss as ecosystem losses instead of export production. The ratio of total particulate biogenic material exported to material in traps may be partly controlled by physical forcing, but the question remains unresolved.

According to Fig. 1 in Legendre and Le Fèvre (this volume), one can expect that most of the export production from the euphotic zone will be composed of larger sized cells whose growth is a consequence of injection of "new" nutrients in association with hydrodynamical singularities (left-hand side of Fig. 1). These may sink to depth as intact cells or within fecal pellets. The fact that the algal remains reaching the sediments consist mainly of spring bloom species is consistent with the prediction of the model. Legendre and Le Fèvre have also explained how different hydrodynamical regimes in frontal areas can lead to either a fully fledged herbivore chain or to a decomposer pathway. For the same amount of phytoplankton produced, the first pathway will export much more particulate material than the second. Moreover, *the end products of the herbivore pathway are likely to be of a siliceous nature while the decomposer pathway may end up producing calcareous pteropods*. These are examples of the significance of physical forcing in controlling the nature and extent of production export.

Alternating cycles of stratification and mixing may take place in the spring. These alternations will transfer organic matter from the surface layer to the permanent thermocline to the extent that production can proceed during periods of stratification. *However, on a scale of thousands of years, it is not obvious how changes in such physical factors as incident light, water temperature, current patterns, or hydrodynamical characteristics would be reflected in sedimentary records*. The inverse problem of interpreting sediment records in terms of prevailing physical and biological conditions cannot therefore be satisfactorily resolved for the time being.

The main advantage of trying to relate export production to physical forcing is that it opens the possibility of modeling export in terms of unambiguously measured variables. One practical application of such models would obviously be the use of remote sensing information to estimate export production, which is central to studying global fluxes. An understanding of the effect of physical forcing might enable us to constrain the range of possible ocean states during geological time.

Recommended Reading

The writings of Riley (1942), Sverdrup (1953), and Margalef (1978) have been crucial in moulding our ideas of the effect of physical forcing on biological processes. The paper of Pingree et al. (1976) and the review by Legendre and Le Fèvre (this volume) will give some insight into more recent studies.

Questions and Research Suggestions

1. What are the relationships between primary production and total export from the euphotic zone? What fraction of the variance in this relationship can be accounted for by physical factors?

2. How do physical conditions control production export? What are the main processes?
3. If both of the above questions can be answered then we need to see if the main transfer functions can be characterized and quantified in order to build up models for remote sensing and for interpretation of sedimentary records.

CHEMICAL CONSIDERATIONS

There are two questions for which one would wish to have answers: *(a)* how does the oceanic content or gradient of biogenic elements force export production and *(b)* were the gradients similar in the past and can they be expected to change in the future?

There is a general correlation between the export of material from the euphotic zone, inferred both from the sediment record and sediment trap data and the concentration of the classical nutrient elements, N and P. This arises from a combination of a larger import of new nitrogen driving export and the presence of larger primary producers in high nutrient areas. In contrast, it appears to be much more difficult to answer the second question. According to Holland (1978), the concentration of the major ions has remained constant in the oceans over geological time. However, it is unlikely that this generalization can be extended to the nutrient elements because all evidence indicates that control of the concentration of the nutrient elements is different from the major ions. Nitrogen (see Codispoti, this volume) is conventionally regarded to be the element limiting the growth of much of ocean phytoplankton. The combined nitrogen reservoir of the oceans is believed to be largely under microbiological control in the form of the two bacterial reactions: dinitrogen fixation and denitrification. In the case of the former, there is no obvious feedback, i.e., the absence of combined nitrogen does not appear in itself to be a strong enough signal to stimulate dinitrogen fixation. In contrast, denitrification is under active feedback control by the organic productivity of the oceans mediated through oxygen and the abundance of organic material in the form of export production.

The question of the nutrient-limiting export production in past oceans is even more difficult to answer. Codispoti (this volume) argues that whereas nitrate has been traditionally regarded as the best candidate for the contemporary ocean, phosphate could just as well have filled the role in the past. Because living organisms share a common biochemical composition, it is likely that the ratios of biogenic elements in organisms (especially N and P) have not changed over geological time. If this is the case, assuming that the controls envisaged by Redfield (1958) applied in the past, then we *may expect that the proportions of the two elements N and P in the past oceans were the same as today.*

The universal acceptance of the classical notion that oceanic productivity is under nitrogen or phosphorus control has been challenged recently by Martin and Fitzwater (1988), who observed that at the stations in the N.E. Pacific subarctic Ocean dissolved iron appeared to be the immediate factor controlling phytoplankton growth. In these areas, local sources of iron appear to be exhausted and run-off is insufficient to meet the demand, so the productivity of these systems may be under the control of the atmospheric input of iron as dust. Martin and Fitzwater further postulate that the reduction of atmospheric CO_2 during the glacial period may have been a consequence of the reported increased amounts of atmospheric dust during that era. The input of iron along with the dust would have elevated ocean productivity, thus drawing down atmospheric CO_2. It is particularly interesting to note that biological dinitrogen fixation is an iron-requiring reaction and could be under the same control; therefore, one would speculate that in this way iron availability may alter the combined nitrogen content of the oceans. If so, iron input would further drive the oceans away from nitrogen limitation.

It is well known that the N:Si nutrient ratio varies widely from (1:0.5 to 1:3) from ocean to ocean (Sverdrup et al. 1946). Impact of variation on this ratio on the biota is not well known. Small diatoms with a large surface area/volume ratio should have a greater silica demand than large diatoms with a small ratio. However, there is no evidence of environmental species succession of diatoms controlled by the ambient N:Si ratios. As the intraspecific degree of silicification of diatom frustules can vary widely, it appears that the species-specific requirements for Si are much more flexible than, e.g., N or P. Furthermore, dissolution rates of silica frustules are also highly variable, which complicates the picture considerably.

Recent measurements (Sugimura and Suzuki 1988) of the dissolved organic carbon (DOC) and nitrogen (DON) content of seawater, using a method incorporating a new high temperature catalytic oxidation, have produced substantially (i.e., 5-fold) higher values that not only question our understanding of the dissolved organic pool in ocean water but also the mode of transfer of nutrients out of the euphotic zone to the deep ocean. Simple mass balance calculations based on previous analytical methods implied that accumulation of nutrients in deep ocean water could only be accounted for by transport of organic material as particles. Expressed another way, the total combined nitrogen content (i.e., dissolved, particulate organic, and inorganic) of surface water, when transported to the deep ocean and on complete decomposition to nitrate, would at most provide only one third (i.e., 10–15 μmol/l) of the deep water nitrate (i.e., 30–45 μmol/l). This gave rise to the need to incorporate particle flux into one-dimensional models of nutrient and oxygen concentration. The elevation of the surface DON values by some 40 μmol/l has taken away the necessary requirement for particulate flux although, as it stands at the present, the observations certainly do not preclude it. The time delay (decades to a hundred or so years) introduced by transfer to the deep water in the soluble phase, rather than the more rapid exit as particles, enabled a much more satisfactory modeling of nutrient distribution (Toggweiler, this volume). The original model, which incorporated rapid loss of particulate material as a nutrient-export term, produced unrealistically high nutrient concentrations beneath the equatorial regions and yielded oceanic production rates many times higher than the conventionally adopted ones. In the more satisfactory model, half of the nutrient (phosphate) exported production was taken to be in the form of long-lived material, with a decomposition rate of 1/200 yr^{-1}. This role could be played by the "newly discovered" DOM. This purported, previously undetermined DOM is also seen to resolve a persistent discrepancy between titration alkalinity and total CO_2 determinations (Brewer 1987). If the analyses turn out to be correct, they will also pose a number of interesting and fundamental microbiological and biochemical problems. How does the material arise: from algal organic exudates or is it composed of refractory organic residues of decomposition, or is it refractory material produced by chemical condensation of reactive low molecular weight biochemicals as speculated by Duursma (1965). If correct, the vertical profiles of the type reported by Sugimura and Suzuki (1988) raise the question of why this material, which may have existed in warm surface water in the presence of an active microbial community for decades or a century, degrades in the cold, sparsely populated deep ocean. If the decomposition in the deep ocean is microbial, it is difficult to believe that the longer period of time available for decomposition is the explanation.

Recommended Reading

The seminal paper is undoubtedly that of Redfield (1958), which lays down a number of the fundamental principles surrounding the control of the distribution of nutrient elements in the oceans. The books by Broecker (Broecker 1974; Broecker and Peng 1982) and the paper by Berger et al. (1987) provide a good summary of more modern work.

Questions and Research Suggestions

1. How widespread is iron limitation in the oceans? What are the interactions between iron, phytoplanktonic production, and nitrogen fixation?

2. If the atmospheric input of iron is shown to be an important controlling factor in the contemporary oceanic production, what is the significance of iron input by vulcanism and aeolian transport of terrestrial material? What effect might early patterns have had upon global production and the distribution of gases such as CO_2 and O_2 in the biosphere?
3. What are the long-term controls on denitrification?
4. How does DON impinge on global production patterns? Does the export of DON from the productive layer represent an inefficiency of the system or can it be viewed as a paradoxical consequence of nutrient retention in the upper ocean?
5. If the new DOM analyses turn out to be correct, we will need to ask how the production and composition of this previously undetermined DOM varies with the seasons.
6. What are the chemical and microbiological events leading up to the formation of DOM? If Sugimura and Suzuki's observations are correct, why does the DOM accumulate in surface water to such an extent?
7. Are there mechanisms of removal of DOM from the water column by sorption onto particles? Does this play a significant role in the export of surface production and is there a molecular selection at this stage?
8. How can we estimate the historical oxygen and nutrient conditions of the oceans?
9. In the present oceans, what is the contribution of deep water nutrients, *in situ* generated combined nitrogen (N-fixation), and atmospherically injected material (iron or nitrogen species) to net production; that is, what drives the system?

BIOLOGICAL FACTORS

Biological factors impinge upon pelagic export through a number of mechanisms: these involve direct sedimentation of phytoplankton cells, usually with some degree of aggregation, and the passage of fecal pellets. The export is not inevitable because certain groups of organisms tend to recycle (and, therefore, to retain) matter within the euphotic zone. *The extent to which these various processes regulate the export of production is closely related to the abundance and species composition of the planktonic community and the ways in which individual species are selected.*

Phytoplankton is very diverse in taxonomic affinity, morphology, motility, and in the exploitation of resources. These differences are the product of, and are actively maintained by, natural selection. The observable diversity attests to the heterogeneity of aquatic environments, particular environments presumably favoring particular preadaptations. These may be superficially evaluated through a simple equation of population growth rate, *(r)*:

$$r = \mathrm{f}(I) \times \alpha \times A \times \mathrm{f}(K) - R - L \qquad (1)$$

where f(*I*) is a function of the incident irradiance available; α is the light-limited efficiency of growth (literally the slope of *r* on *I* where *I* is limiting); *A* is a proportion of photosynthate assimilated; f(*K*) is the resource base (intracellular nutrient pool supplemented by external uptake); *R* is a physiological cost term (respiration, excretion, etc.); and *L* is a term for losses of whole cells, e.g., through grazing or sedimentation. Environments replete with nutrients, and in which saturating irradiances obtain for a large portion of the day, should favor the growth of all species present, but those with highest *r* values would be expected to dominate. High assimilation is favored in small cells (1–10^3 μm^3 in size) with high surface area/volume ratios (>0.3 μm^{-1}) that can take up and turn over nutrients at specific rates in the order of 0.7–1.4 d^{-1} (Lewis 1976). Such species include small diatoms (*Nitzschia, Skeletonema*) and cyanobacterial picoplankton. Clearly, *r* may be slowed as one or another of the vital nutrients is depleted (N or P or a trace element); it then fulfills the constraint imposed by K. Under these circumstances, competitive advantage is expected to move towards species which have superior capacities to scavenge and store scarce nutrients. Such organisms include the larger, motile, and hence self-regulating flagellates such as *Ceratium*, which are therefore selected on the basis of their ability to exploit the available resources, *K* (Margalef 1978).

Physical factors, most particularly extension of the mixed layer relative to the light penetration, impose quite another set of constraints on organisms. They favor those best able to maximize turbulent reentrainment in upwelling eddies and which show some photoadaptation to harvest light efficiently during the short resident periods in the photic zone. Mechanisms to increase α include cell attenuation and pigment enhancement. These are most clearly shown among the larger, chain-forming bloom diatoms (*Chaetoceros, Thalassiosira*). The representation of diatoms in more than one ecological grouping is important, and there may be analogues among other taxonomic classes, e.g., coccolithophorids.

Animals that undergo vertical migrations may actively transport organic matter in the upper several hundred meters, Vinogradov (1970) first proposed this vertical ladder of trophic exchanges. This is treated in detail in Bruland et al. (this volume).

A storm might leave the water column depleted of algae but relatively charged with nutrients. Suitably seeded, this column would favor the growth of *r*-selected species. As the column restratified, growth would again be restricted to the photic zone with *r*-selected species accelerating; in time, however, the depletion of the resource base is such that the more efficient (*K*-selected species) scavengers for the limiting nutrient(s) are selected.

The loss term, *L*, may be contributed chiefly by settling out of the photic zone and/or grazing in the upper layers. In general terms, the species with the highest potential sinking rates are the larger diatoms, which will tend to be lost from the surface layer as mixing weakens (L increases). Sinking may be a positive advantage to these species in escaping high-light environments, and it is well known that sinking rates are mediated by physiological mechanisms (Eppley et al. 1967; Smayda 1970). Of course, sinking cells may die or be consumed elsewhere in the water column. The formation of aggregates of cells, joined by slime or other means, also serves to enhance sinking export. Such a strategy of high sinking requires provision to "re-seed," and although mechanisms have been suggested (Reynolds 1984) there is no consensus on this point. In contrast, slowly sinking small cells and self-regulating flagellates should normally be regarded as being unlikely to settle out as whole cells (Noji et al. 1986). On the other hand, grazing and the consequent formation of fecal pellets and their bacterial mineralization will tend to retard and recycle matter within the photic zone. *Fecal pellets that fall from the layer are liable to be ingested or digested and the contents repackaged several times during descent.*

Both the susceptibility of algae to consumption by grazers and the amounts of material exported vertically are highly dependent upon the species composition and behavior of the zooplankton. Sporadic swarming of opportunist filter feeders, including salps, pteropods, and euphausiids, is frequently observed to make severe impacts upon standing crops of filterable algae through the rapid downward transport of cells in their pellets. In contrast, copepod populations are individually more dispersed over wider areas, and the temporal phasing of population abundance is closely geared to the regular appearance of phytoplankton foods. Because their foods include their own fecal pellets, copepod feeding tends to retain resources in the upper layers and to reduce the export of material to depth. The size of the active population of copepods present at the inception of the phytoplankton growth period is critical to the level of algal biomass that may be attained and to the proportion which will be eventually exported. In this way, an initially ungrazed diatom-dominated population would be expected to export mainly siliceous frustules, whereas a large standing stock of copepods existing before the start of the bloom should transfer the nutrient to the higher trophic level. By the end of the growing season, the particulate export will be dominated by fragments of zooplankton and detritus resulting from their feeding.

It will be seen that the *nature and quantities of material exported* reflect the nature and quantities of material assembled by the planktonic community which, in turn, *are regulated by the characteristics of the physicochemical environment.* A conceptual view of these possibilities is provided in Legendre and Le Fèvre (this volume) and also in Fig. 1. They distinguish between conditions that select in favor of particular algal groups, their fates (sedimentation, grazing, recycling via microbial loops), and the various impacts on the composition of the exported material (from intact siliceous

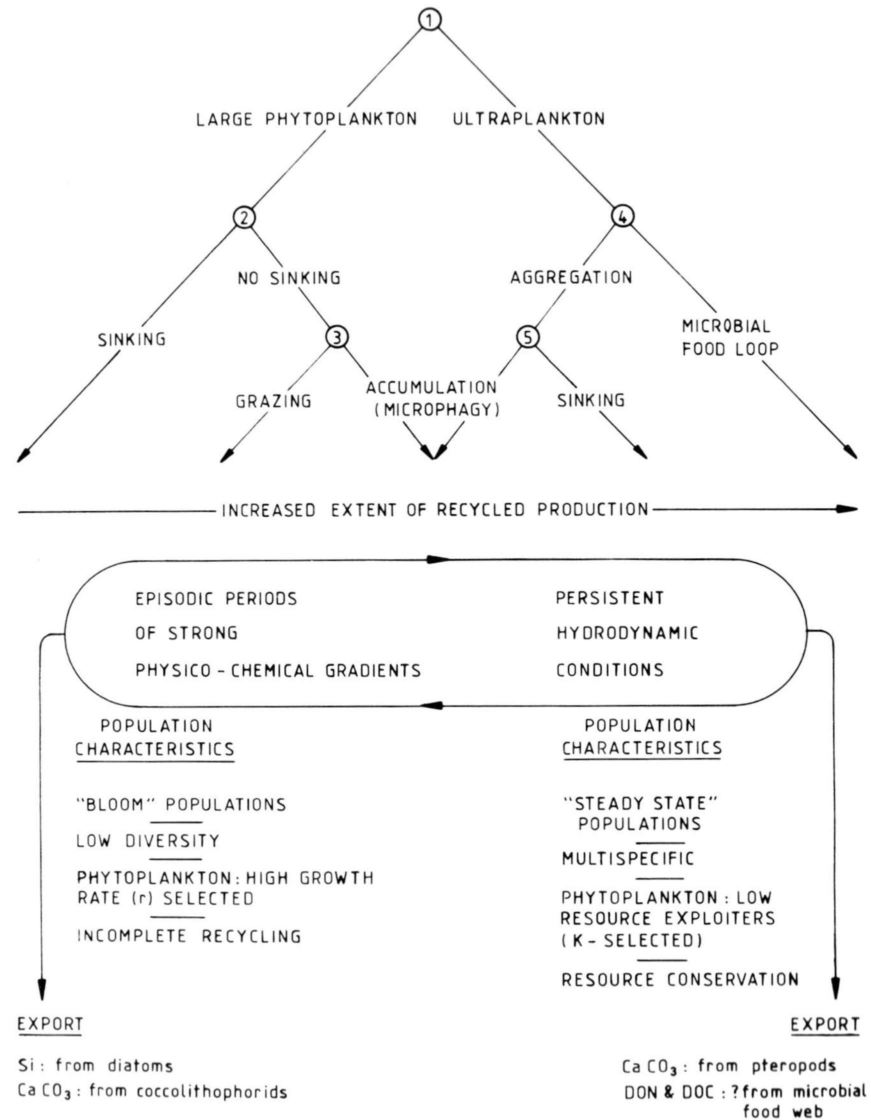

Fig. 1—Summary of the features of the two extreme forms of planktonic community in relation to export production. The model is based upon the bifurcation model of Legendre and Le Fèvre (this volume) and discussions during the group meeting.

diatoms to calcareous debris from pteropod feeders). Though damped in amplitude and lagged in time, the structure and organization of the plankton are major factors in the characterization of the export production of the sea.

Recommended Reading

Textbooks on biological productivity of the oceans rarely deal with export production to any extent. Two important papers are Dugdale and Goering (1967), which formalized the notions of "new" and "recycled" production, and Eppley and Peterson (1979), which considered new production in a global context.

Questions and Research Suggestions

1. Not only diatoms but also coccolithophorids, foraminifera, pteropods, radiolarians, and acantharians should be included in studies of community structure of surface waters and export production.
2. How are particles in surface waters modified by different communities?
3. When and where does aggregation of particles occur? What are the processes involved?
4. Can we obtain a better resolution of the relationship between short-term variability in species composition following nutrient injections and immediate fallout?
5. Can we construct models to predict community structure, production, and export models, not only for the major representatives planktonic food chain but also for the less abundant forms that may make an important contribution to the sedimentary record?
6. Why, in contrast to lakes, does nitrogen fixation appear to play a minor role in the nitrogen budget of the contemporary oceans, even in the extreme nitrogen-deficient central waters of the oceanic gyres? This was, at least, the assessment made in the 1970s. More recently we have realized the possible implications of "microzones" in the oceans, where oxygen concentrations may be low enough for N_2-fixation to occur. We now realize that cyanobacteria are ubiquitous in the sea. Some strains can fix nitrogen. Have we seriously underestimated N_2-fixation in the overall balance of the oceanic N-cycle?
7. What are the processes involved in generating swarms of herbivores which produce immediate fallout? Can the frequency of their occurrence be estimated or ascribed a probability?
8. How do organism seeding strategies fit into sedimentation patterns?

THE BIOLOGY, PHYSICOCHEMICAL ENVIRONMENT, AND EXPORT

In previous sections it has been argued *that pelagic systems are on a continuum between two extremes*. One extreme is characterized by export of large proportions of production on an event scale such as an algal bloom. The other, which behaves more as an equilibrium system, exports very little of the planktonic production. The former is favored by intermittent input of external mechanical energy or strong physicochemical gradients (ergoclines; Legendre et al. 1986). The latter is favored by stable conditions and has developed a complex population structure with efficient remineralizing strategies such as coprophagy, small microbial-based food webs and buffers in the trophic system such as the formation of dissolved organic material, all of which minimize the loss of nutrients from the euphotic zone. Communities associated with temporal or spatial physicochemical gradients are prone to be monospecific; they provide for essential strategies, such as reseeding of large, potentially rapid sinking phytoplankters, and as a result resting stages occur. *The diverse nature of the ocean plankton is sustained by the combination of stable conditions interrupted by episodic events*. If the high energy events are regular in their frequency, then the community may be able to evolve life strategies to prevent exported loss; if they are aperiodic, then such strategies are less likely and export of the bloom will occur.

Figure 1, based upon that of Legendre and Le Fèvre, was an attempt to put together some of our ideas surrounding the export of material from the trophogenic zone. Clearly it is very sketchy, but this properly reflects our very inadequate understanding. This was brought home to us as we considered the export of single elements. In the case of nitrogen, there was a fair measure of agreement that the relationship between total and export production (export production in this case being inferred from new production) followed the guidelines described by Eppley and Peterson (1979). As an exercise we then asked ourselves, as a group, what the diagram would look like for silica; it was our guess that the relationship would be a great deal more complex and we were very unsure of many of the details. We were equally uncertain over the export of calcium carbonate, in part because we knew very little about the factors giving rise to the known organism—the coccolithophorids in particular.

Questions and Research Suggestions

1. When interpreting the analyses of cores an obvious question arises: if monospecific blooms and their export are characteristic of aperiodic physicochemical gradients, do we interpret monospecific layers of

diatoms or coccolithophorids in the sedimentary record as periods of increased intermittent physical disturbance?

2. Conversely, can, for example, the other end of the plankton continuum (the diverse community of the "equilibrium" system) produce similar monospecific deposition by the steady loss of a minor but well preserved component of the community?
3. Are there key species in plankton with mineral skeletons (e.g., diatoms, coccolithophorids, pteropods, or radiolarians) whose dynamics respond predictably to certain combinations of environmental conditions which may therefore serve as indicator species for earlier physicochemical conditions?

OVERALL CONCLUSIONS

1. Biogenic materials exported from the surface layer are those that pelagic ecosystems cannot manage; they therefore reflect whole ecosystem dynamics rather than the activity of the primary producers, i.e., the level of primary production.
2. Although from our experience we have a general knowledge of what types of species assemblages will be found in many situations and the stages in their succession, we are unable to make similar predictions for many of the species commonly preserved in the geological record.
3. Whereas in some cases we can anticipate the effect of physical factors such as temperature, light, hydrodynamic stability, storm events, etc. on production and export, in other cases we find ourselves in disagreement over the sign of the effect because of the complexity of the interactions.
4. The controls on the nitrogen budget of the oceans may be the clue to understanding the productivity of the past oceans.
5. There is an urgency to resolve the present debates over the effect of iron on production rates of oceanic areas and the dissolved organic content of surface oceanic waters because if the recent claims are substantiated, they open up a whole set of new questions and research directions.
6. Clearly, the material that finds its way to the bottom of the oceans, and is incorporated into the sediments, is what one of the hungriest populations on our planet has been unable to digest and as such will only remotely resemble the dishes on the original menu.
7. Finally, as a general conclusion, if we are to relate past oceanic behavior to the remains of surface production found in the sedimentary record, studies of the present-day ocean should be oriented toward studies of the productive system, particle fate, and their relationship to the

physical environment—wherever possible through satellite observations. Attention should be given to evolving such studies to cover the full range of oceanic conditions. At such time, given the deposition rate and composition along with prevailing physical conditions, more reliable information on the productivity of the past ocean may be obtained from comparison with present-day systems.

REFERENCES

Berger, W.H.; Fisher, K.; Lai, C.; and Wu, G. 1987. Oceanic productivity and organic carbon flux. *Scripps Inst. Ocean. Ref.* **87–30**: 1–67.

Bishop, J.K.B.; Edmond, J.E.; Kettner, D.R.; Bacon, M.P.; and Silker, W.B. 1977. The chemistry, biology, and vertical flux of particulate matter from the upper 400 m of the equatorial Atlantic Ocean. *Deep-Sea Res.* **24**: 511–548.

Bishop, J.K.B.; Stepieu, J.C.; and Wiebe, P.H. 1987. Particulate matter distribution, chemistry and flux in the Panama Basin: response to environmental forcing. *Prog. Ocean.* **17**: 1–59.

Blackman, F.F. 1905. Optima and limiting factors. *Annals Bot.* **19**: 281–295.

Brewer, P.G. 1987. What controls the variability of carbon dioxide in the surface ocean? A plea for complete information. In: Dynamic Processes in the Chemistry of the Upper Ocean, eds. J.D. Burton, P.G. Brewer and R. Chesselet, pp.215–231. New York: Plenum.

Broecker, W.S. 1974. Chemical Oceanography. New York: Harcourt Brace Jovanovich.

Broecker, W.S., and Peng, T.-H. 1982. Tracers in the Sea. Palisades, NY: Eldigio Press.

Dugdale, R.C., and Goering, J.J. 1967. Uptake of new and regenerated forms of nitrogen in primary productivity. *Limnol. Ocean.* **12**: 196–206.

Duursma, K. 1965. The Dissolved Organic Constituents of Seawater, eds. J.P. Riley and G. Skirrow, pp.433–475. London: Academic Press.

Eppley, R.W.; Holmes, R.W.; and Strickland, J.D.H. 1967. Sinking rates of marine phytoplankton measured with a fluorometer. *J. Exp. Mar. Biol. Ecol.* **1**: 191–208.

Eppley, R.W., and Peterson, B.J. 1979. Particulate organic matter flux and planktonic new production in the deep ocean. *Nature* **282**: 677–680.

Hargrave, B.T. 1975. The importance of total and mixed layer depth in the supply of organic material to bottom communities. *Symp. Biol. Hung.* **15**: 157–167.

Holland, H.D. 1978. The Chemistry of the Atmosphere and the Oceans. New York: John Wiley.

Legendre, L.; Demers, S.; and Lefaivre, D. 1986. Biological production at marine ergoclines. In: Marine Interfaces Ecohydrodynamics, ed. J.C.J. Nihoul, pp.1–29. Amsterdam: Elsevier.

Lewis, W.M. 1976. Surface/volume ratio: implication for phytoplankton morphology. *Science* **192**: 885–887.

Margalef, R. 1978. Life forms of phytoplankton as survival alternatives in an unstable environment. *Oceanol. Acta* **1**: 493–509.

Martin, J.H., and Fitzwater, S.E. 1988. Iron deficiency limits phytoplankton growth in the north-east Pacific subarctic. *Nature* **331**: 341–343.

Noji, T.; Passow, U.; and Smetacek, V. 1986. Interaction between pelagial and benthal during autumn in Kiel Bight. I. Development and sedimentation of phytoplankton blooms. *Ophelia* **26**: 333–349.

Pingree, R.D.; Holligan, P.M.; Mardell, G.T.; and Head, R.N. 1976. The influence of physical stability on spring, summer and autumn phytoplankton blooms in the Celtic Sea. *J. Mar. Biol. Assn. UK* **56**: 845–873.

Redfield, A.C. 1958. The biological control of the chemical factors in the environment. *Amer. Sci.* **46**: 205–221.

Reynolds, C.S. 1984. The Ecology of Freshwater Phytoplankton. Cambridge: Cambridge Univ. Press.

Riley, G.A. 1942. The relationship of vertical turbulence and spring bloom flowerings. *J. Marine Res.* **5**: 67–87.

Smayda, T.J. 1970. The suspension and sinking of phytoplankton in the sea. *Ocean. Marine Biol. Ann. Rev.* **8**: 353–414.

Smayda, T.J. 1971. Normal and accelerated sinking of phytoplankton in the sea. *Marine Geol.* **11**: 105–122.

Smetacek, V., and Pollehne, F. 1986. Nutrient cycling in pelagic systems: a reappraisal of the conceptual framework. *Ophelia* **26**: 401–428.

Sugimura, Y., and Suzuki, Y. 1988. A high temperature catalytic oxidation method of non-volatile dissolved organic carbon in seawater by direct injection of liquid sample. *Marine Chem.* **24**: 105–131.

Sverdrup, H.V. 1953. On the conditions for the vernal blooming of phytoplankton. *J. Cons. Int. Explor. Mer.* **18**: 287–295.

Sverdrup, H.U.; Johnson, M.W.; and Fleming, R.H. 1946. The Oceans: Their Physics, Chemistry and General Biology. New York: Prentice Hall.

Vinogradov, M.E. 1970. Vertical Distribution of the Oceanic Zooplankton. Jerusalem: Israel Program for Scientific Translation.

Productivity of the Ocean: Present and Past
eds. W.H. Berger, V.S. Smetacek and G. Wefer, pp. 117–137
John Wiley & Sons Limited

Regional Extremes in Particulate Matter Composition and Flux: Effects on the Chemistry of the Ocean Interior

J.K.B. Bishop

Lamont-Doherty Geological Observatory
Columbia University
Palisades, NY 10964, U.S.A.

Abstract. This paper reviews what is known about productivity patterns and vertical flux of organic carbon, opal, and carbonate in the biologically active upper 1000–2000 m of the ocean. Our ability to model particulate carbon flux as a function of depth and primary productivity is much worse than our ability to model particulate carbon flux and regeneration rates as a function of depth once the particulate carbon flux at 100 m is known. A major term missing in equations linking primary production and particle flux appears to be one describing the consumption of particles by macrozooplankton and other large animals.

Simple empirically derived rules based on abundances and chemistry of suspended matter and the mean temperature of the upper 200 m appear sufficient to map the first-order global patterns of production of organic carbon, carbonate, opal, and barium. These may be applicable to both past and present oceanographic conditions.

INTRODUCTION

The vertical distribution of carbon species in the ocean is controlled mainly by biologically mediated processes—photosynthesis, feeding, respiration, and decay—which collectively make up the "biological pump" (Fig. 1). The pump intake is located in surface waters where carbon is transformed from dissolved inorganic forms to particulate organic matter by phytoplankton in the presence of light and nutrients. The impeller—or active mechanism in the pump—is the feeding activity of zooplankton and fish which consume the phytoplankton-produced carbon and package a fraction of it in fecal material which sinks into the deep sea at hundreds of meters per day. Daily vertical migrations of zooplankton and nekton may be an additional way for carbon to be transported from the surface to the deep sea (Angel, this

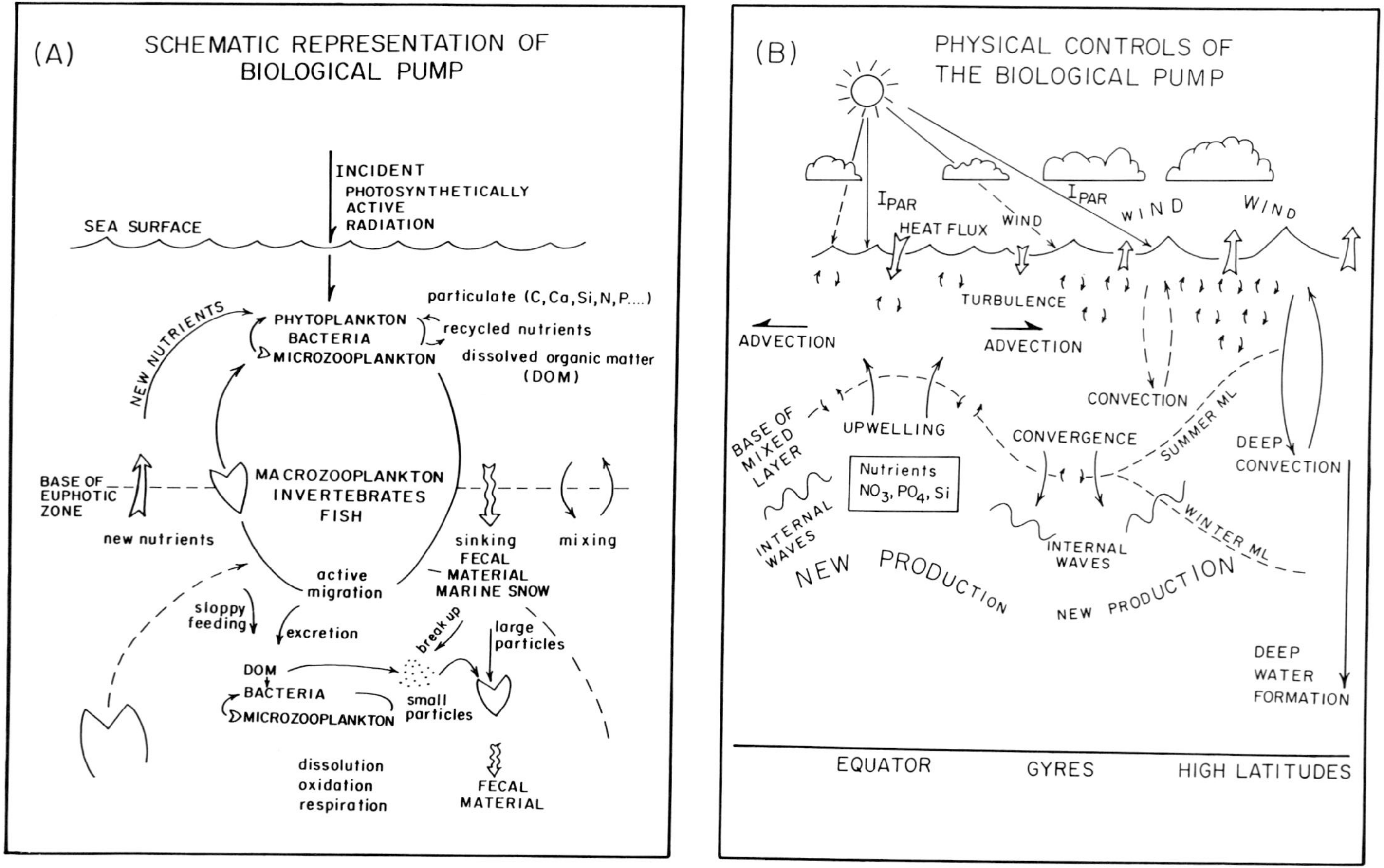

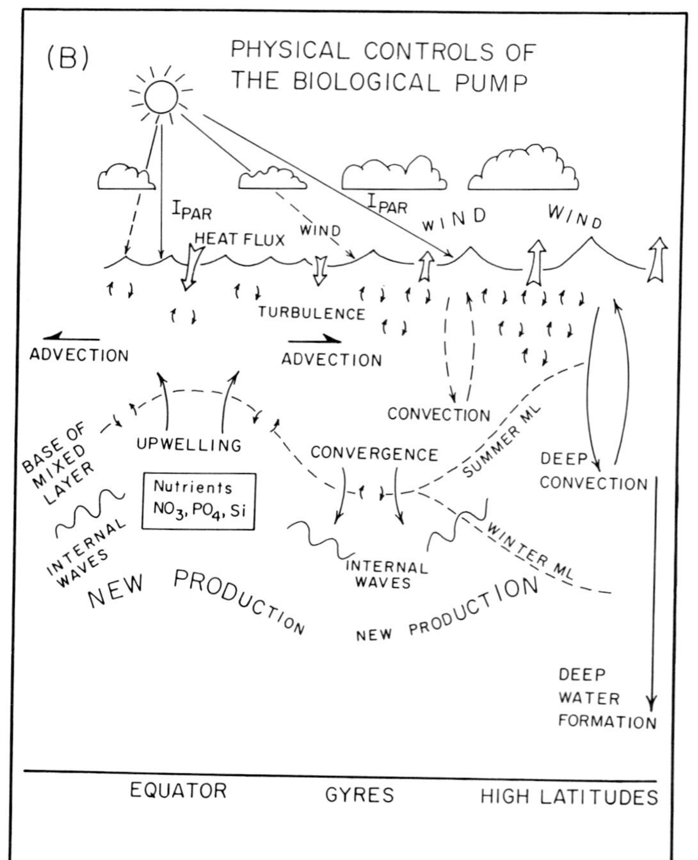

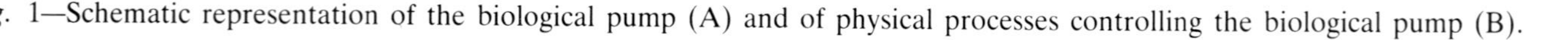

Fig. 1—Schematic representation of the biological pump (A) and of physical processes controlling the biological pump (B).

volume). In some oceanic regions, sinking aggregations of phytoplankton may also contribute to the flux. The discharge of the pump occurs deeper where the organic matter in fecal material, animal tissue, or excreted dissolved organic matter (DOM) is oxidized back to dissolved inorganic carbon species (largely through the activities of bacteria and microzooplankton). The pump forms a major pathway for carbon to move from surface waters into the deep sea. The other pathway for carbon to enter the deep ocean is in polar regions where surface waters are subducted to form deep water. Therefore, the operation and efficiency of the "biological pump" may be an important factor in the removal of anthropogenic CO_2 from the atmosphere to the ocean. The pump is of importance not only to the global carbon cycle but to the cycles of many other chemical species (Ca, Si, N, P, and trace metals such as Ba, Cd, Ni, Zn, and others).

Operation of the pump depends on two dominant processes: (*a*) photosynthesis, which depends on light and nutrient availability in the near surface waters, and (*b*) zooplankton feeding activity. At the base of the euphotic zone, the downward flux (or export flux) of organic matter must, on the whole, be balanced by the upward supply of nutrients (particularly NO_3^-). This is the familiar concept of "new production" (Dugdale and Goering 1967; Eppley, this volume). The fraction of photosynthesis supported by NH_4^+ and urea is recycled within the euphotic zone. A frequently used assumption is that particle flux through the base of the euphotic zone is the sole balancing term for new production and that this balance is achieved on time scales of days. In some oceanic regions, active transport by migrating animals and deep convective mixing may contribute significantly to the flux of organic matter through the base of the euphotic zone, and alter the time scales over which new production is in balance with export flux.

A schematic view of the biological pump and the physical and biological processes which govern its operation is represented in Fig. 1. The global field of photosynthetically active radiation (PAR, about 50% of incident solar radiation)—the primary driving force for the biological pump—varies with latitude, season, and cloudiness. Nutrient supply to the euphotic zone is governed by turbulent, convective, and advective motions of the water column, which are also predominantly forced from above, and by the nutrient concentrations in subsurface waters. Significant effort has been invested in the development of detailed models incorporating the effects of biological and physical processes on the biological pump (e.g., Kamykowski 1987; Toggweiler, this volume) and, in spite of significant progress, they are still in the developmental stages.

The purpose of this paper is to review some of what is known and not known about the biological pump in the upper 1000–2000 m and to test and suggest some simple relationships which can be used as a basis for modeling the cycles of carbon, Si, alkalinity, and Ba—all of which have well-

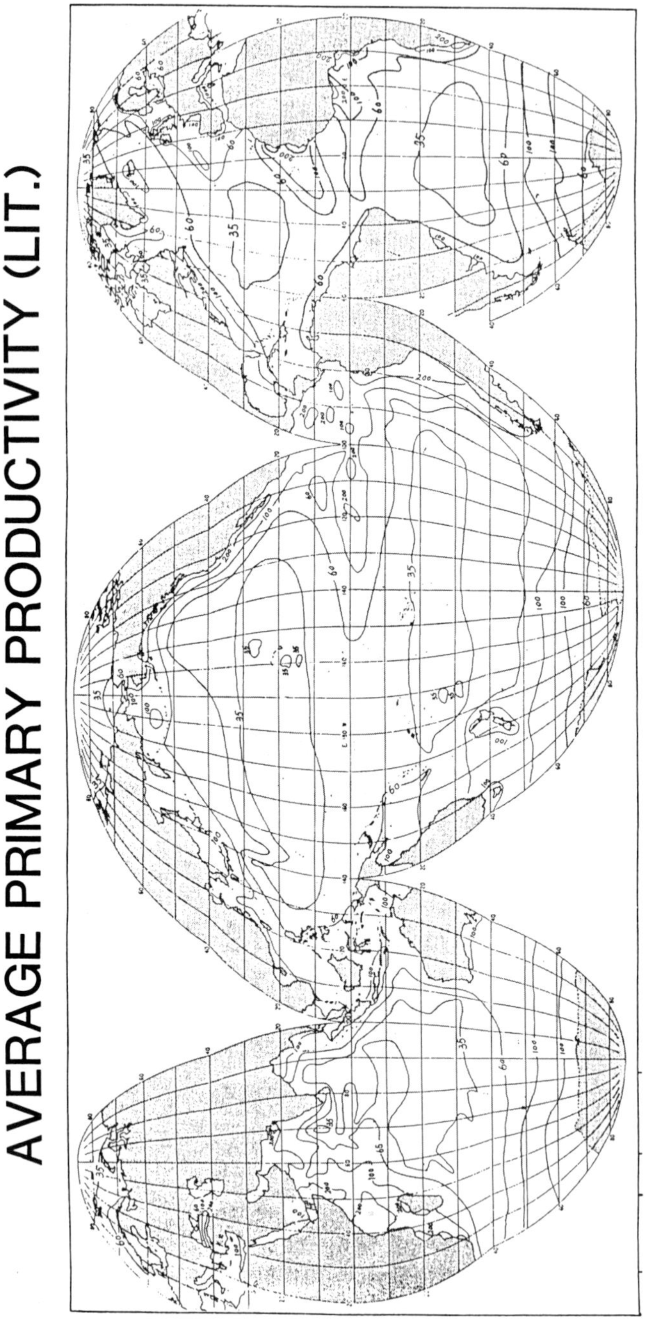

Fig. 2—Average carbon primary productivity of the ocean as compiled from the literature (from Berger et al. 1987).

characterized dissolved distributions in the world's ocean. Some derived relationships may appear to be gross extrapolations but they may permit us to understand the first-order patterns of ocean productivity of these elements past and present. We are presently very far from the goal of understanding the effects of sinking and suspended matter on the chemistry of the ocean interior.

CARBON

Global Patterns of Primary Production

Much of our present understanding of global primary production (the feed for the biological pump) has relied on measurements of primary production measured by the ^{14}C incubation technique and on larger data sets describing the nutrient fields in the upper water column (e.g., Koblentz-Mishke et al. 1970). Figure 2 is a map of mean annual primary productivity constructed by Berger et al. (1987). Large gaps in station coverage exist south of 30°S in all oceans and surprisingly few measurements (1-10 per 10 degree square) were available for large regions of the North Atlantic. Nevertheless, Berger and co-workers constructed a set of simple rules relating primary production to available nutrients (PO_4^{3-} at 100 m), light (latitude), and proximity to continental margins which enabled them to fill in the gaps of station coverage. Whether or not their maps are accurate in detail, they are certainly valuable for understanding patterns of carbon fixation in the ocean and can serve as a starting point for estimates of organic matter delivery to the deep sea. Particulate matter residence times based on $^{234}Th/^{238}U$ are highly correlated with values of primary productivity derived largely from old ^{14}C data and with particle flux rates estimated using sediment traps (Bruland and Coale 1986). The calculated residence times are not as well correlated with contemporary measurements of primary productivity (Bruland, personal communication). Consequently, the patterns shown in the figure may more closely represent that of export production.

Parameterizations of Vertical Carbon Flux

Berger et al. (1987) reviewed sediment trap data published prior to 1986 and various formulations which might be used to model particulate organic carbon flux as a function of depth using maps of primary production. Approximately twice as much data are available now (211 points), mostly for the upper 2000 m from estimates of vertical carbon transport determined by short-term deployments of drifting sediment traps (e.g., Martin et al. 1987) and by large volume *in situ* filtration (LVFS; e.g., Bishop et al. 1987). The new data mostly come from the eastern Pacific but some come from the N.W. Atlantic, including environments influenced by deep convective mixing, persistent upwelling, and late season destratification. The problems associated with sediment trap and LVFS estimates of vertical flux will not be reviewed here since they are adequately described in the literature. The point of this exercise is to show that empirically derived relationships for vertical carbon flux as a function of primary productivity and depth show errors greatly exceeding the factor of 2–3 uncertainty of the data. Significant improvement may be possible by adding a zooplankton consumer term.

Eight empirical relationships describing particle flux (J: units $g\ C\ m^{-2}\ y^{-1}$) are summarized by Equations 1–8 and are evaluated in Figs. 3a and b.

$J = PP/(0.024z + 0.21)$ Suess (1980) (1)

$J = 0.409PP^{1.41}/z^{0.628}$ Betzer, Showers et al. (1984) (2)

$J = 20PP/z$ Berger et al. (1987), simple fit to data <1000 m (3)

$J = 6.3PP/z^{0.8}$ Berger et al. (1987), best fit to data <1000 m (4)

$J = 17PP/z + PP/100$ Berger et al. (1987), refractory carbon model (5)

$J = 9PP/z + 0.7PP/z^{0.5}$ Berger et al. (1987), good fit all data (6)

$J = 1.286PP/z^{0.734}$ Pace et al. (1987), vertex: N.E. Pacific (7)

$J = J_{100}/((z/100)^{0.858})$ Martin et al. (1987), vertex: open ocean (8)

The first 7 empirical relationships tested depend only on primary production (PP), and depth (z), and equations 5, 6, and 7 appear least biased and better behaved compared with equations 1–4; however, it is difficult to prefer any one of the three. Martin et al. (1987) showed that if the particle flux at 100 m (J_{100}) was known, much better estimates of particle flux to deeper waters could be made. The test of this relationship (Eq. 8) using available data did show significant improvement (Fig. 3b). The major implication is that we must be missing an additional term in the equations which use primary production and depth to predict flux. Simply restated, if we had a global map of particle flux at 100 m, then much better estimates of particle flux and regeneration in deeper waters could be made.

Even using the Martin et al. (1987) relationship, some estimated fluxes fall on the high and low sides of the "measured" flux by as much as an order of magnitude. Martin et al. (1987) showed that the depth exponent used in Eq. 8 was as low as 0.3 for their coastal data and as high as 1.0. If some LVFS data from the Warm Core Rings experiment are included, then the upper limit of the exponent may be as high as 1.5. Is the range of the exponent a consequence of differences between zooplankton feeding demand and particulate matter production in the different environments sampled? The answer is probably yes and evidence to support this is found

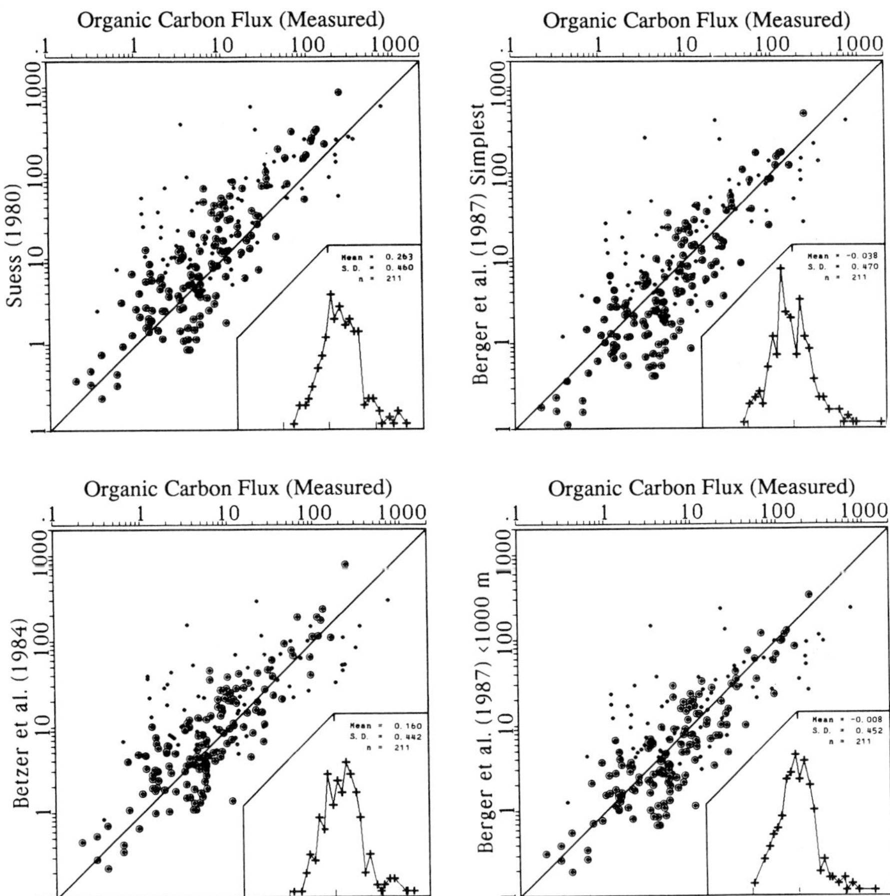

Fig. 3a—Test of empirical relationships (1–4) for carbon flux as a function of primary production and depth. Small symbols (•) denote carbon flux data from LVFS observations, large symbols (◉) denote carbon flux data from sediment trap observations. The inset is a frequency histogram of the ratio of empirically derived carbon flux to measured carbon flux on a log scale from 0.01 to 100. The y axis of the inset is the number of observations on a scale of 0–30.

in Bishop et al. (1987), who found a consistent relationship between particle flux gradient and zooplankton biomass in the Panama Basin. Unfortunately, few data on distributions of zooplankton biomass have been gathered in conjunction with particle flux studies. Simple rules governing the global distributions of zooplankton (and fish) and their impact on particle flux are needed.

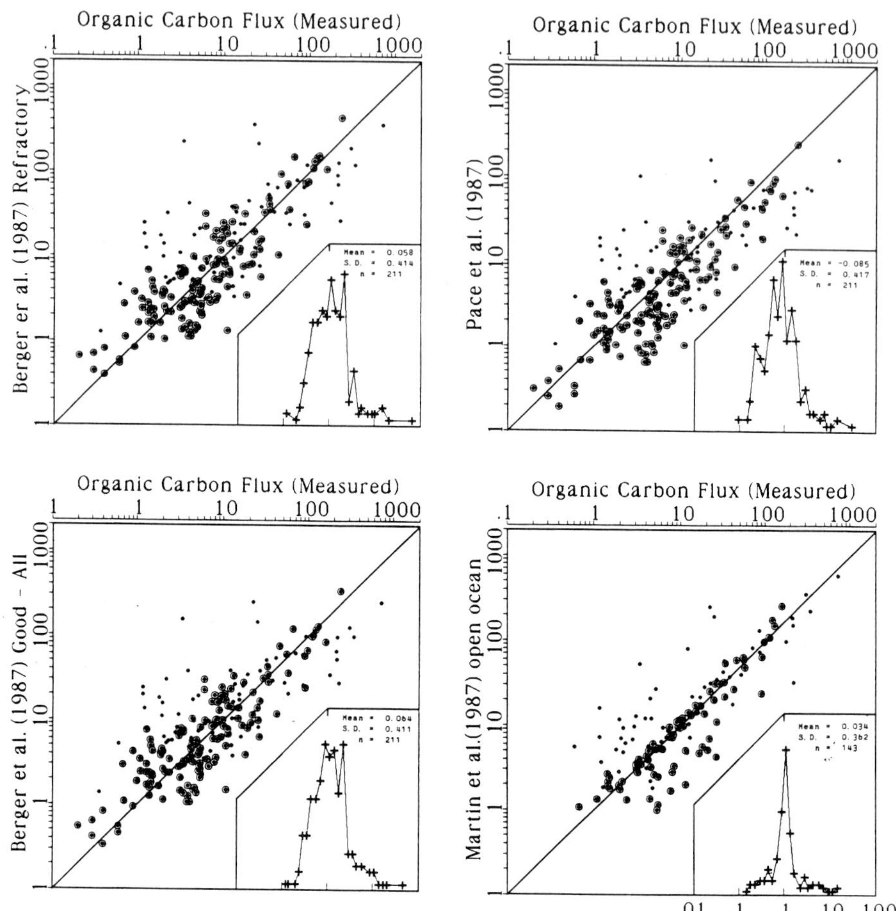

Fig. 3b—Test of empirical relationships (5–7) for carbon flux as a function of primary productivity and depth. Also a test of relationship (8) relating particle flux to known particle flux at 100 m depth. It is clear that knowing the particle flux at 100 m allows a much improved estimation of deeper data at the same station than using primary productivity. One would naturally expect this since flux at one depth is used to calculate flux at greater depths. The y axis scales for the inset are 0–30 for all except (8), where the scale is 0–50.

NITROGEN AND PHOSPHORUS

No maps of global nitrogen and phosphorus fixation appear to be available. It is reasonable to use typical Redfield ratios for conversion between carbon and nitrogen units. Nitrogen measurements are frequently made along with carbon measurements and no great problems appear to exist. An empirical relationship similar to Eq. 7 has been derived to predict nitrogen flux as a function of primary production and depth (Pace et al. 1987).

Phosphorus is a problem. Experimental and field evidence shows that phosphorus is preferentially released from particles relative to nitrogen and carbon (e.g., Bishop et al. 1977; Martin et al. 1979). Once below the euphotic zone, the large particles sampled by traps and pumps exhibit C/P ratios greatly exceeding the 106:1 Redfield ratio. Some reported ratios exceeded 1000:1. On the other hand, small particles filtered from seawater at the same depth frequently show ratios much closer to the Redfield ratio (Bishop et al. 1987). Either there is a methodological bias, or another mechanism apart from the vertical transport of phosphorus in large sinking particles gives the ocean its dissolved phosphate distribution.

SILICA

Global Distributions of Silica Productivity

Although dissolved Si has been extensively measured in the ocean, there have been few attempts to model the global silica cycle. Some still regard Si as having a deep regenerative cycle since profiles of this element frequently show continued increase into the bottom. The work of Berger (1968), Hurd (1972), and Nelson and Goering (1977) provide field and laboratory evidence for shallow silica dissolution. The high deep values for silica in the Pacific may be more related to the fact that the element is dominantly fixed by diatoms living in cold polar and subpolar waters. Such patterns of production are illustrated by Lisitzin (1972; Fig. 4), who used a global primary production map similar to that shown in Fig. 2 and his knowledge of the C/Si ratios of particulate matter filtered continuously from near surface waters during cruises.

Vertical Fluxes of Silica

Little or no sediment trap data exists on the Si flux profiles from the euphotic zone to 1000 m. The few sediment trap data published usually include a single sample within this depth interval. LVFS data from the Panama Basin indicate strong Si dissolution in the upper 1000 m.

CARBONATE

Global Estimates of Carbonate Productivity

According to Lisitzin (1972), the global distributions of coccolith and foraminiferal productivity are maximal between 50°N and 50°S and generally

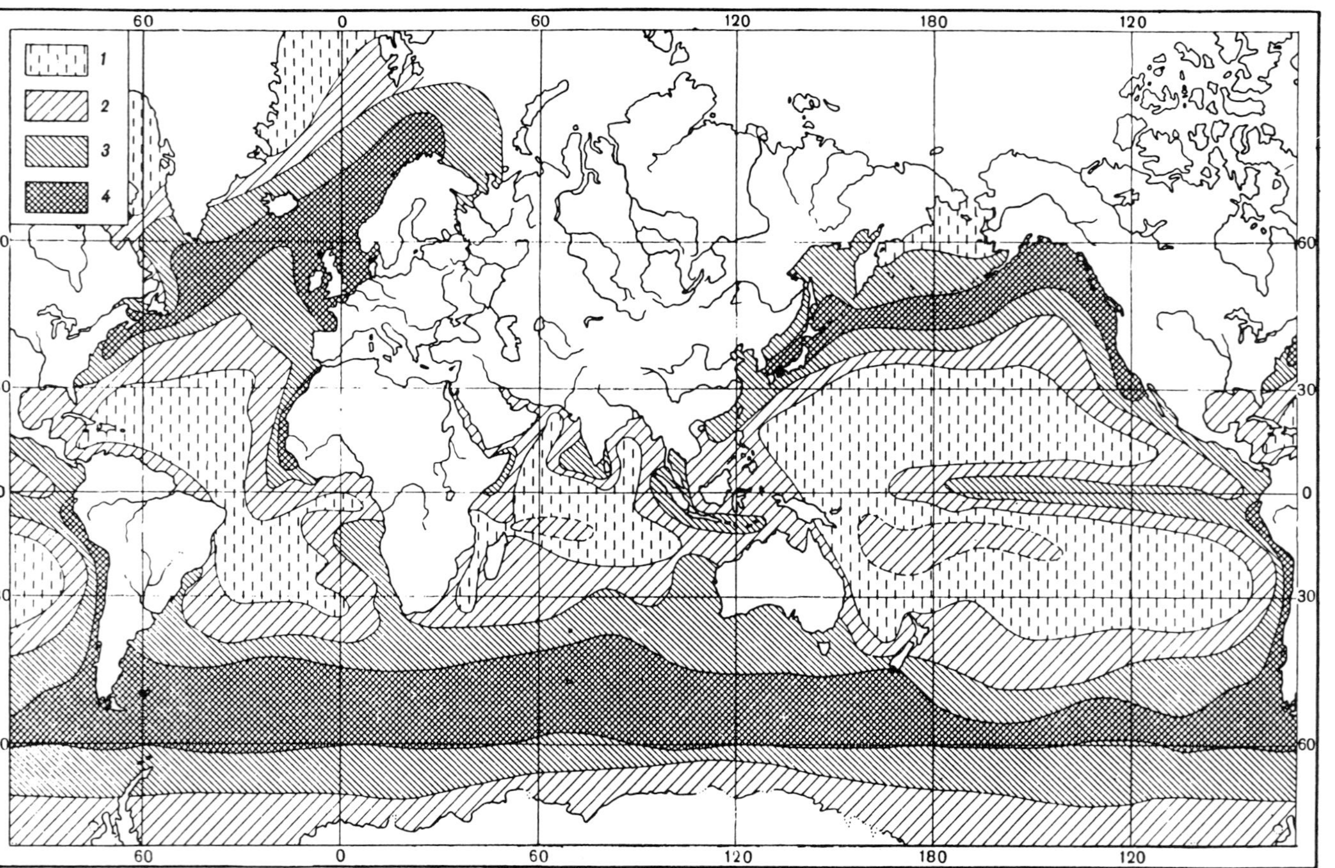

Fig. 4—Annual production of opaline silica in the world ocean (g SiO_2 $m^{-2}y^{-1}$) (from Lisitzin 1972). Key: 1 = <100; 2 = 100–250; 3 = 250–500; 4 = >500. The cross-hatched area centered at 15°S and 180°W is from a printing error in early primary productivity maps and should be <100 (see Berger et al. 1987, p. 49).

coincide with waters warmer than 10°C. His analysis of suspended carbonate distributions in surface waters showed highest concentrations near the Antarctic convergence in the southern hemisphere from 40–60°S, near the polar front in the northern hemisphere, and along the continental margins. Concentrations in equatorial zones were higher than in the gyres. No maps of global carbonate productivity have been drawn.

Vertical Fluxes of Carbonate

There has been no attempt to derive empirical relationships relating vertical fluxes of carbonate to primary productivity. As for silica, few sediment trap data are reported for carbonate flux in shallow waters. The most comprehensive data on fluxes which are published (Betzer, Byrne et al. 1984) appear to be severely biased by entrapment of living pteropods (Harbison and Gilmer 1986). Estimates from LVFS deployments suggest that shallow dissolution of carbonate (especially coccolith carbonate) is a typical occurrence in the eastern equatorial Pacific. This appears to occur in spite of the fact that the water column is supersaturated with respect to calcite from the surface to 1000–2000 m (Bishop et al. 1980, 1987). Evidence for shallow carbonate dissolution in the supersaturated waters of the Atlantic thermocline has been presented by Takahashi et al. (1985). Foraminifera shells have the greatest chance to reach the seafloor unaltered since their residence time in the water column is on the order of days and there is no food value in their empty shells. Foraminifera fluxes may vary by orders of magnitude over time scales of days in productive waters (Bé et al. 1985).

EMPIRICAL SOURCE FUNCTION MAPS FOR ORGANIC CARBON, OPAL, CARBONATE, AND BARIUM

The above sections have focused mainly on rate measurements and empirically derived relationships for estimating particulate carbon flux. This section was motivated by the simple rules suggested by Lisitzin (1972, for carbonate and opal) and most recently by Berger et al. (1987, for organic carbon). All suggested that abundances of particulate matter (and hence patterns of production) could be specified by simple transforms between a well measured quantity (ocean temperature, ocean nutrients, light) and the mean abundances of chemical species in suspended particulate matter. This work was also motivated by evidence suggesting that suspended barium concentrations (which frequently showed maxima at several hundred meters) could be used as an index of organic matter regeneration intensity (Bishop 1988). It must be stressed that parts of this section are speculative, but are deliberately so in order to promote discussion.

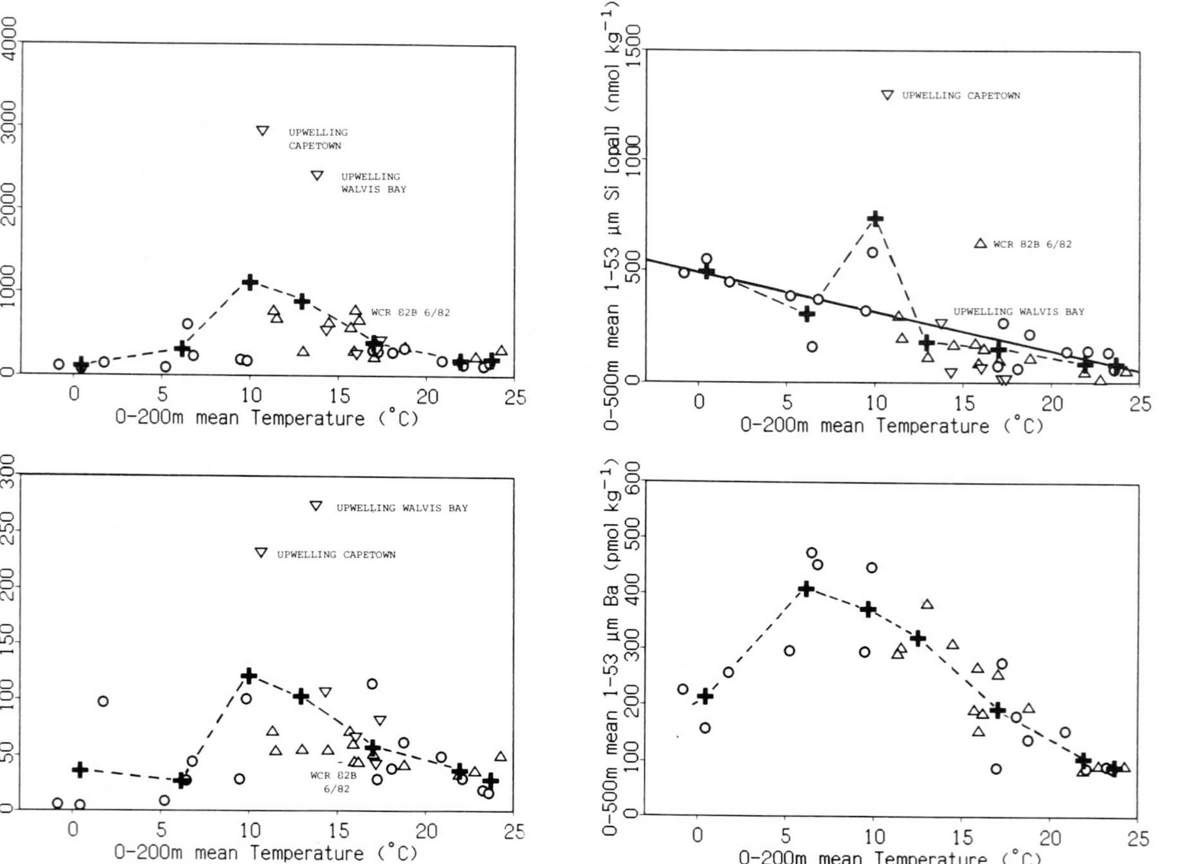

Fig. 5—Average particulate organic carbon, calcium carbonate, opal, and barium in the upper 500 m vs. mean water column temperature at 0–200 m. The (+) symbols denote averages used to derive maps 7–10 from map 6. ○ = GEOSECS data; ▽ = LVFS data from S.E. Atlantic; △ = LVFS data from N.W. Atlantic.

Data used in this analysis came from three sources: (*a*) unpublished Atlantic GEOSECS particulate matter data of Spencer and Brewer, (*b*) LVFS data from the N.W. Atlantic obtained as part of the Warm Core Rings Experiment, and (*c*) LVFS data from the S.E. Atlantic. Only the 1–53 μm LVFS data are used since the GEOSECS samples (10 liter volumes filtered through 0.4 μm Nucleopore filters) and 1–53 μm LVFS data (up to 25,000 liter volumes of water filtered through 1 μm glass or microquartz fiber filters) have been shown to yield equivalent suspended mass concentrations and the GEOSECS data were biased against >53 μm particles. Without going into detail, each kind of data set was processed to yield an estimate of organic carbon, opal, carbonate, and barium and integrated to 500 m (the 500 m limit was chosen to be deep enough to encompass the subsurface barium maximum) and averages were calculated. The organic carbon and opal calculated from GEOSECS chemical data had to be estimated from iodine measurements and gravimetric differences, respectively.

Temperature is inversely related to nutrient concentration in the upper ocean and therefore can be used as an index of nutrient concentration (Kamykowski and Zentara 1986). Profiles determined at each station were integrated to 200 m and the mean temperature calculated. The rationale for the 200 m limit was to be deep enough to minimize the effects of seasonal warming and cooling and yet to be shallow enough so as to adequately proxy nutrient availability to the euphotic zone. Testing other limits of integration yielded less satisfactory results when compared with the suspended matter data.

Figure 5 shows the results of such a correlation exercise. Barium showed a consistent relationship with temperature both in the GEOSECS data and WCRE data with maximum concentrations at 6°C. Silica increased with decreasing temperature. Superimposed on this increase were high values in some upwelling situations. Calcium carbonate and organic carbon tended to peak at intermediate temperatures. Consistently low concentrations for all elements were found in warmest waters. Means and standard deviations were calculated for the data partitioned in 4 degree temperature intervals.

Annual mean ocean temperature data from Levitus (1982) between 100°W and 20°E were used to calculate 0–200 m average temperatures for the Atlantic Ocean (Fig. 6). The empirical relationship relating the mean concentration and variability (standard deviation) of each element to temperature was used to generate corresponding maps (Figs. 7–10).

One question has been asked about the meaning of such maps constructed from few data points. It was suggested that the high values from upwelling situations might be eliminated since they appear to cause a bias of mapped distributions. The answer is that there is no rationale for eliminating particular points. For example, two data points from the S.E. Atlantic

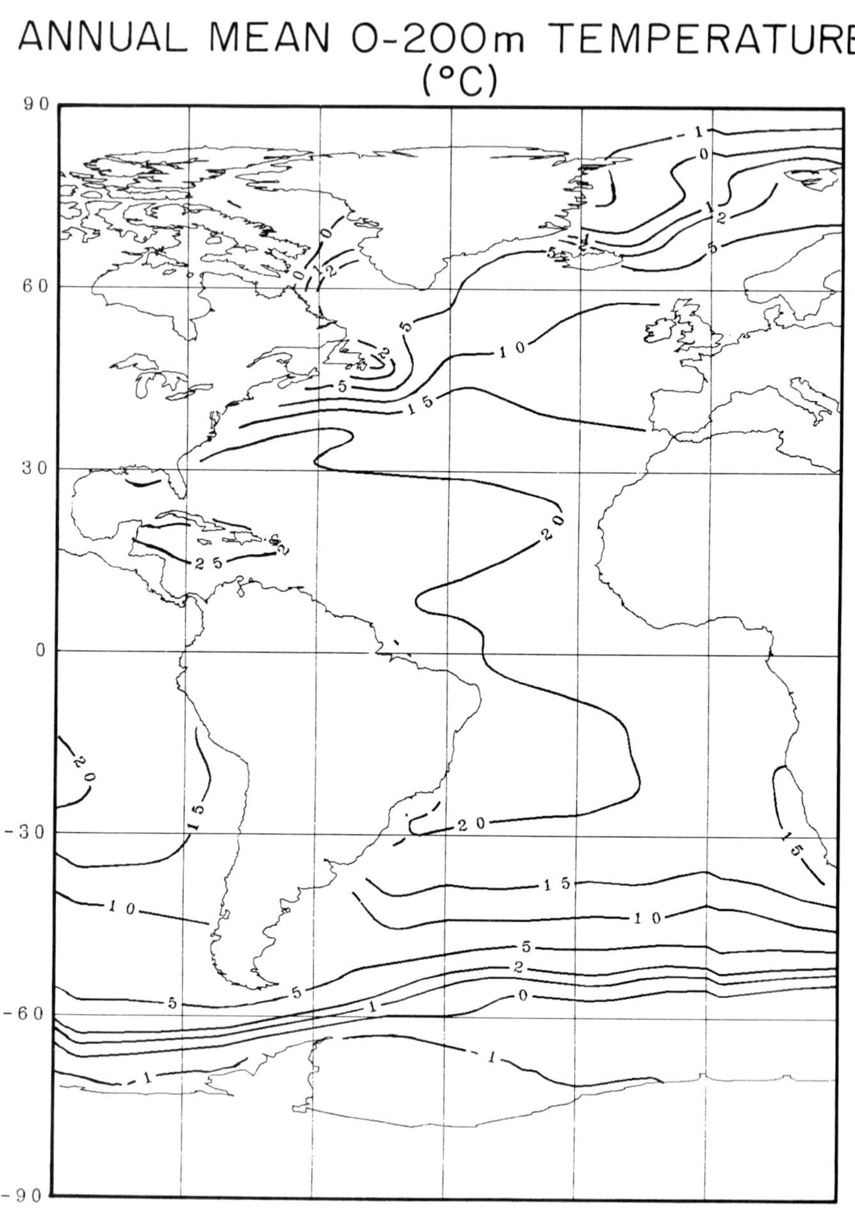

Fig. 6—Annual average 0–200 m ocean temperature (from Levitus et al. 1982).

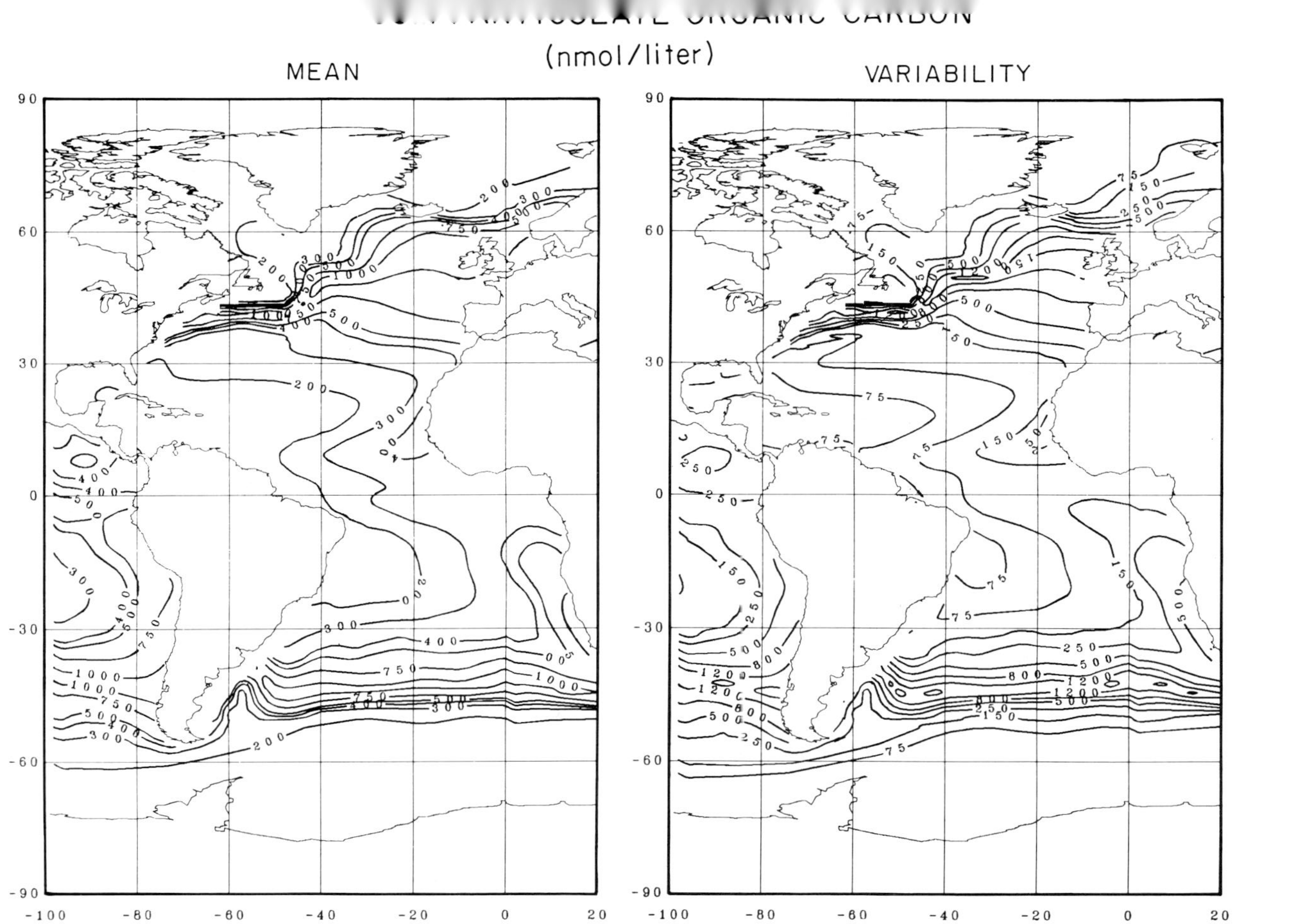

Fig. 7—0–500 m Particulate organic carbon. The left panel displays tne mean concentration derived from ocean temperature (Fig. 6). The right panel is a map of variability (standard deviation) of the data observations about the mean. Lowest concentrations are found in the warmest waters.

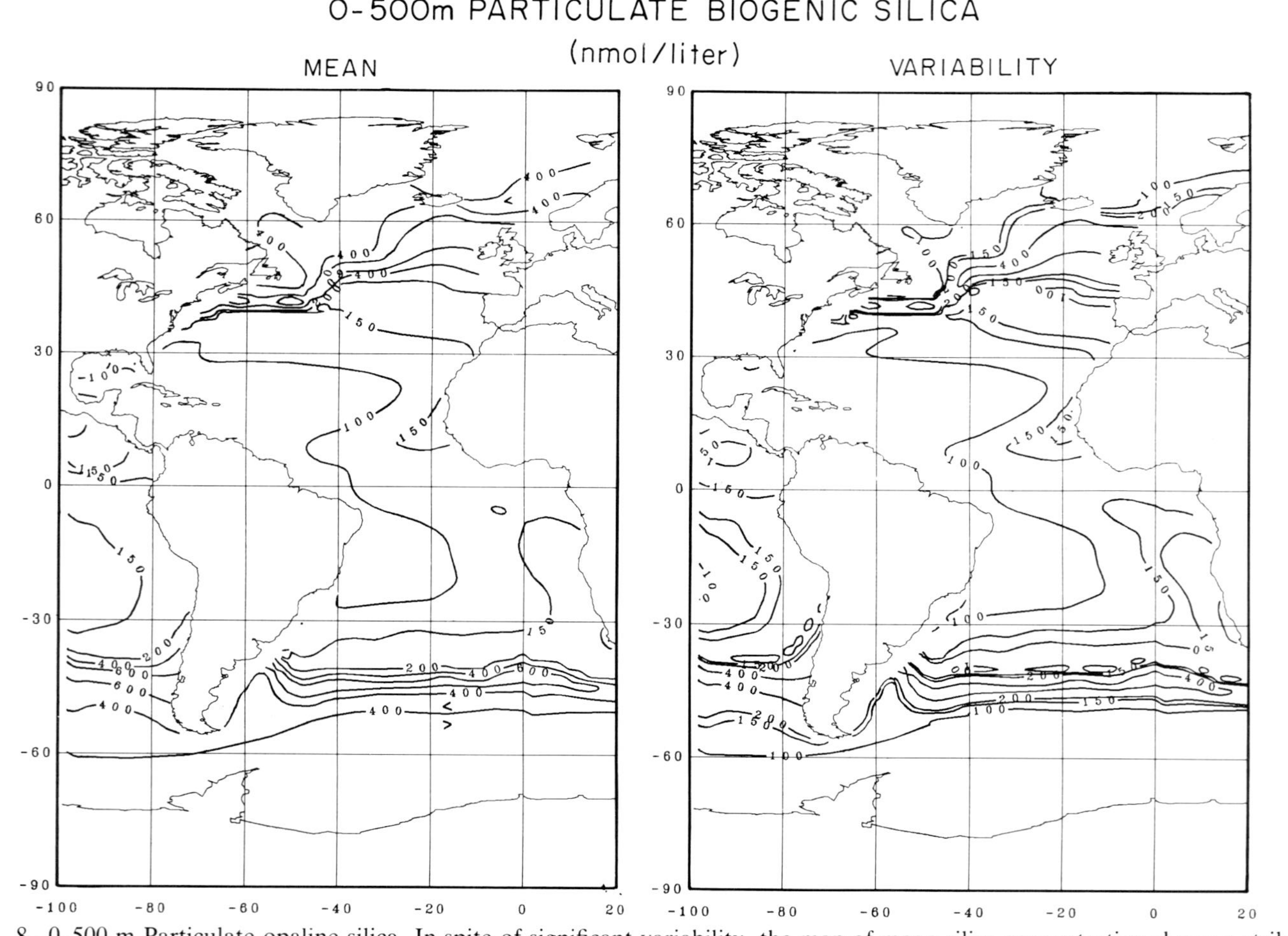

Fig. 8—0–500 m Particulate opaline silica. In spite of significant variability, the map of mean silica concentration shows a striking resemblance to the map of silica productivity drawn by Lisitzin (1972). Highest concentrations appear to occur in the subpolar

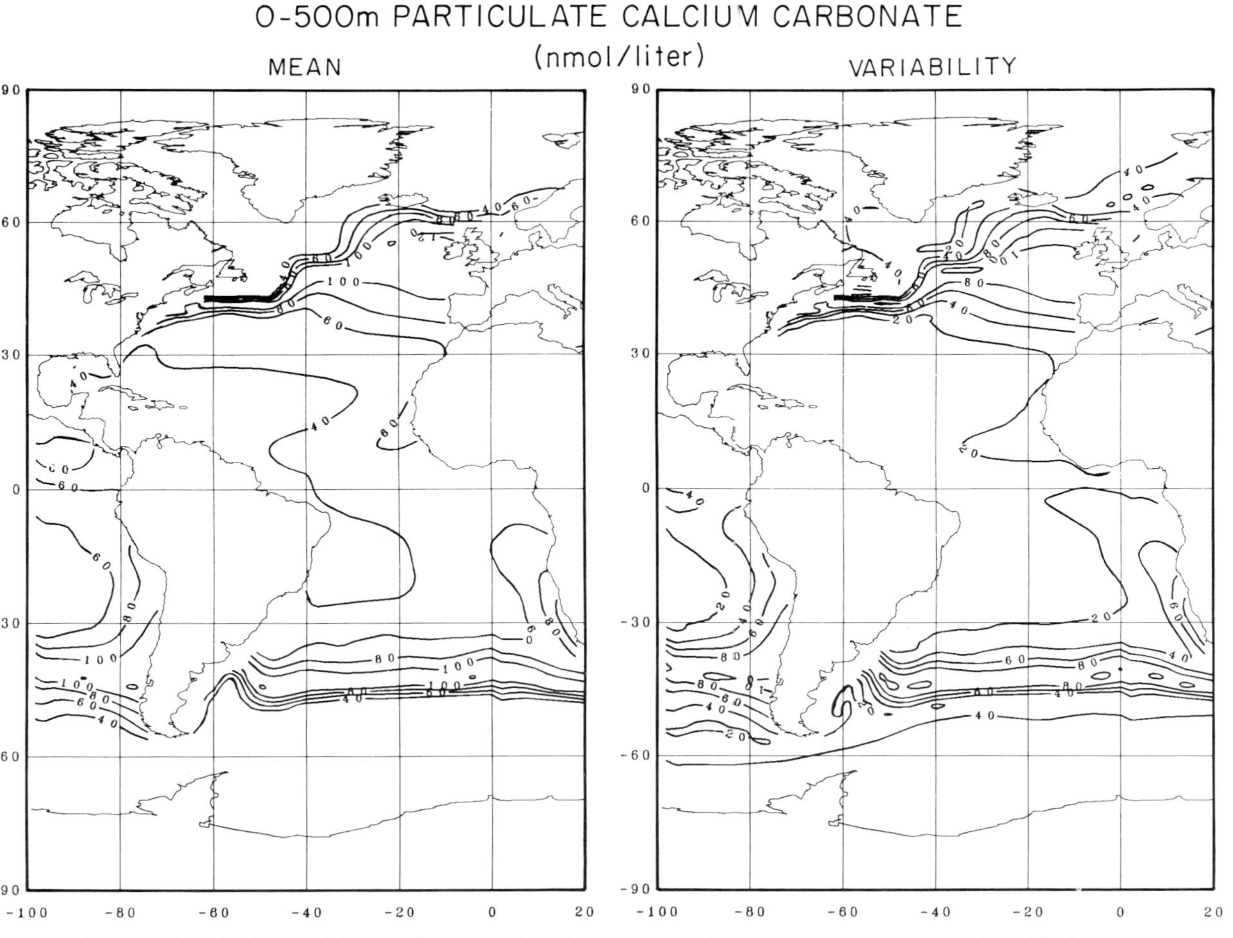

Fig. 9—0–500 m Particulate calcium carbonate. Distributions are highest in waters warmer than 10°C.

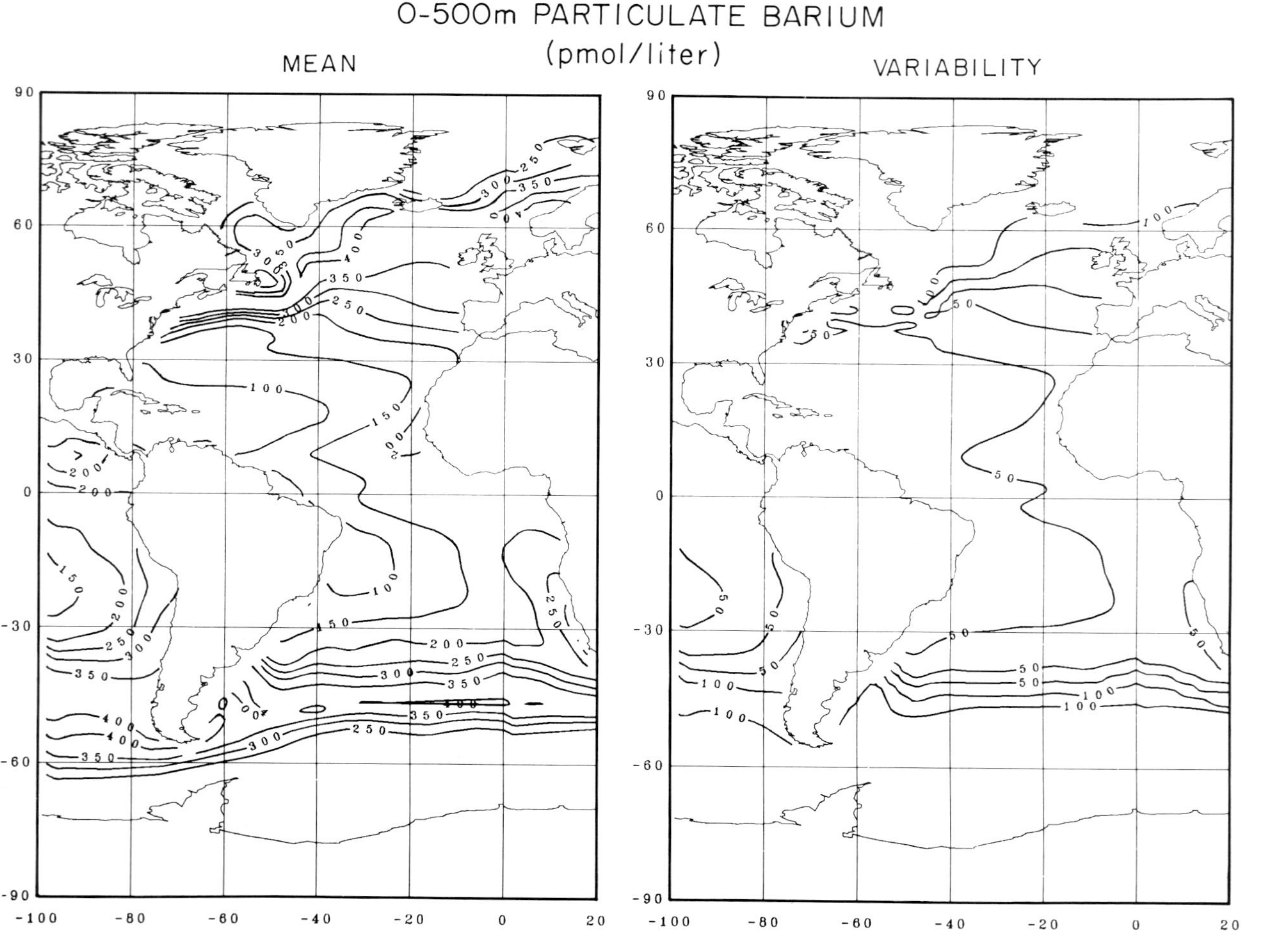

Fig. 10—0–500 m Particulate barium. The distributions of this element show the least variability of all variables mapped and Ba appears to be a good indicator of ocean regions with intense organic production.

appear to be anomalous in carbon and calcium but only one is anomalous in silica. One point anomalous in Si was not anomalous in Ba, Ca, or organic carbon. The important point of this exercise is to illustrate that the suspended particulate matter field does correlate with distributions of ocean physical properties and this information can be used to derive first-order patterns for the production function of each element. By considering the variability as well, we can map regions of the ocean which may contain great variability.

The striking feature of these maps is that the features of the carbon, opal, and carbonate fields reflect closely features described in the previous section. Carbon distributions reflect upwelling and ocean frontal regions. For example, productive regions in the eastern equatorial Pacific (Peru upwelling, Costa Rica Dome) are resolved. Carbonate patterns are confined mostly between 50°N and 50°S and are maximal in the subpolar regions consistent with Lisitzin's simple rule. Some variability may occur in polar waters. Opal patterns are strongest in subpolar zones, remain high in polar regions, and correspond closely with Fig. 4. Barium dominates subpolar and upwelling regions, perhaps indicating global patterns of organic matter regeneration intensity. The warm gyres of the North and South Atlantic contain lowest amounts of all quantities.

Maps of variability (with the exception of barium) show maximal values in the subpolar regions and gyre boundaries. These regions may be naturally variable and therefore may be important areas for the transfer of chemically reactive particles into the deep sea. The regions of great variability are regions of the ocean where more work is required to understand the nature of the variability.

The maps illustrate how knowledge of the suspended particle field leads to simple empirical rules for inferring the production patterns of organic matter, carbonate, opal, and barium. Understanding the workings of the biological pump in the subpolar, frontal, and upwelling regions will be needed to fully evaluate patterns of ocean productivity past and present.

CONCLUSION

Simple rules may exist allowing us to specify the operation of the biological pump on a global basis. Clearly, there is a need to define more of the rules of operation of the pump through cross-disciplinary analysis of existing data and through new field work.

Acknowledgements. The author wishes to thank Drs Derek Spencer and Peter Brewer for providing access to their unpublished GEOSECS Atlantic particulate matter data. Drs W. Berger, P.E. Biscaye, S. Emerson, J. Marra, and V. Ittekkot are thanked for their reviews of the text. Sample collection and analysis were supported by NSF grants OCE88-02773 and OCE85-13420 and ONR contract N00014-80-C-0098. This effort was supported by the NASA cooperative agreement NCC5-29A to Columbia University. L-DGO contribution number 4385.

REFERENCES

Bé, A.W.H.; Bishop, J.K.B.; Sverdlove, M.S.; and Gardner, W.D. 1985. Standing stock, vertical distribution and flux of planktonic foraminifera in the Panama Basin. *Mar. Micropal.* **9**: 307–333.

Berger, W.H. 1968. Radiolarian skeletons: solution at depths. *Science* **159**: 1237–1239.

Berger, W.H.; Fisher, K.; Lai, C.; and Wu. G. 1987. Ocean carbon flux: global maps of primary production and export production. In: Biogeochemical Cycling and Fluxes between the Deep Euphotic Zone and Other Oceanic Realms, ed. C. Agegian. NOAA Symp. Ser. for Undersea Research, NOAA Undersea Research Program, vol. 3(2). Preprint in SIO ref. 87–30.

Betzer, P.R.; Byrne, R.H.; Acker, J.G.; Lewis, C.S.; Jolley, R.R.; and Feely, R.A. 1984. The oceanic carbonate system: a reassessment of biogenic controls. *Science* **226**: 1074–1076.

Betzer, P.R.; Showers, W.J.; Laws, E.A.; Winn, C.D.; DiTullio, G.R.; and Kroopnick, P.M. 1984. Primary productivity and particle fluxes on a transect of the equator at 153 W in the Pacific Ocean. *Deep-Sea Res.* **31**: 1–12.

Bishop, J.K.B. 1988. The barite-opal-organic carbon association in oceanic particulate matter. *Nature* **233**: 241–243.

Bishop, J.K.B.; Collier, R.W.; Ketten, D.R.; and Edmond, J.M. 1980. The chemistry, biology and vertical flux of particulate matter from the upper 1500 m of the Panama Basin. *Deep-Sea Res.* **27**: 615–640.

Bishop, J.K.B.; Edmond, J.M.; Ketten, D.R.; Bacon, M.P.; and Silker, W.B. 1977. The chemistry, biology and vertical flux of particulate matter from the upper 400 m of the equatorial Atlantic ocean. *Deep-Sea Res.* **24**: 511–548.

Bishop, J.K.B.; Stepien, J.C.; and Wiebe, P.H. 1987. Particulate matter distributions, chemistry and flux in the Panama Basin: response to environmental forcing. *Prog. Ocean.* **17**: 1–59.

Bruland, K., and Coale, K.H. 1986. Surface water ^{234}Th/^{238}U disequilibria: spatial and temporal variations of scavenging rates within the Pacific Ocean. In: Dynamic Processes in the Chemistry of the Upper Ocean, eds. J.D. Burton, P.G. Brewer, and R. Chesselet, pp. 159–172. New York: Plenum.

Dugdale, R.C., and Goering, J.J. 1967. Uptake of new and regenerated forms of nitrogen in primary productivity. *Limnol. Ocean.* **12**: 196–206.

Harbison, G.R., and Gilmer, R.W. 1986. Effects of animal behavior on sediment trap collections: implications for the calculation of aragonite fluxes. *Deep-Sea Res.* **33**: 1017–1024.

Hurd, D.C. 1972. Factors affecting solution rate of biogenic opal in sea water. *Earth Planet. Sci. Lett.* **15**: 411–417.

Kamykowski, D. 1987. A preliminary biophysical model of the relationship between temperature and plant nutrients in the upper ocean. *Deep-Sea Res.* **34**: 1067–1079.

Kamykowski, D., and Zentara, S.-J. 1986. Predicting plant nutrient concentration from temperature and sigma-t in the world ocean. *Deep-Sea Res.* **33**: 89–105.

Koblentz-Mishke, O.J.; Volkovinsky, V.V.; and Kabanova, J.G. 1970. Plankton primary production of the world ocean. In: Scientific Exploration of the South

Pacific, ed. W.S. Wooster, pp. 183–193. Washington, D.C.: National Academy of Sciences.

Levitus, S. 1982. Climatological Atlas of the World Ocean. NOAA Prof. Paper 13, US Dept of Commerce.

Lisitzin, A.P. 1972. Sedimentation in the World Ocean. Spec. Publ. 17. Tulsa, OK: Soc. Econ. Paleon. Miner.

Martin, J.H.; Knauer, G.A.; and Bruland, K. 1979. Fluxes of particulate carbon, nitrogen, and phosphorus in the upper water column of the northeast Pacific. *Deep-Sea Res.* **26**: 97–108.

Martin, J.H.; Knauer, G.A.; Karl, D.M.; and Broenkow, W.W. 1987. VERTEX: carbon cycling in the northeast Pacific. *Deep-Sea Res.* **34**: 267–286.

Nelson, D.M., and Goering, J.J. 1977. Near-surface silica dissolution in the upwelling region off northwest Africa. *Deep-Sea Res.* **24**: 31–36.

Pace, M.L.; Knauer, G.A.; Karl, D.M.; and Martin, J.M. 1987. Primary production, new production and vertical flux in the eastern Pacific Ocean. *Nature* **325**: 803–804.

Suess, E. 1980. Particulate organic carbon flux in the ocean—surface productivity and oxygen utilization. *Nature* **288**: 260–263.

Takahashi, T.; Broecker, W.S.; and Langer, S. 1985. Redfield ratio based on chemical data from isopycnal surfaces. *J. Geophys. Res.* **90**: 6907–6924.

Productivity of the Ocean: Present and Past
eds. W.H. Berger, V.S. Smetacek and G. Wefer, pp. 139–153
John Wiley & Sons Limited

Particle Flux in the Ocean: Effects of Episodic Production

G. Wefer

Geowissenschaften, Universität Bremen
2800 Bremen 33, F.R. Germany

Abstract. Present knowledge on seasonal particle sedimentation in deep-sea environments, based on experiments with trap deployment times of one year and longer, is summarized. All experiments showed seasonal flux variations related to biological processes in the surface waters. At some locations, flux maxima are the results of bloom periods of single plankton species. Grazing by zooplankton is an important process in generating particle flux from primary production. In certain cases, phase lags are introduced between production and flux episodes by the expansion of copepod populations. A large variability exists in total flux and composition between the trap experiments. Total annual flux varies over a factor of 300, between about 0.3 g m^{-2} yr^{-1} in the Weddell Sea to more than 100 g m^{-2} yr^{-1} in high production areas such as the Panama Basin, the Bransfield Strait, and the Black Sea.

INTRODUCTION

Sediment trap experiments using time series traps in year-long deployments have demonstrated that particle sedimentation in the open ocean is related to biological processes in the surface ocean waters (Deuser and Ross 1980; Honjo 1982). In the open sea, organisms do not only produce the main portion of the particle flux reaching the deep-sea floor, they also provide the shuttle to transport the particles from the surface to the bottom by means of enlargement of original small particles. Aggregate formation (Alldredge and Hartwig 1986) and fecal pellet production are the main processes in the formation of the larger particles which make up most of the flux.

Biological control of particle sedimentation results in regional variability and a strong seasonality in both composition and amount during the total annual flux. At present, few data are available as time series from stations with year-round deployments. Further experiments are either underway or in the planning stage. This paper summarizes the present knowledge on seasonal particle sedimentation in deep-sea environments obtained from experiments with deployment times of one year or longer.

METHODS

Our knowledge on seasonal sedimentation in the ocean is based on the development of sediment traps, especially time series traps in the manner pioneered by Zeitzschel et al. (1978) and Honjo and Doherty (1987). These traps are attached to moorings deployed at one or more water depths. Where several traps were deployed on the same mooring, only the data from the deepest trap are reported here. In general, the deepest trap is situated well above the seafloor (300–500 m) to avoid the collection of resuspended matter. With the exception of experiments in the Sargasso Sea, Panama Basin, and Equatorial Pacific, which had longer intervals, the sampling periods ranged between two weeks and one month.

In trapping experiments, different preservatives are commonly used depending on the aim of the study. A comparison of the effectiveness of common preservatives is given in Knauer et al. (1984). Between the preservatives and treatments, significant differences were observed. Formalin and mercuric ion treatments appeared to be the most effective.

After recovery of the samples, first-order determinations are undertaken on a split of each sample. The first-order data are fluxes of total mass, carbonate, organic carbon, biogenic silica, and lithogenic components. The data are reported in mg m^{-2} day^{-1} or g m^{-2} yr^{-1}.

ANNUAL SEDIMENTATION PATTERNS

The positions of sediment trap deployments with sampling durations of one year or more are shown in Fig. 1. Data on deployments and annual flux rates per m^2 of total mass, organic carbon, carbonate, biogenic silica, and lithogenic particles are given in Table 1.

In all experiments, large seasonal differences in particle flux are observed (Figs. 2–7). To show the differences in seasonality between the areas studied, the highest and lowest flux rates are listed in Table 2.

It is obvious that the difference between the highest and lowest flux is also dependent on the sampling interval, with longer sampling intervals producing smaller differences and vice versa. Sampling intervals generally varied between two weeks and one month, but were about two months in the Sargasso Sea and Panama Basin experiments and about 100 days at the Equatorial Pacific sites. In addition, the duration of the experiment has an effect on the difference between the highest and lowest flux. The longer an experiment lasts, the higher the difference should be because the greater is the chance to record extremes.

Fig. 1—Positions of sediment trap deployments with sampling durations of more than one year.

In the available data, no differences are obvious between oligotrophic waters and stations from higher productivity areas, such as the equatorial divergence and subpolar zones. The small differences in flux rates from the Sargasso Sea, Panama Basin, and Tropical Pacific are quite likely a result of the long sampling intervals. The relatively small differences between high

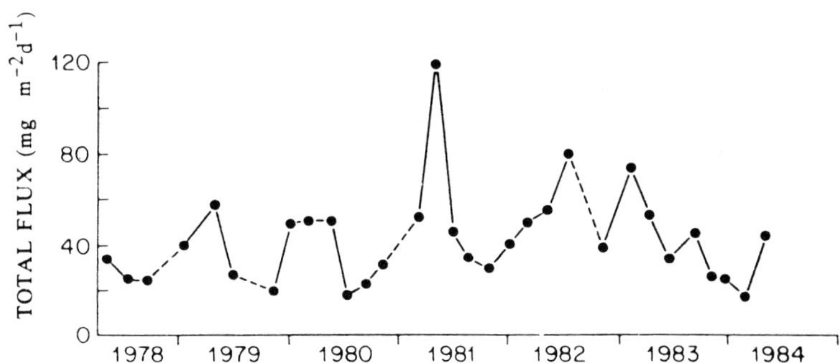

Fig. 2—Six-year record of particle flux to 3200 m water depth 45 miles southeast of Bermuda. Points represent averages of two-month sampling periods (after Deuser 1986).

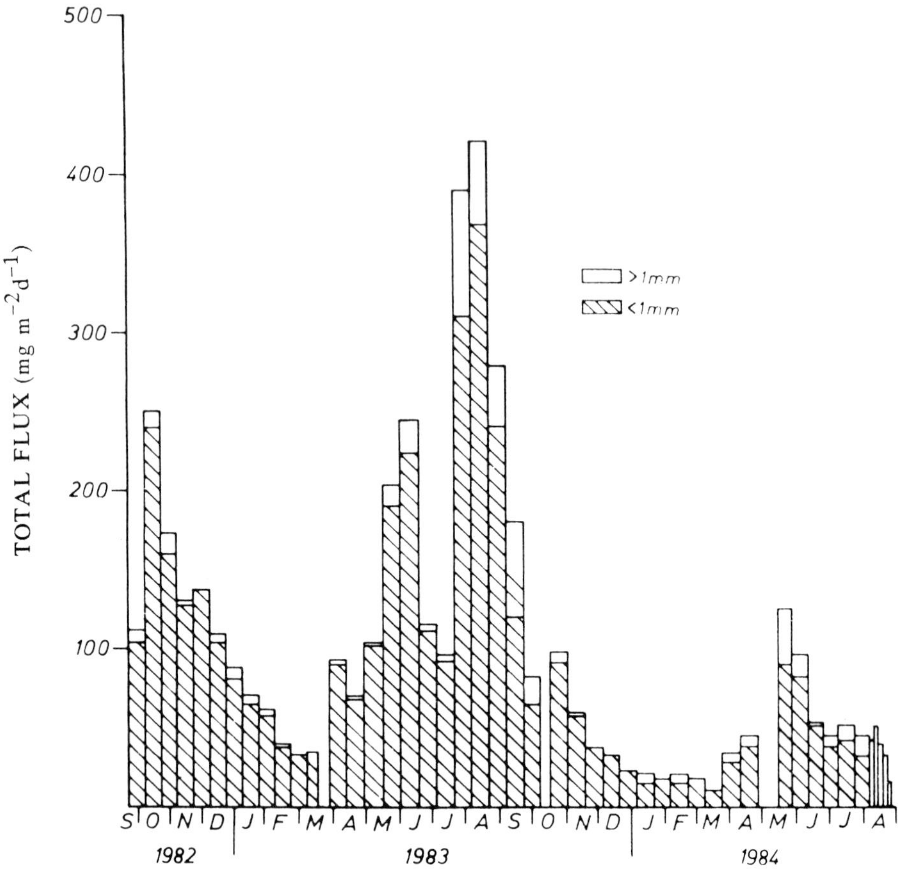

Fig. 3—23-month record of total flux at Ocean Station P in the Gulf of Alaska, water depth 3800 m (redrawn from Honjo 1984).

and low flux rates at the Bear Island station are due to relatively high sedimentation rates during the winter months. These were probably caused by sediment input from downslope transport of shelf material (Honjo et al. 1988).

Where the station was ice-covered for several months (as reported from the Bransfield Strait, Weddell Sea, and Greenland Sea), the difference between flux rates is very great. The reason for this is that the extremely low sediment flux of less than 1 (Bransfield Strait), 0.1 (Greenland Basin), and 0.001 (Weddell Sea) mg m^{-2} day^{-1} was recorded during ice coverage.

All experiments showed that seasonal fluctuations of flux are related to biological processes in the surface waters. At several sites, a close relationship

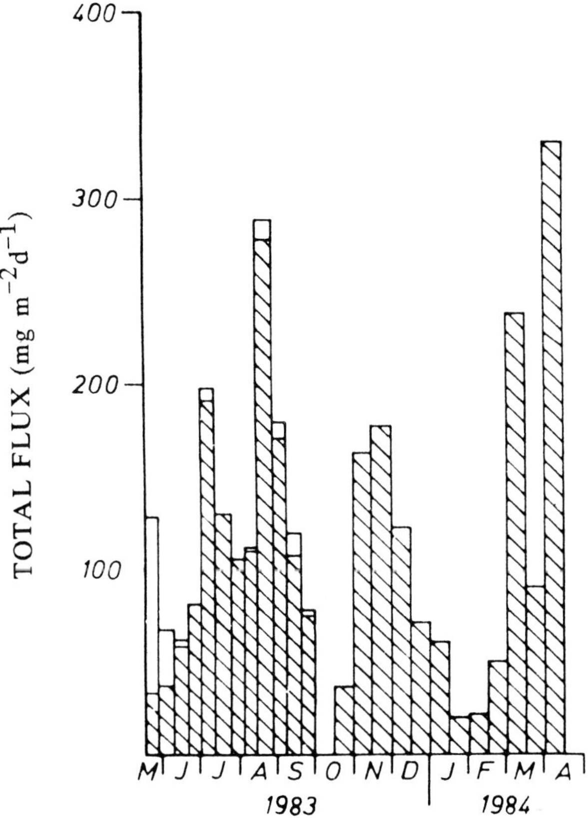

Fig. 4—12-month particle flux record from the Black Sea (redrawn from Izdal et al. 1987).

between the patterns of estimated primary production and particle sedimentation was observed. Examples are the Sargasso Sea (Deuser et al. 1981) and the Black Sea (Izdar et al. 1987; Honjo, Hay et al. 1987). In the Sargasso Sea, variations of total carbonate, silica and organic matter flux are tied to the annual cycle of primary production in the surface water, which peaks in early spring and shows lowest values in late fall (Deuser et al. 1981). In the Black Sea, seasonal variations in flux were related to blooms of diatoms, dinoflagellates, and coccolithophorids (Izdar et al. 1987). The major contributors to the biogenic flux were coccolithophorids, which settled between June and October. Besides the biogenic products of the surface layer, detritus from rivers and shelf sediments were major components of the flux in the Black Sea (Honjo, Hay et al. 1987).

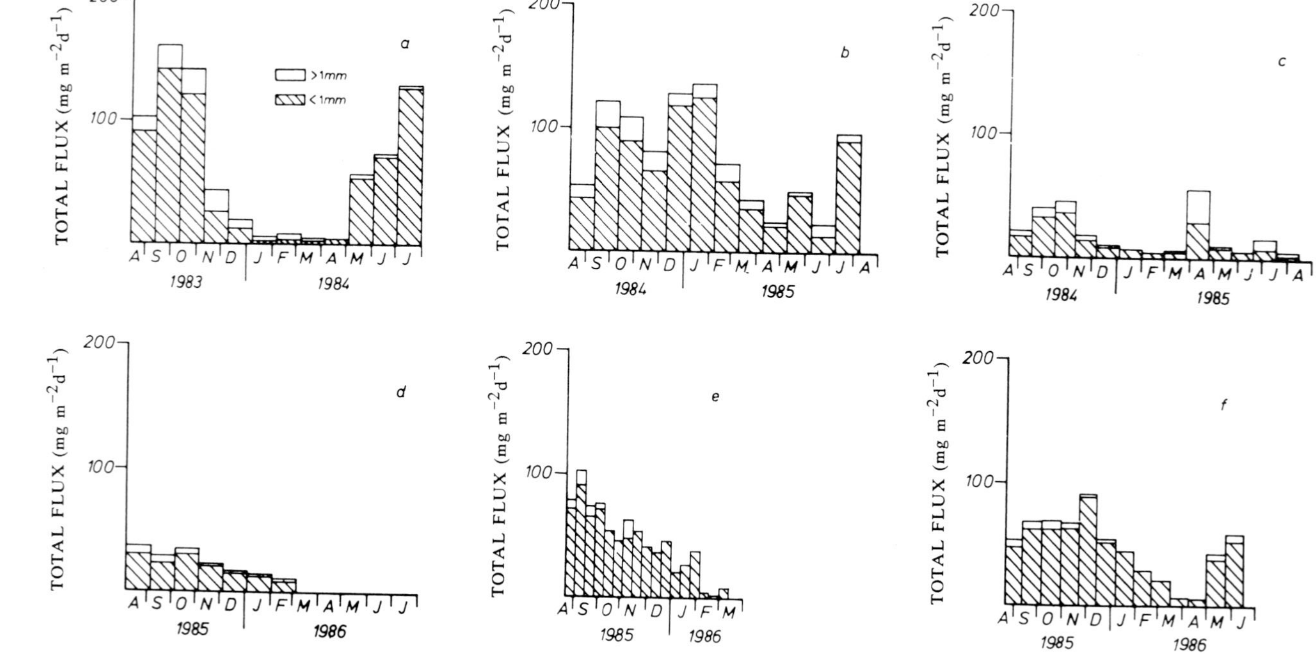

Fig. 5—One-year record of particle flux from the Norwegian–Greenland Sea (redrawn from Honjo, Manganini et al. 1987): (*a*) Lofoten Basin, (*b*) Bear Island, (*c*) Fram Strait, (*d*) Greenland Basin, (*e*) Jan Mayen, (*f*) Aegir Ridge.

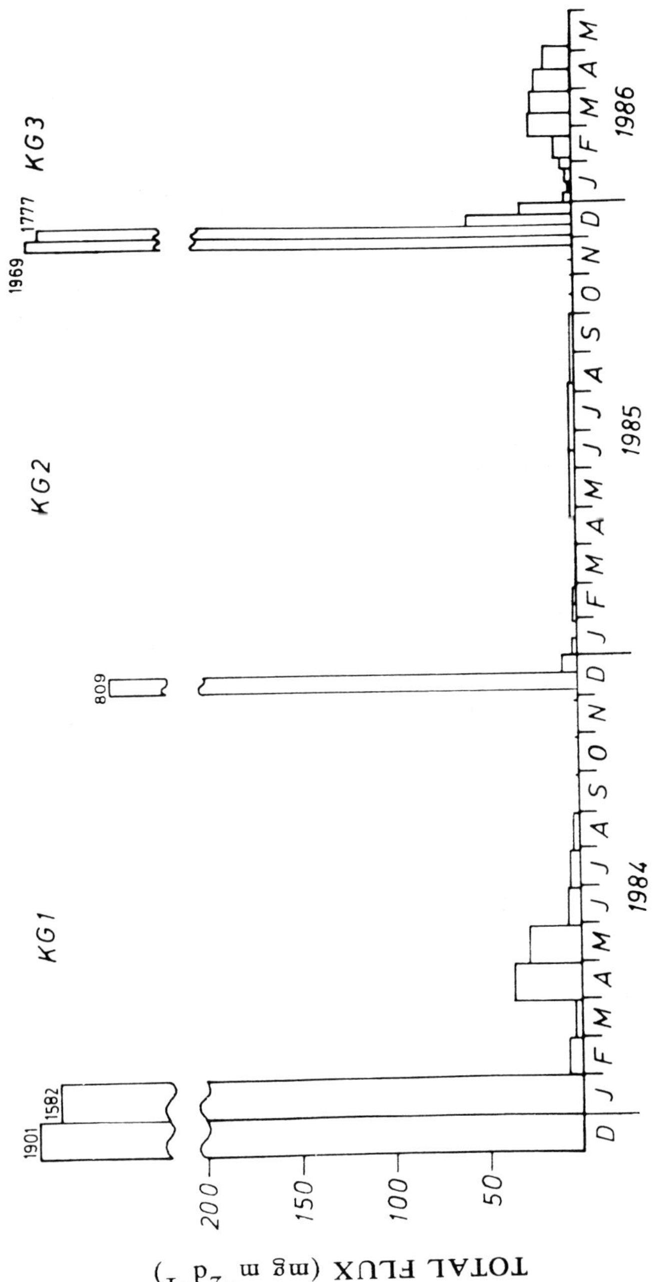

Fig. 6—30-month record of particle flux from the Bransfield Strait (from Wefer et al. 1988, and unpublished data).

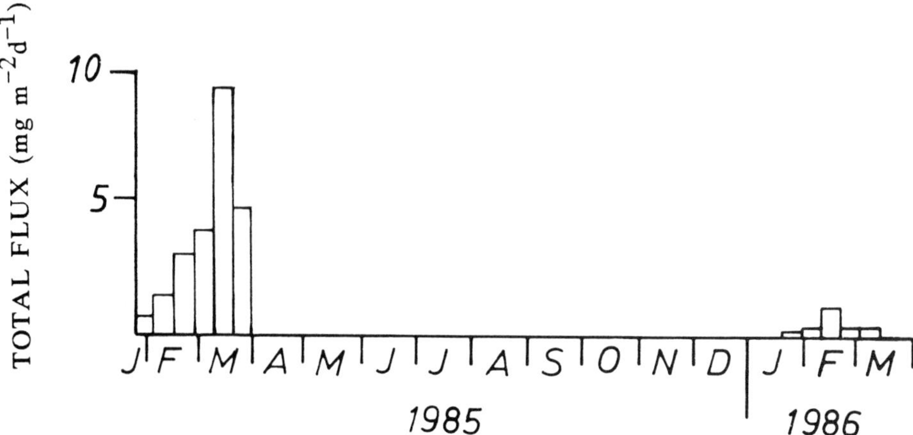

Fig. 7—14-month record of total particle flux from the Central Weddell Sea (from Fischer et al. 1988).

In a recent paper, Deuser (1987) showed that the time lag between changes in mixed layer depth, which is correlated with primary production in the surface water, and particle flux in the Sargasso Sea is less than 1.5 months. In the Black Sea, a time lag of about 2 weeks was observed between the bloom periods of dinoflagellates or coccolithophorids, and increased sediment input at 250 and 1200 m depth (Izdar et al. 1987). In both cases, the calculated time lags can be explained by the average sinking time of the material caught in the traps.

In the Panama Basin, the particulate flux also directly reflected changes in surface primary production (Honjo 1982). During June and July 1980, a pronounced mass flux of 876 mg m^{-2} day^{-1} was observed with carbonate accounting for 62% of the total flux at 3560 m. This high flux was related to the bloom of the coccolithophorid *Umbellicosphaera sibogae*. The sediment collected during this interval was in the form of macroaggregates, which had a size of 0.5 mm and larger, and was cemented together by an organic mucus secreted by *U. sibogae* (Honjo 1982).

Similar relationships between particle flux patterns and biological processes in the surface waters were obtained at two trap sites in the Equatorial Pacific (Dymond and Collier 1988) and three trap sites in the Arabian Sea (Ittekkot et al. 1987). Dymond and Collier (1988) showed that the biogenic particle flux is at least a factor of two lower during periods of intense El Niño Southern Oscillation (ENSO) influence compared to fluxes during other periods. Variations in the total biogenic flux were generally accompanied by compositional changes. During periods with greater nutrient availability, diatom productivity dominated relative to coccolithophorid productivity. At each site in the Arabian Sea, increased particle flux followed the passage of a monsoon (Ittekkot et al. 1987). The annual flux of 54 to 71 g m^{-2} was very large compared with results from Pacific and Atlantic open ocean stations. Seasonality in the particle flux was very pronounced: the flux during the monsoon season was two orders of magnitude larger than during the dry season.

At the Norwegian–Greenland Sea stations, fluxes were highest not during the spring bloom but in the fall between August and October (Honjo, Manganini et al. 1987; Fig. 5). From plankton investigations in the Norwegian Sea it appears that herbivorous zooplankters are responsible for the retention of organic material during the phytoplankton spring bloom. On the Vøring Plateau, the settling of phytoplankton particles (mainly naked flagellates and coccolithophorids) was hampered by recycling the rough copepods during spring, 1986. As a result, a relatively small proportion of the phytoplankton was available for sedimentation (Peinert et al. 1987). The increased sedimentation during fall and early winter is possibly caused by a change in the grazer population from copepods to pteropods (von Bodungen, personal communication). Because of this change, grazing and recycling of fecal pellets might not be as effective as during the period dominated by copepod grazers.

In the Bransfield Strait, pulses in the particle flux are not directly related to pulses in the primary production, although more than 90% of the annual sedimentation is concentrated in a surprisingly short period during the austral summer (Fig. 6). Instead, following phytoplankton growth, vigorous feeding by krill swarms takes place and causes the short sedimentation episodes. During the maximum flux, up to 80–90% of the vertical flux of organic carbon from the surface layer was attributed to krill fecal material (von Bodungen et al. 1987). The small phytoplankton particles are incorporated into krill pellets several hundred μm long and about 150 μm wide which have sinking rates of several hundred meters per day. During settling to the seafloor, the krill feces were altered due to repackaging in midwater depths by zooplankton coprophagy (von Bodungen et al. 1987).

Annual biogenic and lithogenic particle fluxes measured at the Weddell Sea station had the lowest values yet observed in the ocean (Fig. 7). Fischer et al. (1988) suggest that the variability of particle flux in the Northern Weddell Sea is due to spring thaw processes rather than to the availability of larger open sea areas. Significant flux occurred when the sea ice retreated southwards of the mooring site and the flux increased rapidly about 10 weeks after the ice opened. Particle sedimentation ceased almost completely while the ocean was still open.

In general, all areas so far studied over periods of one year or more showed large seasonal differences in particle flux which could be related to biological processes in the surface waters. At some locations, sedimentation

TABLE 1 Logistics of the deployments and annual flux rates for total mass, organic carbon, carbonate, biogenic silica, and lithogenic particles.

Locality	Deployment time	trap depth (m)	depth above seafloor (m)	Flux in g m^{-2} yr^{-1} total	orgC	carbonate	biogenic silica	lithogenic	Reference
Sargasso Sea 45 km (SS) SE Bermuda	6.Apr.1978–11.Aug.1980	3,200	1,000	13.2	0.66	8.2	3.6		Deuser et al. (1981)
Panama Bight 5°22′N (PB) 85°35′W	3.Dec.1979–2.Dec.1980	3,560	300	124.7	4.31	44.71	?	33.1	Honjo (1982)
Station P (SP) 50°N 145°W	Nov.1985–Oct.1986	3,800	520	45.0	1.1	21.9	19.0	0.3	Honjo (1984)
Lofoten Basin 69°30′N (LB) 10°00′E	15.Aug.1983–1.Aug.1984	2,761	400	22.8	1.37	11.4	1.1	6.95	Honjo, Wefer, Manganini, Asper, Thiede (unpublished manuscript)
Bear Island (BI) 75°51′N 11°28′E	12.Aug.1984–10.Aug.1985	1,700	418	28	2.85	6.61	1.96	14.4	Honjo, Manganini and Wefer (1988)
Norway Abyssal Plain 65°31′N (NA) 00°64′E	19.Aug.1985–18.July 1986	2,630	428	17.36	0.59	9.18	1.68	4.26	Honjo, Manganini et al. (1987)

W. Norwegian Sea (NB) 70°00′N 01°58′W	18.Aug.1985– 15.July 1986	2,749	520	16.79	0.53	8.93	1.44	4.65	Honjo, Manganini et al. (1987)
Fram Strait (FS) 78°51.9′N 01°22.0′E	20.Aug.1984– 30.July 1985	2,442	381	7.20	0.41	1.40	0.60	4.00	Honjo, Manganini et al. (1987)
Greenland Basin (GB) 74°35′N 06°43′W	2.Aug.1985– 23.Aug.1986	2,823	622	10.21	0.40	3.28	2.61	3.12	Honjo, Manganini et al. (1987)
Bransfield Strait, Antarctica (KG) 62°15′S 57°31′W	1.Oct.1985– 25.Nov. 1986	1,588	364	107.7	4.3	5.2	38.8	53.5	Wefer et al. 1988
Weddell Sea (WS) 62°26.5′S 34°45.5′W	25.Jan.1985– 19.March 1986	863	3,017	0.37	0.025	0.01	0.29	0.004	Fischer et al. (1988)
Tropical Pacific (TP) Site C 1°N 139°W	12.Dec.1982– 25.Febr.1984	3,495		21.77	1.19	12.31	6.99	n.d.	Dymond and Collier (1988)
	25.Febr.1984– 5.Jan.1985	2,908		41.74	1.89	22.95	15.61	n.d.	
Tropical Pacific (TP) Site S 11°N 140°W	29.Dec.1982– 14.Febr.1984	3,400		13.19	0.59	7.29	3.89	n.d.	Dymond and Collier (1988)

TABLE 2 Ratios of the fluxes between high and low sedimentation (for trap positions see Fig. 1, references for the data are given in Table 1, * denotes flux values calculated from figures).

Locality	highest flux mg $m^{-2}day^{-1}$	lowest flux mg $m^{-2}day^{-1}$	samples per year	remarks
Open Ocean				
SS Sargasso Sea	120*	16*	6	
SP Golf of Alaska	~415*	~10*	24	
LB Lofoten Basin	162	4.5	12	
BI Bear Island	136	24	12	
NA Norway Abyssal Plain	69	6.4	12	
NB West Norwegian Sea	103	2.8	12	
FS Fram Strait	56	4.9	12	
GB Greenland Basin	36.9	0.1	12	several month ice-covered
WS Weddell Sea	~11	< 0.001	20	several month ice-covered
TPC Tropical Pacific site C	77	48	4	
TPS Tropical Pacific site S	75	15.5	4	
High Production Areas				
PB Panama Basin	876	169	5	
KG Bransfield Strait S King George Island	1901	1	12	several month ice-covered
BS Black Sea 1200 m	336	1	24	

maxima are the result of bloom periods of coccolithophorids or dinoflagellates. The time lag between the blooms and arrival at the traps are on the order of weeks and are in agreement with calculations of the time needed for larger particles to settle from the surface to the traps. The relatively short transit times indicate that most of the material caught in the traps settled within fecal pellets or aggregates. Only foraminifera, pteropods, and other zooplankton can produce larger particles with sinking speeds that are high enough to reach the deep sea within a few weeks.

REGIONAL VARIABILITY

With the exception of areas of very high production and polar regions, total flux varies within a surprisingly small range of 7 to 45 g m^{-2} yr^{-1} over most of the ocean (Table 1). Also, the values for organic carbon sedimentation vary little over much of the ocean. Typical organic carbon fluxes to water depths of 1,000 to 3,000 m are between 0.5 and 1.0 g C m^{-2} yr^{-1}, that is,

1–2% of conventional primary production values. The high flux rate of 2.85 g m^{-2} yr^{-1} reported from the Bear Island station is partly caused by old organic matter transported from the Barents Sea shelf to the trap position (Honjo et al. 1988).

Flux values that are higher by an order of magnitude occur in high production areas (primary production larger than about 100 g C m^{-2} yr^{-1}), such as the Panama Basin and the Bransfield Strait (Table 1, Figs. 4,6). The reason for the increase is not organic carbon sedimentation, which remains near 2% of the primary production, but that the high total fluxes are derived from a high carbonate (Panama Basin) or silica (Bransfield Strait) contribution as well as a large flux of lithogenic particles (Table 1).

Considering that the available data comprise a large range of production levels and are from traps set at different depths, the organic carbon proportion of the total flux is surprisingly constant, being in the range of 3.1% (West Norwegian Sea) to 6% (Lofoten Basin) of the total flux (Table 1). An exception is the Bear Island station where 10.2% of the total flux is made up of organic carbon. As previously mentioned, supply from the Barents shelf may be the reason for this high value (Honjo et al. 1988). The Black Sea is exceptional in that organic carbon content varies between 3.5 and 30% in summer and autumn, and between 2 and 3% in winter and spring (Izdar et al. 1987). The trap site is only about 40 km away from the shore. Both coccolithophorids and lateral material transport contribute to the flux of organic matter at this site.

The relationship between total flux and organic carbon sedimentation reflects the type of surface production. Where carbonate production dominates (e.g., Sargasso Sea, Lofoten Basin), the organic carbon content in carbonate producing organisms is crucial. The same is true for silica production (Bransfield Strait, Weddell Sea). Average values for the organic carbon content in organisms with carbonate or silica shells are not available. If we use the carbonate and organic carbon flux at the Panama Basin deployment to 890 m of water depth during June and July 1980 (1595 and 25.1 mg m^{-2} day^{-1}, respectively) (Honjo 1982), for a rough estimate of organic carbon content of the coccolithophorid *U. sibogae*, we obtain 1.5%. If this is representative, then about 15% of the organic carbon which arrived at 3,560 m of water depth over the year can be attributed to coccolithophorid sedimentation. A larger part of the remaining organic carbon was probably resuspended from the slope and laterally transported to the trap site as indicated by the high lithogenic flux (Table 1).

Sites dominated by carbonate flux are the Sargasso Sea, Panama Basin, Black Sea, Lofoten Basin, Norwegian Abyssal Plain, and the West Norwegian Sea. "Dominated" means that carbonate has the highest proportion of the major flux components. The main carbonate producers vary between the areas studied. The most important contributors are coccolithophorids in the

Panama Basin and Black Sea, and planktonic foraminifera in the Sargasso Sea. At the Norwegian Sea stations, pteropods produce a substantial part of the carbonate flux in addition to coccolithophorids. The reasons for the regional differences in carbonate producers are not known.

High silica contents, mainly due to diatoms, are recorded at the trap sites with seasonal ice coverage (Greenland Basin, Bransfield Strait, Weddell Sea). At these sites the content of lithogenic particles is also high. At all areas studied, more than 25% of the total flux was lithogenic material with the exception of the Sargasso Sea. The terrigenous input is explained by resuspension processes in shallower water and lateral transport within the water column to the trap sites. At the Fram Strait site, a portion of the lithogenic material caught in the trap during April (Fig. 5c) is ice-rafted debris. The high contents of lithogenic components demonstrate that most of the trap experiments were influenced by continental margin processes within coastal ocean regions. Quite generally, resuspension processes on the margins apparently play an important role in mass particle flux to the deep-sea floor adjacent to continents. The fact that there was so much lithogenic flux in traps suspended well above the seafloor suggests that much of the material was brought by lateral advection, possibly through subthermocline eddies hugging the margin.

CONCLUDING REMARKS

All sediment trap experiments of one year duration or longer show strong seasonal flux variations, which can be related to biological processes in the surface waters or to more sporadic spikes due to episodic lateral inputs. From these experiments it is concluded that most of the flux to the seafloor is during episodic events. Therefore, composition of particles in the sedimentary record should differ from those measured in the surface ocean. To reconstruct past productivities using assemblages of planktonic organisms (including organic and inorganic tracers) the processes involved in the formation, destruction, and repackaging of particles and aggregates have to be better understood.

Acknowledgements. I thank W.H. Berger, M. Botros, E. Suess, and J. Walsh for a critical reading of the manuscript and helpful comments.

REFERENCES

Alldredge, A.L., and Hartwig, E.O., eds. 1986. Aggregate dynamics in the sea. Workshop report, Office of Naval Research, American Institute of Biological Sciences.

Deuser, W.G. 1986. Seasonal and interannual variations in deep-water particle fluxes in the Sargasso Sea and their relation to surface hydrography. *Deep-Sea Res.* **33**: 225–246.

Deuser, W.G. 1987. Variability and hydrography and particle flux: transient and long-term relationships. In: Particle Flux in the Ocean, eds. E.T. Degens, E. Izdar, and S. Honjo. Mitt. Geol. Paläont. Inst., Univ. Hamburg. Heft 62, pp. 179–193.

Deuser, W.G., and Ross, E.H. 1980. Seasonal changes in the flux of organic carbon to the deep Sargasso Sea. *Nature* **283**: 364–365.

Deuser, W.G.; Ross, E.H.; and Anderson, R.F. 1981. Seasonality in the supply of sediment to the deep Sargasso Sea and implications for the rapid transfer of matter to the deep ocean. *Deep-Sea Res.* **28**: 495–505.

Dymond, J., and Collier, R. 1988. Biogenic particle fluxes in the Equatorial Pacific: evidence for both high and low productivity during the 1982–83 EL NINO. *Geochim. Cosmochim. Acta*, in press.

Fischer, G.; Fütterer, D.; Gersonde, R.; Honjo, S.; Ostermann, D.; and Wefer, G. 1988. Seasonal variability of particle flux in the Weddell Sea and its relation to ice cover. *Nature* **335**: 426–428.

Honjo, S. 1982. Seasonality and interaction of biogenic and lithogenic particulate flux at the Panama Basin. *Science* **218**: 883–884.

Honjo, S. 1984. Study of ocean fluxes in time and space by bottom-tethered sediment trap arrays: a recommendation. In: Global Ocean Flux Study, Proceedings of a Workshop, pp. 305–324. Washington, D.C.: Nat. Acad. Press.

Honjo, S., and Doherty, K.W. 1988. Large aperture time-series oceanic sediment traps; design objectives, construction and application. *Deep-Sea Res.* **35**: 133–149.

Honjo, S.; Hay, B.J.; Manganini, S.J.; Degens, E.T.; Kempe, S.; Ittekkot, V.; Izdar, E.; Konuk, Y.T.; and Benli, H.A. 1987. Seasonal cyclicity of lithogenic particle fluxes at a Southern Black Sea sediment trap station. In: Particle Flux in the Ocean, eds. E.T. Degens, E. Izdar, and S. Honjo. Mitt. Geol. Paläont. Inst., Univ. Hamburg, Heft 62, pp. 19–39.

Honjo, S.; Manganini, S.J.; Karowe, A.; and Woodward, B.L. 1987. Particle fluxes, North-Eastern Nordic Seas: 1983–1986. Woods Hole Oceanog. Inst. Tech. Rep. WHOI–87–17.

Honjo, S.; Manganini, S.J.; and Wefer, G. 1988. Annual particle flux and a winter outburst of sedimentation in the Northern Norwegian Sea. *Deep-Sea Res.* **35**: 1223–1234.

Ittekkot, V.; Manganini, S.J.; Guptha, M.V.S.; Desai, B.N.; Degens, E.T.; and Honjo, S. 1987. Particle flux in the Arabian Sea. *Eos* **68**: 1772 (abstract).

Izdar, E.; Konuk, T.; Ittekkot, V.; Kempe, S.; and Degens, E.T. 1987. Particle flux in the Black Sea: nature of the organic matter. In: Particle Flux in the Ocean, eds. E.T. Degens, E. Izdar, and S. Honjo. Mitt. Geol. Paläont. Inst., Univ. Hamburg, Heft 62, pp. 1–18.

Knauer, G.A.; Karl, D.M.; Martin, J.H.; and Hunter, C.N. 1984. In situ effects of selected preservatives on total carbon, nitrogen and metals collected in sediment traps. *J. Marine Res.* **42**: 445–462.

Peinert, R.; Bathmann, U.; von Bodungen, B.; and Noji, T. 1987. The impact of grazing on spring phytoplankton growth and sedimentation in the Norwegian Current. In: Particle Flux in the Ocean, eds. E.T. Degens, E. Izdar, and S. Honjo. Mitt. Geol. Paläont. Inst., Univ. Hamburg, Heft 62, pp. 149–164.

von Bodungen, B.; Fischer, G.; Nöthig, E.-M; and Wefer, G. 1987. Sedimentation of krill faeces during spring development of phytoplankton in Bransfield Strait,

Antarctica. In: Particle Flux in the Ocean, eds. E.T. Degens, E. Izdar, and S. Honjo. Mitt. Geol. Paläont. Inst., Univ. Hamburg, Heft 62, pp. 243–257.
Wefer, G.; Fischer, G.; Fütterer, D.; and Gersonde, R. 1988. Seasonal particle flux in the Bransfield Strait (Antarctica). *Deep-Sea Res.* **35**: 891–898.
Zeitzschel, B.; Diekmann, P.; and Uhlmann, L. 1978. A new multi-sample sediment trap. *Marine Biol.* **45**: 285–288.

Productivity of the Ocean: Present and Past
eds. W.H. Berger, V.S. Smetacek and G. Wefer, pp. 155–173
John Wiley & Sons Limited

Does Mesopelagic Biology Affect the Vertical Flux?

M.V. Angel

*Institute of Oceanographic Sciences Deacon Laboratory
Wormley, Godalming, Surrey GU8 5UB, U.K.*

Abstract. The biological activities of the midwater fauna and flora at mesopelagic depths are likely to have a significant modifying effect on the fluxes of material from the euphotic zone into the deeper water and the seabed. The components of the flux will be modified by being utilized and repackaged, and there are preliminary data which suggest that up to 80–90% of the organic carbon flux may pass through *in situ* bacterial production. There are mechanisms whereby the size spectra of the suspended particles are modified which will alter the rates of sedimentation and change the availability of the particles to other biological components. There is also active transport of material mediated by the vertical movements of the mesopelagic organisms. There will be substantial changes in these influences which are likely to correlate with the zoogeographical provinces of biological distributions and seasonality cannot be ignored particularly at latitudes > 40°.

INTRODUCTION

The sedimentary record provides a blurred record of past climatic events and the thrust of this workshop is to seek means of improving the precision with which this record can be interpreted. Improved precision can be achieved by increasing, both quantitatively and qualitatively, our understanding of the mechanisms which both create and confuse the sedimentary record. The main sources of oceanic sediments are: (*a*) oceanic sediments produced by *in situ* processes; (*b*) aeolian inputs including both volcanic and desert dusts which are identifiable by their mineralogy; (*c*) cross-shelf inputs of sediments derived from shallow seas or riverine sources; (*d*) ice-rafted debris. All these processes are to some extent sensitive to climatic influences. Predominantly, these inputs enter the deep ocean through the superficial layers and are modified both chemically and biologically during transit through the water column. These water-column processes play a substantial role in finally determining the exact nature of the biogenic material that is eventually deposited on the seabed. On arrival on the sediment interface there are further biological and chemical modifications of the sediments before they become incorporated into the sedimentary record (Fig. 1). This paper will explore the range of modifications which biological processes may impose on the sediment flux during its passage down through the water column, concentrating on those which take place below the euphotic zone, particularly in the mesopelagic zone. It will not deal directly with the initial production and recycling of particles within the euphotic zone, except in considering how these processes may be coupled with the activities of the mesopelagic fauna, nor will it deal with the fate of the material once it reaches the benthic boundary layer.

THE RELEVANT BIOLOGICAL PROCESSES

There are four classes of biological process which may modify the downward flux of organic carbon: (*a*) utilization by grazing including microbial

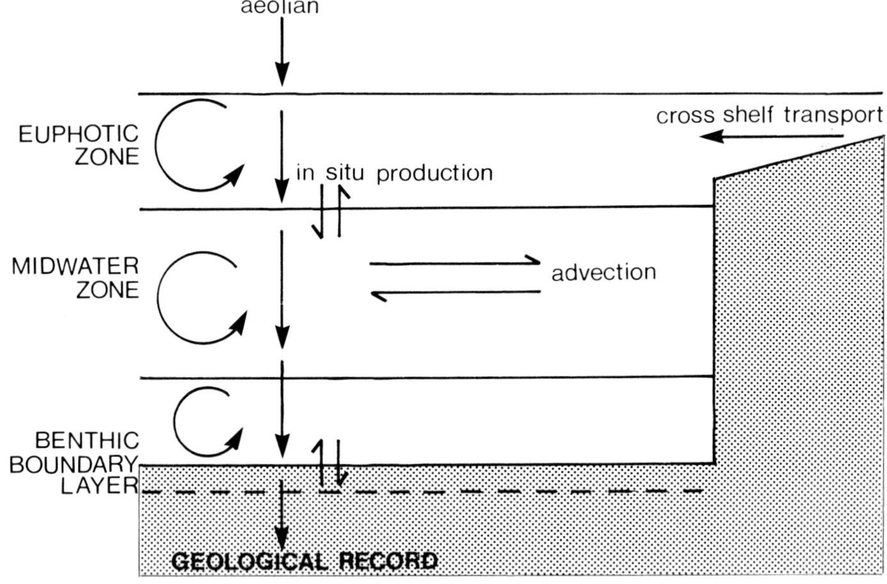

Fig. 1—Schematic representation of the main sources and flux pathways in the deep oceanic water column. The midwater zone is considered to be comprised of two subzones: the mesopelagic, which extends from the base of the euphotic zone to the lower limit to which daylight penetrates in ecologically significant intensity (usually about 800–900 m), and the lower subzone, which is bathypelagic and is further subdivided by some authors at about 2500 m, with the abyssopelagic occupying the deeper part.

degradation, which will reduce the flux; (*b*) *in situ* production based on dissolved organic carbon (DOC), which will increase it; (*c*) modification of the particle-size spectrum by either active or passive processes; and (*d*) the active vertical transport of organic or nutrient material. The dominant influence that any particular assemblage of organisms has on the fluxes will vary with the behavior, abundances, and size spectra of the component species and to some extent on a number of *in situ* environmental factors. For example, microorganisms which inhabit large aggregates (snow) may either be degrading the matrix of the aggregates, leading to their disintegration and the slowing of their descent, or may be increasing the carbon content of the aggregate by production utilizing DOC. In contrast, macroplanktonic organisms may be grazing the particles and/or breaking them up. Large nektonic organisms while performing extensive vertical migrations may actively transport significant quantities of material. The extent of biological activity will be approximately correlated with the distribution of biomass and the community's size spectrum. There are clear zoogeographical divisions described for pelagic communities which correlate closely to the main circulation features of the ocean, and it seems likely that variations in the sedimentary regime will also occur across the boundaries of the main provinces of these distributional patterns (Backus 1985). Before discussing the processes in greater detail, global patterns of biomass are considered.

THE DISTRIBUTION OF BIOMASS

Although secondary production (the production of animal biomass) is correlated with primary production (e.g., FAO 1972), the relationship is far from simple. It is influenced by environmental and physiological factors such as the degree of seasonality in the primary production cycle, the *in situ* temperature, the food web structure, and the assimilation and growth efficiencies of the component species. These latter physiologically related factors have rates which show allometric relationships (i.e., they are size-related). There is a trend for the mean body size of planktonic and nektonic organisms inhabiting the surface layers to increase with increasing latitude, paralleling a trend for standing crops to be higher. The increase is most marked across latitudes of about 40°, where there is substantial change in the effects of winter cooling, and the way in which this boundary might be expected to influence the pattern of sedimentary flux will be discussed below. At higher latitudes the flux will be more variable seasonally and *a priori* it seems likely that the efficiency with which the pelagic communities can exploit this flux will be lower than at lower latitudes. Seasonal migrations are common at high latitudes and in many cases are related to the species entering a state of diapause when they are physiologically inactive (e.g., Miller et al. 1984).

The standing crops of both zooplankton and micronekton decline exponentially with depth (Vinogradov 1972; Angel and Baker 1982). In the North Atlantic the slope of the log-transformed data seems to be fairly constant between different localities (Fig. 2) irrespective of the surface input. A similar relationship has been observed in biomasses of benthopelagic

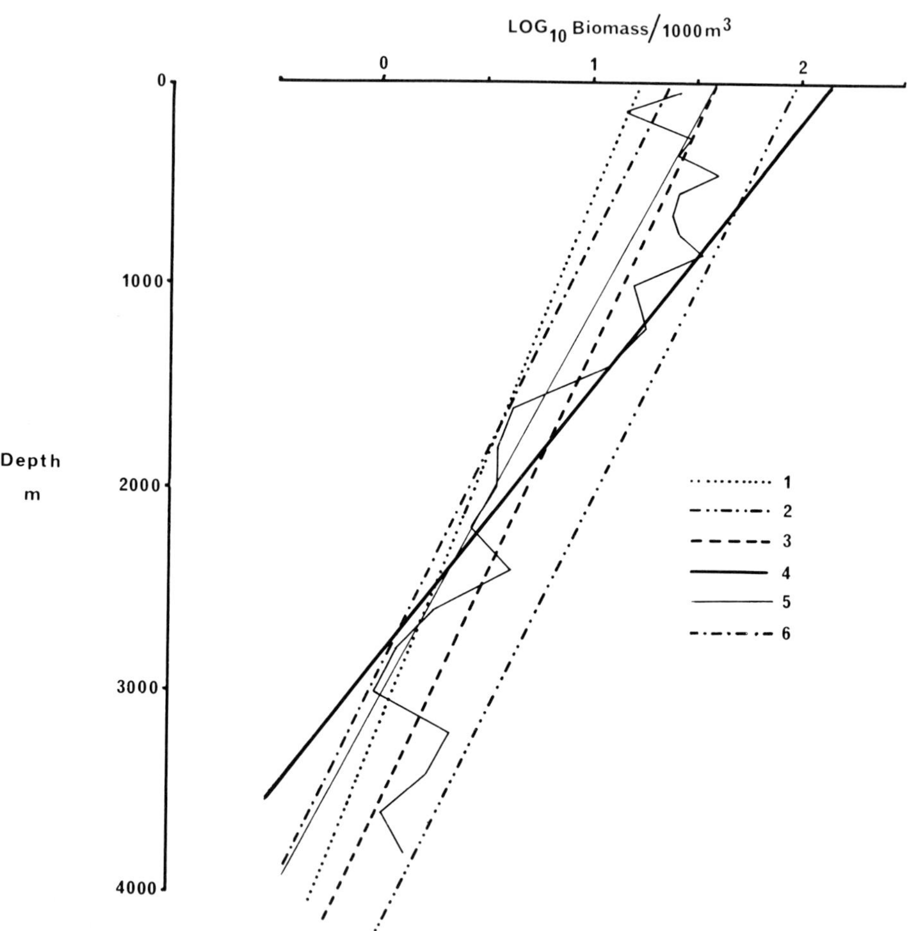

Fig. 2—Regression lines for the relationship between biomass of plankton and micronekton with depth at the stations in the N.E. Atlantic: (1) micronekton at 20°N, 21°W; (2) plankton at 42°N, 17°W; (3) micronekton at 42°N, 17°W; (4) plankton at 49°40′N, 17°W; (5) micronekton at 49°40′N, 17°W; and (6) plankton in the N.E. Atlantic after Wishner (1980). The profile of the raw data for micronekton biomass at 42°N, 17°W is also shown. From Angel and Baker (1982).

communities, which inhabit the benthic boundary layer (Wishner 1980). The populations inhabiting depths below the lower limit of vertical migration must be virtually totally dependent on the organic flux reaching their depth from the surface waters. At such depths (i.e. > 1500 m) mesoscale eddies propagate rather than advect which reduces the lateral transport, so any spatial variability will be generated at shallower depths. The deep-living organisms are long-lived and so their population response time to environmental change is probably longer than the time scale of mesoscale variability. Hence, one possible use of the biomass/depth relationship may be that samples of pelagic communities from depths below 1000–1500 m, at which intersample variability is less than a factor of two (Angel et al. 1982), may provide the simplest way of obtaining comparable semiquantitative data on the long-termed, time-averaged productivity of the overlying water column.

Another characteristic of pelagic communities is that their mean body size increases with depth. This can be seen both in the ratios between the standing crop of macrozooplankton and micronekton sampled simultaneously (Fig. 3) and in the ratio between different size components of the macroplanktonic populations (Fig. 4). The data of Rodriguez and Mullin (1986), on variations in the size spectra of pelagic organisms with both latitude and depth, show similar results. Body size not only affects the size of the particles that an organism can handle and feed on, it also influences the size of the particle it generates (i.e., the fecal pellet), and the larger a pellet is, the faster it sinks.

Another important factor is the proportion of the population which is undertaking diel vertical migrations (DVM) and the vertical extent of these migrations. DVMs tend to be more extensive in planktonic ostracods at middle latitudes than at either lower or higher latitudes (Angel and Fasham 1975), and this is likely to be true for other pelagic organisms. There are also seasonal variations in the expression of migratory behavior. For example, Roe et al. (1984) described clear and extensive diel migrations being undertaken by the deep mesopelagic mysid *Eucopia unguiculata* when the species was breeding in the spring at 44°N, 13°W, whereas at other times of year there was no evidence of such migratory behavior (Discovery Collections, unpublished data). Migration can also develop ontogenetically in some species such that their young stages may be nonmigrant, and as they mature so they begin to migrate to deeper and deeper daytime depths (Angel 1986). DVMs not only cause substantial changes in the vertical distribution of pelagic biomass, they also influence the size structure of the populations (Figs. 4 and 5). DVM links the mesopelagic communities with the euphotic zone. In macroplankton the lower limit of the migrations is around 900 m (Angel and Baker 1982) whereas in micronekton the lower limit can be as deep as 1600–1700 m (Angel 1988). An obvious connotation

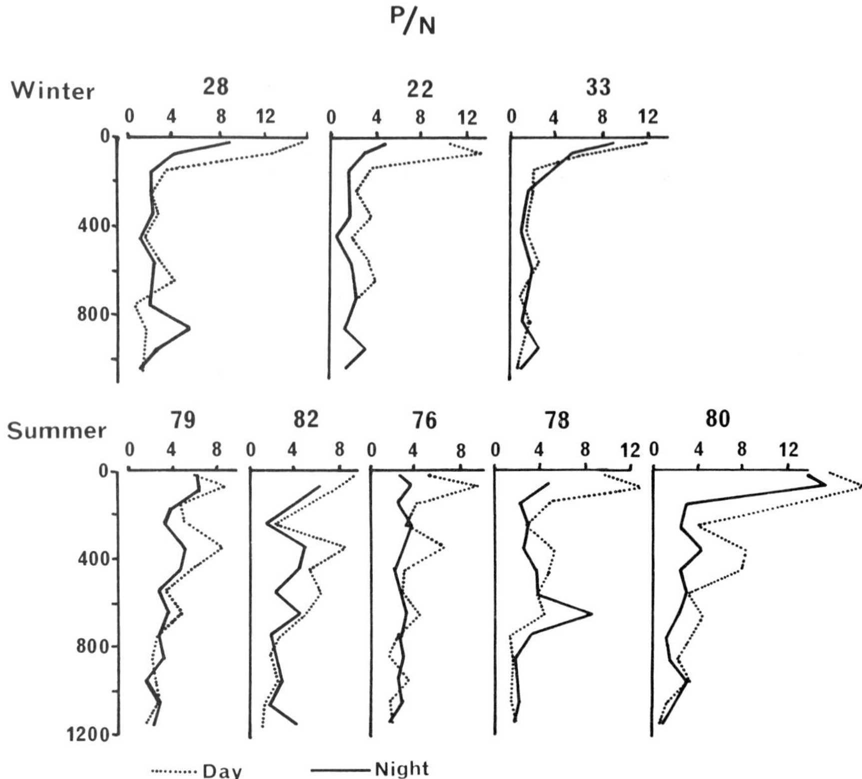

Fig. 3—Depth profiles of the ratios between standing crops of macroplankton/micronekton sampled simultaneously at three wintertime stations (28, 22, and 33) and five summertime stations (79, 82, 76, 78, and 80) both by day and by night, in the vicinity of the Azores Front, all close to 30°N, 30°W (Angel 1988).

of this is that organisms inhabiting depths deeper than the lower limit of diel migration cannot be herbivores and must be either detritivorous or carnivorous. Any model describing the fluxes of material out of the euphotic zone must take into account the coupling between the processes operating in the euphotic and the mesopelagic zones.

FACTORS INFLUENCING THE SIZE SPECTRUM OF PARTICLES

Away from the continental margins, sources of particles in the water column are concentrated in the euphotic zone. The main mass of the particles is generated by phytoplankton and 30–60% of this production is by picoplankton

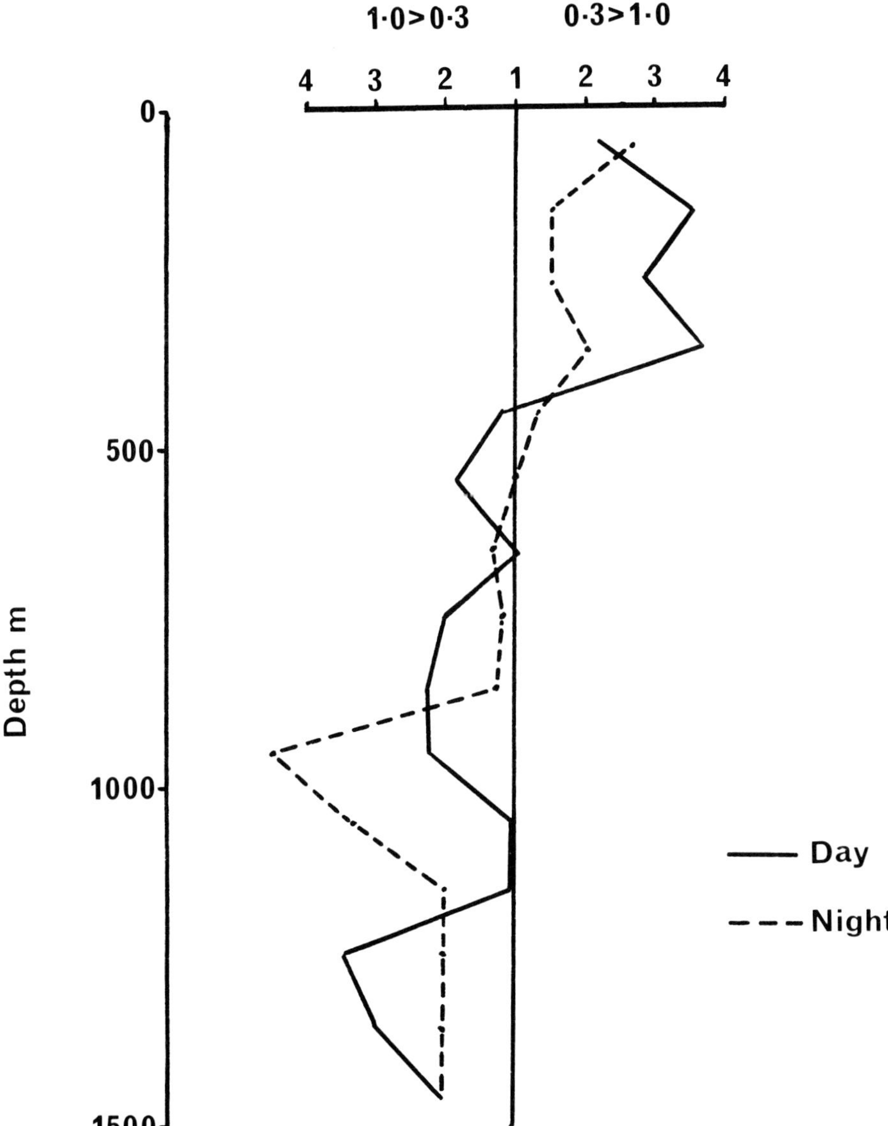

Fig. 4—Day and night depth-related changes in the ratio between the biomass of macroplankton retained by a 1 mm mesh and that passing through the 1 mm mesh and retained by a 0.32 mm mesh, at a station in the N. Atlantic at 31°30′N, 25°30′W.

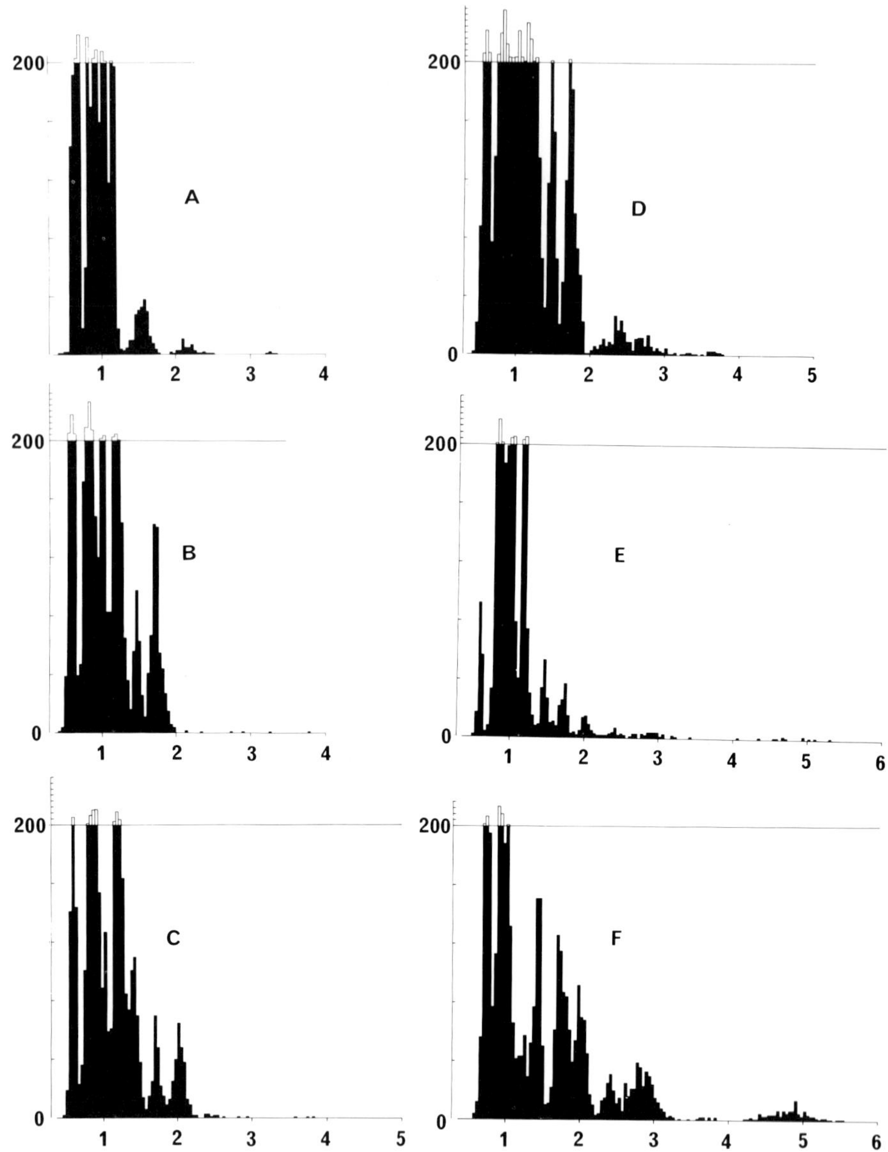

Fig. 5—Depth-related changes in the size spectra of planktonic ostracod populations. Carapace lengths are in millimeters. The numerical scale for abundance > 200 changes to 100 intervals. A = daytime, 100–50 m; B = night, 100–50 m; C = daytime, 300–200 m; D = night, 300–200 m; E = daytime, 400–500 m; F = daytime, 600–700 m. Note that the average size increases with depth (compare A with E and F) and that it is the large specimens that are migrating up at night (compare C with D, and A with B). Data from Ocean Acre near Bermuda.

(i.e., phytoplankton with cell diameters of < 2 μm). The primary production either becomes incorporated into larger particles by biological processes, through scavenging by other particles, or else sinks out directly. Direct sedimentation by phytoplankton cells probably only occurs in those cells which are large enough to have sinking rates fast enough to overcome the effects of turbulent mixing which will otherwise keep them in suspension. There is extensive recycling of material within the euphotic zone. One reason may be the recently observed habit that some planktonic grazers have of breaking up fecal material without apparently ingesting it (Lampitt, personal communication). Another factor is that the empirical relationship between sinking rates and particle size (Fowler and Knauer 1986 and Fig. 6) may overestimate *in situ* sedimentation rates. Recent data on the residence times of fecal pellets in the euphotic zone imply that they persist there longer than expected (Alldredge et al. 1987). The application of Lagrangian modeling techniques to the study of the fate of particles within the euphotic

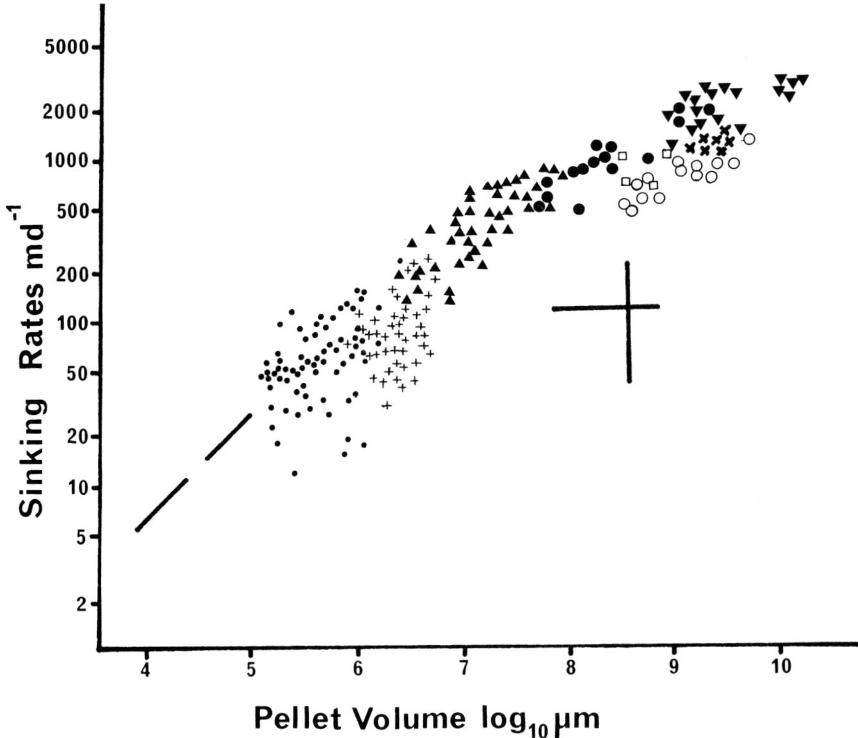

Fig. 6—Relationship between fecal pellet size and sinking rates (further modified by Fowler and Knauer 1986).

zone has highlighted the importance of turbulent mixing (Woods and Onken 1982).

Since large percentages of the primary production are generated by picoplankton, which have almost negligible sinking rates, aggregation mechanisms are clearly important. Aggregation may result from organisms grazing the particles and generating larger particles by releasing fecal pellets. In addition, several microphagous feeders secrete mucus structures on which they trap their food (e.g., thecosomatous pteropods and larvaceans); these are frequently discarded and then act as foci for the passive generation of aggregates. Aggregation may also occur through the scavenging of small particles by larger ones, and here too, biological processes play a role because microorganisms may generate the "glues" which help to stick the aggregates together. Sediment trap data imply that 10–20% of the primary productivity sinks out of the euphotic zone. Size spectra of the suspended particles, as well as the contents of the sediment traps, suggest that the main sink is via larger particles and aggregates. However, as will be discussed below, there is circumstantial evidence that a flux of around 1–3% of primary production may occur within the gut contents of migrating micronekton (Angel 1986), and hence give rise to the maximum in flux rates often observed in midwater.

Fluxes measured by time series of sediment traps show a clear seasonal signal. Even in subtropical environments such as the Sargasso Sea, sediment trap records from depths of 3800 m show the flux varies threefold (Deuser 1986). At higher latitudes the fluxes probably vary by orders of magnitude as shown by Lampitt (1985) in the North Atlantic using time-lapse photography of the seabed. The detrital material photographed by Lampitt was sampled and was found to contain quite substantial quantities of undegraded chlorophyll, moreover the floristic composition of the phytoplankton remains was identical to the community occurring at the same time in the euphotic zone (Billett et al. 1983). Both pieces of evidence indicate that there had been little if any off-shelf transport contributing to the flocs. The degradation products of the chlorophylls were also of interest because they implied that equal proportions of the chlorophyll had been broken down during passage through the guts of herbivorous plankton, and by microorganisms.

In few cases has any attempt been made to conduct biological sampling in conjunction with trap studies which would enable the sources of the fecal material in the traps to be identified unequivocally. A notable exception is the study by Bishop et al. (1987) in the Panama Basin. However, this is a region where there is a strong oxygen minimum, which will have curtailed some of the mesopelagic biological activity and limited vertical migration. Also, there was considerable lateral advection at a depth of around 100 m so the conditions observed in the Panama Basin are unlikely to be typical of most oceanic areas. Intuitively the groups of animals that might be expected to make significant contributions to the fecal flux will be those which are most important in terms of biomass. Table 1 gives examples of the community structure of the micronekton sampled at three stations in the vicinity of the Azores Front in the central region of the North Atlantic. These data show that there are several taxonomic groups which were important in terms of biomass and might be expected to produce significant quantities of faster-sinking feces. Their fecal material, however, is still to be identified in sediment trap samples.

In the VERTEX experiment, particle traps set at 30 m yielded no particulate carbon flux, but were full of zooplankton. The traps set at 120 m registered a particle flux of 38 mgC m^{-2} day^{-1} of which 30% was in the form of fecal pellets produced by 200–2000 μm zooplankton (Small et al. 1987). In this case it seemed that zooplankton can be more significantly coupled to carbon flux out of the bottom layers of oligotrophic euphotic

TABLE 1 Integrated standing crops for the top 1100 m of the water columns of the main groups of micronekton in order of abundance (mls displacement volume.m^{-2}) in the N. Atlantic at four stations across the Azores Front in the vicinity of 30°N, 30°W, showing both summer and winter conditions. Displacement volumes can be converted into carbon using the data of Wiebe et al. 1975 in which 1 ml of displacement volume is equivalent to 50 mgC.

Station number[a]	10380[b]		10376		10379		10228	
Position	30°N 34°W		33°N 33½°W		35°N 33°W		33°N 31½°W	
Time	16–19 June '81		26–29 May '81		11–14 June '81		1–3 Nov. '80	
Season	Summer		Summer		Summer		Winter	
Water type	W. Atlantic Water		Front		E. Atlantic Water		W. Atlantic Water	
	Day	Night	Day	Night	Day	Night	Day	Night
Fish	1.10	1.47	1.77	3.72	2.01	2.10	1.79	1.57
Siphonophora	0.45	0.41	1.84	2.38	1.32	1.87	0.88	0.91
Chaetognatha	0.32	0.32	1.33	1.36	1.24	1.26	1.89	1.42
Decapoda	0.42	0.42	0.76	0.97	0.47	0.73	0.86	0.65
Euphausiacea	0.16	0.28	0.29	0.49	0.32	0.55	0.23	0.28
Mysidacea	0.05	0.05	0.18	0.23	0.22	0.23	0.21	0.33
Cephalopoda	0.12	0.08	+	+	0.11	0.08	0.11	0.12
Amphipoda	0.07	0.03	+	+	0.07	0.05	0.06	0.02

[a] At all the stations there were occasional aggregations of the salp *Pyrosoma atlantica*, which were migrating between daytime depths of around 600–700 m and nighttime depths in the euphotic zone where they often congregated within the thermocline. When large numbers were caught, *Pyrosoma* dominated the communities in terms of their displacement volumes.

[b] At station 10380, in Western Atlantic Water, there was a thermostad of 18°C water and the community had a close affinity with the Sargasso Sea region where Deuser (1986) conducted his long-term sediment trap observations.

zones which are stratified, than to the flux out of unlayered eutrophic euphotic zones; there was no examination of micronektonic species.

THE INFLUENCE OF MICROORGANISMS ON FLUXES AT MESOPELAGIC DEPTHS

Bacteria are present in the water column either freely suspended or attached to particles. Alldredge and Silver (1988) review the status of knowledge of marine snows, which they define as large aggregates with longest dimensions > 500 μm. In deep water, abundances of the large aggregates range from 0.5–7500 m^{-3}. Such aggregates are often seeded with bacteria in the euphotic zone, for example, with the gut flora of the organisms producing fecal material which gets incorporated into the snow. As the snow aggregates sink, the numbers and types of microorganisms they contain change, both through the aggregates scavenging further particles as they sink and through *in situ* growth within them. The bacteria within the aggregates tend to be considerably larger than those occurring freely suspended in the surrounding water. The presence of the bacteria results in the development of a microfauna which grazes them, and the utilization of the aggregates by detritivorous macroplankton may be more to exploit their microflora and fauna rather than for the relatively refractory organic material the aggregates contain. The carbon source for the attached bacteria may either be the organic matrix of the aggregate or DOC. Logan and Hunt (1987) examined the sinking rates and porosities of aggregates and concluded that because the aggregates are relatively porous, flow rates through an aggregate sinking at a rate of 60 md^{-1} (or 0.07 cm s^{-1}) will be around 10^{-2} to 10^{-4} cm s^{-1}. This through flow has the potential to enhance bacterial uptake of DOC by as much as 60% as compared with freely suspended bacteria. This may account for attached microorganisms being up to three orders of magnitude more abundant within snow aggregates than freely suspended in the surrounding water. Alldredge and Silver (1988) conclude from these observations that snow aggregates form local "hotspots" of metabolic activity which are islands of chemosynthetic productivity at certain depths within the aphotic zone. However, Cho and Azam (1988) have published some preliminary results which suggest that the freely suspended bacteria are the more important.

Cho and Azam (1988) examined the abundances of freely suspended bacteria at mesopelagic depths and studied their production by thymidine uptake. They used two models to estimate what proportion of the new production enters the microbial loop in the deeper water. They conclude that 60–98% of the carbon flux passes through this loop, which implies that most of the carbon flux enters the DOC pool within the water column and does not sediment out directly in particulate form. This has startling implications for many of the assumptions that are made in studies of particle flux based on sediment trap data. For example, because the flux appears to occur mainly in the form of large particles, an implicit assumption is made that once material enters the downward escalator provided by sinking particles very little gets off. These data, should they be supported by further observations, show that there is likely to be continual interchange between the dissolved and the particulate material. While it is relatively easy to conceive mechanisms for the fine particles, it is less easy to postulate how this interchange occurs between the larger particles unless there is extensive intervention by deep-living plankton and nekton. Bishop (personal communication) has data on certain types of highly labile polyunsaturated acids for which 90% of the flux is by active transport.

At present, most geochemical models of trace metal distributions within oceanic water columns relate these distributions to suspended clay fractions. Data from the vicinity of 31°30′N 25°W collected by an *in situ* pump system (Simpson, personal communication) have shown that the surface area of the freely suspended bacteria exceeds that of the clay particles by around two orders of magnitude. Given the reactiveness of their surfaces, it would seem likely that these bacteria will provide the dominant influence on these trace metal profiles.

Microbial activity is strongly influenced by temperature so the hydrographic regime may induce regional variations in the impact of the microorganisms within the water column. For example, in the North Atlantic the Mediterranean outflow maintains much higher water temperatures within the water column at depths of 500–1500 m than occur in other oceans. Hence, the relative influence of the microorganisms may be relatively greater in the North Atlantic than in other oceanic areas. Another example of the possible influence of the hydrographic regime is beneath the equatorial upwelling region. There the pelagic and benthic biomasses in the deep underlying waters are much lower than beneath higher latitude areas where the annual primary productivity is comparable. A number of factors may contribute to this, but the combination of the smaller size of the herbivorous plankton, the reduction in the degree of vertical migration, and the greater metabolic activity of the microorganisms in the near-surface layers probably result in much greater recycling occurring in the equatorial regions.

THE SIGNIFICANCE OF DIEL VERTICAL MIGRATION

There have been surprisingly few attempts to quantify the movements involved in vertical migration. Estimates based on net samples will underestimate the vertical exchanges because (*a*) there can be considerable avoidance of nets by day and (*b*) nets fail to discriminate asynchronous movements. Longhurst (1976) suggested that the exchange amounted to around 10% of the pelagic standing stock. Data from the Azores Front region in the North Atlantic (Angel 1988) suggest that as much as 25% of the standing stock of plankton and micronekton in the top 1100 m of the water column moves in and out of the surface 200 m of the water column between day and night. These data also show that there were important and significant differences in the proportions of the individuals migrating in any given taxonomic group as well as differences in the ranges of the migrations.

The hypothesis most favored at present as providing the most likely explanation for the function of DVM is that by day migrants seek depths at which visually hunting predators can no longer detect them, and at night they return to the shallow, more productive depths to feed. Implicit within this hypothesis is that feeding by the migrants will tend to be more intensive by night than by day. Examination of the stomach contents of migrators usually reveals that the animals feed constantly but at different rates and on different items as the availability of food changes. There are records of migrators caught at depth which have only shallow-living species within their guts, and these are thought only to feed at shallow depths. Thus, if gut residence times are long relative to the time taken to complete the downward migration, there is likely to be a net export of organic material from the euphotic zone down into deeper water. There are relatively few data available either for the rates of the migrations (see Angel 1986) or for gut residence times (see Fowler and Knauer 1986). In euphausiids, gut residence times increase with increased body size, but in one of the larger species, *Meganyctiphanes norvegica*, the gut residence time for fully grown adults is only 30 minutes, which is much shorter than the 1–2 hours the downward migration takes. Thus the organisms which will be contributing most to any active export of organic material will be the larger nektonic fish and decapod crustacean species. In the Azores Front the daily mass flux of these two groups alone in terms of carbon is around 70 mgC m^{-2} d^{-1}, whereas the total mass flux of the migrating plankton and micronekton was 200–700 mg C $m^{-2}d^{-1}$. Furthermore, these estimates of active transport ignore the possibility of predation occurring at depth.

As well as any possible influence of the migrators on carbon fluxes, another interaction that needs to be considered is the impact of the migrations on the distribution of nutrients about the nutricline. Nitrogen which has been assimilated is excreted as dissolved nitrogen (DN), mostly in the form of ammonia. The excretion is continuous although the rate is related to the *in situ* temperature. Migrants which feed only at night at shallow depths will only take up nitrogen at shallow depths, but they will continue to excrete nitrogen while they are at greater depths during the day, and so they will be net exporters. However, for those species which feed continuously no matter where they are within the water column—which seems to be the rule rather than the exception from the data of Roe et al. (1984)—it is not intuitively obvious as to which direction the net export will occur. Longhurst and Harrison (1988) have published a paper in which they suggest that there is a net export of nitrogen as a result of vertical migration, although the cooler ambient temperatures deeper down will slow the excretion rates. Detritivores feeding on snow particles may be exploiting a food resource which is enriched by microbial activity at shallow depths, although C/N ratios of particles sampled using sediment traps tend to decrease with depth (see Fowler and Knauer 1986). There is a strong likelihood that where there is extensive DVM undertaken by the pelagic communities the nutrient supply to the euphotic zone will be influenced.

In the Azores Front region where there is extensive vertical migration, Fasham et al. (1985) concluded that vertical diffusion would support 50% of the primary productivity within the Deep Chlorophyll Maximum (DCM). However, in the Eastern Tropical Pacific where vertical migration is reduced because of the oxygen minimum, King and Devol (1979) estimated that only 30% of the production was new. It is tempting to postulate that the relatively low proportion of "old" production (i.e., that portion which is based on recycled nutrient) in the Azores Front region is the result of the export of nitrogen by the vertical migrants. Hayward (1987) concluded from his studies of the distribution of nitrogen in the oligotrophic region of the central Northern Pacific that there is no large-scale correlation between the shape of the nitricline and the primary productivity. He argues that nondiffusive nutrient input must be responsible for increasing the nutrient supply in the euphotic zone. It is postulated here that variations in the extent to which vertical migration occurs may be the factor which perturbs the nutrient supply away from the pattern expected from a simple diffusion model. If so, the suggestion of Longhurst and Harrison implies that it may be variations in the export of nitrogen rather than in the input which is the factor causing the unexpected variability in the productivity from area to area.

Biomass profile data from the Azores Front are now used to estimate how much nitrogen may be released by the excretion of the pelagic community. Table 2 summarizes the information from five stations taken across the front and from an eddy which became detached from the front during the course of the sampling. It has been assumed that migrants (*a*) spent eight hours above 100 m in the euphotic zone; (*b*) that the dry weight of both plankton and nekton is 5% of their displacement volumes, and (*c*) that the mean rate of nitrogen excretion (mostly in the form of ammonia) is 5 μg.mg dry weight^{-1} d^{-1}. Table 2 shows that the macroplankton is likely to be responsible for the majority of the recycling by the size ranges sampled; no assessments were made of the microzooplankton standing crops. In the front the migrants were probably releasing more DN than the nonmigrants, but only about 30% of the total excreted DN in the Eastern Atlantic Water,

TABLE 2 An analysis of the distribution of macroplanktonic and micronektonic biomass in the top 100 m of the water column at five stations in the vicinity of the Azores Front, all close to 30°N 30°W (for full details see Angel 1988), and its possible influence on the availability of nitrogen within the euphotic zone. The migrants which feed in deeper water during their daytime sojourn will be transferring some of the nitrogen they release from below the nutricline, although the exact proportion cannot be estimated at present. It is assumed that migrants stay within the surface layers for 8 hours, hence their biomass estimate has been reduced by a third in subsequent calculations.

Station Number Water type	10376 Front	10380 W. Atlantic Water	10378 W. Atlantic Water (M)	10379 E. Atlantic Water	10382 Eddy
Biomass of nonmigrants					
(g dry wt m^{-2})	0.13	0.19	0.15	0.22	0.23
% of biomass plankton	88	95	92	88	90
Biomass of migrants					
(g dry wt m^{-2})	0.60	0.054	0.15	0.45	0.28
% of biomass plankton	73	92	75	86	84
% of biomass in top 50m of water column					
during day	33	28	23	33	43
during night	28	47	35	56	38
Estimated nitrogen excretion (mg m^{-2} day^{-1})					
by nonmigrants	0.65	0.95	0.75	1.10	1.15
by migrants	1.0	0.09	0.26	0.75	0.47
total	1.65	1.04	1.01	1.85	1.62

and only about 10% in the Western Atlantic Water. In the front the total estimated nitrogen source via excreted DN amounted to about 1.65 mgN m^{-2} d^{-1}. This was about one-third of the nitrogen requirement of the production in the DCM. Fasham et al. (1985) estimated that half the required nitrogen was being supplied by turbulent mixing, which leaves a balance of 15–20% to be provided by microzooplankton excretion.

The important point to make is that a fraction of the nitrogen excreted by the vertical migrants will probably have originated from their feeding below the nutricline. Apparently even nonmigratory species exhibit periods of active upward swimming which are interspersed with passive sinking. Thus they may oscillate vertically over distances of several meters or even tens of meters. These movements are unsynchronized and so will be undetected by the sampling regime. These movements in at least some of the nonmigrants will probably transverse the nutricline and so provide a transport mechanism. The size of the fraction transported by vertical movements cannot be estimated at present but it may be sufficiently large to decouple the productivity from the character of the nutricline in water columns particularly where the nutricline lies below the 1% light level, as observed by Hayward (1987) in the North Pacific. Thus it can be argued that the vertical movements of migrating plankton, and to a lesser extent micronekton, may result in the redistribution of nitrogen about the nutricline.

THE INFLUENCE OF THE SUBTROPICAL CONVERGENCE

Changes which occur across the boundary between the regions where stratification is permanent in the near-surface waters and where it occurs only seasonally result in major changes in the pelagic communities (McGowan and Walker 1985). The substantial changes in the pattern of the production cycle which are observed there are accompanied by equally large differences in the degree to which DVM is expressed. On the low latitude side of the boundary, migration can always be seen regardless of the season. On the high latitude side, migration occurs seasonally with its maximum expression occurring in the spring and late summer, usually when there are peaks in the production cycle. In winter there tends to be a seasonal migration down into the deeper water (Angel 1986). The result of all these changes in the North Atlantic is that there is a clear change in the structure and composition of the benthic communities across 42°N (Merrett 1987), which indicates that the whole sedimentary regime is substantially modified. In the midwater communities the major changes seem to be in the high abundances of gelatinous plankton at high latitudes, particularly at depths around 1000 m (Angel and Baker 1982). The results of the CLIMAP experiment suggested that this boundary has been relatively stable during the recent change between glacial and interglacial period (Cline and Hays 1976). This boundary should be persistent enough geologically to generate a clear signal in the sediment record as a discontinuity in the sequence.

REFERENCES

Alldredge, A.L.; Gotschalk, C.C.; and MacIntyre, S. 1987. Evidence for sustained residence of macrocrustacean fecal pellets in surface waters off Southern California. *Deep-Sea Res.* **34**: 1641–1652.

Alldredge, A.L., and Silver, M.W. 1988. Characteristics, dynamics and significance of marine snow. *Prog. Ocean.* **20**: 41–82.

Angel, M.V. 1986. Vertical migrations in the oceanic realm: possible causes and probable effects. In: Migration: Mechanisms and Adaptive Significance, ed. M.A. Rankin. *Contr. Mar. Sci. (Suppl.)* **24**: 45–70.

Angel, M.V. 1988. Vertical profiles of pelagic standing crop in the vicinity of the Azores Front and their implications to deep ocean ecology. *Prog. Ocean.* **20**, in press.

Angel, M.V., and Baker, A. de C. 1982. Vertical distribution of the standing crop of plankton and micronekton at three stations in the northeast Atlantic. *Biol. Ocean.* **2**: 1–29.

Angel, M.V., and Fasham, M.J.R. 1975. Analysis of the vertical and geographical distributions of the abundant species of planktonic ostracods in the northeast Atlantic. *J. Mar. Biol. Assn. U.K.* **55**: 709–737.

Angel, M.V.; Hargreaves, P.; Kirkpatrick, P.; and Domanski, P. 1982. Low variability in planktonic and micronektonic populations at 1000 m depth in the vicinity of 42°N, 17°W; evidence against diel migratory behaviour in the majority of species. *Biol. Ocean.* **1**: 287–319.

Backus, R.H. 1985. Biogeographic boundaries in the open ocean. In: Pelagic Biogeography, eds. A.C. Pierrot-Bults, S. van der Spoel, B.J. Zahuranec, and R.K. Johnson. *UNESCO Tech. Papers Mar. Sci.* **49**: 9–13.

Billett, D.S.M.; Lampitt, R.S.; Rice, A.L.; and Mantoura, R.F.C. 1983. Seasonal sedimentation of phytoplankton to the deep-sea benthos. *Nature (Lon.)* **302**: 520–522.

Bishop, J.K.B.; Stepien, J.C.; and Wiebe, P.H. 1987. Particulate matter distributions, chemistry and flux in the Panama Basin: response to environmental forcing. *Prog. Ocean.* **17**: 1–59.

Cho, B.C., and Azam, F. 1988. Major role of bacteria in biogeochemical fluxes in the ocean's interior. *Nature (Lon.)* **332**: 441–443.

Cline, R.M., and Hays, J.D., eds. 1976. Investigation of late Quaternary paleoceanography and paleoclimatology. *Geol. Soc. Amer. Mem.* **145**: 1–464.

Deuser, W.G. 1986. Seasonal and interannual variations in deep-water particle fluxes in the Sargasso Sea and their relation to surface hydrography. *Deep-Sea Res.* **33**: 225–246.

FAO. 1972. Atlas of the Living Resources of the Seas. Rome: Food and Agricultural Organization of the United Nations.

Fasham, M.J.R.; Platt, T.; Irwin, B.; and Jones, K. 1985. Factors affecting the spatial pattern of the deep chlorophyll maximum in the region of the Azores Front. *Prog. Ocean.* **14**: 129–165.

Fowler, S.W., and Knauer, G.A. 1986. Role of large particles in the transport of elements and organic compounds through the oceanic water column. *Prog. Ocean.* **16**: 147–194.

Hayward, T. 1987. The nutrient distribution and primary production in the central North Pacific. *Deep-Sea Res.* **34**: 1593–1628.

King, F.D., and Devol, A.H. 1979. Estimates of vertical eddy diffusion through the thermocline from phytoplankton nitrate uptake rates in the mixed layer of the eastern tropical Pacific. *Limnol. Ocean.* **24**: 645–651.

Lampitt, R.S. 1985. Evidence for the seasonal deposition of detritus to the deep-sea floor. (Porcupine Bight, NE Atlantic) and its subsequent resuspension. *Deep-Sea Res.* **32**: 885–897.

Logan, B.E., and Hunt, J.R. 1987. Advantages to microbes of growth in permeable aggregates in marine systems. *Limnol. Ocean.* **32**: 1034–1048.

Longhurst, A.R. 1976. Vertical migration. In: The Ecology of the Seas, eds. D.H. Cushing and J.J. Walsh, pp. 116–137. Oxford: Blackwell.

Longhurst, A.R., and Harrison, W.G. 1988. Vertical nitrogen flux from the oceanic photic zone by diel migrant zooplankton and nekton. *Deep-Sea Res.* **35**: 881–889.

McGowan, J.A., and Walker, P.W. 1985. Dominance and diversity maintenance in an oceanic ecosystem. *Ecol. Monogr.* **55**: 103–118.

Merrett, N.R. 1987. A zone of faunal change in assemblages of abyssal demersal fish in the eastern North Atlantic: a response to seasonality in production? *Biol. Ocean.* **5**: 137–151.

Miller, C.B.; Frost, B.W.; Batchelder, H.P.; Clemons, M.J.; and Conway, R.E. 1984. Life histories of large, grazing copepods in subarctic ocean gyre: *Neocalanus plumchrus, Neocalanus cristatus* and *Eucalanus bungii* in the Northeast Pacific. *Prog. Ocean.* **13**: 201–243.

Rodriguez, J., and Mullin, M.M. 1986. Relation between biomass and body weight of plankton in a steady state oceanic system. *Limnol. Ocean.* **31**: 361–370.

Roe, H.S.J.; Angel, M.V.; Badcock, J.; Domanski, P.; James, P.T.; Pugh, P.R.; and Thurston, M.H. 1984. The diel migrations and distributions within a mesopelagic community in the Northeast Atlantic. *Prog. Ocean.* **13**: 245–511.

Small, L.F.; Knauer, G.A.; and Tuel, M.D. 1987. The role of sinking fecal pellets in stratified euphotic zones. *Deep-Sea Res.* **34**: 1705–1712.

Vinogradov, M.E. 1972. Vertical Migration of the Oceanic Plankton. Israel Program for Scientific Translation.

Wiebe, P.H.; Boyd, S.; and Cox, J.L. 1975. Relationships between zooplankton displacement volume, wet weight, dry weight and carbon. *Fish. Bull.* **73**: 777–786.

Wishner, K.F. 1980. The biomass of the deep-sea benthopelagic plankton. *Deep-Sea Res.* **27**: 203–216.

Woods, J.D., and Onken, R. 1982. Diurnal variation and primary production in the ocean – preliminary results of a Lagrangian ensemble model. *J. Plankton Res.* **4**: 735–756.

Productivity of the Ocean: Present and Past
eds. W.H. Berger, V.S. Smetacek and G. Wefer, pp. 175–191
John Wiley & Sons Limited

How Much Shelf Production Reaches the Deep Sea?

J.J. Walsh

Department of Marine Science
University of South Florida
St. Petersburg, FL 33701, U.S.A.

Abstract. Moored arrays of fluorometers, sediment traps, and current meters on the continental margins of the Atlantic and Pacific Oceans suggest that 30–47 g C m^{-2} yr^{-1} might exit the shelf-break as particulate organic carbon (POC), falling towards slope depocenters (32.5×10^6 km^2). Such an annual carbon loading of $1.0–1.5 \times 10^9$ tons C yr^{-1} is consistent with both neglected pCO_2 fluxes across the air-sea interface of continental shelves and a terrestrial source of CO_2 from deforestation. Apparently 5–16 g C m^{-2} yr^{-1} of the phytodetritus survive oxidation during their descent to the top of the 25–50 m bottom boundary layer on the mid-slope, where this near-bottom flux is equivalent to the total POC rain at a depth of 2000 m in the deep sea. Estimates of bottom metabolism, carbonate dissolution, and sediment accumulation suggest that ~50–100 percent of the near-bottom carbon flux may be consumed during early diagenesis, with evolution of additional biogenic CO_2 as a coastal sink within the bicarbonate cycle. The physical flow regimes in eastern and western boundary currents are consistent with both the supply of nutrients for "new" production and the lateral export of organic matter to the deep sea. Anthropogenic nutrient sources could presumably account for at most $0.4–0.5 \times 10^9$ tons C yr^{-1} of the estimated shelf export. Spatial inhomogeneity and seasonality of biological processes at the margins of the ocean preclude further refinement of these estimates; however, until additional field and theoretical studies are conducted on element fluxes between these regions and the deep sea.

INTRODUCTION

Compilation of numerous measurements of the gradient of the partial pressure of CO_2 across the air-sea interface of the global ocean over the last 15 years (1972–1986) provides a basis for estimating the net annual exchange of carbon between its atmospheric and oceanic reservoirs. Using a spatially varying gas exchange coefficient of 0.037–0.111 g-at CO_2 m^{-2} yr^{-1}

μatm^{-1}, as a function of latitudinal change in wind speed, and averaging over seasonal changes of photosynthesis, Takahashi et al. (1986) have updated estimates of the source and sink terms of atmospheric CO_2 at the surface of the deep sea (Table 1).

Earlier estimates, based on a constant gas exchange coefficient of 0.062 g-at CO_2 m^{-2} yr^{-1} μatm^{-1} (T. Takahashi, personal communication) and the GEOSECS observations of 1972–1978, had indicated a mean annual influx of 2.1×10^9 tons C yr^{-1} from the atmosphere to the sea. Their revised net uptake rate of 2.6×10^9 tons C yr^{-1} by the oceanic food web is now half that released to the atmosphere each year by burning of fossil fuels and cement production (Marland and Rotty 1984), and similar to the yearly atmospheric increment of CO_2 (Keeling et al. 1982).

Their considerations of the global CO_2 budget (Table 1), however, ignore both the continental margins in carbon uptake by the sea and inputs from the terrestrial reservoir of carbon. From the latter, 0.8–2.0 $\times 10^9$ tons C yr^{-1} may be released to the atmosphere during tropical deforestation. These estimates are based on changes in land use (Houghton 1986) and the $\delta^{13}C$ content of tree rings (Stuiver et al. 1984). The 7.8% of the ocean, not

TABLE 1 Seasonal mean of net CO_2 flux across the sea surface of open ocean regions, where + denotes outgassing (after Takahashi et al. 1986).

Region		Areal extent (10^6 km^2)	Exchange coefficient (g-at m^{-2} yr^{-1} μatm^{-1})	pCO_2 gradient (μatm)	Carbon flux (10^8 tons C yr^{-1})
Pacific	(10°N-10°S)	34.4	0.037	+50	+ 7.68
Atlantic	(10°N-10°S)	14.8	0.037	+30	+ 1.92
Indian	(10°N-10°S)	13.6	0.037	+10	+ 0.60
Pacific	(10°-40°N)	42.1	0.059	− 0.4	− 0.12
Pacific	(10°-40°S)	47.0	0.059	− 8	− 2.64
Atlantic	(10°-40°N)	23.1	0.059	−16	− 2.64
Atlantic	(10°-40°S)	20.4	0.059	+ 3	+ 0.48
Indian	(10°-20°N)	5.0	0.059	+25	+ 0.84
Indian	(10°-40°S)	30.2	0.059	−10	− 2.16
Pacific	(40°-60°N)	13.7	0.089	+14	+ 2.04
Atlantic	(40°-80°N)	13.9	0.089	−33	− 4.92
Antarctic	(40°-50°S)	31.2	0.111	−38	−15.84
Antarctic	(50°-70°S)	43.5	0.111	−20	−11.64
Total (92.2%)		332.9	0.069	− 9.4	−26.40

considered in Table 1, is the area of the continental shelves. Here, annual primary production is the highest of the sea (> 200 g C m^{-2} yr^{-1}) and the biological drawdown of atmospheric CO_2 per unit area may be the greatest. Comparison of the seasonal CO_2 cycles above weather ships in the midocean, i.e., 66°N, 2°E and 50°N, 145°W, with those at coastal stations between 40° and 70°N, for example, suggests that the seasonal amplitude of atmospheric CO_2 is 3–5 μatm greater at the edge of the North American continent than over the open ocean (Komhyr et al. 1985).

At the latitudes of the southeastern Bering Sea, the pCO_2 of shelf surface waters in March during the PROBES program was 339 μatm, similar to the atmosphere, but it was reduced to < 125 μatm by the end of the spring bloom in June (Codispoti et al. 1986). In contrast, open waters of the Gulf of Alaska appear to be a net source of CO_2 (Table 1), where the seasonal amplitude of primary production is weak. Smaller seasonal trends of dissolved CO_2 are found in the nutrient-rich, but light-limited waters south of the Antarctic Convergence, where summer pCO_2 values can be 35% less than those of the atmosphere (Takahashi et al. 1986).

Within North Atlantic waters of the Norwegian–Greenland–Labrador Seas, however, where an intense spring bloom evidently occurs (Esaias et al. 1986), Takahashi et al. (1986) found during the TTO program that the seasonal trend of dissolved CO_2 was similar to that of the Bering Sea, with the pCO_2 as little as 160 μatm in summer, compared to atmospheric values of ~340 μatm in the winter surface mixed layer. If the annual net flux of atmospheric CO_2 per unit area into North Atlantic or Antarctic waters (Table 1) is applied to the neglected area of the shelves (26×10^6 km^2), a range in net atmospheric carbon input to the ocean margins might be another $0.7–1.3 \times 10^9$ tons C yr^{-1}, thus offsetting terrestrial inputs of CO_2 to the atmosphere. An average particulate carbon flux of 42.4 g C m^{-2} yr^{-1}, caught at 100 m depth within several coastal regions of the VERTEX program (Martin et al. 1987), suggests that 1.1×10^9 tons C yr^{-1} might indeed exit the shelf-break as the biogenic expression of such a net CO_2 flux into shelf regions of the ocean.

A net gaseous flux of ~1×10^9 tons C yr^{-1} across the air-sea interface of the shelves, or similar particulate fallout at 100 m, implies that as much as 25% of the annual carbon fixation on shelves (at least 5.2×10^9 tons C yr^{-1}, assuming 200 g C m^{-2} yr^{-1}; see Walsh 1984) must be transferred each year to deep sea storage, either in the waters or on the floor. Decomposition of organic debris in the open ocean, for example, may liberate CO_2 from 50% of the mass of sinking particles within < 300 m of the upper water column and 90% within 1500 m (Martin et al. 1987). Refractory shelf detritus, reaching the slope bottom and escaping consumption by the benthos, could accumulate in the sediments as organic carbon rather than being stored as dissolved inorganic carbon within the open sea. About half of the organic carbon in surficial sediments of the oceans is found on the continental slopes (S. Emerson, personal communication).

Based on recent estimates of benthic respiration at various depths within the Atlantic and Pacific Oceans (Hinga 1985; Smith 1987), about 25% of the ocean's total bottom respiration may also occur within slope sediments. These regions represent only about 9% of the total ocean's area, however, suggesting that the organic debris, fueling such bottom catabolism, is focused here by transport from other regions. Indeed, estimates of vertical particle flux and benthic oxygen demands in the eastern Pacific (Smith 1987) imply lateral injection and/or undersampled seasonality of food supplies to the deep sea from the continental margins.

Based on short-term sediment trap moorings 100 m above the bottom and benthic chambers deployed at 1300 m, 40 km off San Diego, for example, the particulate carbon input can be 79% of the bottom metabolism in March, but only 15% in June. In contrast, at a mooring located 315 km offshore on the continental rise in a bottom depth of 3815 m, this ratio was 0.16 in April and 0.08 in June (Smith 1987). Trapping efficiency of the particle interceptor trap declines with increasing current speed, such that the horizontal advection of 6–8 cm sec^{-1} found at these stations could have led to an underestimation of the vertical particle flux by 50% (Smith 1987). Additional lateral input within a bottom boundary layer of < 100 m thickness would also have been undetected on the California slope.

To further quantify the extraction of CO_2 from the atmosphere by the biological pump—first as particulate carbon, then its export, and subsequent survival and transformations within shelf and slope sediments (Walsh et al. 1981; Emerson 1984; Walsh et al. 1985; Carpenter 1987; and Shaffer 1987)—a group of physicists, geochemists, and biologists conducted the first field experiment of the Shelf Edge Exchange Processes (SEEP) Program in 1983–84 within the Mid-Atlantic Bight. This multidisciplinary SEEP-I study of particle formation, transformation, transport, sedimentation, and storage was conducted along two sections across the shelf-break south of Martha's Vineyard and Long Island during July 1983 to October 1984 (Walsh et al. 1988).

Moored arrays of current meters, thermistors, transmissometers, fluorometers, and sediment traps provided time series data, ranging from 1 to 12 months duration, on the distributions of currents, temperature, light transmission, fluorescence, organic carbon, carbonate, and particulate ^{210}Pb within these shelf and slope waters of the Mid-Atlantic Bight. SEEP-I shipboard and aircraft measurements also focused on the birth and death processes of plankton during the 1984 spring bloom, when seasonal fluxes of carbon, carbonate, and ^{210}Pb to slope waters and sediments were maximal. Because of constraints on citing references in the proceedings of a Dahlem

Workshop, I refer to Walsh et al. (1988) for citation of individual publications by my SEEP colleagues; a summary of their results follows.

MID-ATLANTIC SHELF EXPORT

The consensus appears to be that the primary production of the spring bloom was neither quickly exported nor immediately consumed on the Mid-Atlantic shelf. For example, an oxygen budget was constructed for the shelf water column over 40 days between the east and west SEEP-I lines, south of Martha's Vineyard and Long Island, respectively. The budget suggested that a maximal consumption of 50% of the daily photosynthetic evolution of O_2 had occurred from all pelagic and benthic respiration by 4 April 1984, the end of the 40-day period.

The concomitant net particulate production of carbon by phytoplankton on the shelf was estimated by ^{14}C techniques to be 0.6 g C m^{-2} day^{-1} during March and 1.3 g C m^{-2} day^{-1} during April. At measured mean daily ingestion rates by all copepods of 0.10 g C m^{-2} day^{-1} during March and 0.50 g C m^{-2} day^{-1} during April within shelf waters south of Long Island, they could consume 17–38% of the daily fixation of particulate algal carbon. Such a grazing stress in terms of carbon is greater than the equivalent respiratory sink for oxygen evolution, since losses of the assimilated algal carbon to zooplankton reproduction must also be budgeted. In this same shelf region, estimates of benthic secondary production and all respiration requirements suggest a daily total carbon demand of only 0.15 g C m^{-2} day^{-1}, or 12–25% of the March–April primary production, respectively.

At a single measured fecal pellet flux of 0.12 g C m^{-2} day^{-1}, however, little or no input of ungrazed phytoplankton need directly be consumed by the benthic food web. Computed estimates of the seasonal input of sinking pellets range from 0.02 g C m^{-2} day^{-1} in March and 0.11 g C m^{-2} day^{-1} in April, assuming an assimilation efficiency of 70% for the zooplankton herbivores. Such estimates suggest that the direct ingestion of sinking phytoplankton, in contrast to fecal pellets, by the benthos might be, respectively, 22% and 3% of the daily primary production over these two months.

Without microbial respiration measurements in the 1984 water column, the spring oxygen budget may not be an accurate assessment of the metabolic demands of the bacterioplankton within slowly warming shelf waters of 3–9°C. Assuming that a 50% consumption of daily photosynthesis by the rest of the food web is correct, however, these microorganisms would, by difference, require carbon inputs of at least 11% (March) and 9% (April) of the particulate algal production. If the daily March–April oxygen demands over the water column are partitioned as 10% to phytoplankton, 30% to zooplankton, and 30% to benthos, one can assign the remaining 30% of the aerobic metabolism to bacterioplankton, i.e., about 15% of the gross photosynthesis.

With a bacterial secondary production of only ~2% of the rate of photosynthesis measured at this time of year, one then arrives at a total (production and respiration) carbon demand for the bacterioplankton of about 17% of the daily particulate carbon fixation by phytoplankton. Such food requirements of the bacterioplankton may perhaps be equally met from algal release of dissolved organic carbon (DOC) rather than from bacterial colonization of particles. The mean DOC release rate is ~16% of the daily particulate fixation of carbon by phytoplankton in the Mid-Atlantic Bight, for example, as measured by the ^{14}C technique, which presumably approximates net photosynthesis.

I have constructed a simple carbon budget (Fig. 1) to summarize our present knowledge of the shelf and slope food webs of the Mid-Atlantic Bight. An input of 0.16 g C m^{-2} day^{-1} from DOC exudates leads to a total net carbon fixation of 1.16 g C m^{-2} day^{-1} by phytoplankton for transfer of organic matter to the rest of the shelf food web (Fig. 1). About half may be removed directly within the water column, i.e., 0.30 g C m^{-2} day^{-1} by the zooplankton and perhaps 0.17 g C m^{-2} day^{-1} by particle-related bacteria, if they attach to living phytoplankton. The other free-floating forms of the bacterioplankton might derive an additional 0.16 g C m^{-2} day^{-1} from phytoplankton exudates of DOC, for a total microbial sink of 0.33 g C m^{-2} day^{-1}.

Since little algal biomass accumulates in Mid-Atlantic surface waters, the other half of the carbon produced by the spring bloom evidently sinks to the shelf bottom. Here perhaps 0.06 g C m^{-2} day^{-1} of the primary production is directly taken up by the benthic communities, which might ingest a further 0.09 g C m^{-2} day^{-1} as fecal pellets, for a total benthic consumption of 0.15 g C m^{-2} day^{-1}. The unconsumed algal production of ~0.47 g C m^{-2} day^{-1} (Fig. 1) presumably moves downstream, in time and space, to be either utilized by the summer shelf food web (Rowe et al. 1986) or exported to the slope (Walsh et al. 1981).

Assessment was made by the near-bottom arrays of fluorometers and current meters near the shelf-break on the west SEEP-I line of how many fluorescent particles were left behind in the spring water column. The product of the fluorometer and current meter data suggested a somewhat smaller seaward flux of 0.30 g C m^{-2} day^{-1} over 40 days, compared to 0.47 g C m^{-2} day^{-1} from the carbon budget, without consideration of the size composition of the phytoplankton. Analysis of the size classes of shelf and slope phytoplankton communities (Walsh et al. 1988) suggests, in fact, that 0.30 g C m^{-2} day^{-1} of large, sinking diatoms may exit the shelf surface waters, compared to an export of 0.05 g C m^{-2} day^{-1} from slope surface waters, in the form of both fecal pellets and algal cells larger than

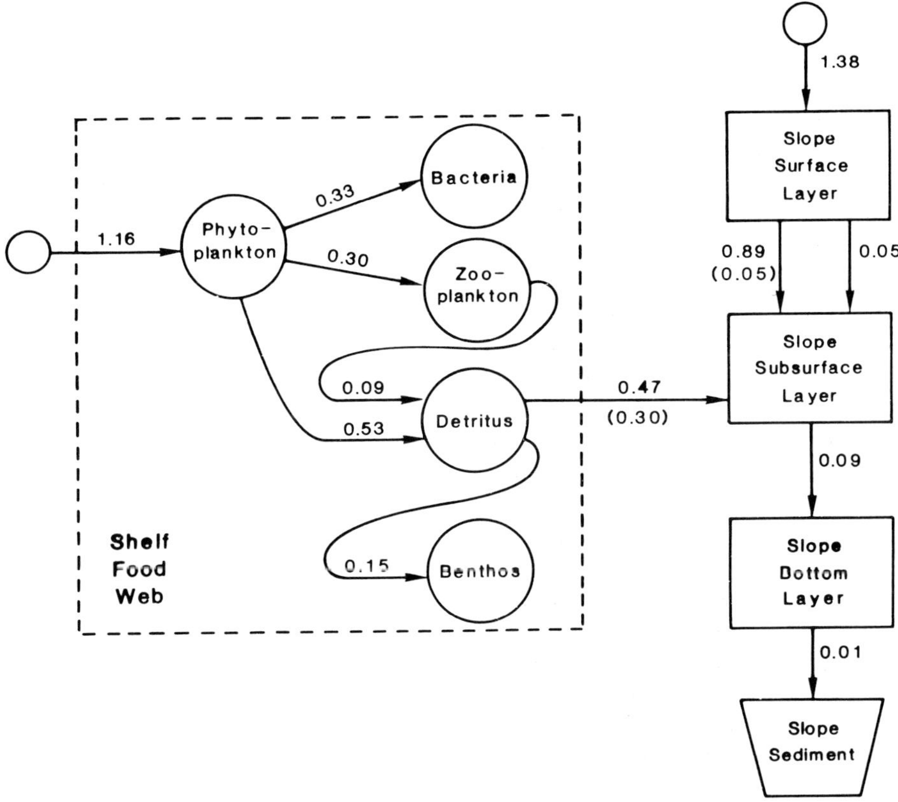

Fig. 1—The daily carbon flow (g C m^{-2} day^{-1}) within shelf and slope food webs of the Mid-Atlantic Bight during March-April, 1984. The numbers in parentheses are additional estimates of export production, assuming size fractionation of phytoplankton groups as large diatoms and small picoplankton.

picoplankton; these size-fractionated algal fluxes are the numbers in parentheses of Fig. 1.

Assuming a primary production of 1 g C m^{-2} day^{-1} during a spring bloom of 100 days, and 0.5 g C m^{-2} day^{-1} for the rest of the year, an annual carbon fixation of 232.5 g C m^{-2} yr^{-1} is a reasonable estimate of autotrophic removal of CO_2 on the Mid-Atlantic shelf. Assuming further that an export of 0.30–0.47 g C m^{-2} day^{-1} occurred only during the spring bloom, with full utilization of primary production by this shelf food web over the rest of the year, a flux of 30–47 g C m^{-2} yr^{-1} might exit the shelf-break. This is in reasonable agreement with Martin et al.'s (1987) measurements for other coastal regions. What evidence for such a particle transport process exists within the physical and geochemical results of SEEP-I, obtained over a longer period of 12 months?

CROSS-SHELF TRANSFER MECHANISMS

Based on freshwater discharge, salt budgets, $^{16/18}O$ ratios, and nutrient fluxes required to sustain shelf primary production, the long-term, two-way exchange rate of shelf and slope waters at the Mid-Atlantic shelf-break is estimated to be 0.2–0.4 m^2 sec^{-1}. Warm core rings, baroclinic instabilities, and wind events are all causative factors in the mixing process, parameterized by this exchange coefficient. Within the shelf-break front, moreover, greater diabathic fluxes of dissolved substances, e.g., salt and perhaps nutrients, may occur in the summer when isopycnal surfaces extend continuously across the frontal zone.

Over a 20–40 m euphotic zone, offshore Ekman transport, in response to upwelling favorable winds, together with such a mean exchange rate at the shelf-break would lead to average cross-isobath surface flows of 0.5–2.0 cm sec^{-1}. Observed diabathic currents during SEEP-I, measured within both the surface (10 m below) and bottom (5 m above) Ekman layers on the 80 m, 125 m, and 150 m isobaths of the outer shelf during each of the fall–winter (9/9/83–1/15/84) and spring–summer (4/19/84–10/17/84) deployments, were indeed offshore at long-term means of 0.6–1.2 cm sec^{-1}. During winter–spring, large cross-isobath excursions of the shelf-break front over distances of 10–20 km were also correlated with the alongshore wind stress in an Ekman sense, when episodes of diabathic flow of as much as 10–20 cm sec^{-1} were observed.

During an upwelling favorable wind event, particles situated near the bottom on the landward side of the inclined shelf-break front could be resuspended and advected seaward in the surface Ekman layer. After this wind event, the biogenic and lithogenic particles could then fall through the inclined front to bottom waters of the outer shelf. Once seaward of and beneath the front, however, near-bottom particles could no longer be mixed up again to the surface. They would instead be further exported to the slope within a bottom Ekman layer during a downwelling favorable wind event.

Resuspension of particles, trapped beneath the shelf-break front, by both tides and internal waves would actually be enhanced during periods of downwelling. In these events, the bottom of the front is steepened and baroclinic tides are amplified. On the outer shelf, the semi-diurnal tide generates mean near-bottom, cross-shelf velocities of ~4 cm sec^{-1} amplitude. However, at this and even higher frequencies of internal waves, originating within slope waters and dissipating on the outer shelf bottom, transitory near-bottom oscillations of 10–30 cm sec^{-1} amplitude were observed.

These flows constitute a bottom stress of as much as 6 dynes cm^{-2} on the outer shelf, i.e., equivalent to the storm-induced friction velocities on the inner shelf. To place these forces in perspective, a bottom stress of ~1 dyne cm^{-2} would be required for resuspension of particles of 37 μm diameter, i.e., larger diatoms and silt-sized sediments. High-frequency water motions on the upper slope may similarly lead to resuspension and preferential downslope transport of near-bottom particles once they penetrate the shelf-break front. Surviving shelf particles enter a near-bottom physical regime on the continental slope which actually favors their downslope transport and deposition.

Flows within the bottom Ekman layer on the inner shelf were frequently greater than 20 cm sec^{-1}, which is probably sufficient to resuspend and transport sediment. However, in water depths > 70 m, the bottom stress contributed by surface waves is a minor factor in resuspension of particles, and thus is not discussed above. Similarly, resuspension events at the shelf-break in SEEP-I were poorly correlated with direct changes in wind stress but were related instead to low salinities, i.e., seaward movement of the foot of the shelf-break front, where the kinetic energy of semi-diurnal tides and internal waves is focused. Resuspension by high-frequency forcing of tides and internal waves on the outer shelf, as well as by surface waves on the inner shelf, coupled with offshore advection within surface and bottom Ekman layers in response to low-frequency wind events, is a feasible net export mechanism, if the particles settle out of the water column before the next onshore movement of the shelf-break front.

An offshore displacement of the shelf-break front over 15–20 km at speeds of 5–20 cm sec^{-1} (5–20 km day^{-1}), measured at the 125 m isobath, suggests that a shelf particle might have a seaward journey of 0.8–4.0 days during a downwelling event. For net seaward export, however, it must sink out of the bottom Ekman layer onto the outer shelf during such an offshore excursion, before return of the shelf-break front towards the coast. With no further resuspension, a sinking velocity of 10–20 m day^{-1} over a 10–30 m nepheloid layer would strip particles from the bottom water column within 0.5–3.0 days. Assuming no motility of algae or detritus, this is a likely scenario for those particles whose diameter exceeds 48–68 μm, i.e., approaching equivalent Stokes settling velocities.

Such a size range is somewhat larger than that which can be resuspended by a bottom stress of 1 dyne cm^{-2}, given earlier. However, the diameters of solitary diatom species, e.g., *Coscinodiscus* and *Biddulphia*, found in slope sediments during SEEP-I fall in that range. When colonial diatoms of individual diameters of 10–20 μm form chains, e.g., *Chaetoceros, Thalassiosira*, and *Skeletonema* as in March–April, 1984, the resultant diameters of the colony can also be in the range of 48–68 μm, or larger. It would appear that both diatoms and other fine-grained (< 63 μm) particles could sink out of the shelf-break water column during surface or bottom excursions of the front, to be incorporated within a near-bottom nepheloid layer of the upper slope.

In this region, the net diabathic current on the upper slope (200–1150 m), between 67° and 70°W, was seaward at 7 m above bottom. Over a Mid-Atlantic continental slope of 10% inclination, a downslope flow of 1 cm sec^{-1} is equivalent to a depth change of about 100 m day^{-1}. This apparent vertical displacement rate is similar to that estimated from time-lapse photography for the arrival of phytodetritus on the Irish slope after the spring bloom (Lampitt 1985). Similar flocs of near-bottom organic debris have been detected visually at mid-slope (989–1750 m) within a camera sled traverse along the SEEP-I east transect during 4–5 May, 1986 (B. Hecker, personal communication).

The thickness of the bottom boundary layer on the Mid-Atlantic slope was > 7 m and < 50 m, with an average downslope current of 3.8 cm sec^{-1} at the 200 m isobath, 1.4 cm sec^{-1} at the 500 m isobath, and 1.8 cm sec^{-1} at the 1150 m isobath. An apparent vertical displacement of 140–380 m day^{-1} would ensure rapid delivery of particles to the lower slope after their bioturbation and resuspension on the upper slope. Here, high-frequency, asymmetric oscillating currents occurred, with downslope flows which were 2–3 times greater than the upslope flow. On the lower slope, a depositional flow regime prevailed, where near-bottom currents > 20 cm sec^{-1} were observed less than 1% of the time. Particles, escaping oxidation in the water column and sediments, would presumably accumulate in this slope depocenter.

MID-ATLANTIC SLOPE IMPORT

On an annual basis, however, ^{210}Pb budgets of the slope sediments suggest that only 20–60% of organic debris from the adjacent shelf ecosystem may exit the shelf-break in the SEEP-I region, with the remainder entrained downstream towards Cape Hatteras. One constraint on the magnitude of export of organic matter from the shelf is the amount of ^{210}Pb which might accumulate in the slope sediments on an equal area basis, over longer time scales than the spring bloom. Based on sediment inventories of ^{210}Pb from all six SEEP-I cores taken between the 500 m and 2000 m isobaths, the mean ^{210}Pb accumulation rate in surface sediments of the slope might only be 1.4 dpm cm^{-2} yr^{-1}, the same as that caught with sediment traps moored 50 m off the bottom at the 500 m and 1250 m isobaths.

From the three SEEP-I cores on the 500 m, 870 m, and 1130 m isobaths of the upper slope, however, the mean ^{210}Pb accumulation was instead

1.9 dpm cm^{-2} yr^{-1}, similar to prior estimates over the whole Mid-Atlantic Bight. Since the downslope transport of particles apparently occurs in a bottom boundary layer of < 50 m thickness, the near-bottom sediment traps in SEEP-I may not have caught all the lateral export of particulate ^{210}Pb and organic carbon from the shelf. We explore the consequences of possible undersampling in a simple budget.

Assume that the shelf width (~100 km) is twice that of the slope (~50 km), with an atmospheric input of ^{210}Pb over the entire region of 1.0 dpm cm^{-2} yr^{-1}. If all of the atmospheric ^{210}Pb fallout over the shelf were exported to the slope, with *none* buried within the relict shelf sediments, then the total input of ^{210}Pb focused in the slope sediments should be 3.0 dpm cm^{-2} yr^{-1}, i.e., 2.0 dpm cm^{-2} yr^{-1} from the shelf waters and 1.0 dpm cm^{-2} yr^{-1} from the slope waters. A measured ^{210}Pb accumulation rate of 1.4–2.2 dpm cm^{-2} yr^{-1} in slope sediments then accounts for 100% of the local atmospheric input to slope waters (1.0 dpm cm^{-2} yr^{-1}). Another 20–60% of the atmospheric input of ^{210}Pb from shelf waters (equivalent to 0.4–1.2 dpm cm^{-2} yr^{-1}) is also presumably stored in slope sediments. The remaining 40–80% of the shelf ^{210}Pb is apparently absorbed on detritus which exits at Cape Hatteras, where perhaps 50% of the shelf water and 33% of the unconsumed phytoplankton from the spring bloom may also be exported.

Fresh phytodetritus appears to be a major export in the SEEP-I region, since C/N ratios of < 7 were measured in the organic fraction within the near-bottom sediment traps at 1000–2500 m, compared to larger ratios found both in the slope water column and shallow traps at 150–450 m depths. A seasonal peak of the calcium carbonate flux was also found in April at 500 m on the upper slope. Finally, observations of shelf diatom species were made within the bottom layer at 400–1000 m on the slope. These SEEP-I data all suggest that fast-sinking (~100 m day^{-1}) macroaggregates of diatoms and coccolithophores, in contrast to thecate dinoflagellates and picoplankton, constituted the bulk of the carbon caught in the upper slope traps.

Continued oxidation of the spring organic matter within slope sediments explains total carbon accumulation rates in the SEEP-I region of less than 1% of the annual primary production on the shelf. During summer months the particulate ^{210}Pb flux in the traps at 150 m depth on the middle slope (1250 m) was a seasonal minimum. Monthly values ranged from < 0.2 dpm cm^{-2} yr^{-1} in September to > 2.0 dpm cm^{-2} yr^{-1} in May, with a seasonally averaged flux of 0.8 dpm cm^{-2} yr^{-1}. From near-surface sediment traps at 150 m to 450 m depths, where particulate matter first enters the slope water column, the daily organic carbon flux can be roughly related to the annual ^{210}Pb flux by 25 mg C m^{-2} day^{-1} $\simeq$ 1 dpm ^{210}Pb cm^{-2} yr^{-1}.

Less organic carbon flux was found for the same input of ^{210}Pb within subsurface sediment traps, i.e., implying more oxidation of the organic matter during the longer residence time in the deeper water column. Without decomposition of the detrital carbon, a ^{210}Pb accumulation rate of 1.4–2.2 dpm cm^{-2} yr^{-1} within slope sediments and the surface ^{210}Pb:C relationship imply an organic carbon burial rate of 12.8–20.1 g C m^{-2} yr^{-1}. The seasonally averaged carbon flux actually caught in sediment traps 50 m off the bottom was instead about half, 11.3 and 5.3 g C m^{-2} yr^{-1} at the 500 m and 1250 m isobaths respectively; only 2.2 g C m^{-2} yr^{-1} were caught 50 m above bottom at depths of 2300 and 2750 m on the continental rise. This slope fallout at the top of the benthic boundary layer is 2–5% of a mean annual primary production of 230 g C m^{-2} yr^{-1}, estimated previously for the Mid-Atlantic Bight, and 12–27% of the particulate flux of 42.4 g C m^{-2} yr^{-1} estimated to sink below 100 m in coastal regions (Martin et al. 1987).

A previous estimate of sediment accumulation of carbon over slope regions of the whole Mid-Atlantic Bight was similar to the near-bottom trap flux at 500 m, about 9.9 g C m^{-2} yr^{-1} (Walsh et al. 1985). In just the SEEP-I region, however, where *a priori* we thought carbon export would be minimal, revised estimates of carbon burial on this slope are now instead 1.0–2.5 g C m^{-2} yr^{-1}, i.e., at most 1% of the annual primary production and 9–47% of the particulate flux just above the bottom. A similar range in the percentage of carbon buried to near-bottom input fluxes was found in previous studies (Emerson 1984) of the Mid-Atlantic slope and rise (8%) and of the California borderlands, e.g., the Santa Catalina Basin (45%).

The SEEP-I study area, however, represents a region of minimal export of particles from the Mid-Atlantic Bight. For example, 250 km farther south on the 1000 m isobath, the annual fluxes of ^{210}Pb and organic carbon, within another 10-month time series at 25 m above bottom in 1985–86, were respectively two- to threefold greater than those at equivalent depths in the SEEP-I experiment. In this second time series, 15.5 g C m^{-2} yr^{-1}, or 7% of the annual shelf production was caught closer to the bottom at mid-slope off New Jersey, implying as much as 34 g C m^{-2} yr^{-1} might have sunk to the 500 m isobath here, using a depth relationship from the SEEP-I experiment.

ANTHROPOGENIC TRANSIENTS

The amount of shelf production that reaches the deep sea is thus clearly a function of shelf width, slope depth, and changes of the nutrient supply rate. Time scales of interest range from wind events to glaciations, with particular emphasis now being placed on the 100-yr period of the Industrial Revolution (Walsh et al. 1981). In shallow waters, such as the Baltic Sea, for example, where the production of the spring bloom is similar to that of the Mid-Atlantic Bight, as much as 2 g C m^{-2} day^{-1} is caught in March with traps on the 20 m isobath (Smetacek 1984). This near-bottom settling flux is about tenfold the seasonal peak of particulate carbon input found at 500 m on the Mid-Atlantic slope.

The annual carbon flux to the bottom in Kiel Bight during 1976 was ~104 g C m^{-2} yr^{-1} at 2 m above the 20 m isobath (Smetacek 1984). As a result of eutrophication since 1900, nutrient input to the Baltic has perhaps increased tenfold (Shaffer 1987). In response, the annual primary production of the Baltic Sea has at least doubled from 90 to 175 g C m^{-2} yr^{-1}, similar to the North Sea (Walsh 1984), and anoxia has occurred. The ensuing burial flux may have increased from ~2 g C m^{-2} yr^{-1} in 1900 to 10 g C m^{-2} yr^{-1} by 1980 (Shaffer 1987), i.e., a maximum of perhaps 10% for the present ratio of burial to input flux. Farther offshore, where advective and diffusive resupply of oxygen may be greater, increased nutrient loading may instead lead to larger rates of respiration rather than of burial.

Phytoplankton biomass (Venrick et al. 1987) and primary production (Laws et al. 1987) in the central North Pacific gyre at ~28°N, 155°W, may also have doubled between 1968 and 1985, from means of 11.2 mg chl m^{-2} and 57 g C m^{-2} yr^{-1} to 24.3 mg chl m^{-2} and 100 g C m^{-2} yr^{-1}. In contrast to coastal eutrophication, increased surface wind stress and vertical mixing during the winters of 1980–1985 may have led to enhanced rates of nutrient input to the central North Pacific (Venrick et al. 1987). Increased algal biomass, at ~95–120 m near the base of the euphotic zone, may have led to higher primary production, although the net drawdown of atmospheric CO_2 for this region is only ~0.1×10^9 tons C yr^{-1} (Table 1). Long-term time series of particle fluxes are required to assess whether increased export of "new" production occurs in response to enhanced input of oceanic nutrients.

Nutrient increments, however, have not been observed within the open ocean during this time period (Peng and Broecker 1984; Venrick et al. 1987). Furthermore, the 1985 PRPOOS estimates of primary production are within the range of carbon fixation measured in the North Pacific during previous studies of 1968–1980 (Laws et al. 1987). A similar period of increased wind stress also occurred in the North Pacific during 1960–1965, suggesting that such possible trends of carbon fixation in the open ocean may be cyclic rather than indicative of a long-term response to increased nutrient loading, with the consequences of anthropogenic stimulation confined to shelf ecosystems.

TERRESTRIAL SOURCES

The higher ratio of burial to near-bottom input fluxes found farther landward on the California borderlands may reflect greater concentrations here of refractory terrestrial carbon, associated with nutrient loading from the adjacent land ecosystems (Emerson 1984). The amount of such refractory organic matter in Mid-Atlantic slope sediments was also traced in SEEP-I with biomarkers, e.g., lignin phenols or sterols from higher vascular plants. The abundance of vanillin and β-sitosterol, in comparison with total organic carbon content of the SEEP-I surficial sediments, suggested that as much as 49% of the organic carbon in these slope sediments could be from the shelf and coastal regions.

The $\delta^{13}C$ signature of these sediments is mainly marine (Walsh et al. 1985), however, suggesting a planktonic source of carbon as well. The absence of most lignin phenols from the sediment trap moored 50 m above bottom on the 1250 m isobath implied that near-bottom, downslope sediment movement, rather than atmospheric deposition and zooplankton scavenging within slope waters, was the transport mode for shelf export of terrestrial carbon. An average age of 2000 yr for surficial sediments on the Mid-Atlantic slope may reflect this terrestrial source (Walsh et al. 1985), masking recent bomb signals of ^{14}C-labeled plankton carbon, which should be focused here (Peng and Broecker 1984).

Alternatively, preferential oxidation of the labile plankton carbon, fixed as a consequence of coastal eutrophication, could have increased within slope depocenters. The input flux of organic matter on the slope has a C/N ratio of < 7, yet the C/N ratio of Mid-Atlantic sediment is 8 as a result of early diagenesis (Walsh et al. 1985). On the Washington slope, however, where biomarkers indicate that the terrestrial fraction is $< 10\%$ of the organic carbon (Emerson 1984), accelerated rates of carbon storage and oxygen depletion have not been detected; the annual flux of nutrients from the Columbia River has also changed less than twofold from 1966 to 1981 (Carpenter 1987).

Comparison of $CaCO_3$ fluxes and inventories in the SEEP-I sediments suggests that half of the input of calcareous matter may be dissolved as a result of biogenic production of CO_2 within the bottom boundary layer. Such biogenic CO_2, evolved during diagenesis on the lower slope, would be below the main thermocline of the ocean. Here, isopycnal transport might allow horizontal dispersion of this source of CO_2 towards the deep sea to be possibly sequestered within the inorganic bicarbonate cycle as a partial coastal sink for emissions of fossil fuel.

GLOBAL IMPLICATIONS

To place the particulate part of such a coastal sink in a global perspective, the depth and seasonally averaged vertical flux of organic carbon of 21 mg C m^{-2} day^{-1} caught in the six SEEP-I sediment traps on the slope was threefold that of 7 mg C m^{-2} day^{-1} from another six traps on the adjacent rise. Farther seaward on the same daily basis, the seasonal maximum of sinking carbon within deep-sea sediment traps at 3200 m in the Sargasso Sea was only 3–4 mg C m^{-2} day^{-1} (Deuser 1986). This offshore flux is more than an order of magnitude less than that of the seasonal maximum of 46 mg C m^{-2} day^{-1} found throughout the April–May water column at the SEEP-I moorings on the continental rise.

Since the surface area of the continental slopes is ~10% that of the open ocean, the total particulate carbon export from the continental margins which eventually arrives at 2000 m may be equivalent to that from the open ocean. For example, over a world slope area of 32.5×10^6 km^2, the mean particulate carbon flux within the SEEP-I water column of 7.7 g C m^{-2} yr^{-1} would yield an input of 0.25×10^9 tons C yr^{-1} to the bottom boundary layer, extending from the 200 m to the 2000 m isobaths. Within light-limited polar regions, e.g., on the slopes of the Barents and Weddell Seas, the annual carbon input to near-bottom sediment traps above the continental margin is somewhat less, 2.9–4.3 g C m^{-2} yr^{-1} (Wefer, this volume), than the SEEP-I average. Carbon export from the more productive eastern boundary currents should be greater than the SEEP-I estimate; however, it should compensate for the lower export from polar seas.

Using a very high estimate of oceanic primary production of 130 g C m^{-2} yr^{-1} from the central North Pacific, rather than previous estimates of 50–100 g C m^{-2} yr^{-1}, Martin et al. (1987) calculated that 0.46 $\times 10^9$ tons C yr^{-1} might instead arrive intact at a depth of ~2000 m in the open sea. Excluding the Panama Basin, where lateral inputs of organic matter are significant, year-long sediment trap moorings at a mean depth of 2842 m yield an average particle flux of 0.80 g C m^{-2} yr^{-1} for 10 open ocean (bottom depth > 2000 m) sites (Wefer, this volume). This mean open ocean export of carbon is tenfold less than the SEEP-I average, and over the 3.1×10^8 km^2 area of the deep sea, also yields a particulate input of 0.25×10^9 tons C yr^{-1}.

From the GEOSECS, VERTEX, PRPOOS, TTO, PROBES, SEEP, and other studies, it would appear that the export of organic matter, in a combination of dissolved and particulate forms, from the margins of the ocean might now be as much as $\sim 1 \times 10^9$ tons C yr^{-1}. About 75% of the particulate carbon is oxidized in the water column between depths of 100 and 1950 m. At least another 20% of the particulate carbon is now oxidized in the benthic boundary layer, with 1–5% buried in slope sediments.

The total input of the shelf/slope biological pump (1×10^9 tons C yr^{-1}) to the adjacent open ocean is about half of the most recent estimate of the net transfer of CO_2 from surface waters to just the deep sea ($\sim 2.6 \times 10^9$ tons C yr^{-1}). A possible tenfoid increase of man's nitrogen loading within the coastal zone to 6×10^7 tons N yr^{-1} by 1980 (Walsh 1984), and C/N ratios of 6/1 to 8/1 for various decomposition stages of phytodetritus, imply that about $0.36–0.48 \times 10^9$ tons C yr^{-1} of the coastal export might be derived from anthropogenic sources. Man now fixes about half as much nitrogen from the atmosphere, for use in agriculture, as that extracted by other terrestrial and marine organisms. If the uses of fertilizer and/or soil leaching rates increase, one might expect larger exports of organic matter from the continental margins to the deep sea. Further quantification of this process awaits additional multidisciplinary studies of the coupling of element fluxes between the margins and the interior of the sea.

Acknowledgements. This research was supported by the Department of Energy, the National Aeronautics and Space Administration, and the National Science Foundation of the United States.

REFERENCES

Carpenter, R. 1987. Has man altered the cycling of nutrients and organic C on the Washington continental shelf and slope? *Deep-Sea Res.* **34**: 881–896.

Codispoti, L.A.; Friederich, G.E.; and Hood, D.W. 1986. Variability in the inorganic carbon system over the southeastern Bering Sea shelf during spring 1980 and spring–summer 1981. *Cont. Shelf Res.* **5**: 133–160.

Deuser, W.G. 1986. Seasonal and interannual variations in deep-water particle fluxes in the Sargasso Sea and their relation to surface hydrography. *Deep-Sea Res.* **33**: 225–246.

Emerson, S. 1984. Organic carbon preservation in marine sediments. In: The Carbon Cycle and Atmospheric CO_2: National Variation, Archean to Present, eds. E.T. Sundquist and W.S. Broecker. *Geophys. Monog.* **32**: 78–87. Washington, D.C.: Amer. Geophys. Union.

Esaias, W.E.; Feldman, G.C.; McClain, C.R.; and Elrod, J.A. 1986. Monthly satellite-derived phytoplankton pigment distribution for the North Atlantic ocean basin. *EOS* **67**: 835–837.

Hinga, K.R. 1985. Evidence for a higher average primary productivity in the Pacific than in the Atlantic Ocean. *Deep-Sea Res.* **32**: 117–126.

Houghton, R.A. 1986. Estimating changes in the carbon content of terrestrial ecosystems from historical data. In: The Changing Carbon Cycle: A Global Analysis, eds. J.R. Trabalka and D.E. Reichle, pp. 175–193. Berlin: Springer.

Keeling, C.D.; Bacastow, R.B.; and Whorf, T.P. 1982. Measurements of the concentration of carbon dioxide at Mauna Loa observatory, Hawaii. In: Carbon Dioxide Review: 1982, ed. W.C. Clark, pp. 377–385. London: Oxford Univ. Press.

Komhyr, W.D.; Gammon, R.H.; Harris, T.B.; Waterman, L.S.; Conway, T.J.; Taylor, W.R.; and Thoning, K.W. 1985. Global atmospheric CO_2 distribution and variation from 1968–82 NOAA/GMCC flask sample data. *J. Geophys. Res.* **90**: 5567–5596.

Lampitt, R.S. 1985. Evidence for the seasonal deposition of detritus to the deep-sea floor and its subsequent resuspension. *Deep-Sea Res.* **32**: 885–898.

Laws, E.A.; Di Tullio, G.R.; and Redalje, R.G. 1987. High phytoplankton growth and production rates in the North Pacific subtropical gyre. *Limnol. Ocean.* **32**: 905–918.

Marland, G., and Rotty, R.M. 1984. Carbon dioxide emissions from fossil fuels: a procedure for estimation and results for 1950–82. *Tellus* **36**: 232–261.

Martin, J.H.; Knauer, G.A.; Karl, D.M.; and Broenkow, W.W. 1987. VERTEX: carbon cycling in the northeast Pacific. *Deep-Sea Res.* **34**: 267–285.

Peng, T.H., and Broecker, W.S. 1984. Ocean life cycles and the atmospheric CO_2 content. *J. Geophys. Res.* **89**: 8170–8180.

Rowe, G.T.; Smith, S.L.; Falkowski, P.G.; Whitledge, T.E.; Theroux, R.; Phoel, W.; and Ducklow, H. 1986. Do continental shelves export organic matter? *Nature* **324**: 559–561.

Shaffer, G. 1987. Redfield ratios, primary production, and organic carbon burial in the Baltic Sea. *Deep-Sea Res.* **34**: 769–784.

Smetacek, V.S. 1984. The supply of food to the benthos. In: Flows of Energy and Materials in Marine Ecosystems: Theory and Practice, ed. M.J. Fasham, pp. 517–548. New York: Plenum.

Smith, K.L. 1987. Food energy supply and demand: a discrepancy between particulate organic carbon flux and sediment community oxygen consumption in the deep ocean. *Limnol. Ocean.* **32**: 201–220.

Stuiver, M.; Burk, R.L.; and Quay, P.D. 1984. $^{13}C/^{12}C$ ratios and the transfer of biospheric carbon to the atmosphere. *J. Geophys. Res.* **89**: 11731–11748.

Takahashi, T.; Goddard, J.; Sutherland, S.; Chipman, D.W.; and Breeze, C.C. 1986. Seasonal and geographic variability of carbon dioxide sink/source in the oceanic areas: observations in the North and Equatorial Pacific Ocean, 1984–1986 and global summary. DOE Technical Report MRETTA 19X–89675C, pp. 1–52.

Venrick, E.L.; McGowan, J.A.; Cagaw, D.R.; and Hayward, T.L. 1987. Climate and chlorophyll. I. Long-term trends in the central North Pacific Ocean. *Science* **238**: 70–72.

Walsh, J.J. 1984. The role of the ocean biota in accelerated ecological cycles: a temporal view. *Bioscience* **34**: 499–507.

Walsh, J.J.; Biscaye, P.E.; and Csanady, G.T. 1988. The 1983–84 Shelf Edge Exchange Processes (SEEP)-I experiment: hypotheses and highlights. *Cont. Shelf Res.* **8**: 435–456.

Walsh, J.J.; Premuzic, E.T.; Gaffney, J.S.; Rowe, G.T.; Harbottle, G.; Stoenner, R.W.; Balsam, W.L.; Betzer, P.R.; and Macko, S.A. 1985. Organic storage of CO_2 on the continental slope off the mid-Atlantic Bight, the southeastern Bering Sea, and the Peru coast. *Deep-Sea Res.* **32**: 853–883.

Walsh, J.J.; Rowe, G.T.; Iverson, R.L.; and McRoy, C.P. 1981. Biological export of shelf carbon is a neglected sink of the global CO_2 cycle. *Nature* **291**: 196–201.

Standing, left to right:
Jörn Thiede, Gerold Wefer, Paul Bienfang, Venu Ittekkot, Michael Sarnthein, Richard Lampitt

Seated, left to right:
Geoffrey Eglinton, John Walsh, Ken Bruland, Jim Bishop

Productivity of the Ocean: Present and Past
eds. W.H. Berger, V.S. Smetacek and G. Wefer, pp. 193–215
John Wiley & Sons Limited

Group Report
Flux to the Seafloor

K.W. Bruland, Rapporteur
P.K. Bienfang
J.K.B. Bishop
G. Eglinton
V.A.W. Ittekkot
R. Lampitt
M. Sarnthein
J. Thiede
J.J. Walsh
G. Wefer

INTRODUCTION

The principal goal of our group was to discuss the processes resulting in the modification of the export flux of material settling from the euphotic zone through the oceans' interior to the benthic boundary layer. These processes must be better understood in order to predict the geographic and seasonal variability of the flux of biogenic material to the seafloor.

During transit from the bottom of the euphotic zone to the deep-sea floor, more than 90% of the sinking particulate organic material degrades and a portion of the carbonates and opal dissolves. During this transfer, organic carbon is altered and new organic material can be added by horizontal inputs from shelf/slope boundaries and by bacterial particle production.

The interior of the ocean is filled by water masses with complex hydrographies which modify, and are altered by, the export products of the surface water masses on their way to the seafloor. Because of the Earth's continuing climatic evolution, the modern ocean is not always an ideal analog for understanding processes controlling past productivity. Nevertheless, a major step in better understanding present and past patterns of ocean productivity can be achieved through a thorough examination of the modern scenario, with all its regional and temporal variability.

Our working group was composed of a mixture of biologists, chemists, geochemists, and geologists and a variety of topics were addressed including particle interactions, the contribution due to episodic events, the role of dissolved organic matter (DOM), and the contributions of terrigenous and shelf organic carbon inputs to the interior of the oceans. We attempted to identify and to address problem areas in which our understanding is limited by a lack of data or realistic models. The limited time frame of the meeting and the subjective scientific interests of the working group members forced the exclusion of a number of processes which nevertheless were considered important. These include, among others, the influence of hydrothermal venting and submarine volcanism on particle fluxes and patterns of small- to intermediate-scale regional variability of particle fluxes in the oceans' interior.

PARTICLE INTERACTIONS

Sediment trap and large volume *in situ* filtration studies have demonstrated that particle sedimentation in the oceans is controlled primarily by biological processes in the upper kilometer of the ocean, which in turn appear to be forced by physical processes influencing the surface layer (Honjo 1984; Deuser 1986, 1987; Bishop et al. 1987). Such biological control of particle sedimentation results in a marked regional variability, with a given oceanic regime also exhibiting a strong seasonality in annual total flux and composition.

Bishop (this volume) has reviewed various parameterizations developed to describe the vertical carbon flux. Most of these are empirical relationships describing the particulate organic carbon flux as a function of depth and primary production in the overlying surface waters. However, the parameterization providing the best fit, with respect to available field data, was that of Martin and co-workers in the VERTEX Program (Martin et al. 1987), which was based upon knowing the particulate carbon flux, or export flux, at 100 meters depth [$J = J_{100}/(Z/100)^{0.86}$]. In other words, knowing the particle flux at 100 m allows a much better prediction of fluxes deeper in the water column. Nevertheless, the range of the exponent appears to vary between 0.3 and 1.5 in individual profiles, and the variation appears to be due to differences in the activities and distributions of particle consumers which are not included in the empirical model. Generally, all the empirical relationships predict a relatively rapid decrease in the export flux within the upper 1000 m of the water column, with only 10–20% of the initial export flux remaining at a depth horizon of 1000 m. Throughout the remainder of the deep water column, the modification of the particulate flux appears to be substantially less; however, significant particle transformations do still occur.

A Conceptual Model of Material Transformations

A schematic figure of material transformations (emphasizing C and N) in the oceans' interior is presented in figure 1. This conceptual model has two dissolved pools: the first includes the dissolved inorganic forms while the second includes dissolved organic matter, and also any colloidal material. There is a fine suspended particle fraction and two large particle fractions: one composed of rapidly sinking fecal material and the other of amorphous aggregates, or marine snow. The model includes the active role of bacteria, zooplankton, and micronekton in modifying the carbon flux. The transformations between these various forms are indicated by the different arrows and are defined in the figure caption. The majority of the transformations are mediated by the activities of organisms. This representation of elemental pathways is of similar complexity to a recently published model of the pelagic food web (Vezina and Platt 1988).

This system is unlikely to be found at steady state conditions; instead it is a dynamic system responding to perturbations in the large particle flux and the impact of varying zooplankton and micronekton activities. Perturbations in the particle flux can be associated with short events such as episodic pulses due to blooms or longer annual or interannual variations (e.g., ENSO events). The rates of the various transformations vary spatially, in both the vertical and horizontal sense, and temporally. However, the rules governing the magnitude and variability of the various transformation rates are poorly understood.

Modification of particles may be conveniently defined as either physical (aggregation, disaggregation) or chemical (assimilation, remineralization, dissolution, etc.), and within this framework the biotic effects may be described as internal, if dominated by the microbial communities associated with the particles, or external, if dominated by planktonic organisms spending the majority of their time detached from particles.

Recent results from the VERTEX Program suggest that internal microbial activity in the trap-collected material appears insufficient to account for the strong decrease in near-surface organic carbon fluxes (Karl et al. 1988). Thus, it has been suggested that either the material is fragmented to be consumed by the microbiota associated with the small particle reservoir, or it is consumed by the zooplankton community. This latter scenario is supported by zooplankton/particle flux studies in the Panama Basin (Bishop et al. 1987) and the subtropical Atlantic (Lampitt, in preparation). In these cases, a substantial proportion of the flux reduction is required by the macrozooplankton, leaving little for the microzooplankton and microbiota. This appears to be in contrast to the results of Cho and Azam (1988), which indicated that up to 98% of the carbon flux passes through the microbial loop. Although these results appear to conflict, they may be explained by factors as simple as regional or temporal variation in biological processes. As it stands, they illustrate our lack of understanding of the food web in the oceans' interior.

Several mechanisms have been proposed for the formation of marine snow or amorphous aggregates, including production of mucus-feeding structures by organisms such as salps and larvacians, and purely physical-chemical coagulation. Models of physical coagulation are poorly tested for typical marine conditions, i.e., dilute suspensions of heterogeneous particle sizes and specific gravities in a viscous electrolyte. A major unknown in the

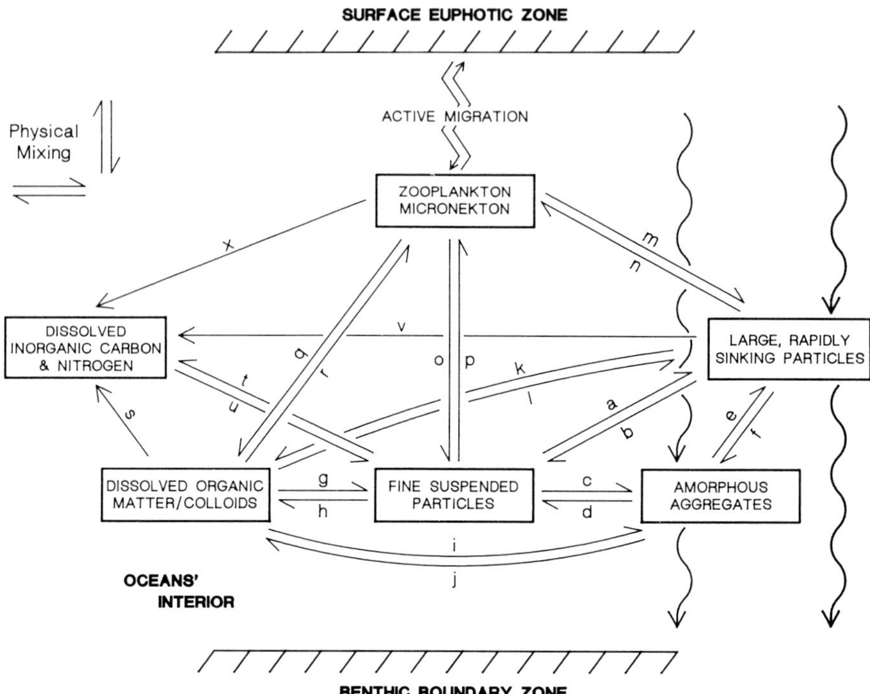

Fig. 1—Particle transformations in the oceans' interior. The different processes represented in the figure will be of different importance spatially and temporally. What we lack presently is a global understanding of the rules giving rise to these regional and seasonal variations. The following transformation rates are presented:

a) packaging of small suspended particles (both living cells and detritus) by animal grazers into large rapidly sinking fecal pellets,
b) fragmentation of fecal pellets, either internally or externally mediated, into the fine suspended particle pool,
c) aggregation of fine suspended particles into amorphous aggregates or marine snow,
d) fragmentation of amorphous aggregates into the small suspended particles by either biotic or abiotic means,
e) packaging of amorphous aggregates into fecal material by grazers.
f) degradation of fecal material into amorphous aggregates,
g) assimilation of DOM by free-living bacteria or aggregation of colloids into suspended particles,
h) excretion or release of DOM by metabolic activity associated with fine suspended particles,
i) assimilation of DOM by bacteria associated with amorphous aggregates,

physical coagulation models is the so-called "sticking coefficient": the fraction of interparticle contacts that result in coagulation. In recent models this value ranges from 0.1 to 0.9 without any supporting measurements (McCave 1984). In addition, the mechanical strength of aggregates is not well measured, making it difficult to model the disaggregation effects of physical processes such as shear. Thus, it is difficult to specify the importance of purely physical processes on the vertical particulate flux (McCave 1984).

Although there are few data to support it, the prevailing mood is that activities of biological organisms are more important for the packaging/aggregation and fragmentation of particles in the oceans' interior than are purely physical-chemical processes. The impact of the zooplankton may have been seriously underestimated in many oceanic areas due to the inability of nets to capture fragile gelatinous organisms which are destroyed, or fast swimmers which escape. A better understanding of the biotic processes and their rates is sorely needed (Alldredge and Hartwig 1986).

It has been suggested that lateral processes, possibly involving DOM, can also play an important role in supporting the oxygen utilization rates that are inferred for the midwater column. The DOM would serve as a carbon source for bacterial growth in the oceans' interior. Thus, DOM could support particle production leading to a source of particulate organic matter at mid-depths. If the new high dissolved organic carbon (DOC) results observed by Sugimara and Suzuki (1988) are confirmed, the advective transport of DOM, followed by bacterial production at mid-depths, could prove to be extremely important to particle transformations in the oceans' interior. Besides simple advection and diffusion, active transport by migrating species may also affect the vertical distribution of DOM.

---

j) excretion or release of DOM by metabolic activity associated with amorphous aggregates,
k) excretion or release of DOM from fecal pellets,
l) assimilation of DOM by bacteria associated with fecal pellets,
m) *in situ* injection of fecal pellets or other aggregates by actively migrating zooplankton or micronekton,
n) grazing upon fecal pellets by zooplankton or micronekton with incorporation into biomass or loss due to active migration,
o) injection of small particles from actively migrating zooplankton or micronekton,
p) grazing upon small particle fraction by zooplankton with incorporation into biomass or loss due to active migration,
q) excretion of DOM by zooplankton,
r) assimilation of DOM as a food source by zooplankton,
s) oxidation of DOM to DIC and DIN either abiotic or mediated by bacteria,
t) metabolic respiration of the small particulate fraction to DIC and DIN,
u) assimilation of DIC or DIN by bacteria into biomass in the fine particle pool,
v) respiration of large particle fraction of POM to DIC and DIN,
x) excretion of DIC and DIN by zooplankton and micronekton.

The Influence of Vertical Migration of Biota on the Flux of Dissolved and Particulate Organic Matter

Many midwater organisms exhibit vertical migrations on diurnal or seasonal scales (Angel 1986 and this volume). Diurnal migrations from depths of 600 m (zooplankton) or 1200 m (micronekton) during the day, to the surface waters at night, are likely to be of importance to flux studies. Rates of population movement are generally in the range of 40–120 m/hr (but have been noted up to 300 m/hr) and depending on the region and season, daily carbon movement into and out of the top 200 m may often exceed 50 mg/m^2, a value on the same order or only slightly lower than total primary production. If any of the processes of feeding, excretion, defecation, or mortality are discontinuous in a migration cycle, there is likely to be an influence on net material flux. From evidence gleaned from gut content analyses, there are many examples of migrators continually feeding throughout their migration cycle. However, without knowledge of diurnal changes in gut residence time it is not possible to derive feeding rates, and one should conclude that feeding is likely to be most intense in the surface waters where food is most plentiful. Solid material and dissolved material is released from the migration cycle by defecation and mortality, while dissolved material is released by excretion of metabolic end products (ammonia, urea, and CO_2). If such release occurs continuously throughout the daily migration cycle, or even predominantly at lower depths, there will be a net downward flux of material. In certain cases, this net downward flux may exceed fluxes due to the passive settling of detrital particles (Angel, this volume). Data on all of these processes are extremely sparse and firm conclusions cannot be made. At present, however, such migratory movements cannot be reasonably ignored in flux studies.

Tracers

Estimates of some of the transformation rates depicted in Fig. 1 have been made with the use of particle-reactive radioisotopes. The use of multiple isotopes of thorium with half-lives of 24 days (^{234}Th), 1.9 years (^{228}Th) and 8×10^5 years (^{230}Th) has allowed estimations of both the aggregation or packaging rate and the fragmentation rate of particulate thorium (Bacon, in preparation). These results have supported a model of continual exchange between the pools of particles in the deep sea, with the fine particle pool and large particle pools having a mean life of 2 years and 24 days, respectively. These rates are consistent with the transport, removal, and sedimentation of particulate aluminum (Spencer 1984) and other particle-reactive trace metals (Whitfield and Turner 1987) within the oceans' interior.

However, these metal radiotracers will not properly trace all the various classes of organic and inorganic compounds. There is a great need for developing other tracers of these material transformations in the oceans' interior. We need tracers of labile organic components, as well as tracers of the more refractory ones. Ideally these tracers will provide information on the compound's source, as well as the rates and extent of subsequent alteration.

An example of such an approach is that of Wakeham and Canuel (1988), who analyzed lipids in large particles (from sediment traps) and small particles (from *in situ* filtration), and who observed that large particles in the mesopelagic zone displayed extensive alteration by animals, while the small particles at the same depth contained a remarkable abundance of labile organic compounds attributable to phytoplankton sources. This suggested that the required rapid delivery of fresh material to the small particle pool was not derived from the large particles sampled in the sediment traps. Based on this evidence, Wakeham and Canuel argued for a second large particle pool containing a larger proportion of apparently undegraded algal cells, perhaps bound loosely together in marine snow aggregates, which rapidly carries them from surface waters to depths of 1000–2000 m. Disintegration of these particles at depth would contribute relatively undegraded algal material to the suspended particle pool sampled by *in situ* filtration. Large populations of picoplankton have been documented in the deep sea (Silver et al. 1986).

Conte and Bishop (1988) studied large and small particles collected by large volume *in situ* filtration and found similar lipid behavior. The large particles showed extensive alteration by zooplankton (even in the euphotic zone) and the small particles showed labile w3-polyunsaturated fatty acid enrichment. These acids (thought to be produced only by phytoplankton) were shown to decay on the time scale of several days, and their presence in the small particles could only be explained by processes occurring at the time of sampling. Conte and Bishop suggested that over 90% of the total export flux of phytoplankton-produced w3-polyunsaturated fatty acids was carried by migratory zooplankton and released into the small particle pool in the mesopelagic zone, where it appeared to support enhanced bacterial activity (as shown by the abundances of odd and branched-chain fatty acids). The flux of w3 fatty acids carried by large, fast-sinking particles was only 10% of the total w3 flux into subeuphotic zone waters. Active transport apparently accounted for 20–30% of the export flux of bulk particulate organic carbon (POC).

There is a great need for a better understanding of the various roles of organisms in modifying the export flux within the oceans' interior, and one should be cautioned that not all elements or compounds will behave similarly during their transit. Nor will all the component processes be equally present in time and space.

EPISODIC DEPOSITION

Episodic events embrace phenomena occurring over time scales ranging from days to thousands of years. Examples of episodic events over this time frame include upwelling, zooplankton swarms, storms, monsoons, El Niño events, moving ice fronts, volcanism, glaciation, and changes in sea level or ocean circulation. Episodic events are thought to be important because their destabilizing effect would increase export production and deliver a signal to the sedimentary record. The variety of episodic events would likely elicit dissimilar responses both temporally and spatially.

Do Episodic Events Dominate the Sedimentary Record?

Sediment trap time series of one year duration or longer show strong seasonality of the particulate flux to the deep sea, which can be related either to biological processes in the overlying surface waters or to more sporadic spikes of the flux due to episodic lateral inputs (Wefer, this volume). In addition, interannual variations of particle flux, such as those due to El Niño Southern Oscillation (ENSO) events, are also apparent (Dymond and Collier 1988). Variations in export flux can be due to changes in primary production and the degree of coupling between this and secondary consumption. Examples of decoupling have been found during intense spring blooms of diatoms at high latitudes, when phytodetritus rapidly accumulates on the seafloor (Billet et al. 1983; Lampitt 1985; Rice et al. 1986). Examples of episodic zooplankton grazing are swarms of krill in the Antarctic (von Bodungen et al. 1987) or swarms of salps off the coasts of North America, both of which involve migrating opportunistic grazers whose activities can lead to episodic export fluxes of fecal material or episodic export associated with their active migration.

These episodic events of settling material reflect the uncoupling of various components of the biological system within the euphotic zone, which otherwise tends toward tight coupling. Our notion is that in the normal or "equilibrium" case, the biological system of oceanic surface waters is quasi-conservative, thereby restricting gravitational losses of elements. This hypothesis means that with the normal case, only small fractions of the primary production of biogenic material exits the surface layer to the oceans' interior. In contrast, episodic perturbations represent opportunities for uncoupling the production/consumption components of the pelagic food web, increasing significantly the export flux and, thus, the input of biogenic material to the sediments.

Episodic events are thus potentially important contributors to the sedimentary record; most of the flux to the deep sea may occur at these restricted times (and/or sites) and the mean composition of particles and transport processes may therefore differ from those measured on most occasions in the open ocean. For example, the episodic deposition of phytoplankton, as a result of the spring bloom in the North Atlantic, appeared to overwhelm the "biological pump" and be transferred through the oceans' interior relatively untouched. Analysis of lipids in phytodetritus from the blooms collected as fluff at the seafloor suggested that the bulk of the material arrived rapidly at depth without passage through animal guts (Rice et al. 1986). Consumption and modification of this material by the benthos presumably followed. It is important to recognize that the sedimentary record may be greatly biased towards these relatively rare uncoupled events rather than the "norm" of leakage from the coupled biological system of the euphotic zone, which is subsequently further processed and modified in the oceans' interior.

To What Extent is the Transport to Depth of Intact Phytoplankton, Resulting from Spring Blooms, Significant on a Basin or a Global Scale?

For significant downward fluxes of micro-algae (less than 20 μm diameter) to occur, sinking rates must be increased by formation of larger particles. Algae of less than 2 μm diameter, such as those found in picoplankton-dominated regions of the open ocean, might sink a few meters during a year and, thus, relative to their life span, are not subject to direct sinking losses.

Export production in the open ocean presumably occurs mainly in the form of aggregates, fecal material, and/or body tissue of diurnal migrators. In the open North Pacific at Station P, a seasonal flux of algal remains (diatom frustules, etc.) occurs in the summer when ontogenetic migrators (e.g., *Neocalanus*) have left the surface water column (Takahashi 1986). In the open North Atlantic at Station S, a seasonal flux of carbon similarly occurs in response to nutrient addition during the early spring, when seasonal cohorts of crustacean herbivores are presumably out-of-phase with the primary producers (Deuser 1986, 1987). Tight coupling of picoplankton and protozoan herbivores in the open ocean, however, generally results in sinking fluxes at 1500–2000 m depth of less than a few percent of the primary production.

At the boundaries of the central gyres (i.e., shelf-breaks, equatorial divergence regions, and subpolar ocean divergences) upwelling and enhanced vertical mixing are sufficient energy subsidies to allow dominance of larger phytoplankton groups such as nonmotile diatoms and motile coccolithophores. At times of the year when light/nutrient conditions are optimal and grazing stress is minimal, spring and fall blooms of phytoplankton biomass are followed by aggregations of algal debris with apparent sinking rates of 150–200 m/day (i.e., similar to copepod fecal pellets). These episodes of sinking phytoplankton may represent much larger fractions of the daily primary production at the continental margins. At continental margins with bottom depths of less than 2000 m, as much as 5% of the annual primary production may arrive at the bottom boundary layer, compared to ca. 0.5% in the deep sea (Walsh, this volume).

Consumption of phytodetritus by benthic biota on the continental slope leads to burial estimates of less than 1% of the annual primary production in some regions. However, horizontal heterogeneity of particle export at the synoptic length scale (ca. 1000 km) may lead to local depocenters with much higher accumulation rates. In terms of oxygen utilization/CO_2 evolution beneath the main thermocline, the biological pumps in the interior of the deep sea and at the continental margin may be quite similar, with oxidation of 0.3–0.5 $\times$ 10^9 tons C yr^{-1} at a depth of 2000 m, both above the slopes and throughout the deep sea (Walsh, this volume). Further research must be conducted to test such rudimentary budgets, which imply that the flux of algal carbon to slope regions, with subsequent lateral exchange, is of the same importance as the direct flux of particulate organic carbon to the deep sea.

Although most evidence for episodic blooms of micro-algae is found in continental margin areas, these events may be more significant in the open ocean than was previously recognized. Recent satellite CZCS imagery (Feldman et al. 1986; McClain et al. 1987) has shown that one of the largest seasonal signals for ocean color is a basin-wide phenomenon in the North Atlantic. This variability is clearly associated with the spring bloom, which begins around 30°N latitude in the early spring and proceeds to 60–70°N by late June. Although this appears to be a basin-wide phenomenon, at present we have little information on the fate of these phytoplankton, regarding whether they are grazed and recycled in the upper 1000 m or whether a substantial portion of the bloom can aggregate and rapidly sink 4000–5000 m to the bottom boundary layer of the basin.

DISSOLVED ORGANIC MATTER

DOM constitutes by far the largest reservoir of reduced carbon in the oceans. DOM can be an intermediate stage during degradation of particulate organic matter (POM) to dissolved inorganic carbon (DIC). In spite of its reservoir size and its strategic position in the degradation pathway, very little is known about the role of DOM in the carbon cycle of the ocean.

Perhaps the best illustration of our general lack of knowledge on DOM concerns the uncertainty of the absolute concentration. Recent work by Sugimura and Suzuki (1988), using a high temperature combustion technique, suggests that oceanic DOC concentrations are at least twice as high and perhaps 4–5 times higher than those obtained by conventional wet chemistry techniques upon which current global ocean inventories are based. A similar discrepancy appears with the determination of dissolved organic nitrogen (DON) (Suzuki et al. 1985).

DOM, POM, and the biota are intimately related and must exchange molecules and degradation products on a time scale ranging from seconds to thousands of years. POM and the microbiota are virtually indistinguishable and inseparable in gross terms; for some, particles composing POM will be living organisms while others will be dead organic matter, perhaps heavily colonized by bacteria. Understanding the nature, timing, and extent of molecular exchange and interplay between DOM, POM, and the biota should be a key objective in our efforts to understand the fate of ocean productivity.

Sugimura and Suzuki's rapid micro-method for the determination of DOC provides a new window through which to examine DOM (Toggweiler, this volume). Essentially, the method consists of the injection of 200 μl of filtered seawater into a high temperature catalytic oxidation system, with in-line CO_2 measurement. The system is claimed to give complete oxidation of DOM, as documented by oxidizing pure compounds and biopolymers added to seawater. An important feature of the method lies in the ability to carry out the oxidations over a temperature range from 50° to 700°C.

Sugimura and Suzuki (1988) have found that the susceptibility to oxidation of single organic compounds and of DOM varies greatly, with some being 100% oxidized at relatively low temperatures (e.g., a monosaccharide, xylose, at 220°C; a humic acid fraction at 300°C), while others are only completely oxidized at high temperatures (e.g., the polysaccharide, xylon, at 510°C; the protein, albumin, at 580°C; the nitrogen-containing polysaccharide, chitin, at 620°C). These data are surprising and, if validated, lead to the following important conclusions. The catalytic oxidation temperature profiling data cannot provide a measure of degradation of DOM in the oceans. Biopolymers such as albumin, xylon, and chitin are rapidly degraded by appropriate enzymes, in contrast to the high oxidation temperature profile shown in the Sugimura and Suzuki process. There may be some parallel with photooxidation reactions; for example, it is conceivable that humic acids, which exhibit a low temperature oxidation profile, might readily undergo photolytic reactions. However, in general, the molecular characteristics important in determining the temperature profile in the platinum-catalyzed oxidations are unlikely to be similar to those operating in the microbiological and photochemical milieu of the oceans.

Sugimura and Suzuki (1988) also report molecular size fractionation studies with DOM. In one experiment, they oxidized DOM size fractions before and after persulfate oxidation (the classical method for DOC) and showed that each size fraction no longer expressed a "low temperature" component. Most original DOM size fractions showed a temperature of oxidation onset ca. 100°C, while after persulfate oxidation this onset was typically ca. 300°C; persulfate oxidized samples, however, still contained considerable DOC. These are important observations but, as discussed earlier, they cannot be interpreted in terms of biodegradability in natural seawater conditions. For example, the large quantity of high temperature oxidizable DOC at the surface cannot be directly equated with environmentally (biologically) refractory DOM. However, such caveats in no way detract from the potential importance of the Sugimura and Suzuki method, i.e., as a new, rapid micro-tool for a better evaluation of DOC and DOM in the oceans. The method opens the way for exploring the DOM as follows:

1. quantitation on an extensive scale at sea,
2. qualitative assessment with the temperature profiling feature and size fractionation,
3. elemental composition of C, N, etc., and
4. pulse labelling experiments with ^{13}C and ^{14}C, for example in microbial and enzymatic studies.

A high priority should be placed on independent testing and confirmation of the results of Sugimura and Suzuki (1988) and Suzuki et al. (1985). Not only would these new values increase our current estimates of the DOC and DON reservoirs in the world's ocean by a factor of two or three, they would also substantially diminish the importance of the particulate flux (and that due to active migration) or organic carbon and nitrogen in the export flux from the surface layers. Historically, oxygen consumption in the deep sea has been primarily considered a result of the "J flux," due to oxidation of sinking POM (Martin et al. 1987). If Sugimura and Suzuki's results are confirmed, then a major component of the oxygen utilization must be from the oxidation of DOM mixed to depth on time scales of tens to hundreds of years. Three-dimensional circulation models that require such high DOM values and subsequent mixing and advection in order to predict the correct nutrient field are already being proposed (Toggweiler, this volume).

It is now timely to intensify attempts to improve the characterization of DOM and POM at the molecular level in parallel with a redefinement of the bulk DOM concentrations. A better molecular understanding will have immediate bearing on the sources of organic matter and particle interactions, such as DOM-POM microbial processes, and will assist in the rationalization of stable isotope data, biomarker fluxes, and DOC/POC export fluxes. Such organic chemistry studies will require new methodologies and will be demanding and time consuming.

THE CONTRIBUTION OF ORGANIC CARBON FROM LAND AND SHALLOW WATERS

Organic matter associated with the material flux to the seafloor is derived from both internal (marine production) and external (riverine, aeolian) sources. Regional variation of terrigenous inputs via rivers, in the form of particulate and dissolved impacts on the coastal boundaries of the ocean, can be traced by surface salinities. Deposition of organic matter from the atmosphere, however, is much more broadly dispersed and is not as well known. It also appears that new particles can be generated within the oceans' interior by the conversion of DOM (some of which has a terrigenous source) into POM. The material arriving at the seafloor is thus a mixture of inputs from all these sources. At the seafloor this material is subjected to early diagenetic processes occurring at the sediment-water interface. The use of organic matter preserved in the sedimentary record for paleoproduction studies depends on how well we can differentiate the inputs from these various sources, and in relating it to sea surface biological processes.

One of the problems associated with the estimation of actual flux from the land to ocean sediments (via rivers or aeolian transport) stems from the uncertainty concerning its fate upon reaching the sea. A recent estimate suggests that about 35% of the POC from rivers can be oxidized in the sea and the fluxes of refractory POC from rivers is surprisingly close to most current estimates of organic carbon burial rates in the deep sea (Ittekkot 1988). Up to 60% of the delivery of POC occurs from the Asian continent, from rivers with extensive deep-sea fans. However, there is a paucity of data on the nature and cycling of organic matter in tropical and subtropical sea areas into which this major flux enters.

Little organic matter appears to be stored in shelf sediments compared to those of estuaries and marshes. Instead, cross-shelf transport of particles is a phenomenon which frequently affects particle sedimentation in slope and deep-sea areas (Walsh, this volume). Questions to be resolved are (*a*) what is the nature of this organic matter and (*b*) what are the time scales under which this lateral transport occurs. Characterization of this flux is an extremely significant aspect, since labile organic carbon associated with this advected material might produce conditions of oxygen minima in midwaters and be subsequently buried in depocenters of organic carbon on the slopes, where this oxygen minimum impinges.

How much carbon of terrestrial origin is part of the marine accumulation of organic carbon is an open question and is subject to refinement of methodologies. Bulk parameters, such as C/N, ^{15}N, or ^{13}C, may provide information at time scales different than those of biomarkers, e.g., lignin. Information on stable carbon isotopes and lignin-derived materials appears to put limits on the amount of terrestrial organic carbon which can become part of marine sediments (Prahl and Muehlhausen, this volume). These recent studies show that gaps exist in our knowledge of the land-ocean carbon balance, on the one hand, and on the cycling of organic matter in the sea, on the other. Such gaps include:

1. What is the fate of the terrestrial organic matter reaching the sea? How much of which fraction is being oxidized in the water column and surficial sediments? Which part of the terrestrial organic matter is being preserved in modern marine sediments? According to estimates based on chemical characterization of the labile fractions, only part of the river-derived POM and DOM gets oxidized (Ittekkot 1988).
2. The potential for oxidation of the terrestrial organic matter in the sea suggests that river-influenced coastal oceans may be a heterotrophic environment thriving on nutrients released from riverine organic matter. There are reports of coastal embayments being heterotrophic in character. However, uncertainty in the primary production data has made a reliable quantification of these processes difficult.
3. It is possible that the present values for marine organic carbon burial are a gross underestimate. Suggested deposition sites for carbon burial are the slope environments and deep-sea fans, where 50% or more of the organic carbon in surficial sediments of the sea are located.
4. In order to quantify accurately the fluxes from various sources in the midwater and to understand the processes controlling these fluxes, the use of specific classes of organic geochemical compounds is needed. Included in these classes should be compounds that are indicators of both marine and terrestrial sources.

TRACERS OF BIOGENIC SOURCES AND PALEOPRODUCTIVITY

Biomarker Molecules

By biomarker molecules we mean organic compounds with unique chemical structures specific enough to permit them to be related to particular source organisms. Generally, in the geologic perspective, this is taken to mean environmentally resistant compounds which may persist into the geologic record. However, within the context of the modern ocean, one can also consider readily degraded, or labile, compounds whose presence can provide information on the freshness or extent of alteration of the organic material.

Biomarkers thus provide information of source organisms and their biomass, and also the extent of various processes such as zooplankton grazing, microbial alteration, and photochemical degradation.

There is an enormous scope for the use of fairly resistant biomarkers for assessing the production of different groups of organisms, such as dinoflagellates, cyanobacteria, etc. They are particularly important as markers of those primary producers that do not have opal or calcite tests or shells that are preserved in the fossil record. We need to look extensively for the ultimate biological sources of compounds found in marine sediments and to understand the microbial processes operative in the water column and surface sediments in altering these biomarkers. There is a great deal of "ground truthing" in the modern ocean that is required before full use of biomarkers as indexes of paleoproduction can be realized.

Examples of some common, relatively resistant, polymeric organic materials are presented in Table 1, with regard to provenance or production. It is apparent that four biopolymers should provide a means of tracking terrestrial inputs, since they are relatively difficult to degrade and have been thought to have relatively little interaction with other substances. However, no techniques currently exist for determining sporopollenin. Enzymatic, oxidative, and hydrolysis techniques are available for the remaining three. The most studied is lignin (Hedges et al. 1984), but already it is apparent that substantially degraded or incorporated lignin may be the norm in the marine environment and, hence, estimation of terrestrial input must try to take this into account.

TABLE 1 Examples of polymer organic material and their source.

	Terrestrial (land)	Marine
Lignin[b]	vascular tissues	absent
Cutin[b]	leaf cuticles[a]	absent (a little?)
Suberin[b]	root cuticles	absent (a little?)
Sporopollenin[b]	pollen grains[a]	absent
Humic*	debris	debris (DOM)

[a] Such polymers are especially subject to aeolian transport, as minute fragments of leaf tissue and as pollen grains. All organic matter, including polymers, are likely to be aeolian transported to some extent, for example, as a result of incorporation in/on clay minerals or soils.

[b] All of these polymers are known to be resistant to biodegradation, but certain organisms do have effective enzymes. An approximate order of degradability might be cutin>suberin>lignin>sporopollenin. Photooxidation must also be important as part of the breakdown process.

* Humic acids are not biopolymers. They are formed by heteropolymerization in the aquatic environment. There is evidence that they become partially degraded in photic zones and then repolymerize, but with other molecular fragments. Such breakdown and polymerization processes may generate both partially degraded and source-heterogenous humic acid.

Additionally, microbial activity contributes an organic fraction comprised of low molecular weight compounds contributing polymeric material to marine organic matter. The contribution of this fraction of organic matter is largely unknown, but may be considerable.

For the biomarker compounds themselves, we consider the nonpolymers with a molecular size less than 1000 Daltons, which may be extracted with organic solvents. A few examples are given in Table 2, but the field is large and rapidly developing, mainly as an outgrowth of the organic geochemical approach.

A number of questions arise in the potential use of the approaches presented in Tables 1 and 2:

TABLE 2 Examples of biomarkers.

	Terrestrial (land)	Marine
Epicuticular Waxes[a]		
Alkanes, alkanols, alkanoic acids, etc., A-ring functionalized triterpenoids	ubiquitous (higher plants)	identical homologues virtually absent; same classes of compounds are present but generally much lower chain length.
Steroids[b]		
B-sitosterol	abundant	less abundant?
Dinosterol and Dinostanol	absent (except lacustrine)	abundant dinoflag. (bloom-related)
Diatomsterol	absent (except lacustrine)	abundant in diatoms
Diols		
Alkan- a, w-diols	uncommon	abundant cyanobacteria
Alkenones/Alkenoates[b]	absent (except lacustrine)	abundant prymnesiophytes (water temperature ind.)
Loliolides		
Oxidative degradation of carotenoids	probably small riverine input	measure of input of carotenoids

[a] Aeolian transport has been demonstrated to be an important contributing pathway from continents to the oceans for these epicuticular waxes. The particles may conceivably be minute fragments of epicuticular waxy crystallites, perhaps slightly larger portions of cuticle, or adsorbed in/on clays.

[b] Degradability for gross fractions such as "steroids" has been inferred from water column and sediment trap data. Additionally, feeding experiments, mainly with zooplankton (including coprophagy), have demonstrated the high resistance to degradation or removal by the organisms of dinosterol, dinostanol, and the alkenones. Such experiments provide significant support for making use of these compounds as "conservative markers" (Bristol and Plymouth work).

1. Can we link the molecular approach (biomarkers and biopolymers) to other measures of the fate of organic carbon in the water column? Microscopic recognition and quantitation of particles is clearly desirable, as are also data for the mineral and elemental compositions. However, the most opportune are perhaps the new capabilities for microscale stable isotope ($\delta^{13}C$) and ^{14}C studies. For $\delta^{13}C$, commercial instruments are nearing release for capillary GC-MS determination at the ng level for individual compounds. This will allow a vast burgeoning of ecological studies involving paths and fluxes of carbon specific to particular species of organisms, carbon sources and sites. For ^{14}C, wider availability of new AMS techniques at the mg sample size present outstanding opportunities for less selective analyses (e.g., of organic fractions), which will assist our understanding of the longer time scales for organic matter in the oceans.
2. What component part do small biomarker molecules make up as intact or fragmentary incorporates of higher molecular weight DOM and POM? This is a little studied area which awaits detailed effort using new methods for selective degradation and estimation. This seems especially important in view of the growing evidence that POM in the ocean is typically ingested, microbially attacked, fragmented/packaged, etc., many times before final burial. Surely this means that this heteropolymeric debris must contain trapped, bound, and cross-linked intact and altered biomarker molecules? Assessment of such previously unrecognized compounds would enable a better quantitation of the contributions from terrestrial and marine sources. In this regard, we know from laboratory experiments with single compounds and with model polymers that steric hindrance and functional group stereochemistry have dramatic effects in controlling the accessibility of a molecular framework to attack by reagent ions, radical molecules, and, in particular, enzymes. Such considerations must be important controls on the lifetimes of biomarkers and the degradability of organic matter in the oceans.
3. Oxic/anoxic conditions (biomarkers) or other indicators of the history of exposure of organic matter to anoxic conditions are particularly desirable. How do we explain the apparent anomaly of the persistence of small quantities of biomarkers in sediments deposited under oxic conditions? Are they sited within particles and are hence inaccessible to bacterial attack?

Traditional Tracers of Paleoproduction

What properties of sediments at the seafloor are most appropriate to document and quantify variations in export productivity, and how can this be interpreted in the geologic record? Traditionally, a number of variables in the magnitude and composition of the sedimentary flux have been used to answer this question: variables that were not discussed extensively in the working group which, however, will continue to be important in deciphering the productivity signal in the future. These are listed in Table 3 and briefly discussed in the text.

TABLE 3 Tracers of ocean productivity.

1. Organic soft tissue (flux rates)
2. Biogenic silica: diatoms and radiolaria
3. Planktonic carbonates: foraminifers and coccoliths
4. The "rain ratio" between bulk organic carbon and $CaCO_3$
5. C/N ratios and $\delta^{13}C$
6. Trace metals such as Cd, Ba, Cu, and inorganic tracers
7. Benthic response markers

Soft tissues or POM. Since the landmark empirical approach of Suess (1980) there have been a number of attempts to quantify further the flux rates of organic tissue or POM as related to water depth (Berger et al. 1987). There is an ever rapidly increasing wealth of sediment trap data which, however, has hardly been evaluated or even documented in the literature, which could be used to improve these preliminary attempts on flux versus water depth ratios for various basins of the oceans. Important sources of bias, such as the production in the inner ocean and lateral advection of organic matter from shelf and slope regions, are difficult to quantify. Likewise, the role of the addition of lithogenic particles, as a factor of accelerating the flux of organic tissues to the seafloor and potentially improving its chances of preservation, needs to be evaluated.

Biogenic silica. As shown by time series sediment trap data, flux rates of diatoms provide on a temporal basis a close approximation of productivity (Takahashi 1986). However, it is difficult to quantify the bias caused by opal dissolution, in particular on the species composition which occasionally suggests very clear signals of high productivity such as a dominance of *Ethmodiscus rex* in Quaternary diatom oozes along the equator. Much less is known about the relationship between productivity and species composition of radiolarians. Famous events of the fossil record, such as the radiolarian oozes of the Eocene, still wait to be deciphered in terms of their relationship to ocean productivity.

Plankton carbonate. The $CaCO_3$ fluxes of both coccoliths and planktonic foraminifera are subject to (*a*) rapid biogenic crushing in the surface and intermediate waters due to the grazing activity of crustacean zooplankters and (*b*) dissolution below the lysocline in the deep ocean. Whereas dissolution processes have been largely elucidated, those involving biogenic crushing are poorly known. Hence, it is difficult to quantify which proportion of planktonic foraminifera shells are preserved as full tests or are fragmented or crushed to a fine-grained ooze, which does not allow their identification as foraminifers any more. Furthermore, the original species composition can greatly differ from that observed in surficial sediments, a difference which has great implications for the indicator species and any transfer function used to estimate paleoproductivity (Mix, this volume). Variations in test or shell size may have an important effect on the results.

"Rain ratio" of planktonic organic carbon and $CaCO_3$. Based upon recent trap data, Berger and Keir (1984) suggested that the flux ratio ("rain ratio") between carbon fixed in organic tissues and carbon fixed in $CaCO_3$ increases with increasing export production. They showed that this shift in the rain ratio has great implications for the CO_2 budgets of the surface and deep ocean. Currently, no test of variations in the rain ratio in the sedimentary record has been reported. This lack of data is partly due to the bias of planktonic $CaCO_3$ accumulation rates by dissolution on the seafloor. However, slope sediments obtained from above the $CaCO_3$ saturation level, and carefully investigated for dissolution effects, may provide an interesting record of long-term changes in carbon fluxes from the surface to the deep ocean. The development of other "rain ratios," such as organic carbon to opal ratios or the use of rain ratios of specific organisms, may also prove useful.

C/N ratios and $\delta^{13}C$. The C/N ratio and the $\delta^{13}C$ of organic matter have been used to differentiate between inputs from terrestrial and marine sources. Usually C/N ratios of around 6–8 are taken to represent organic matter derived from marine sources, whereas higher ratios are indicative of a terrestrial source. A major problem associated with this approach is the uncertainty concerning the amount of ammonium ions fixed into clay minerals (Müller 1977), which are measured along with organic nitrogen in conventional C/N analyzers. With the $\delta^{13}C$ approach, values of -18 to $-21‰$ are taken to indicate a marine planktonic source, and lighter values of $-25‰$ and less to indicate a land source. Here again caution is called for in interpreting the $\delta^{13}C$ values in terms of source inputs. This is due to the fact that $\delta^{13}C$ values are found to exhibit wide variations within materials derived from the same source. For example, values of marine plankton tend to vary widely depending on factors such as the temperature and biochemical composition of the material examined. Knowledge of the biochemical nature of the sample and, in particular, separation of the sample into chemical components may aid considerably in interpreting the $\delta^{13}C$ signal (Prahl and Muehlhausen, this volume).

Trace metals and inorganic chemical tracers. Cadmium has been shown to provide a close proxy of the nutrient content of the ocean (Boyle et al. 1976; Bruland 1980). Boyle (1986) showed that past changes in the nutrient content of the ocean deep waters are well documented in the Cd/Ca ratio of benthic foraminifera and may help to reconstruct the carbon-nutrient balance in the ocean during past climatic regimes.

Barium is a nutrient-like trace metal that has been extensively mapped in the world's oceans. Recent studies of the water column abundances of particulate barium (Bishop, this volume) showed a striking correlation with estimates of annual mean primary production. This correlation does not appear to result from the active fixation of barium as barite by phytoplankton during photosynthesis, but instead from inorganic precipitation of barite within microenvironments (broken diatoms, fecal material, etc.) in the presence of decaying organic matter and silica surfaces (Bishop 1988). This explains the apparent correlation with primary production and suggests its potential utility as a sedimentary indicator of export flux. The fact that Ba appears to get its nutrient-like behavior not through the process of active uptake by phytoplankton reopens questions of mechanisms giving nutrient-like behavior to other trace metals. Copper, uranium, and other heavy metals may form additional proxies for export productivity in the ocean (Brongersma-Sanders 1983). The proximity or distance of these deposits to submarine vents may, of course, form a bias for the deduction of the productivity-related organic particle flux.

CONCLUSIONS

The particulate flux of biogenic material is composed of both living and nonliving particles with a wide range of sizes. There is a broad range of settling velocities for biogenic particles depending on their size, shape, chemical composition, and density. It appears that larger particles, such as fecal pellets and micro- and/or macroaggregates of detritus and microorganisms, comprise the bulk of the vertical particulate flux. Knowledge of the dynamics of production, consumption, and repackaging of living and detrital particles and organic aggregates within the oceans' interior is primitive, yet models of biogenic flux will have to deal with this problem.

Dissolved organic compounds contribute to the export flux of carbon from the euphotic zone but are not measured as part of the particulate flux. The same mixing mechanisms which bring nutrients to the surface also remove dissolved organic compounds. It is not well known at the present time just how big the dissolved flux might be, relative to the particulate flux, but it is easy to show how the fluxes might be comparable.

Microbiologists maintain that DOC provides the main substrate for bacterial growth in the ocean. Thus, DOC mixed or otherwise delivered to the oceans' interior can provide the substrate for bacterial respiration and growth, or production of particulate organic material.

An additional process that appears to be of increasing importance is the role of actively migrating zooplankton in contributing to the vertical flux of both particulate and dissolved organic matter. Active migrators can diurnally migrate over depths of 1000 m or more. This can result in the rapid and active transfer of material to depth: material that is missed by sediment traps and large volume *in situ* pumps, which are designed to estimate fluxes of passively sinking particulate material.

The above-mentioned processes have to be better understood in order to use assemblages of different kinds of organisms and organic and inorganic tracers to reconstruct past productivities of the ocean. Besides paleomass flux rates, we need information on the organisms contributing to this flux and the mode of sedimentation (episodic versus more continuous flux). A question which remains open is how much of the organic matter, accumulated on the continental slope and in the deep sea, is supplied from the highly productive continental shelf regions and the continents themselves. Today the species composition of planktonic organisms, organic carbon accumulation rates, C/N ratios, stable isotope data, and chemical tracers are used to distinguish the source of organic matter and to calculate paleoproductivity. Biomarkers appear to be extremely promising tracers that will provide more information on the source of organic matter and better define the extent of alteration or modification. A great deal of ground truthing in the modern ocean is required prior to, and in concert with, using these biomarkers as paleoproductivity indexes.

REFERENCES

Alldredge, A.L., and Hartwig, E.O., eds. 1986. Aggregate Dynamics in the Sea. Workshop Report, Office of Naval Research, Amer. Institute of Biological Sciences.

Angel, M.V. 1986. Vertical migrations in the oceanic realm: possible causes and probable effects. In: Migration: Mechanisms and Adaptive Significance, ed. M.A. Rankin. *Marine Sci.* **24** (Suppl.): 45–70.

Berger, W.H.; Fisher, K.; Lai, C.; and Wu, G. 1987. Ocean carbon flux: global maps of primary production and export production. In: Biogeochemical Cycling and Fluxes between the Deep Euphotic Zone and Other Oceanic Realms, ed. C. Agegian. NOAA Symposium Series for Undersea Res., vol. 3 (preprint).

Berger, W.H., and Keir, R. 1984. Glacial-Holocene changes in atmospheric CO_2 and the deep-sea record. In: Climate Processes and Climate Sensitivity, eds. J.E. Hansen and T. Takahashi. *Geophys. Monog.* **29**: 337–351. Washington, D.C.: Amer. Geophys. Union.

Billet, D.S.M.; Lampitt, R.; Rice, A.L.; and Mantoura, R.F.C. 1983. Seasonal sedimentation of phytoplankton to the deep-sea benthos. *Nature* **302**: 520–522.

Bishop, J.K.B. 1988. The barite-opal-organic carbon association in oceanic particulate matter. *Nature* **233**: 241–243.

Bishop, J.K.B.; Stepien, J.C.; and Wiebe, P.H. 1987. Particulate matter distributions, chemistry, and flux in the Panama Basin: response to environmental forcing. *Prog. Ocean.* **17**: 1–59.

Boyle, E.A. 1986. Paired carbon isotope and cadmium data from benthic foraminifera: implications for changes in oceanic phosphorus, oceanic circulation and atmospheric carbon dioxide. *Geochim. Cosmochim. Acta* **50**: 265–276.

Boyle, E.A.; Sclater, F.R.; and Edmond, J.M. 1976. On the marine geochemistry of cadmium. *Nature* **263**: 42–44.

Brongersma-Sanders, H. 1983. Unconsolidated phosphorites, high barium and diatom abundances in some Nambian shelf sediments. In: Coastal Upwelling, Its Sedimentary Record, Part A, eds. E. Suess and J. Thiede, pp. 421–437. New York: Plenum.

Bruland, K.W. 1980. Oceanographic distribution of cadmium, zinc, nickel and copper in the North Pacific. *Earth Planet. Sci. Lett.* **47**: 176–198.

Cho, B.C., and Azam, F. 1988. Major role of bacteria in biogeochemical fluxes in the ocean's interior. *Nature* **332**: 441–443.

Conte, M.H., and Bishop, J.K.B. 1988. Active vertical transport of fatty acids in an oligotrophic Gulf-stream warm core ring. *Science*, in press.

Deuser, W.G. 1986. Seasonal and interannual variations in deep-water particle fluxes in the Sargasso Sea and their relation to surface hydrography. *Deep-Sea Res.* **33**: 225–246.

Deuser, W.G. 1987. Variability and hydrography and particle flux: transient and long-term relationships. In: Particle Flux in the Ocean, eds. E.T. Degens, E. Izdal, and S. Honjo. *Mitt. Geol. Paläont. Inst. Univ. Hamburg* **62**: 179–193.

Dymond, J., and Collier, R. 1988. Biogenic particle fluxes in the equatorial Pacific: evidence for both high and low productivity during the 1982–83 El Niño. *Geochim. Cosmochim. Acta*, in press.

Feldman, G.; Esaias, W.; McClain, C.; Evans, R.; Brown, J.; and Carle, M. 1986. Production of a global ocean color data set. *EOS* **68**: 1722.

Hedges, J.I.; Turin, H.J.; and Ertel, J.R. 1984. Sources and disturbances of sedimentary organic matter in the Columbia River drainage basin, Washington and Oregon. *Limnol. Ocean.* **29**: 35–46.

Honjo, S. 1984. Study of ocean fluxes in time and space by bottom-tethered sediment trap arrays: a recommendation. In: Global Ocean Flux Study, Proceedings of a Workshop, pp. 305–324. Washington, D.C.: National Acad. Press.

Ittekkot, V. 1988. Global trends in the nature of organic matter in river suspensions. *Nature* **332**: 436–438.

Karl, D.M.; Knauer, G.A.; and Martin, J.H. 1988. Downward flux of particulate organic matter in the ocean: a particle decomposition paradox. *Nature* **332**: 441–443.

Lampitt, R.S. 1985. Evidence for the seasonal deposition of detritus to the deep-sea floor (Porcupine Bight, N.E. Atlantic) and its subsequent resuspension. *Deep-Sea Res.* **32**: 885–897.

Martin, J.H.; Knauer, G.A.; Karl, D.M.; and Broenkow, W.W. 1987. VERTEX: carbon cycling in the Northeast Pacific. *Deep-Sea Res.* **34**: 267–285.

McCave, I.N. 1984. Size spectra and aggregation of suspended particles in the ocean. *Deep-Sea Res.* **31**: 329–352.

McClain, C.R.; Esaias, W.; Feldman, G.; Meeson, B.; and Evans, R. 1987. A review of the Nimbus-7/coastal zone color scanner processing and archiving activity. *EOS* **68**: 1322.

Müller, P.J. 1977. C/N ratios in Pacific deep-sea sediments: effect of inorganic ammonium and organic nitrogen compounds sorbed in clays. *Geochim. Cosmochim. Acta* **41**: 765–776.

Rice, A.L.; Billet, D.S.M.; Fry, J.; John, A.W.G.; Lampitt, R.S.; Mantoura, R.F.C.; and Morris, R.J. 1986. Seasonal deposition of phytodetritus to the deep-sea floor. *Proc. Roy. Soc. Edinb.* **88B**: 265–279.

Silver, M.W.; Gowing, M.M.; and Davoll, P.J. 1986. The association of photosynthetic picoplankton and ultraplankton with pelagic detritus through the water column (0–2000 m). In: Photosynthetic Plankton, eds. T. Platt and W.K.W. Li. *Can. Bull. Fish Aquat. Sci.* **214**: 583.

Spencer, D.W. 1984. Aluminum concentrations and fluxes in the ocean. In: Global Ocean Flux Study, Proceedings of a Workshop. Washington, D.C.: National Acad. Press.

Suess, E. 1980. Particulate organic carbon flux in the oceans: surface productivity and oxygen utilization. *Nature* **288**: 260–263.

Sugimura, Y., and Suzuki, Y. 1988. A high temperature catalytic oxidation method of non-volatile dissolved organic carbon in seawater by direct injection of liquid samples. *Marine Chem.*, in press.

Suzuki, Y.; Sugimura, Y.; and Itoh, T. 1985. A catalytical oxidation method for the determination of total nitrogen dissolved in seawater. *Marine Chem.* **16**: 83–97.

Takahashi, K. 1986. Seasonal fluxes of pelagic diatoms in the subarctic Pacific, 1982–83. *Deep-Sea Res.* **33**: 1225–1251.

Vezina, A.F., and Platt, T. 1988. Food web dynamics in the ocean. I. Best estimates of flow networks using inverse methods. *Mar. Ecol. Prog. Ser.* **42**: 269–287.

von Bodungen, B.; Fischer, G.; Nothig, E.-A.; and Wefer, G. 1987. Sedimentation of krill faeces during spring development in Branfield Strait, Antarctica. In: Particle Flux in the Ocean, eds. E.T. Degens, E.I. Izdar, and S. Honjo. *SCOPE/UNEP Sonderband Heft* **62**: 243–257.

Wakeham, S.G., and Canuel, E.A. 1988. Organic geochemistry of particulate matter in the eastern tropical Pacific Ocean: implications for particle dynamics. *J. Marine Res.* **46**: 183–213.

Whitfield, M., and Turner, D.R. 1987. In: Aquatic Surface Chemistry, ed. W. Stumm, pp. 457–494. New York: Wiley.

Productivity of the Ocean: Present and Past
eds. W.H. Berger, V.S. Smetacek and G. Wefer, pp. 217–233
John Wiley & Sons Limited

Control of Benthic Fluxes by Particulate Supply

C.E. Reimers

Scripps Institution of Oceanography, A-015
University of California, San Diego
La Jolla, California 92093, U.S.A.

Abstract. More than 80% of the particulate supply to the benthos in the deep sea consists of organic, opaline, and calcareous materials. Since less than 10% of these materials are buried, fluxes of dissolved substances from the benthos back to the water column must be strongly dependent on particulate supply.

Two simple kinetic models for the degradation/dissolution of biogenic particles illustrate the likelihood that benthic fluxes are dominated by the most recently settled and therefore most reactive detritus. Observations indicate that the majority of the organic material reaching the deep seafloor (and possibly the other biogenic fractions as well) degrades over time scales of a few months very near the sediment-water interface. The organic matter buried several centimeters down in the sediment mixed layer is a small, slowly degrading residue of the incoming flux. If the degradation rate of a significant portion of this residual material is the only part of the total carbon oxidation which is redox dependent, bottom water oxygen concentrations can significantly affect carbon burial rates with little detectable impact on the overall carbon reflux.

INTRODUCTION

The deep seafloor, the interface between water and sediment, is more active biologically than either zone it joins. Fluxes of biogenic particulate material through the bathypelagic water column may change dramatically on seasonal time scales (Deuser et al. 1981; Honjo 1984), but the rapidly settling organic, opaline, and calcareous materials, which form these fluxes, are regenerated chiefly after they reach the sediment-water interface (Noriki and Tsunogai 1986; Jahnke and Jackson 1987). These findings suggest that in the deep ocean, the benthic biological community responds to particulate supply, and that benthic fluxes are controlled by particulate supply.

The methods that have been used most often to quantify benthic regeneration rates in the deep sea include comparisons between rates of accumulation in deep sediment traps and in underlying sediments, direct measurements over time of the utilization (or release) of reaction substrates (or end products), and kinetic models of pore water or solid phase depth-concentration profiles (e.g., Dymond 1984; Berelson et al. 1987). Throughout this paper a corollary of the "particulate supply rule" will be stressed, which, simply stated, is that the most recent inputs to the seafloor exert the greatest control on reflux rates. An examination of two kinetic models helps illustrate its meaning.

Consider first Jørgensen's (1978) or Westrich and Berner's (1984) first-order kinetic model to describe organic matter decomposition near the surface of marine sediments. According to the model, organic matter reaching the seafloor is composed of several groups of compounds (G_i s) that differ in their reactivity with respect to decomposition. The overall decomposition rate is expressed as

$$-dG_T/dt = \sum_{1}^{n} k_i G_i \,, \tag{1}$$

where i is the fraction number ranked in order of decreasing reactivity (total = n), G_i is the concentration of carbon per unit mass of total solids in a pool of organic compounds with similar decomposition kinetics, G_T is the total decomposable organic carbon, and k_i is the first-order decomposition rate constant. This model predicts that recently deposited detritus (if it carries a "significant" supply of labile organic matter) will tend to dominate the overall benthic organic matter degradation rate.

Similar models can be made for opaline silica and $CaCO_3$ dissolution. Because of the variety in their form, size, and surface and structural chemistries, siliceous and calcareous microfossils may be more or less "reactive" (Schink et al. 1975; Berner 1980). Highly reactive forms are rarely preserved in sediment assemblages, but their presence in collections from deep sediment traps indicates that they do reach the sediment-water interface before dissolving (e.g., Pisias et al. 1986). Therefore, the total seafloor dissolution rate is best represented by a rate expression dependent on the concentration of the different reactive material(s) and the degree of undersaturation. For silica or $CaCO_3$ such an expression may be written as

$$-dB_T/dt = \sum_{1}^{n} k_i B_i (1-\Omega)_i^m \,, \tag{2}$$

where B_i is the concentration of each form of reactive solid per unit mass solid, k_i is a rate constant which contains a surface area per unit volume term, m is an exponent ≥ 1,[1] and $(1 - \Omega)_i$ is the degree of undersaturation for the fraction under consideration.

It is implicit in Eqs. 1 and 2 that the rate constants for decay or dissolution are a function only of the solid reactant. This assumption, not universally accepted, has led to debate over the importance of a variety of environmental conditions (e.g., boundary layer flow, bioturbation intensity, and bottom water oxygen content) for controlling certain sediment/seawater reflux rates. Each of these factors will be discussed in this review by emphasizing deep-sea examples and in many instances the recycling of organic matter. Probably the most persistent "environmental factor" controversy is whether or not overall rates of organic matter oxidation at the seafloor are dependent on the oxygen content of water overlying the sediment. This is important because cycles of alternating high and low organic carbon concentrations are common in marine sediment sections and have been interpreted as records of both paleoproduction (or paleoflux) and bottom water oxygen content. Arguments for the latter state that anaerobic decomposition is ineffective, that a decrease in bottom water oxygen concentration decreases the depth to which oxygen diffuses into sediments, and that the consequence of a shallower redox boundary is a higher probability that organic carbon will escape oxidation (by burial in anoxic sediments) (Emerson 1985). Clearly for this mechanism to operate, there must be some organic materials which degrade much slower in the absence, than in the presence, of oxygen.

THE REACTIVITY OF THE PARTICULATE FLUX

Most marine scientists do not think of the material reaching the deep seafloor as being highly reactive. This is because this material must settle through the water column for periods of weeks to months, during which time it is assumed to loose its most labile constituents. However, if settling phytodetritus is well packaged (Schrader 1971) and reaches the seafloor unscathed, and if reactions are catalyzed at the sediment-water interface by biological activity, a large part of the particulate flux to the seafloor could decompose quickly after arrival. This would imply that (*a*) degradation reactions should be concentrated in the uppermost portion of the sediment

[1] Empirical results of laboratory dissolution rate studies with biogenic calcite, aragonite, and opal suggest that m is 4–4.5 for carbonates but is equal to 1 for silica (Keir 1980; Hurd 1972). The higher order dependence for calcium carbonate minerals has been explained as resulting from surface reaction-controlled dissolution decelerated at low degrees of undersaturation by inhibitor ions (primarily phosphate), which slow the detachment of Ca^{2+} and CO_3^{2-} from exposed points or "kinks" on the crystal surface (Berner and Morse 1974). Opaline silica is noncrystalline and not an ionic solid. Thus, although its dissolution is also perceived as surface reaction-limited and inhibited by adsorbed cations (Hurd 1973), the inhibitory effect of cation surface attachment is not overcome at high degrees of undersaturation. The overall result is a first-order dependence on the degree of undersaturation.

column, (*b*) benthic fluxes should respond to input changes associated with seasonal productivity cycles or rare climate events, and (*c*) material buried several centimeters down in the sediment mixed layer should represent a small residue of the incoming flux. Observations which support these conclusions are summarized below.

The first examples are time-lapse photographic and current surveys in the north equatorial Pacific at depths > 4000 m by Gardner et al. (1984) and in the northeast Atlantic at depths up to 4000 m by Lampitt (1985). They show that detritus from surface phytoplankton blooms reaches the seafloor sporadically. The phytodetritus consists of "fluffy," rapidly sinking aggregates which, according to Lampitt (1985), disappear from the sediment-water interface in less than two months under the combined influence of bottom currents and the benthic biological community. Gardner et al. (1984) note that phytodetritus, as well as fecal pellets, decompose without sediment resuspension and, in many instances, without biological disturbance in less than a month. Animal trails also are short-lived.

A second data set which suggests that phytodetritus degrades rapidly after its arrival on the seafloor includes the time series studies of sediment community oxygen utilization by Smith and Baldwin (1984) and Smith (1987). Smith has shown that at five deep stations on a transect across the eastern North Pacific the benthic oxygen flux varies by factors of 2–4 with a yearly periodicity. The data are unfortunately too spotty within given years to demonstrate completely shorter-term seasonal trends. However, at one station at 3800 m depth, off the Patton Escarpment, an observed doubling of the oxygen flux between February and June of 1981 strongly suggests the arrival and rapid decay of reactive organic matter from a spring bloom. The question which remains is: Could the labile component always be the major fraction of the deep-sea particulate flux when it is apparent that the bulk of the sediment mixed layer is composed of much less reactive materials?

Intuitively the answer is yes. In simple terms, what degrades away rapidly is not there to contribute to the average age and composition of the mixed layer or sediment below the mixed layer. To appreciate this point, consider Emerson et al.'s (1987) recently determined depth profiles of organic ^{14}C age through and below the sediment mixed layer, at three deep-water locations. The average sedimentation rates at the three sites were between 0.9 and 2.0 cm ky^{-1}. Emerson et al. found that the mean, mixed layer, organic ^{14}C ages were between 2,000 and 4,700 y B.P., respectively. Emerson et al. substituted these ages into an expression derived from four steady state mass balance equations, each for either ^{14}C or ^{12}C in two organic carbon fractions (one degradable and one nondegradable), to estimate an average first order rate constant, $\bar{k}$ for the degradable fraction (Emerson et al. 1987). They concluded the bulk of the carbon in the sediment mixed layer has a residence time ($1/\bar{k}$) < 1000 y. In Appendix 1, I present the same model used by Emerson et al. (1987) but add to it a second, more rapidly degradable, carbon fraction. Simplifying what now becomes six mass balance equations by the assumption that all fractions have an equal initial $^{14}C/^{12}C$ ratio (Emerson et al. 1987 assume this as well), yields

$$\lambda t = 1 - \left[\frac{1 + \frac{R_2}{R_1}\left(\frac{k_1}{k_2+\lambda+s}\right) + \frac{R_3}{R_1}\left(\frac{k_1}{\lambda+s}\right)}{1 + \frac{R_2}{R_1}\left(\frac{k_1}{k_2+s}\right) + \frac{R_3}{R_1}\left(\frac{k_1}{s}\right)} \right], \tag{3}$$

where t (y) is the ^{14}C age, s (y^{-1}) is the sedimentation rate normalized by the mixed layer thickness, λ (y^{-1}) is the radioactive decay constant, and R_1, R_2, and R_3 are the particulate flux rates of the rapidly degrading, slowly degrading, and nondegradable carbon fractions, respectively (Appendix 1).

Utilizing Eq. 3, ratios of highly degradable to slowly degradable carbon in the particulate flux reaching the sediment can be predicted assuming variable rate constants, the ^{14}C age, and the size of the refractory carbon fraction. Solutions are presented in Fig. 1 for conditions comparable to the Emerson et al. (1987) data. The trends in Fig. 1 illustrate that the deep-sea flux of rapidly degrading organic matter could be more than 100 times greater than the flux of slowly degrading material.

The solutions presented in Fig. 1 assume that the highly degradable carbon fraction (C_1) is mixed into the top 10 cm of sediment. This may never occur considering what is known of deep-sea sediment bioturbation rates. Instead, as noted earlier, the highly degradable materials probably are concentrated in the uppermost sediment. Under such circumstances, my model reverts to one with only two components (as Emerson et al. (1987) presented), except that the flux into the mixed layer is identified not as the flux to the seafloor, rather instead as a small fraction of that flux which passes through the surface sediment zone. Lines of evidence indicating that this last representation is appropriate are: (*a*) the 20–50% younger ^{14}C ages measured in the top few centimeters of all the cores analyzed by Emerson et al. (1987), and (*b*) the 100-fold difference between the total organic carbon inventory in the mixed layer of the abyssal ocean (17–33 Gt) (Emerson et al. 1987) and the carbon flux to abyssal depths (~2,500 Gt C) (Martin et al. 1987)[2].

[2] From Martin et al. (1987):
Open ocean new production, F_{100} = 6 Gt y^{-1};
carbon flux to an average seafloor depth (z=4000 m) = $F_{100}(z/100)^{-.858}$;
mixed layer age = 10,000 y.
Therefore, carbon flux to the seafloor over 10,000 y = 2,500 Gt.

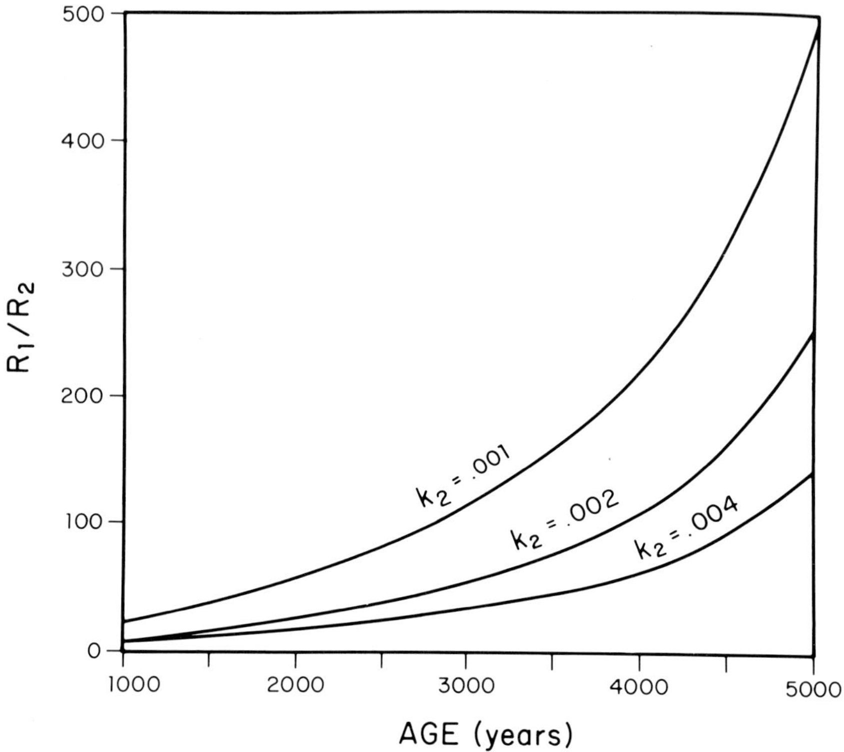

Fig. 1—The variation in the flux ratio of rapidly degrading organic carbon, R_1, to slowly degrading carbon, R_2, as a function of the mean ^{14}C age of the organic carbon in the sediment mixed layer. The conditions assumed were $k_1 = 10\ y^{-1}$, $R_3 = .001(R_1 + R_2)$, and $s = 10^{-4}\ y^{-1}$ (see Appendix 1). The three lines represent solutions for different values of k_2. They suggest, for example, a ^{14}C mixed layer age of 3000 years allows flux ratios that are between approximately 30 and 110.

PARTICLE FLUX-BENTHIC FLUX COMPARISONS

In the previous section it was illustrated that the majority of the particulate organic carbon flux to the seafloor probably degrades so rapidly that it is not buried very far or very long within surface sediments. Comparisons of sediment burial rates of opal and $CaCO_3$ (at locations below the lysocline) indicate that often >90% of the particulate fluxes of these materials are also recycled (Dymond 1984; Noriki and Tsunogai 1986). Are the residence times of these materials at the seafloor as short as the bulk of the organic matter flux? Also, can it be determined whether most of the dissolution/decomposition actually occurs in near-surface sediments or within the water column following resuspension (Dymond 1984)?

By Eq. 2, it is clear that the fate of opal and $CaCO_3$ materials on the deep-sea floor depends both on the reactivities of these materials and the degree to which bottom waters and pore waters are undersaturated with respect to each mineral. Little is known about the *in situ* dissolution kinetics of the different biogenic skeletal materials that make it to the deep sea, except that aragonite dissolution is favored over calcite and some weakly silicified species of diatoms, radiolarians, and silicoflagellates are preferentially dissolved over more robust forms (Pisias et al. 1986). The ideal test for a tight functional dependence between benthic dissolution rates and biogenic opal and $CaCO_3$ fluxes would be to relate a continuous time series of well-characterized, particle flux measurements to repeated benthic flux measurements. No such experiment has yet been reported. In Fig. 2 a time series of biogenic, particulate flux (sediment trap) measurements from MANOP Site C, a location below the lysocline in the equatorial Pacific, are related to benthic reflux rates derived from benthic chamber measurements made two years following the sediment trap deployments. The benthic reflux rates (which are plotted to the right of the particulate flux series) appear to reflect the average of the particulate inputs. However, since they were not measured contemporaneously, it can not be determined whether this is because the seafloor regeneration reactions at this site reflect time-averaged particulate supplies or because the benthic flux measurements were made at a time when the supply was average.

Two observations relevant to the above problem arise from pore water data. First, at Site C the benthic chamber fluxes compare to within 20–40% of diffusive fluxes calculated from pore water profiles for oxygen, alkalinity, and silica (Berelson and Hammond, personal communication), indicating the majority of the benthic regeneration of organic matter, $CaCO_3$, and opal occurs in communication with pore waters. Second, pore water gradients are steepest just below the sediment surface (Jahnke et al. 1982). This indicates maximum reaction rates there, which could be in response to the most recent particulate fluxes.

ENVIRONMENTAL FACTORS AFFECTING BENTHIC FLUXES

Benthic fluxes have two preconditions: (*a*) the production or consumption of dissolved solutes by sediment-water reactions, and (*b*) the transport of solutes across the sediment-water interface. The presumption that rates of sediment-water exchange are limited primarily by the nature of the mass flux implies solute transport is not limiting, and environmental factors (other than degree of saturation) have relatively little effect on overall reaction kinetics. For the deep sea, this is reasonable as a general rule. There are specific cases, however, where transport or environmental factors may suppress or enhance sediment-seawater reactions. Some of these cases and the factors at work are discussed individually below.

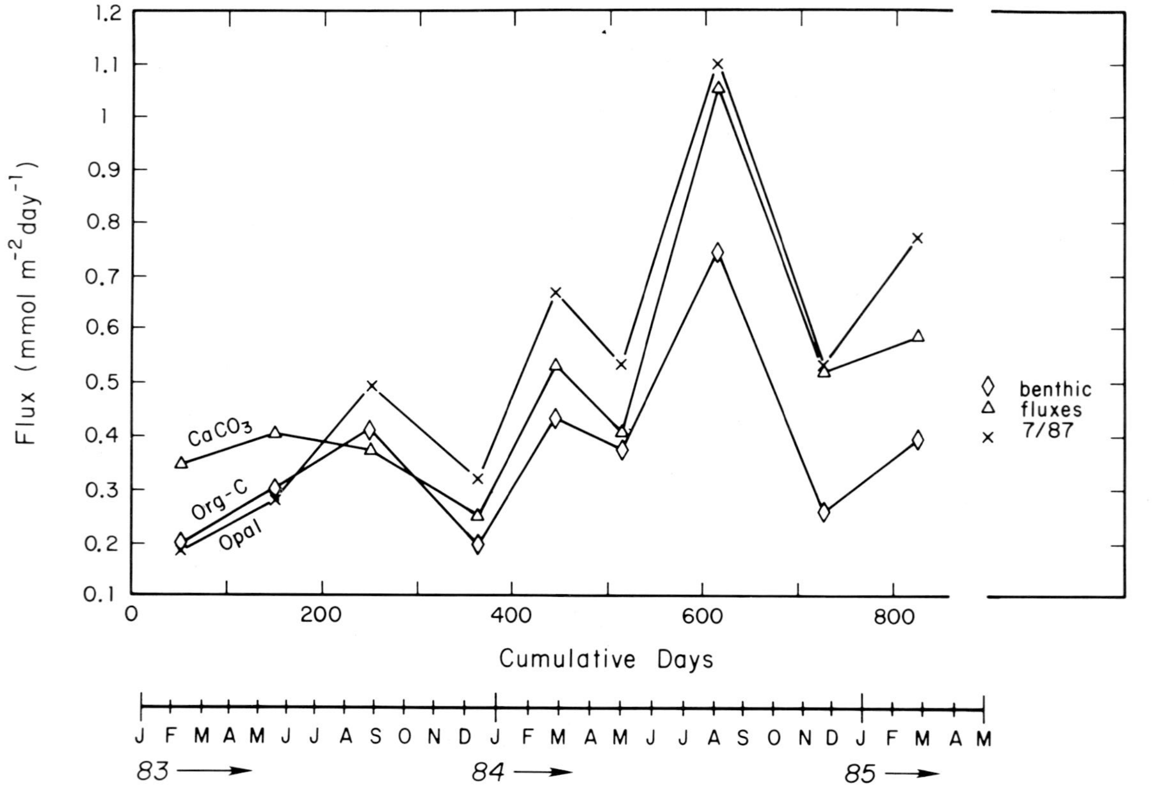

Fig. 2—A comparison between two years of particulate fluxes of organic carbon, opal, and $CaCO_3$ (points connected by solid lines), and benthic fluxes measured two years later (points to the right). All data are from MANOP Site C in the equatorial Pacific. The particulate fluxes were measured with sediment traps and each point represents a collection period of approximately 100 days (Dymond and Collier 1988). The benthic reflux rates were estimated from benthic chamber fluxes of dissolved oxygen, silica, alkalinity, and ΣCO_2 (Berelson and Hammond, personal communication). These benthic flux measurements were made in July 1987.

Boundary Layers at the Sediment-Water Interface

Once particles come to rest on the seafloor, they immediately become components of the sediment-pore water system by virtue of a transition zone from advective to diffusive fluid transport called "the diffusive sublayer." Using a convention established by Santschi et al. (1983) and Boudreau and Guinasso (1982), the equivalent "stagnant film" thickness of the layer, z (cm), is related to a chemical flux, F (g $cm^{-2}sec^{-1}$) by

$$F = \frac{D}{Z}(C_\infty - C_0) , \tag{4}$$

where D (cm^2sec^{-1}) is the molecular diffusion coefficient of the chemical solute in seawater, C_∞(g cm^{-3}) is the bottom water solute concentration, and C_0 (g cm^{-3}) is the solute concentration at the sediment interface. Z is fixed by the hydrodynamics of flow over the bottom and is an inverse function of the bottom stress. At one site on the deep seafloor of the eastern Pacific, Z was calculated as equal to 575 ± 50 μm by an alabaster dissolution model (Santschi et al. 1983).

Although diffusive sublayers are ubiquitous features, they will only limit benthic exchange when the diffusive impedance of the sublayer is greater than the kinetic impedance of reactions within the sediment (Boudreau and Guinasso 1982). That is to say, if reaction rates within the sediment are very fast, C_0 may approach a maximum (e.g., saturation) or minimum (e.g., zero) concentration supportable by both reaction and diffusion. Under these extremes, $D(C_\infty - C_0)$ will be constant and the resulting flux a function of Z (and therefore the bottom stress). The only deep-sea cases in which this situation is suspected are: (*a*) during the transport of metal ions from seawater to the solid surfaces of manganese nodules; (*b*) during oxygen uptake by sediments in oxygen minimum zones, i.e., if C_0 $(O_2) \simeq O$; and possibly (*c*) after initially rapid calcite dissolution, i.e., if $C_\infty \ll C_{saturation}$ (for CO_3^{2-}). Calcite dissolution would not be boundary layer limited, however, if the moles of $CaCO_3$ dissolved in the surface sediment were also less than or equal to the moles of organic carbon oxidized. This is because the CO_2 generated from organic carbon oxidation reacts with excess CO_3^{2-} to form bicarbonate, keeping $C_0(CO_3^{2-})$ below saturation values (Emerson and Bender 1981).

These arguments suggest that boundary layers, in general, do not play a large role in determining deep-sea benthic flux rates.

Bioturbation

To understand the multiple effects bioturbation processes have on deep-sea benthic fluxes, one must separate and compare (*a*) the pore water sensitive and non-pore water sensitive reactions, (*b*) the relative impact mixing has on particles with long and short seafloor residence times, (*c*) deep-sea environments where the numbers of macrofauna are high from those where they are not, and (*d*) the influences of biogenic structures that act as passive particulate traps from those of surrounding seafloor features. This task is so large and the information so scanty, that only a few illustrative highlights can be presented here.

First example. One of the arguments made against the significance of diffusive sublayer impedance is that the mixing activities of benthic organisms may promote solute transport at rates far greater than diffusive fluxes (Berner 1980). This will happen in bathyal environments if organisms turnover near-surface sediments so rapidly that diffusion gradients cannot stabilize, or if irrigated burrows are maintained in high enough densities to bring the concentrations of pore water solutes close to those of overlying waters (Aller 1980). The resulting "bio"-solute transport increases benthic exchange rates of solutes involved in reactions normally slowed by increased pore water concentrations (e.g., SiO_2 and $CaCO_3$ dissolution, see Eq. 2). Pore water bioturbation has little net effect on the destruction rates of solids such as organic matter, which are degraded primarily near the sediment-water interface and appear to be rate-limited by their own chemistry (Eq. 1) (Aller 1980; Reimers and Smith 1986).

Second example. In the abyssal oligotrophic Pacific where the sedimentation rate is only ~1 mm ky^{-1}, macrofauna numbers are very low. Meiofauna and nanobiota living primarily in the surface 1–3 cm of sediment tend to dominate the benthic community (Snider et al. 1984). Therefore, particle mixing to depths greater than 3 cm can be considered a rare event and unimportant for reactions involving biogenic materials with short seafloor residence times. In contrast, the radionuclides $^{239,240}Pu$ and ^{137}Cs (sorbed onto clays and produced since 1945), and ^{14}C (in nearly refractory organic carbon pools) are measurable to greater than 10 cm depth in the sediment column (Druffel et al. 1984). Thus, even though deep mixing is rare, it enhances the net flux of chemically stable, long residence-time tracers into abyssal sediments.

Third example. There is one means, usually overlooked, whereby bioturbation may slow down benthic dissolution rates. It is an indirect consequence of the many relict biogenic structures on the seafloor. Relict structures, which include long troughs, burrows, and mounds, act as traps for surface-derived, reactive, biogenic debris (Aller and Aller 1986). If this debris were to otherwise remain rafting over the sediment or were to settle out onto an exposed surface, its destruction would be rapid. Concentrated in traps, localized solute concentrations may rise above those in surrounding sediments, retarding solid dissolution rates. Conversely, debris-filled biogenic features may produce anoxic or suboxic microenvironments imbedded in a more oxygenated background (Aller and Aller 1986). This could result in the mobilization of redox sensitive elements, such as Fe and Mn, and cause heterogeneous benthic fluxes over small (i.e., < 1 m) spatial scales.

Oxic vs. Anoxic Diagenesis and the Role of Bottom Water Oxygen Content

There is considerable disagreement and confusion in marine science literature concerning the influences of redox state on benthic rates of organic matter decomposition (and conversely burial). Two opposing positions are frequently stated. These are either (*a*) organic matter is degraded faster and more completely under oxic rather than suboxic or anoxic conditions because aerobic bacteria can metabolize a wider range of organic compounds than anaerobic bacteria (due to the higher energy yielded from oxygen compared to alternative electron acceptors, and because aerobes may have a wider suite of extracellular enzymes able to degrade resistant structures), or (*b*) because most marine particulate organic matter is not structurally complex (in contrast to dissolved organic matter and many terrestrially derived compounds, e.g., lignin), rates of organic matter decomposition by aerobic and anaerobic microorganisms will be very similar. For those who advocate the first position, an additional point of contention is whether it makes a difference how much oxygen there is in waters overlying the seafloor. This argument depends on (*a*) whether the O_2-zero redox boundary in surface sediment pore waters moves up or down with respect to the sediment surface with decreasing or increasing overlying oxygen concentrations, respectively, and (*b*) whether biological mixing will in effect transport more particulate organic matter below a shallow than a deep redox front, so that in the low bottom water oxygen case, more organic carbon is ultimately buried.

Oxygen profiles, measured in surface sediment pore waters, indicate that redox boundary depth does vary broadly and that shallow depths are common in sediments deposited under waters with low oxygen concentrations (Reimers 1987 and unpublished). Therefore, if sedimentary organic matter oxidation was fueled only by oxygen, there would be reason to believe bottom water O_2 conditions could, together with the carbon rain rate and bioturbation, determine the organic matter reflux. This has been argued by Emerson (1985) for deep-sea environments with sedimentation rates < 30 cm ky^{-1} and used by McCorkle et al. (1985) as a precondition for predicting either the bottom water oxygen value or the carbon particulate flux, given the other, and pore water $\delta^{13}C$ values of ΣCO_2.

How can the influence of environmental redox conditions on benthic organic carbon reflux rate be tested? In recent papers, Henrichs and

Reeburgh (1987) and Canfield (1988), after a thorough literature review, have compiled a wealth of data comparing aerobic and anaerobic decomposition rates both in laboratory experiments with fresh marine algae and in natural marine sediments with a wide range of sedimentation rates and partially degraded, organic matter pools. To quote their conclusions: the data "suggest that the most labile fractions of algae are decomposed rapidly and at comparable rates under both oxic and anoxic conditions" (Henrichs and Reeburgh 1987), and "over a wide range of organic matter predecomposition (during settling to the seafloor and sediment burial), the ability of aerobic and anaerobic microorganisms to decompose organic matter is very similar" (Canfield 1988). Similar and comparable here means within factors of 2–3.

Implicit in Canfield's analysis was an assumption that sediments with the same sedimentation rate receive similar types and rain rates of organic matter. The crux of his argument is a result showing overlapping integrated carbon oxidation rates (i.e., the total carbon oxidized by both oxygen and/or sulfate) as a function of sedimentation rate, no matter what the overlying bottom water oxygen content (Fig. 3).

To reconcile Fig. 3 with observations that more organic carbon often becomes buried in bathyal and abyssal marine sediments when bottom water conditions become less oxygenated (Emerson 1985) requires only one more step in thinking. Recalling Eq. 1 and arguments made earlier, most of the organic matter raining to the deep seafloor can be assumed to be labile and probably to degrade in uppermost sediments on time scales < 1 year. By the results shown in Fig. 3, the degradation of this major and labile fraction of organic matter may be assumed to be independent of redox conditions. However, within the remaining fraction normally found slowly degrading in or below sediment mixed layers, there must be some organic pools (G_i where $1<i<n$) which can only be degraded in the presence of oxygen (i.e., $k_{1<i<n}$ goes to zero as O_2 goes to zero). Since the supply of these pools is small compared to the total particulate flux (see Fig. 1), their decomposition, or lack thereof, does not significantly affect the overall benthic organic carbon decomposition rate. The fate of these "resistant when anaerobic" materials does, however, influence the carbon burial rate, which also is small compared to the particulate carbon flux (Canfield 1988).

CONCLUDING REMARKS

The deep-water chemical cycles of many minor sediment-seawater constituents (e.g., trace metals, phosphorus, iodine, and light and heavy rare earth elements) are intricately tied to the particulate and benthic fluxes of opal, $CaCO_3$, and organic matter by adsorption, reaction, and redox-sensitive solubility. In the deep sea, these three biogenic materials typically sum to $>80\%$ of the total mass flux (Noriki and Tsunogai 1986; Dymond and

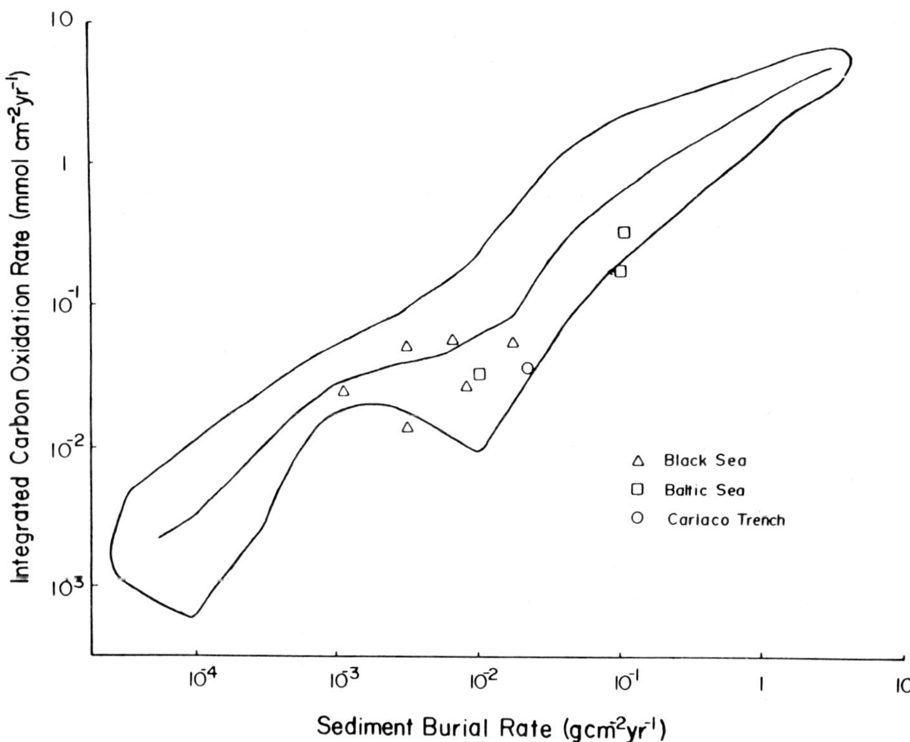

Fig. 3—Total rates of organic carbon oxidation by oxygen and/or sulfate are shown as a function of sedimentation rate. The symbols are from euxinic environments. These fall within a field of data (indicated by the outline) from marine sediments overlain by water with varying oxygen concentrations. The particulate carbon flux is assumed to covary with sedimentation rate. Reproduced with permission from Canfield (1988).

Collier 1988). However, within each opal, $CaCO_3$, or organic matter flux exists a broad spectrum of fractions with varying reactivities.

The focus of this paper has been to illustrate how the reactivities and magnitudes of biogenic particulate fluxes might determine regional and temporal patterns in benthic flux and burial rates. It was concluded that in the deep sea, reactivities are generally high and environmental variables (other than degree of saturation) have only secondary impact on reflux. The most informative study that could test for the kinetic limits of chemical reactions on the seafloor would be paired time series particulate and benthic flux measurements.

Acknowledgements. The author thanks Robin Keir and George Jackson for helpful discussions of the subjects in this paper. Financial support for her salary when writing this paper was provided from NSF grant OCE86-08497. Erwin Suess, Wolf Berger, George Jackson and Gert De Lange provided helpful reviews of the manuscript.

REFERENCES

Aller, J.Y., and Aller, R.C. 1986. Evidence for localized enhancement of biological activity associated with tube and burrow structures in deep-sea sediments at the HEBBLE site, western North Atlantic. *Deep-Sea Res.* **33**: 755–790.

Aller, R.C. 1980. Quantifying solute distributions in the bioturbated zone of marine sediments by defining an average microenvironment. *Geochim. Cosmochim. Acta* **44**: 1955–1965.

Berelson, W.M.; Hammond, D.E.; and Johnson, K.S. 1987. Benthic fluxes and the cycling of biogenic silica and carbon in two southern California borderland basins. *Geochim. Cosmochim. Acta.* **51**: 1345–1364.

Berner, R.A. 1980. Early Diagenesis: A Theoretical Approach. Princeton, NJ: Princeton Univ. Press.

Berner, R.A., and Morse, J.W. 1974. Dissolution kinetics of calcium carbonate in seawater. IV. Theory of calcite dissolution. *Am. J. Sci.* **274**: 108–134.

Boudreau, B.P., and Guinasso, N.L., Jr. 1982. The influence of a diffusive sublayer on accretion, dissolution, and diagenesis at the sea floor. In: The Dynamic Environment of the Ocean Floor, eds. K.A. Fanning and F.T. Manheim, pp. 115–148. Lexington: Lexington Books.

Canfield, D.E. 1988. Sulfate reduction and oxic respiration in marine sediments: implications for organic carbon preservation in euxinic environments. *Deep-Sea Res.,* in press.

Deuser, W.G.; Ross, E.H.; and Anderson, R.F. 1981. Seasonality in the supply of sediment to the deep Sargasso Sea and implications for the rapid transfer of matter to the deep ocean. *Deep-Sea Res.* **28**: 495–505.

Druffel, E.R.M.; Williams, P.M.; Livingston, H.D.; and Koide, M. 1984. Variability of natural and bomb-produced radionuclide distributions in abyssal red clay sediments. *Earth Planet. Sci. Lett.* **71**: 205–214.

Dymond, J. 1984. Sediment traps, particle fluxes, and benthic boundary layer processes. In: Global Ocean Flux Study, Proc. of a Workshop, pp. 260–284. Washington, D.C.: National Academy Press.

Dymond, J., and Collier, R. 1988. Biogenic particle fluxes in the equatorial Pacific: evidence for both high and low productivity during the 1982–83 El Nino. *Glob. Biogeochem. Cyc.* **2**: 129–137.

Emerson, S. 1985. Organic carbon preservation in marine sediments. In: The Carbon Cycle and Atmospheric CO_2: Natural Variations Archean to Present, eds. E.T. Sundquist and W.S. Broecker. *Geophys. Monog.* **32**: 78–88. Washington, D.C.: Amer. Geophys. Union.

Emerson, S., and Bender, M. 1981. Carbon fluxes at the sediment-water interface of the deep-sea: calcium carbonate preservation. *J. Marine Sci.* **39**: 139–162.

Emerson, S.; Stump, C.; Grootes, P.M.; Stuiver, M.; Farwell, G.W.; and Schmidt, F.H. 1987. Estimates of degradable organic carbon in deep-sea surface sediments from ^{14}C concentrations. *Nature* **329**: 51–53.

Gardner, W.D.; Sullivan, L.G.; and Thorndike, E.M. 1984. Long-term photographic, current, and nephelometer observations of manganese nodule environments in the Pacific. *Earth Planet. Sci. Lett.* **70**: 95–109.

Henrichs, S.M., and Reeburgh, W.S. 1987. Anaerobic mineralization of marine sediment organic matter: rates and the role of anaerobic processes in the oceanic carbon economy. *Geomicrobiol. J.* **5**: 191–237.

Honjo, S. 1984. Study of ocean fluxes in time and space by bottom-tethered sediment trap arrays: a recommendation. In: Global Ocean Flux Study, Proc. of a Workshop, pp. 305–324. Washington, D.C.: National Academy Press.

Hurd, D.C. 1972. Factors affecting solution rate of biogenic opal in seawater. *Earth Planet. Sci. Lett.* **15**: 411–417.

Hurd, D.C. 1973. Interactions of biogenic opal, sediment and seawater in the central equatorial Pacific. *Geochim. Cosmochim. Acta* **37**: 2257–2282.

Jahnke, R.; Heggie, D.; Emerson, S.; and Grundmanis, V. 1982. Pore waters of the central Pacific Ocean: nutrient results. *Earth Planet. Sci. Lett.* **61**: 233–256.

Jahnke, R.A., and Jackson, G.A. 1987. Role of sea floor organisms in oxygen consumption in the deep North Pacific Ocean. *Nature* **329**: 621–623.

Jørgensen, B.B. 1978. A comparison of methods for the quantification of bacterial sulfate reduction in coastal marine sediments. II. Calculation from mathematical models. *Geomicrobiol. J.* **1**: 29–47.

Keir, R.S. 1980. The dissolution kinetics of biogenic calcium carbonates in seawater. *Geochim. Cosmochim. Acta* **44**: 241–252.

Lampitt, R.S. 1985. Evidence for the seasonal deposition of detritus to the deep-sea floor and its subsequent resuspension. *Deep-Sea Res.* **32**: 885–897.

Martin, J.H.; Knauer, G.A.; Karl, D.M.; and Broenkow, W.W. 1987. VERTEX: carbon cycling in the northeast Pacific. *Deep-Sea Res.* **34**: 267–285.

McCorkle, D.C.; Emerson, S.R.; and Quay, P.D. 1985. Stable carbon isotopes in marine porewaters. *Earth Planet. Sci. Lett.* **74**: 13–26.

Noriki, S., and Tsunogai, S. 1986. Particulate fluxes and major components of settling particles from sediment trap experiments in the Pacific Ocean. *Deep-Sea Res.* **33**: 903–912.

Pisias, N.G.; Murray, D.W.; and Roelofs, A.K. 1986. Radiolarian and silicoflagellate response to oceanographic changes associated with the 1983 El Nino. *Nature* **320**: 259–262.

Reimers, C.E. 1987. An *in situ* microprofiling instrument for measuring interfacial pore water gradients: methods and oxygen profiles from the North Pacific Ocean. *Deep-Sea Res.* **34**: 2019–2035.

Reimers, C.E., and Smith, K.L., Jr. 1986. Reconciling measured and predicted fluxes of oxygen across the deep sea sediment-water interface. *Limnol. Ocean.* **31**: 305–318.

Santschi, P.H.; Bower, P.; Nyffeler, U.P.; Azevedo, A.; and Broecker, W.S. 1983. Estimates of the resistance to chemical transport posed by the deep-sea boundary layer. *Limnol. Ocean.* **28**: 899–912.

Schink, D.R.; Guinasso, N.L., Jr.; and Fanning, K.A. 1975. Processes affecting the concentration of silica at the sediment-water interface of the Atlantic Ocean. *J. Geophys. Res.* **80**: 3013–3031.

Schrader, H.J. 1971. Fecal pellets: role in sedimentation of pelagic diatoms. *Science* **174**: 55–57.

Smith, K.L., Jr. 1987. Food energy supply and demand: a discrepancy between particulate organic carbon flux and sediment community oxygen consumption in the deep ocean. *Limnol. Ocean.* **32**: 201–220.

Smith, K.L., Jr., and Baldwin, R.J. 1984. Seasonal fluctuations in deep-sea sediment community oxygen consumption: central and eastern North Pacific. *Nature* **307**: 624–625.

Snider, L.J.; Burnett, B.R.; and Hessler, R.R. 1984. The composition and distribution of meiofauna and nanobiota in a central North Pacific deep-sea area. *Deep-Sea Res.* **31**: 1225–1250.

Westrich, J.T., and Berner, R.A. 1984. The role of sedimentary organic matter in bacterial sulfate reduction: the G model tested. *Limnol. Ocean.* **29**: 236–249.

Appendix: Solution of the Emerson et al. (1987) model with three carbon fractions.

The carbon in a sediment mixed layer is assumed to be a three component mixture:

$$C_T = C_1 + C_2 + C_3 . \tag{A}$$

C_1 represents carbon which degrades rapidly on the seafloor with a residence time on the order of one month; C_2 degrades much slower than C_1; and C_3 is refractory.

Assuming a steady state mass balance for the $^{12}C(')$ and $^{14}C(^*)$ isotopes of each fraction:

$$R_1' = k_1 C_1' + sC_1' \tag{B1}$$

$$R_1^* = k_1 C_1^* + sC_1^* + \lambda C_1^* \tag{B2}$$

$$R_2' = k_2 C_2' + sC_2' \tag{B3}$$

$$R_2^* = k_2 C_2^* + sC_2^* + \lambda C_2^* \tag{B4}$$

$$R_3' = sC_3' \tag{B5}$$

$$R_3^* = sC_3^* + \lambda C_3^* \tag{B6}$$

where R_1', R_1^*, . . ., etc. are the particulate fluxes of carbon arriving at the sediment-water interface, and $k_1 C_1'$, . . ., etc., λC_1^*, . . ., etc. and sC_1', . . ., etc. are the removal rates by reaction, radioactive decay, and sedimentation, respectively.

The radioactivity of the carbon in the mixed layer of the sediment is given by:

$$\frac{C_T^*}{C_T'} = \left(\frac{R_1^* + R_2^* + R_3^*}{R_1' + R_2' + R_3'}\right) e^{-\lambda t} \tag{C}$$

where t is an average or effective time since the carbon was fixed in the surface ocean.

By solving equations B1–B6 for C_1', C_1^*, . . ., etc., summing these and substituting into equation C:

$$\frac{\left(\dfrac{R_1^*}{k_1+\lambda+s}+\dfrac{R_2^*}{k_2+\lambda+s}+\dfrac{R_3^*}{\lambda+s}\right)}{\left(\dfrac{R_1'}{k_1+s}+\dfrac{R_2'}{k_2+s}+\dfrac{R_3'}{s}\right)}=\left(\frac{R_1^*+R_2^*+R_3^*}{R_1'+R_2'+R_3'}\right)e^{-\lambda t} \qquad \text{(D)}$$

Equation D may then be simplified assuming λt is small so that $e^{-\lambda t} \sim 1 - \lambda t$, and assuming $\dfrac{R_1^*}{R_1'}=\dfrac{R_2^*}{R_2'}=\dfrac{R_3^*}{R_3'}$.

The result of this simplification is Eq. 3 in the text. To derive the results in Fig. 1 it was assumed that:

$$k_1 = 10\,y^{-1},$$

$$k_2 = .004 \text{ to } .001\,y^{-1},$$

$$\lambda = 1.24 \times 10^{-4}\,y^{-1},$$

$$s = 10^{-4}\,y^{-1}, \text{ and}$$

$$R_3 = .001(R_1 + R_2)\,.$$

The last assumption is based on burial rates of truly refractory carbon at depth in oxic sediments from the central North Pacific Ocean (Reimers, unpublished).

Productivity of the Ocean: Present and Past
eds. W.H. Berger, V.S. Smetacek and G. Wefer, pp. 235–253
John Wiley & Sons Limited

Responses of Benthos to Changing Food Quality and Quantity, with a Focus on Deposit Feeding and Bioturbation

P.A. Jumars and R.A. Wheatcroft

School of Oceanography, WB-10
University of Washington
Seattle, WA 98195, U.S.A.

Abstract. Deposit feeding is the process that accounts for most bioturbation of particles most of the time because there is selective pressure on deposit feeders to feed on generally nutrient-poor sediments at high rates and to separate sites of ingestion and egestion. Conversely, burrowing has substantial energetic cost, and there is strong selective pressure to minimize the amount and distance of sediments moved. The most rapid advances in understanding deposit feeding and thus bioturbation have and will come from examining the selective pressures of changing food quality and quantity on individuals. Comparatively little predictive value has been or can be gleaned from empirical, ecosystem level examinations of detrital processing. In particular, predictive models are needed of deposit feeder size, abundance, and activity as functions of food quality and quantity variation over a broad range of time scales. The roles of deposit feeder digestion in geochemical reactions are conspicuously underexplored either with models or measurements.

Community level responses to varying organic input are best known in cases of near-shore, anthropogenous organic inputs, where a succession from small, shallowly burrowing to large, deeply burrowing species is usually seen after loading ceases. In the deep sea, population responses of meiofauna to seasonal and episodic inputs of fresh phytodetritus have been noted, but the importance of such inputs to macrofauna is unknown. Exceptionally deep burrowers are characteristic of regions in which sedimentation has buried substantial organic carbon. Tranquil deep-sea sites under oligotrophic oceans are characterized by very sparse and small individuals of motile, deposit-feeding infauna, but inference of vertical flux of organics from standing stocks and community composition of benthic animals is confounded by the as yet poorly understood effects of horizontal sediment transport.

Depth of mixing by animals is surprisingly invariant and may be set by rapidly mounting costs of burrowing with increasing sedimentary overburden. Bioturbation intensity also varies less than one might expect, because animal density falls off with decreasing sedimentation rate and organic flux, but individuals in food-poorer regions presumably must feed at greater rates to meet similar metabolic needs per gram of body mass. We also suggest that bioturbation intensity is much greater in the horizontal than in the vertical and accounts for surprisingly rapid disappearance of surface traces. Edges of communities (ecotones) and regions of unsteady sedimentation (in particular turbidite zones) are understudied in terms of the insights they could provide into benthic organism responses to varying food supply.

INTRODUCTION

Separating dependent from independent variables in the system of organisms, sediments, and boundary-layer fluids is certainly difficult and perhaps unwise. In systems with feedbacks, that separation must be arbitrary. At the risk of being both unwise and arbitrary, but for the sake of organizational simplicity, we adopt the point of view of benthic organisms and ask how they respond to changes in the quality and quantity of food supplied to them. We then switch perspectives to ask how these responses may affect the sedimentary record. We do not begin with the latter perspective because without an understanding of the selective pressures driving animal responses, organisms would appear both capricious and arbitrary in their effects on the sedimentary record. We endeavor to express and support strong opinion rather than to present a balanced review.

Our focus on feeding in general and on deposit feeding in particular may appear excessive, but there are several reasons for it. Feeding is the animal activity that under nearly all circumstances is responsible for the overwhelming majority of biogenous movement of sediments. A common misconception, due to such prominent structures as fossilized escape burrows across strata of strongly contrasting lithologies, is that organisms spend a great deal of time burrowing. Even when they do move from one place to another, there is strong selective pressure on organisms not in life-threatening situations to minimize the amounts of sediment displaced; burrowing is the most expensive means of locomotion, including flight (joules per body length moved, cf. Trevor 1978). Because water is so much easier to move than denser solids, it is no accident that apparent eddy diffusion coefficients of fluid bioturbation exceed by an order of magnitude or more those of particulate bioturbation (Aller 1982). Animals do erect tube and burrow structures within the seabed and move about within them: the anthropomorphic analogy is to digging subway tunnels rarely but using them frequently once built. The prime exceptions are a few: comparatively rare subsurface burrowers that combine the costs of ingesting sediments with the costs of burrowing, eating their ways along (e.g., heart urchins and large ophelliid polychaetes). Even these exceptions are better understood by focusing on feeding rather than burrowing, for there is still strong selective pressure to minimize the amounts of sediment moved but not ingested. Furthermore, the kinetics of deposit feeding are counterintuitively rapid due to the short gut residence times (30 min to 2 hr being often seen in shallow-water representatives) and large gut volumes (typically 30–80% of total body volume) imposed by the generally low food quality of sediments.

To relate these ideas about animal activities to the usual biodiffusion coefficient (D_b in units of length squared per unit of time) Wheatcroft et al. (in preparation) have decomposed it into a step length and a frequency. Because step length is squared, this term dominates D_b. In burrowing, the typical distance moved by a sediment particle is small: a fraction of a burrow or body radius. In feeding, by contrast, gut length (more precisely mean distance from point of ingestion to point of egestion) is the appropriate length scale. Not only is the typical displacement distance by feeding thus much greater, we suggest that feeding is also the more frequent activity, assuring its dominance over burrowing in determining D_b.

This bioturbation formalism can be applied to either conservative or reactive chemical species (Aller 1982). To date, however, reaction terms inside animal guts have not been examined in the detail they deserve. What makes the deposit feeder gut a unique sedimentary environment is the otherwise rare combination of low gut oxygen levels (produced by microbial activity) with mechanical agitation. This combination is rare outside animal guts and is effective at speeding reactions over rates normally encountered in more tranquil anaerobic environments outside animal guts (Kristensen and Blackburn 1987).

A secondary purpose of ours is to challenge a virtually automatic focus on successional time scales when dealing with benthic biological response to varying organic input. The reason for this automatic focus is the relative abundance of data on these scales characteristic of anthropogenous organic inputs and scarcity of data on both larger and smaller spatial and temporal scales. In particular we question the notion that mechanistic, *predictive* understanding of aggregate phenomena, such as bioturbation, can be achieved without a focus on the selective processes that operate on individuals. We will examine both steady and unsteady responses over a much broader range of scales than those encompassed by most studies of organic loading—to highlight the lack of data and potential utility of models on these shorter and longer scales.

BIOLOGICAL STRUCTURE AND FUNCTION

Concepts of Food Quality and Quantity

Semantic difficulty runs rampant in discussions of organism response to food differences. To be useful, however, the amount of food supplied or available to an individual, population, or community must be expressed per unit of time. Any unit of biomass requires a rate of supply in order to maintain itself or to grow. Fundamentally, an individual requires a given number of mol s^{-1} of some limiting nutrient in order to survive. Confusion often results, however, because at steady state the standing stock of food and its rate of supply can be related through a simple constant, leading to the assertion (true under steady state but not in general) that biomass of food is a perfectly valid indicator of food availability. Another special case in which standing stock of food can serve as an empirically, if not conceptually, satisfactory surrogate for rate of food supply to an individual is when the individual's utilization rate is swamped by the rate of supply, i.e., by advection or microbial growth. What is not always apparent is that growth is still rate-limited, but in this case it is limited by the animal's processing rate rather than by the rate of supply afforded by the external environment.

The latter point may seem subtle, but it must be treated explicitly if one is to determine an appropriate geometry for measurement of rate of supply. Rate of supply is often, and in general perhaps best, expressed in terms of mol m^{-2} s^{-1}, where the area is a horizontal plane representing the sediment-water interface. In reality, however, feeding volumes of organisms have more complex geometries. Among the simplest are those of surface deposit feeders having circular or cardioid areas reached by the feeding appendages, but even here some finite and not necessarily areally uniform thickness of sediments is accessible. In the absence of horizontal movement, organic matter sedimenting outside the geometry swept out by feeding appendages would remain unavailable; in the presence of horizontal sediment transport, due either to physical forces or to bioturbation, material deposited outside can become available, and conversely material initially deposited inside can become unavailable. Feeding geometries and transport interact, further complicating the situation; protruding animal tubes can create areas of scour (reduced deposition until some steady state is reached), and many animals build pits that increase deposition rates locally.

Flux geometry is perhaps most critical to suspension feeders. Passive suspension feeders—those that do not pump water past their capture appendages—depend upon the geophysically driven flux of material reaching the appendages. In the bottom boundary layer, horizontal fluxes of suspended particles typically exceed vertical fluxes by at least an order of magnitude. It is doubtful whether a suspension feeder could exist anywhere on the ocean floor if there were no turbulence or horizontal flow to provide a supply in excess of the net vertical sedimentation rate. An important corollary is that the same net vertical flux of organic matter from surficial waters may support varying proportions of suspension-feeding versus deposit-feeding animals (including the foraminiferans discussed by Altenbach and Sarnthein, this volume); a faunal change need not signal a change in vertical flux. Even under steady state transport, both suspended particle concentration and horizontal velocity show very strong gradients near the bed, producing diverse and often strongly vertically varying flux profiles (concentration × horizontal velocity = horizontal flux). At first it would appear that active suspension feeders (such as clams) would be sensitive only to ambient concentration. Reinforcing the point made above this insensitivity to flux should be observed only when supply overwhelms demand. Interactions of the flux rates and geometries produced by pumping with the flux rates and geometries produced by the ambient flow become critical.

It is thus apparent that rate of food supply—food quantity—cannot be defined accurately without taking into account to what organisms this rate is being supplied. For example, a benthic bacterium may depend upon vertical sedimentation of organic particulate carbon for its supply of chemical energy and vertical turbulent and molecular diffusive exchange of inorganic nitrogen to supply building blocks for proteins. A critical and unanswered question for both suspension feeders and deposit feeders in the deep sea is the extent to which they feed on a renewable resource in the form of such bacteria attached to particles versus the extent to which they compete with bacteria for newly arriving and labile organic material. In the former case, their food supply is strongly buffered against short-term oscillations in rate of supply (with bacterial growth rate determining that supply), while in the latter these animals should be strongly keyed to vertical and horizontal sedimentation events of the most organically labile, new materials. Identifying precisely the food resources assimilated by deposit feeders is a key problem in shallow water as well as in the deep sea (Lopez et al. 1989); the problem is not straightforward because, at the high processing rates characteristic of deposit feeders, a minor sedimentary organic constituent might be a major source of assimilated energy or mass even if it were assimilated with only moderate efficiency. The problems raised by this lack of knowledge are pervasive: if a downstream individual assimilates a different component of sedimentary organic carbon from that assimilated by individuals of other species found upstream, then horizontal transport is all the more effective as a source of food.

Any biologically useful definition of food quality must be even more closely tailored to the organism in question, as is obvious from Aesop's fable of the dog in the manger. One component of sedimentary food quality is the proportion of potentially digestible material in the food. A more specific definition is the volumetric or gravimetric concentration of the growth rate-limiting component, yet only for a handful of shallow-water species are there data to suggest what such components are (Tenore et al. 1984). This handful of results gives some hope that growth in deposit feeders may be limited more generally by either available (labile organic) nitrogen or available caloric energy, depending upon the ratio in which these two key ingredients are provided. To an organism, however, time is energy (or mass that contains energy), and food quality must deal in net return to the animal per unit of time. Consistent descriptions would be the functions that describe net rate of gain to a metazoan versus residence time of a food in its sensory field (detection, pursuit, and capture) and gut or that describe net rate of gain to a bacterium versus time exposed to its exoenzymes. From such plots (Fig. 1) it is readily apparent that food quality and quantity are

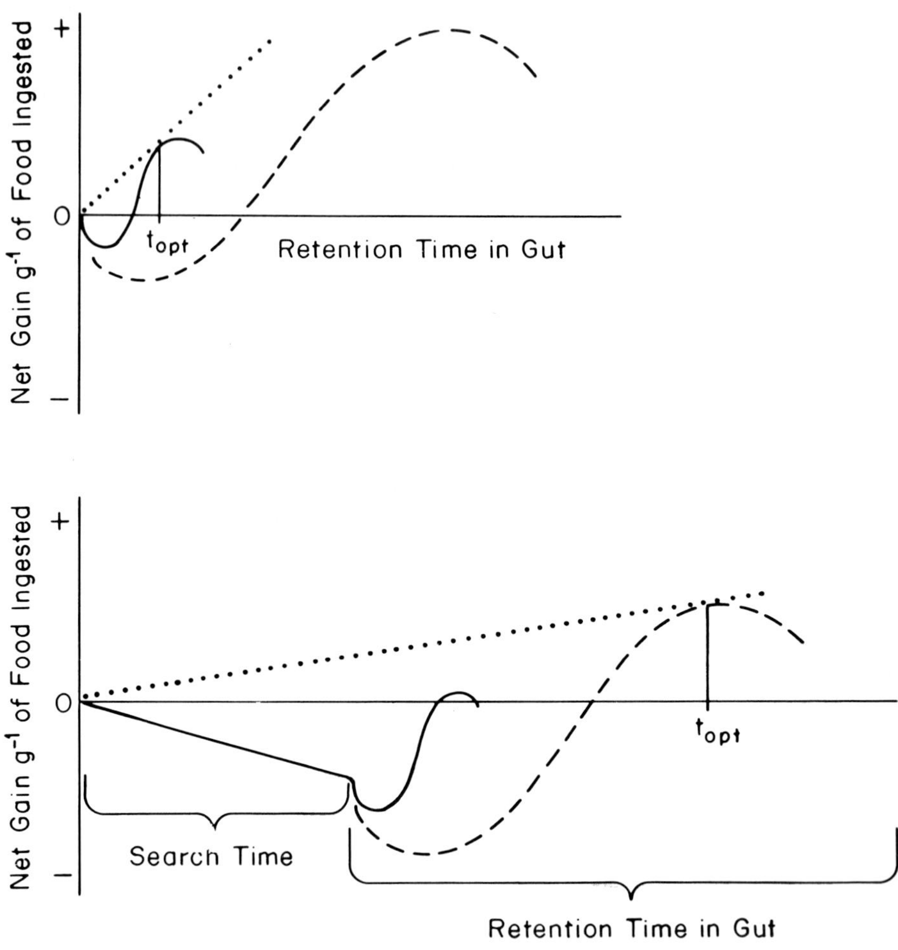

Fig. 1—Interaction of food quality (described by the solid line for a food type giving rapid return after ingestion and the dashed line for a food type giving slower but eventually greater return) with food quantity. The upper graph is for the case of unlimited food while the lower graph is for a lower abundance (supply rate), requiring a finite search time to locate a given item. The dotted tangents through the origins represent the maximal rates of gain possible (at the optimal retention time, t_{opt}), showing that the favored food type should change with food abundance. Modified from Sibly (1981).

not independent; in general, it is not possible even to ordinate food quality without specifying rate of supply, since rate of supply and quality interact in determining the foraging strategy that can maximize net rate of gain of mass or energy.

Clearly, there is a long way to go in determining food supply rate and food quality for deep-sea organisms. Sediment trap data provide apparent vertical fluxes expressed in total weight, carbon content, or nitrogen content per unit of area per unit of time. Over the range of C:N ratios seen in the deep sea, it seems safe, despite considerable scatter, to say that the trend is toward higher food quality the lower the ratio. Similarly, there is a tendency for food quality and quantity to covary. Much supply to the benthos seems to be from unsteady events in overlying waters, with consequent pulses of relatively labile material. Given this tendency towards covariation of quality and quantity and the present state of knowledge, we will subsequently not spend much effort (beyond a quick look at individuals' responses) to resolve quality from quantity. As the utility of specific biomarkers increases, however, this distinction will become crucial.

Responses of Individuals and Populations to Time-varying Food Supply

Feeding rates of active (spending energy to obtain new water parcels by swimming or pumping) planktonic suspension feeders and benthic bivalves have been much studied with respect to the rate of food supply. Foraging theory and empirical observations agree (Fig. 2) in illuminating individual responses to short-term changes ($\leq$ mean gut residence time) in suspended food concentration, which acts as a surrogate for food supply rate since food is maintained at a given level while the response is measured. Data are consistent with the idea that suspension feeders act in such a way as to maximize their individual net rates of gain of energy or limiting nutrient. A plot very much like that of Fig. 2 can, in fact, be generated from Fig. 1 by means of continually shifting the origin in order to change the time between ingestion events.

Exercising Fig. 1 in this manner also demonstrates the predictive value of examining biological phenomena at the level of ecological organization where selection takes place, i.e., on individuals that constitute populations. As food becomes scarcer (time between ingestion events increases), a shift toward a new, longer optimal retention time will increase digestive efficiency. The ecosystem result will be greater retention of nutrients by organisms and more efficient recycling. In the past, such phenomena have been considered to be intrinsic features of "mature ecosystems" (Odum 1969), but it is now plain (Miller et al. 1984; Penry and Jumars 1987) that predictive theory at the level of the individual is much better poised to provide an understanding of them.

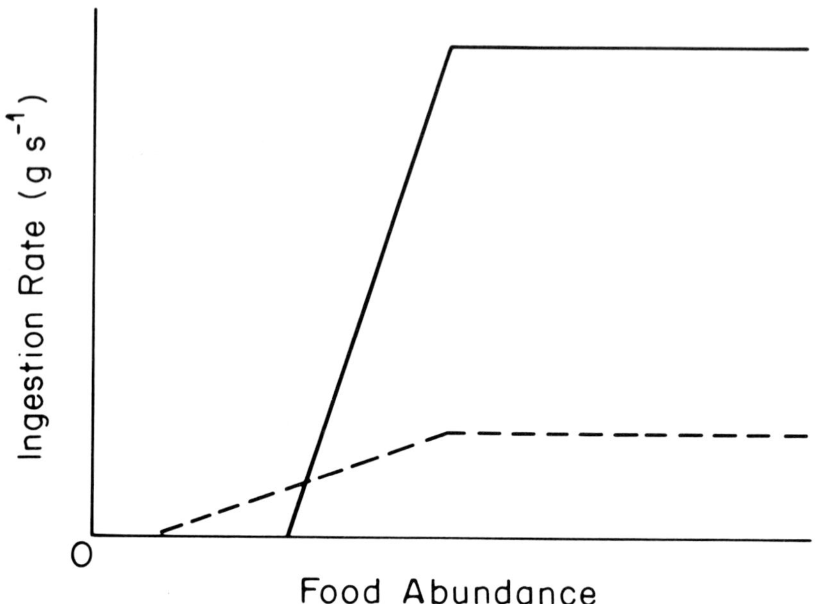

Fig. 2—Ingestion rate and food supply rate (as indicated by food concentration in the case of suspension feeders) on the two different types of food of Fig. 1, offered separately. The plateaus occur when search time of Fig. 1 drops to zero.

At first it might appear that deposit feeders are never limited by food quantity, since mud or sand is always available and nearly invariably has some organic coating. This intuition is wrong, as can be seen by doing the thought experiment of allowing an animal to feed in the absence of advection and positing its digestion to be highly efficient. If feeding rate exceeds resupply rate, food content of the sediments can be driven so low that feeding provides no net gain. While one could argue that only quality has changed, it seems more consistent to argue that rate of supply is a primary determinant of this food quality. Whether one argues on the basis of quantity or quality, the empirical observation is that surface deposit feeding rates of sedentary deposit feeders rapidly decline (order of one gut residence time) in the absence of advective resupply and removal of accrued fecal material. Again, further resolution is unlikely until the foods that deposit feeders digest and assimilate are better identified and quantified (Lopez et al. 1989).

Experiments that vary only the concentration of one pure food material covalently bonded to artificial sediments have been conducted using deposit feeders. Sufficient quantity has been provided to preclude change of quality by the animal. The general result with the handful of species thus far examined is for more rapid feeding (g dry weight sediments ingested s^{-1})

as food quality (protein g^{-1} dry weight of sediment) is increased, as would be predicted from a plot like that of Fig. 1 tailored to this situation (Taghon and Jumars 1984). There is a hint from work with more complex mixtures of food that at some (perhaps unnaturally) high concentrations, feeding rate turns down (Taghon 1988); this downturn may, however, be due to diffusional constraints in such a thick stew of nutrients.

That faster feeding individuals among one species of a deposit feeder actually do gain mass at a greater rate (or lose it at a lower rate) has now been established (Taghon 1988). Given the same food quality and quantity, the faster feeding individuals showed greater net gains in weight. Presumably some other selective pressure, such as greater exposure to predation during feeding, prevents selection from proceeding to produce a population of uniformly high feeding rate. This established link between short-term feeding behavior and potential reproductive gain is a critical one in optimal foraging theory, so Taghon's study and new ones like it are extremely important. The idea underpinning optimal foraging theory is that the individual gaining energy at a greater rate has a greater excess over its own metabolic needs to place into its somatic growth or defenses and into enhancing the robustness or numbers of its offspring. Thus, this individual under steady food supply is better able to sequester available material, while under fluctuating food supply is better able to channel food into population growth. In the former case, investment in rarer, larger individuals is likely; in the latter, rapid population growth is the expected channel for available resources. The natural time scale for these population phenomena is the generation time, whose minima range from the order of an hour in bacteria to decades in the largest animals. There is now evidence that meiofaunal populations at shelf and deep-sea depths respond by rapid population growth to seasonal or episodic inputs of labile organic matter (Gooday 1988; Cammen, in preparation; Fleeger, in preparation). It is not clear whether larger, longer-lived organisms—particularly deposit feeders that by definition frequently ingest material of low bulk food value—depend similarly on this labile fraction or are keyed instead, or also, to slower but steadier yields from more refractory material. The finding of a few species that do depend upon the most labile material should not draw attention away from the question of whether these species are typical or unusual.

Succession

Changes in species composition due to organic enrichment in benthic communities are better documented, at least for intertidal and shallow subtidal cases. There is a repeatable succession from small, dense (number m^{-2}) shallowly burrowing species with rapid population increase toward large, sparse, more slowly growing, deep burrowers (McCall 1977; Pearson and Rosenberg 1978). Exactly this kind of succession in population dynamics would be predicted, as the rate of food supply from the enrichment diminished, causing a transition from advantage in allocating resources toward population growth to advantage in allocation to individual maintenance and maintenance of a constant population size. In ecological terms, populations undergo *K* selection as food supply diminishes and becomes more steady and predictable. Since metazoans of organic-rich sediments must pump in their own oxygen, from the surface (or from neighbors' ventilation tubes), there is a general correlation between body size and depth of burrowing. If the severely disturbed area is large enough, new recruits near its center must come from the water column, so it is logical that shallow dwellers arrive and (since they on the average are smaller) mature first. The time scale for recovery from an organic enrichment depends upon the area covered by the enrichment and upon the intensity and time variation of the organic input. Clearly no steady state can be reached in less than the generation time of the longest-lived species in the latest stage of succession.

Most of the data on responses to organic loading come from anthropogenous events, and their relevance to nature is difficult to discern. Natural organic loadings rarely drive the sediments to a condition in which metazoans are absent and rarely occupy the size scales of these anthropogenic events. Nonetheless, there do seem to be parallels in nature. Floods of rivers sometimes cause similar conditions. Natural anoxia of continental shelf sediments and bottom waters from intense blooms has been observed, and successional recovery does appear similar to that observed in sediments affected by organic wastes (Boesch, in preparation). As expected, recovery from disturbances of large physical scale in which all fauna are removed is fastest among those species that have planktonic larvae. Recent results call into question, at least for disturbances of small spatial scale, the slow pace of recovery previously measured or assumed from deep-sea disturbances (Smith et al. 1986).

In most shallow-water environments a seasonal cycle is imposed on longer-scale trends. Seasonal reproductive periodicity certainly is seen in some benthic populations (e.g., Tyler 1986), but there is not a single published, deep-sea time series of quantitative samples that addresses the degree to which seasonality of reproduction in some deep-sea species is reflected in seasonally changing community structure. As a consequence, we do not know whether seasonality in population density and structure in the deep sea is an oddity or the norm.

Given that most sequences of deep-sea environments preserved on land come from bathyal depths and that many are of turbidite origin, successional patterns after major burial events are grossly underexplored. We have no idea how long complete recovery from a turbidity flow might take. Turbidity currents do bury organic-rich sediments and appear to allow deeply burrowing species to make a net profit from deposit feeding at a depth below the sediment-water interface, where the costs of producing burrows otherwise would not repay the costs of making them (Griggs et al. 1969). Animals in effect "mine" the "coal seams" produced by the event. There is not a piece of benthic sampling gear in existence that is well designed to capture large and deep burrowers, but shallow-water representatives (e.g., thalassinid shrimp) easily can exceed 2 m depth below the sediment-water interface (Dworschak 1983), and we and other deep-sea biologists and geologists routinely see open tubes and burrows coming out the bottoms of 50 cm-long box cores taken from turbidites. An interesting question is whether these deep burrowers must be preceded by a sequence of shallower burrowers. Alternatively, they might find such events through the nonsteady state migration of pore waters (Sorensen et al. 1987). In either case it is not clear how the deep burrower grows large enough to reach the organic-rich deposit unless seams outcrop. It would appear profitable in view of all these unknowns to study living faunas and trace fossils in turbidites having varying vertical and horizontal structures and oxygenation histories.

Evolution

Clearly the most underexplored time scale for paleoceanographic purposes is the evolutionary one (from generation time upward). The level of resolution in most stratigraphic records gives little information on shorter-term phenomena (Schindel 1980). The key issue is whether one can identify characteristics of individuals, populations, and communities that reflect the local mean and variation of organic input. The goal from a biologist's perspective would be to predict the sizes, abundances, and types of organisms expected to evolve at or invade a site given its vertical inputs and horizontal transports of food. That goal is very distant, but as it is approached it will provide a strong means of converting paleontological evidence into paleoproductivity estimates. Despite the distance of the goal, it is clear that effects of millenial increases in productivity over broad reaches of the ocean floor do not parallel anthropogenous loadings. Equatorial deep-sea benthic faunas are neither dramatically low in diversity nor do they comprise exclusively small, shallow-burrowing individuals.

This empirical approach of comparing present-day faunas from disparate vertical and horizontal flux regimes, assuming that these faunas have evolved to some near-equilibrium with their present food supplies, can provide other insights as well. For the deep sea, however, this effort is hampered by lack of published information on geographic distribution of infaunal species. Community structure appears similar under the oligotrophic gyres of the North and South Pacific and quite different from that of the equatorial zone

(Hessler, personal communication), but the number of distinct communities and the characteristic lifestyles of their members are very poorly known, and one must invoke many arguments by analogy with shallow-water results.

In a compilation of feeding rates in 19 shallow-water, deposit-feeding species, Cammen (1980) corrected for the fact that metabolic rates generally, and deposit feeding rates in particular, scale as body weight or volume raised to a power near 2/3 to 3/4. He then found that ingestion rates (organic matter ingested per unit of time per size-corrected gram of flesh) are roughly constant across a thousandfold range in organic content of ingested sediments. That is, there is no free lunch. Species that live in organic-poor sands process much greater sediment volumes per unit of time than do species that live in organic-rich muds. Note that this among-species comparison is not at all in conflict with foraging theory predictions that an individual exposed to normal within-environment variation will feed faster on higher-quality foods; this response is simply behavioral noise in Cammen's empirical relation. Unfortunately, there is no *in situ* measurement of feeding rate in any deep-sea deposit feeder for comparison with Cammen's empirical findings for shallow water, but direct extrapolation of Cammen's results would predict feeding rate (volume or mass per unit of time) to be relatively high in deep-sea deposit feeders because organic content of deep-sea sediments is usually low. The only evidence thus far of this selective pressure is the relative increase in volume of guts of deeper-dwelling (water depths) species when closely related taxa are compared (Allen and Sanders 1966). Such an adaptation allows a greater flow rate of sediments without a decrease in residence time or an increase in residence time without a decrease in flow rate. Furthermore, there is less body (versus gut) volume to support.

Biomasses of macroscopic metazoans and their numerical abundances decrease roughly exponentially with water depth. Below continental shelf depths the overwhelming majority of metazoan macroscopic organisms are deposit feeders. This seeming overdescription is needed to hide the fact that much of the total biomass in the deep sea is in poorly known and often large-bodied protozoans. There is some suggestion, from observations reported by Altenbach and Sarnthein (this volume), that protozoans may comprise more suspension feeders. While some foraminiferans are known to be mobile, they almost surely account for a much smaller amount of movement of sedimentary particles than do deposit-feeding metazoans. Body size of deep-sea infauna characteristically is small and is smallest where the flux of food is smallest. This general tendency makes the few larger animals among the infauna all the more important (Smith et al. 1986), but their abundances are poorly known because rare, deep-burrowing animals are the least well sampled components of the deep-sea fauna.

In looking at these patterns one sees a peculiar melange of theory and empirical pattern analysis, with little direct connection between the two. One exception that soon should emerge will be the issue of body size. One might expect deep-sea deposit feeders to be larger than their shallow-water counterparts, because metabolic cost per gram of flesh goes down with body size. Body size models such as those of Sebens (1987), however, show that this cost must be balanced against gain, i.e., one must know how rate of collection or digestive production scales with body size. Hessler and Jumars (1974) suggest that small deposit feeders can better focus food collection on interfacial sediments and thus get food of higher quality than can a bigger animal with a bigger collector. Penry and Jumars (1987) suggest that small body size avoids diffusional constraints on rates of digestion and absorption. Body size models have the potential to explain and to predict how biomass will be distributed among body sizes and how fast each of the individuals will process sediments. There are certainly strong patterns to explain (Schwinghamer 1985); the challenge is to couple mechanistic, predictive theory to them.

Motility patterns also change with water depth and distance from shore. Frequent and intense sediment transport precludes sessility, as does a food flux so low that an animal cannot encounter sufficient food to sustain itself by staying put. Hence, an intermediate-depth maximum in sedentariness of the fauna is expected and observed (Jumars and Fauchald 1977). One also sees the anticipated demise of active benthic suspension feeders and decreasing abundance of passive suspension feeders with depth, except where horizontal fluxes of seston are unusually high. Feeding type changes can be much more subtle, however, as a look at the life habits among the tube-dwelling polychaete worm family Ampharetidae demonstrates (Fig. 3). In the high-energy intertidal, ampharetids build no special devices for capturing sediments and show no special adaptations for getting rid of their fecal matter; the former are supplied in abundance while the latter are removed regularly by waves and currents. In lower-energy subtidal to bathyal and abyssal sites, ampharetids make pits that enhance sedimentation locally by inducing settlement or retention of horizontally transported material, and they show much more care in keeping such feeding pits free of their own feces (Nowell et al. 1984). At the deepest, most food-poor abyssal sites, ampharetids disappear altogether. One gets the impression that it may be possible in the next decade to build models that predict what lifestyles will occur under what rates of organic rain and horizontal transport.

A striking anomaly in the typical pattern of abundance change with depth has recently been discovered at the base of the continental rise (approx. 4800 m) off Nova Scotia (Thistle et al. 1985). This region, studied under the HEBBLE (High Energy Benthic Boundary Layer Experiment) program,

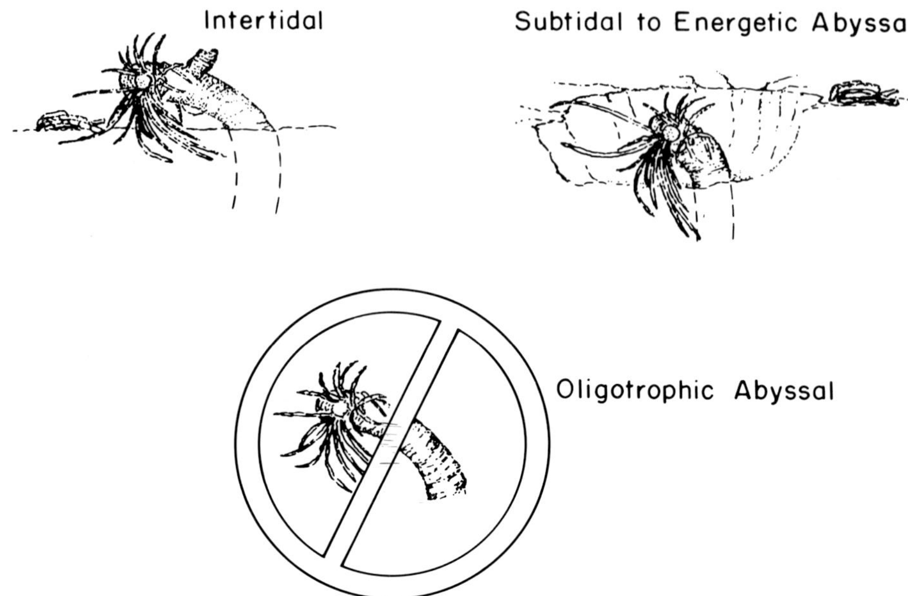

Fig. 3—Sketches of ampharetid polychaete lifestyles as the flux of food from local production, waves, and currents decreases. Intertidal ampharetids deposit their fecal wastes in the feeding area, relying on current and wave energy to remove them. Where wave and current energy is reduced, ampharetids often make pits that enhance sedimentation, and they take care to deposit fecal wastes outside the feeding area (helping to maintain the pit). At most food-poor abyssal sites, they and other sedentary tube builders are rare or absent.

is characterized by high horizontal sediment transport rates in "storms" that occur roughly five times per year and are now thought to represent geostrophically trapped edge waves along the rise. Bacterial and faunal abundances are an order of magnitude larger than otherwise would be expected at this depth. The storm events apparently are effective at stimulating bacterial production sufficient to fuel this enhanced standing stock. The numerically dominant macroscopic infaunal species uses the food-gathering method of the second panel of Fig. 3, and a tube of the adult size provides shelter within the sediments from storms of the magnitudes observed. The high organism abundances are not consistent with the idea that the deep-sea fauna is fed with little lag directly by the labile components of phytodetritus falling vertically. If the latter notion were true, then it would not be apparent why areas with large horizontal fluxes of sediments should yield such high standing stocks. Epifaunal species are extremely rare, however, and so part of the standing-stock increase might be explained by reduced predation from epifauna.

SEDIMENTARY CONSEQUENCES OF BIOLOGICAL STRUCTURE AND FUNCTION

Stratigraphic Resolution

If one accepts the argument that deposit feeding is the principal component of bioturbation, then can one rationalize the biological patterns noted above with observed stratigraphic resolution? Moreover, can one suggest particularly good places (outside anoxic basins) to look for improved resolution? Unfortunately, the most obvious answer—to look where suspension feeders that do not rummage through the sediments for food dominate—is not a good one. Metazoan suspension feeders simply are too rare in the deep sea ever to dominate soft bottoms. The best one can hope for is to find variations in the abundances, sizes, and activity levels of deposit feeders.

Guinasso and Schink (1975) provide a nondimensional index (G) of mixing intensity that varies inversely with stratigraphic resolution:

$$G = \frac{D_b}{L_b S} \tag{1}$$

where S is sediment accumulation rate (thickness per unit of time) and L_b is the thickness of the layer in which bioturbation occurs. Variations in L_b and D_b are quite limited so that the best stratigraphic resolution to date has been observed in areas with highest values of S (Schindel 1980; Shiffelbein 1984). A modest increase in L_b with increasing flux of organic carbon and sedimentation rate (Berger, personal communication) is consistent with the idea that deeper burrowing costs can be repaid when fluxes and concentrations of organic carbon are higher at greater sediment depths. It is tempting to speculate that the relative invariance of L_b is due to rapidly increasing costs of deeper burrowing as the sedimentary overburden increases. L_b does not scale with the most deeply burrowing species found at a site; depth of deepest burrowing is nearly invariably larger. Note also that this mixing relation is developed only for regions free of obvious complications from turbidites. The range in G, nonetheless, is smaller than one would expect if D_b were independent of S. Over the full range of gradual sediment accumulation rates seen in the deep sea, one expects animal size and abundance to vary directly with S, but with a great deal of scatter due to variations in horizontal fluxes. The greatest apparent deep-sea values for D_b in regions of continuous sedimentation have indeed been measured in nearshore regions of high productivity. However, since individual feeding rate (evolutionary time scale) by Cammen's (1980) arguments varies inversely and animal abundance varies directly with sedimentary organic content, the range in D_b is not as large as one might expect. Unfortunately, information to make a more informed guess as to where abundances, kinds, and activities of deposit feeders might permit stratigraphic resolution best to persist is still lacking. Clearly, if we better knew how to predict D_b and L_b from biological information we could better choose sites of higher potential stratigraphic resolution.

We would not expect similar invariance of the equivalent parameters to D_b and L_b with respect to pore water solutes. Animals pump water to meet respiratory needs for oxygen. Where sediments are richer in organic matter and animals are larger, volumetric pumping rates and pumping depths will have to be greater. In oxygenated red clays sparsely populated by small animals, molecular diffusion without pumping may meet metabolic oxygen demand.

Faunas of calcareous and siliceous oozes are particularly understudied. Digestive enzymes are highly pH sensitive, so it would be reasonable to anticipate some differences in digestive kinetics between the terrigenous-sediment feeders so far examined in feeding-rate relationships (Cammen 1980) and deposit feeders on calcareous oozes. Casual (personal) observations of equatorial Pacific radiolarian oozes and animals in them would suggest mechanical difficulty in ingestion and gut passage by deposit feeders. Perhaps Antarctic shelf sediments in areas of massive siliceous sponges also warrant stratigraphic attention for similar reasons, particularly if the sponges mechanically (by spicules fallen to the sediments) or competitively (by intercepting food for deposit feeders) decrease deposit-feeder activity or abundance.

Enough is known, however, to rule out some possibilities. It might be tempting to try, for example, to use apparently large D_b (or large G observed in a fossil sequence) as an index of paleoproductivity. The HEBBLE observations, however, suggest that the correlation of G is good with secondary productivity at the bed or with horizontal flux just above it but not with overlying water productivity. Interpretation is difficult because physical transport as well as biological mixing contribute to the apparent D_b values measured at the HEBBLE site, but the animal abundances and D_b values are comparable to those seen under much more productive overlying waters (DeMaster et al. 1985).

Surface Traces

Surface trace abundance has been suggested to be an index of animal abundance and hence a potential fossil index of paleoproductivity. To evaluate this suggestion, we recently have developed a model of surface trace concentration (fraction of surface area covered by traces) that demonstrates that at the highest animal abundances trace concentration should be low; traces can be erased both by epifaunal movement and by horizontal mixing due to infauna (Wheatcroft et al. 1988). Since trace concentration is simply the areal production rate of new traces (ones that do not cover previous traces) times their residence times, there is not a monotonic relationship between animal abundance (surrogate for production further down the food web) and trace abundance.

We also suggest that bioturbation is generally much more intense in the horizontal than in the vertical. Feeding and defecation by surface deposit feeders (typically comprising one-half of all deposit feeders at deep-sea sites) move material horizontally to a far greater extent than it moves material vertically. The same will be true of horizontally burrowing subsurface deposit feeders. It is the horizontal component of bioturbation that is effective in erasing surface traces, and so disappearance rates of traces may be much faster than anticipated from D_b values estimated from vertical mixing of tracers.

Ecotones and Unsteadiness

We have focused implicitly on what occurs within a given community of benthic organisms and on how what goes on differs among benthic communities. It is also worthwhile to consider the edges of communities, the so-called ecotones. Inefficiency of utilization is often associated with rapid change, and productivity changes may also be associated with changes in the borders of high- and low-productivity regimes. Such regions of change often produce the most spectacular evidences of bioturbation, e.g., of exchange between brown clays and calcareous oozes (Nelson 1985). These regions have not been popular sites for estimation of mixing parameters because so many variables change downcore. The very fact that the sediments remain mottled rather than appearing completely mixed, however, gives impetus to further exploration. These sites would also appear to warrant attention from the standpoint of being places where one might expect some inefficiency in processing of sediments by an imperfectly adapted community. In fact, we do not know whether such border regions in the benthos contain "ecotone specialist" species—as is found in the oceanic zooplankton between major water masses—or whether one simply sees a mixture of two neighboring faunas.

Acknowledgements. The opportunity for us to explore these relations experimentally from the perspectives of both organisms and sediments was provided by NSF Grant OCE-8608157 and ONR contract N0014-87-C-0160, respectively. Wolf Berger, Steve Emerson, Erwin Suess, and especially Barry Hargrave gave valuable comments on an earlier draft.

REFERENCES

Allen, J.A., and Sanders, H.L. 1966. Adaptations to abyssal life as shown by the bivalve *Abra profundorum* (Smith). *Deep-Sea Res.* **13**: 1175–1184.

Aller, R.C. 1982. The effects of macrobenthos on chemical properties of marine sediment and overlying water. In: Animal-Sediment Relations, eds. P.L. McCall and M.J.S. Tevesz, pp.53–102. New York: Plenum.

Cammen, L.M. 1980. Ingestion rate: an empirical model for aquatic deposit feeders and detritivores. *Oecologia (Berl.)* **44**: 303–310.

DeMaster, D.J.; McKee, B.A.; Nittrouer, C.A.; Brewster, D.C.; and Biscaye, P.E. 1985. Rates of sediment reworking at the HEBBLE site based on measurements of Th-234, Cs-137 and Pb-210. *Marine Geol.* **66**: 133–148.

Dworschak, P.C. 1983. The biology of *Upogebia pusilla* (Petagna) (Decapoda, Thalassinidae). 1. The burrows. P.S.Z.N.I: *Marine Ecol.* **4**: 19–43.

Gooday, A.J. 1988. A response by benthic foraminifera to the deposition of phytodetritus in the deep sea. *Nature* **332**: 441–443.

Griggs, G.B.; Carey, A.G., Jr.; and Kulm, L.D. 1969. Deep-sea sedimentation and sediment-fauna interaction in Cascadia Channel and on Cascadia Abyssal Plain. *Deep-Sea Res.* **16**: 157–170.

Guinasso, N.L., Jr., and Schink, D.R. 1975. Quantitative estimates of biological mixing rates in abyssal sediments. *J. Geophys. Res.* **80**: 3032–3043.

Hessler, R.R., and Jumars, P.A. 1974. Abyssal community analysis from replicate box cores in the central North Pacific. *Deep-Sea Res.* **21**: 185–209.

Jumars, P.A., and Fauchald, K. 1977. Between-community contrasts in successful polychaete feeding strategies. In: Ecology of Marine Benthos, ed. B.C. Coull. Columbia: Univ. South Carolina Press.

Kristensen, E., and Blackburn, T.H. 1987. The fate of organic carbon and nitrogen in experimental marine sediment systems: influence of bioturbation and anoxia. *J. Marine Res.* **45**: 231–257.

Lopez, G.R.; Taghon, G.L.; and Levinton, J.S., eds. 1989. Ecology of Marine Deposit Feeders. New York: Springer, in press.

McCall, P.L. 1977. Community patterns and adaptive strategies of the infaunal benthos of Long Island Sound. *J. Marine Res.* **35**: 221–266.

Miller, D.C.; Jumars, P.A.; and Nowell, A.R.M. 1984. Effects of sediment transport on deposit feeding: scaling arguments. *Limnol. Ocean.* **29**: 1202–1217.

Nelson, C.S. 1985. Bioturbation in middle bathyal, Cenozoic nannofossil oozes and chalks, southwest Pacific. In: Initial Reports of the Deep Sea Drilling Project, eds. J.P. Kennett and C.C. von der Borch, vol. 90, pp.1189–1200. Washington: US Govt. Printing Office.

Nowell, A.R.M.; Jumars, P.A.; and Fauchald, K. 1984. The foraging strategy of a subtidal and deep-sea deposit feeder. *Limnol. Ocean.* **29**: 645–649.

Odum, E.P. 1969. The strategy of ecosystem development. *Science* **164**: 262–270.

Pearson, T.H., and Rosenberg, R. 1978. Macrobenthic succession in relation to organic enrichment and pollution of the marine environment. *Ocean. Mar. Biol. Ann. Rev.* **16**: 229–311.

Penry, D.L., and Jumars, P.A. 1987. Modeling animal guts as chemical reactors. *Am. Natural.* **129**: 69–96.

Schiffelbein, P. 1984. Effect of benthic mixing on the information content of deep-sea stratigraphic signals. *Nature* **311**: 651–653.

Schindel, D.E. 1980. Microstratigraphic sampling and the limits of paleontologic resolution. *Paleobiol.* **6**: 408–426.

Schwinghamer, P. 1985. Observations on size-structure and pelagic coupling of some shelf and abyssal benthic communities. In: Proceedings of the 19th European Marine Biology Symposium, Plymouth, Devon, UK, 16–21 September 1984, ed. P.E. Gibbs, pp. 347–359. Cambridge: Cambridge Univ. Press.

Sebens, K.P. 1987. The ecology of indeterminant growth in animals. *Ann. Rev. Ecol. Syst.* **18**: 371–408.

Sibly, R.M. 1981. Strategies of digestion and defecation. In: Physiological Ecology: An Evolutionary Approach to Resource Use, eds. C.R. Townsend and P. Calow, pp.109–139. Sunderland, MA: Sinauer.

Smith, C.R.; Jumars, P.A.; and DeMaster, D.J. 1986. *In situ* studies of megafaunal mounds indicate rapid sediment turnover and community response at the deep-sea floor. *Nature* **323**: 251–253.

Sorensen, J.; Jorgensen, K.S.; Colley, S.; Hydes, D.J.; Thomson, J.; and Wilson, T.R.S. 1987. Depth localization of denitrification in a deep-sea sediment from the Madeira Abyssal Plain. *Limnol. Ocean.* **32**: 758–762.

Taghon, G.L. 1988. The benefits and costs of deposit feeding in the polychaete *Abarenicola pacifica*. *Limnol. Ocean.* **33**, in press.

Taghon, G.L., and Jumars, P.A. 1984. Variable ingestion rate and its role in optimal foraging behavior of marine deposit feeders. *Ecology* **65**: 549–558.

Tenore, K.R.; Hanson, R.B.; McClain, J.; Maccubbin, A.E.; and Hodson, R.E. 1984. Changes in composition and nutritional value to a benthic deposit feeder of decomposing detritus pools. *Bull. Mar. Sci.* **35**: 299–311.

Thistle, D.; Yingst, J.Y.; and Fauchald, K. 1985. A deep-sea benthic community exposed to strong near-bottom currents on the Scotian Rise (western Atlantic). *Marine Geol.* **66**: 91–112.

Trevor, J.H. 1978. The dynamics and mechanical energy expenditure of the polychaetes *Nephthys cirrosa, Nereis diversicolor*, and *Arenicola marina* during burrowing. *Est. Coast. Shelf Sci.* **6**: 605–619.

Tyler, P.A. 1986. Studies of a benthic time series: reproductive biology of benthic invertebrates in the Rockall Trough. *Proc. Roy. Soc. Edinb.* **88B**: 175–190.

Wheatcroft, R.A.; Smith, C.R.; and Jumars, P.A. 1988. Dynamics of surficial trace assemblages in the deep sea. *Deep-Sea Res.*

Productivity of the Ocean: Present and Past
eds. W.H. Berger, V.S. Smetacek and G. Wefer, pp. 255–269
John Wiley & Sons Limited

Productivity Record in Benthic Foraminifera

A.V. Altenbach and M. Sarnthein

Geologisch-Paläontologisches Institut der Christian-Albrechts-Universität 2300 Kiel, F.R. Germany

Abstract. Single-celled organisms contribute about 50–70% of the total metabolic turnover in the deep sea, and benthic Foraminifera constitute a major portion of the benthic biomass. Thus, foraminiferal biomass provides a record of the nutrient flux and, ultimately, of sea surface productivity. The same holds true for distinct species assemblages, where *Uvigerina* sp. and *Globobulimina* sp. are characteristic of high nutrient fluxes. The distribution pattern of different species is based on characteristic habitat preferences. In contrast to many uvigerinids, *Cibicidoides wuellerstorfi* is representative of a purely epibenthic lifestyle. The $\delta^{13}C$ record of benthic Foraminifera reflects two major signals. Whereas the global record of deep-water ventilation dominates the epibenthic test composition, the record of ocean productivity prevails in the $\delta^{13}C$ of endobenthic forms. The question of how to isolate the two factors in the endobenthic record quantitatively has for the most part remained unsolved.

INTRODUCTION

Foraminifera are rhizopod protozoans characterized by granulated, anastomosing pseudopods and a secreted, agglutinated, or calcareous test. These single-celled organisms comprise one of the most widespread groups, living in nearly all marine and brackish environments. Also, Foraminifera represent one of the most important microfossil groups, based on a detailed paleontological record of more than 40,000 species. The high abundance and the diversity of modern and fossil assemblages allows statistical analyses of small sediment samples. In spite of their importance in biostratigraphy and paleoecology, they are ignored in most marine benthic studies (Lipps 1983; Gooday 1986). When considering the current knowledge of the biology of distinct species and their distribution in space and time, however, important interrelations with marine primary productivity are detectable.

In this paper we document the impact of ocean productivity and carbon flux on benthic Foraminifera and their test compositions. In a number of cases, parameters such as oxygen depletion in benthic microenvironments and physical properties of the sediments will covary, depending on the flux rates of organic matter. Almost the entire bioenergetic activity in the deeper parts of the world ocean is dependent upon this flux. The major exception may be the chemosynthetic activity in the vicinity of hydrothermal vents, which could contribute to the bioenergetic activity independent of organic flux.

FORAMINIFERAL SHARE IN BIOMASS AND METABOLISM OF THE MARINE BENTHOS

Single-celled organisms play a crucial role in the global nutrient cycle of the biosphere. We estimate their contribution to the total metabolic turnover to range from 50% to 70%. Alongside the production performance of single-celled algae and bacterial decomposition during remineralization, Protozoa make up a large percentage of the consumers. Animals of their wide size ranges are important for the food chain, where foraminifers and other large protozoa form an intermediate link by consuming single-celled organisms and, in turn, by being consumed by the meio- and macrobenthos. In the marine benthos, the foraminifers not only build up an important portion of the biomass (Gooday 1986), their turnover rate may even eclipse that of the meio- and macrobenthos (Gerlach et al. 1985). First tentative estimates of the turnover rates in the deep sea indicate an annual reproduction cycle (Altenbach 1985). Recent studies have also focused on benthic Foraminifera which occupy empty tests of planktonic Foraminifera, the abundance of forms with proteinaceous tests (Allogromiina), the importance of Komokiacea (Textulariina), and the barite crystal-bearing xenophyophores (Xenophyophorea, Protozoa). Some of these rhizopods seem hitherto unrecognized or, at least, underrepresented in benthic studies (for discussion see Gooday 1986). Thus, their contribution to total turnover rates is still unknown.

On the marine shelf and slope the regional distribution of biomass of benthic Foraminifera varies strongly (Table 1). In the deep sea, Foraminifera are major components of the total biomass, especially if the abundance of agglutinated Foraminifera is fully recognized (Khusid 1974; Gooday 1986). Foraminiferal biomass exceeds that of any other taxon in the abyssal Pacific (Smith et al. 1976). In the central North Pacific, foraminiferal biomass was estimated to be two orders of magnitude higher than that of other meiofauna (Snider et al. 1984). In the Arctic Ocean, benthic Foraminifera represent 53% of the benthic biomass, with bivalves forming only 27%, sponges 7%, and polychaetes 5% (Paul and Menzies 1974). Further research is necessary for better estimation of deep-sea benthic biomass, although the dominance of Foraminifera appears well substantiated.

TABLE 1 Maximum values of benthic foraminiferal biomass reported from different regions in the world ocean.

Area of Investigation	Refs.	Water depth (m)	Primary product $g\ C\ m^{-2}\ yr^{-1}$	Carbon flux $g\ C\ m^{-2}\ yr^{-1}$	Maximum biomass $g\ m^{-2}$
Puerto Deseado, Patagon	(1)	3–9	?	–	2.5_{++}
Baltic Sea, Kiel Bight	(2)	26	158	–	2.6
North Sea, Helgoland	(3)	34	100–150	–	40.0
Celtic Sea	(4)	128–138	100?	–	0.8
Peru-Chile Trench	(5)	330	200–500	17.7–32.0	70.0
Falklands	(6)	720	100?	4.6	29.9
Off NW Africa	(7)	980	212	10.1	5.3_{+}
Okhotsk Sea	(8)	1020	100–200	3.8–9.5	21.5
Arctic, Alpha Cord	(9)	2038	< 1?	< 0.005	0.8
Peru-Chile Trench	(5)	2100	200–500	6.4–11.7	50.0
Off NW Africa	(7)	2822	212	5.6	1.2_{+}
NW Pacific	(8)	3100	100–200	2.0–5.1	8.0
South Sandwich Trench	(6)	4720	100–200	1.6–4.0	2.0
Peru-Chile Trench	(5)	6900	200–500	3.3–6.0	12.0
Kurile Trench	(10)	7000–8000	100–200	1.2–3.3	9.0

Values based on extrapolation of plasm volumes from test volumes. Exceptions: $_{+}$ = wet weight from measurement of org. C content of protoplasm (fraction > 250 μm); $_{++}$ = wet weight from subtracting ash weight from dry weight after combustion. Additional tables in Murray (1973: 202) and Gooday (1986: 1365).

Values on primary productivity from cited authors or from Berger et al. (1987). Carbon flux based on Sarnthein et al. (1988, equation 3).

(1) Boltovskoy and Lena (1969)
(2) Wefer and Lutze (1976)
(3) Gerlach et al. (1985)
(4) Murray (1973)
(5) Khusid (1974)
(6) Basov (1974)
(7) Altenbach (1985)
(8) Basov and Khusid (1983)
(9) Fetter (1973)
(10) Saidova (1967)

Recent findings from sediment trap studies and deep-sea photography indicate that the deep-sea benthos receives a large proportion of its nutrient supply from short-term flux pulses (Gooday 1988). This possibly provides insight into the role Foraminifera play in the deep-sea benthos. In shallow seas, the input of a single plankton bloom is consumed completely in one or two weeks. The extremely fast response of Foraminifera is reflected by their concomitantly rapid growth phase (Lipps 1983; Altenbach 1985). The potential turnover rate of single-celled organisms strongly depends on the ratio of cell surface to cell volume, a ratio which is maximized in rhizopods.

With their large biomass and high potential growth rates, benthic deep-sea Foraminifera possibly have a more significant effect on the structure and dynamics of the deep-sea benthic ecosystem than any of the macrofaunal taxa that are the usual objects of deep-sea ecological studies (Bernstein et al. 1978).

PRODUCTIVITY RECORDS FROM FORAMINIFERAL ASSEMBLAGES

Numerous examples of modern and ancient sediments demonstrate that high accumulation rates of organic matter on the seafloor underlie regions of high primary production and are demarcated by distinct species or communities of benthic Foraminifera. Upwelling productivity in particular can induce distinct assemblages of benthic Foraminifera and a large population size (Sen Gupta et al. 1981; Lipps 1983; Lutze and Coulbourn 1984; Caralp 1984; Boersma 1986; Lutze et al. 1986).

Off Northwest Africa, the species distribution of benthic Foraminifera traces distributional patterns of primary production downslope to the deep sea (Lutze and Coulbourn 1984). A major downslope running, i.e., vertical border of the biocenoses parallels the northern edge of maximal upwelling productivity. Note that it is not related to any physical or chemical parameter defining a deep-ocean water mass or other deep-sea physiographic barriers. Extensive grain-size studies (Koopmann 1981) show that sediment winnowing, and hence current strength at the seafloor, does not vary across boundaries between foraminiferal biocenoses (cf. Jumars and Wheatcroft, this volume). The *Cibicidoides wuellerstorfi*/*C. kullenbergi* biofacies dominates in areas with low productivity and the resulting low flux rates of carbon, while the *Uvigerina peregrina*/*Globobulimina turgida* biofacies is present in areas with high productivity and flux rates (Altenbach 1985). Accordingly, the biomass total of benthic foraminiferal assemblages is positively correlated with the estimated local annual flux rate of organic carbon into the deep sea (Fig. 1, below 500 m). Biological studies have shown that concentrations of organic carbon in the surface sediment correlate only loosely with large benthic biomass values and high turnover rates in the deep sea. Better correlations are achieved with organic matter derived from the primary production cycle, such as chlorophyll pigments (Fig. 2). This example of a linear correlation ($r = 0.83$) seems promising since it provides a first step towards the understanding of foraminiferal nutrient requirements in the deep sea (see also Gooday 1988).

A more complex example of the nutrient control of benthic foraminiferal assemblages was found in the Arctic. Oceanographic data indicate that near-surface Atlantic waters extend via the Svalbard current to almost 83/84°N below perennial sea ice cover. On the bottom beneath this current, where

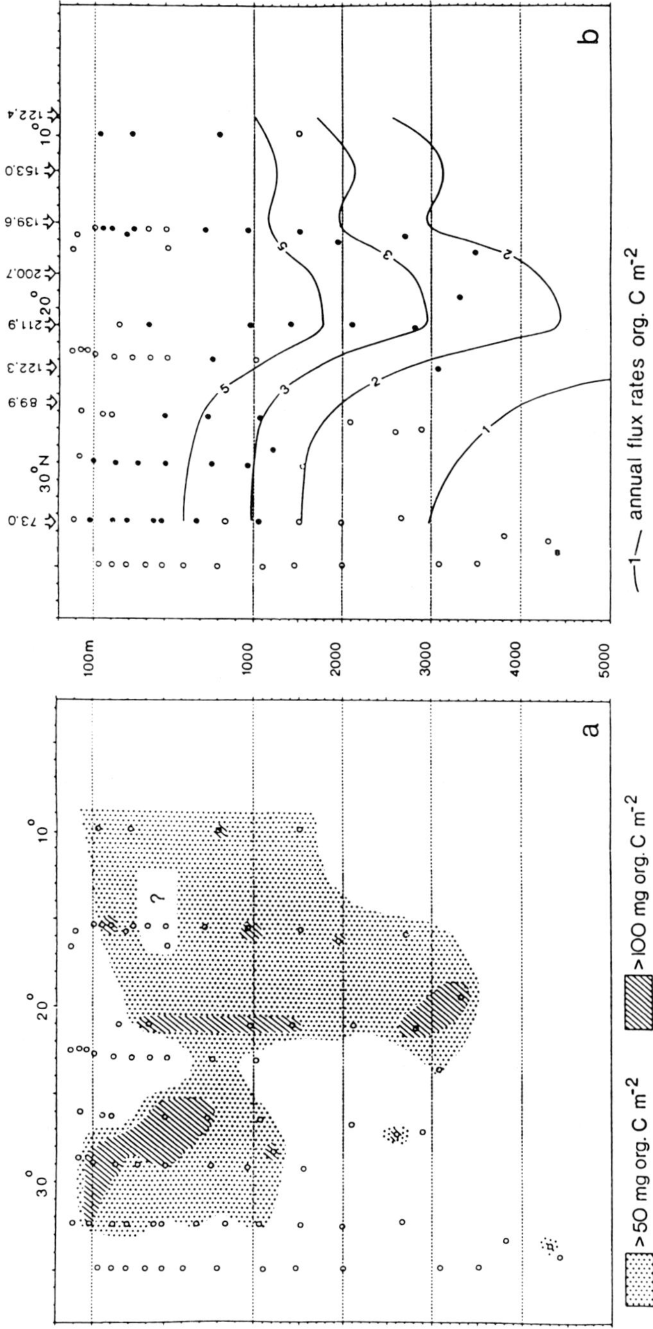

Fig. 1—Three-step vertically exaggerated depth scale of the continental margin of N.W. Africa from 10°N to 35°N (modified from Altenbach 1985). Description of marked stations in Lutze (1980) and Lutze and Coulbourn (1984). (*a*) Biomass of benthic Foraminifera (fraction > 250 μm) in milligrams organic C per square meter. (*b*) Annual flux rates of particulate organic C (g m^{-2}) to the sediment surface.

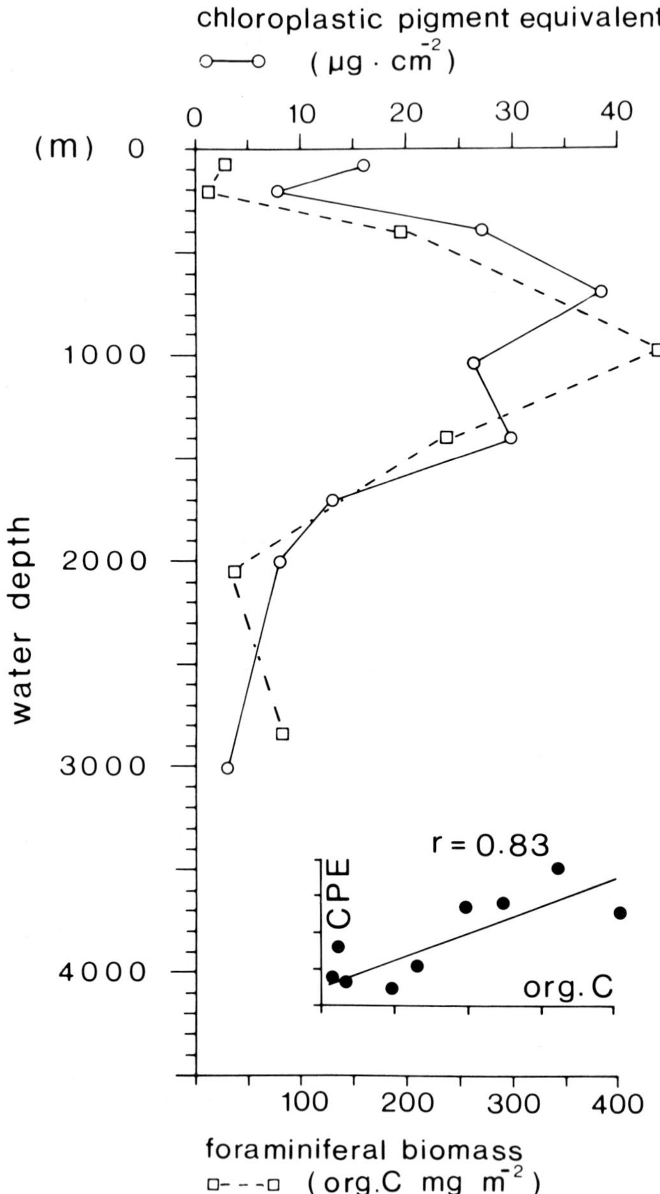

Fig. 2—Correlation of chloroplastic pigment equivalents (CPE) in surface sediments vs. foraminiferal biomass (organic C, fraction > 250 µm) on the N.W. African continental margin at 21°N. The correlation coefficient is significant at the 0.01 level (from Altenbach 1985).

the overlying surface layer is populated primarily by North Atlantic phyto- and zooplankton, benthic foraminiferal communities are similar to those offshore from the ice-free Norwegian coast. Further north, a specific Arctic plankton community exists, from which a reduced flux rate and a different composition is derived, resulting in a totally different benthic community (Thiede 1988). These preliminary results suggest that near-surface currents may transport nutrients far into areas where productivity is low, inducing locally enhanced input of organic matter and a specific response in the benthic community.

The pulsation of high productivity in space and time appears to be the primary cause of benthic foraminiferal fluctuations off northwest Africa. At present, the benthic Foraminifera patch is located right below the Cap Blanc upwelling area between 12°N and 22°N, i.e., the South Atlantic Central Water (Labracherie 1980). The patch expanded up to 27°N during the last glacial stage, indicating an equivalent expansion of the region of upwelling productivity and carbon flux. In this way changes in benthic foraminiferal assemblages have continuously recorded fertility changes in surface water during the last 130,000 years (Lutze et al. 1986). Boersma (1986) observed similar trends in Tertiary sediments of the Tasman Sea, where foraminiferal assemblage patterns were primarily influenced by invigorated oceanic circulation and increased rates of upwelling productivity.

Among the primary parameters most often considered in modern statistical analyses of foraminiferal ecozonations is oxygen content in the bottom water mass and at the water-sediment interface. Increasing flux rates of organic matter will not result only in increasing food abundance in the deep sea, oxygen concentrations and substrate type will also closely depend on the input of organic matter. These factors must overlap in the benthic ecological network and hence will be difficult to distinguish (for discussion see Van Der Zwaan 1982; Woodruff 1985). In order to arrive at a better understanding of these links, we need improved information concerning the requirements for food and/or oxygen of single species and especially of those species dominating in the deep sea.

HABITAT PREFERENCE AND ADAPTATION OF SINGLE SPECIES TO DIFFERENT FOOD SOURCES AND CONCENTRATIONS

Besides food supply and other factors, such as predation and bioturbation, competition for food is an important factor for growth and structure of foraminiferal assemblages in the shallow seas (Lee 1974) and, most probably, in the deep-sea as well (Gooday 1988). Different sources of food are available in different foraminiferal habitats and environments. In the deep sea, some benthic Foraminifera are positively correlated in abundance with subsurface macrobenthic deposit feeders while other foraminiferal species are correlated with surface deposit feeders (Bernstein et al. 1978). Likewise, the distinction of microhabitats in the sediment, on the sediment surface, and on emergent substrates can reveal specific food preferences and explain distinct morphological features in epifaunal and infaunal species (Corliss 1985).

Excellent examples are found in *Uvigerina peregrina* and *Cibicidoides wuellerstorfi*, which greatly differ in their main distributional patterns in the deep sea. In the record of deep-sea cores, the abundance of these species usually follows negatively cross-correlated fluctuation patterns. According to various authors (Van Der Zwaan 1982; Lutze and Coulbourn 1984; Altenbach 1985; Boersma 1986), *U. peregrina* requires a high nutrient input whereas *C. wuellerstorfi* is frequent in areas or stages of low supply of organic matter. Recent information on the lifestyles of these species reveals that *C. wuellerstorfi* belongs to the epifauna and inhabits the "trees" on the seafloor, i.e., protruding substrates such as macrobenthic hard parts and exposed rocks (Lutze and Thiel 1987). Vital specimens were never found on the sediment surface. On the contrary, our *in vivo* observations on *U. peregrina* suggest that this species lives within the uppermost sediment layer. Note that different species and subspecies of the *U. peregrina* group may require different habitats (Lutze 1986; Lutze et al. 1986). The dependence on nutrients of these species may be reasonably established by their lifestyles and habitats. As a suspension feeder, *C. wuellerstorfi* does not depend on high vertical flux rates and easily tolerates rates of less than $2\ g\ C\ m^{-2}\ yr^{-1}$ (Altenbach 1988). In the case of favorable current conditions, a low particle concentration is sufficient to satisfy the necessary nutrient particle input per unit of time and, hence, the nutritional needs of the species. In contrast, *U. peregrina* lives in sediment zones in which the metabolic turnover rates in the deep sea reach a maximal value. The microhabitat of this species is characterized by high concentrations of bacteria, exoenzymes, and meiofauna and is typical of sediments enriched in organic carbon and depleted in oxygen, such as in areas below upwelling productivity. Feeding on solutes (dissolved organic carbon: DOC) may contribute significantly to the nutrition of foraminiferal plasm below the sediment surface (DeLaca et al. 1981).

THE PRODUCTIVITY RECORD OF CARBON STABLE ISOTOPES IN BENTHIC FORAMINIFERA

The dissolved oxygen and CO_2 content of a surface water mass, which is not upwelled, is in balance with the atmosphere. Hence, a deep-ocean water mass originating from this surface water is enriched in dissolved oxygen and characterized by dissolved atmospheric CO_2 with a carbon isotopic composition of a little over 1‰ $\delta^{13}C$ (vs. PDB standard). During its passage through the deep reaches of the ocean, dissolved oxygen is utilized by the oxidation of particulate organic matter sinking from the sea surface, i.e., the water mass "ages" (Duplessy et al. 1980). Based on the addition of isotopically light organic carbon averaging about −20‰ (a result of metabolic carbon isotopic fractionation; Fontugne and Duplessy 1986), the total dissolved CO_2 in the deep water becomes isotopically lighter (Fig. 3). This supply of organic matter largely originates from the primary production in the overlying ocean surface layer, which in turn is strongly controlled by upwelling intensity, i.e., the availability of nutrients in low- and mid-latitudes. Thus, when interpreting the $\delta^{13}C$ composition of a bottom water mass, which serves as a measure for total dissolved CO_2, two factors must be considered that control the CO_2 concentration: (*a*) the local surface ocean fertility and flux of carbon and (*b*) the general, global renewal rate of oxygenated deep water. Besides the general ^{12}C enrichment/^{13}C depletion of deep ocean water, the carbon flux from high productivity belts also results in a local ^{13}C depletion of the bottom water within the nepheloid layer. Kroopnick (1971) observed ^{13}C depletion by 0.35‰ in the bottom water up to 50 m above the seafloor below the belt of upwelling productivity in the equatorial Pacific. $CaCO_3$ tests of benthic Foraminifera provide a good opportunity to document variations in $\delta^{13}C$ near the seafloor through time (e.g., Shackleton 1977). However, there are strong interspecific differences in $\delta^{13}C$ composition as summarized by Grossman (1984). For example, *C. wuellerstorfi* is close to $\delta^{13}C$ equilibrium with the $\Sigma\ CO_2$ of the ambient deep water mass, whereas many coeval benthic species, such as *Pyrgo* sp. and the *Uvigerina peregrina* group, are not (Woodruff et al. 1980). Traditionally, this $\delta^{13}C$ departure is mainly ascribed to metabolic processes summarized as "vital effects" and is regarded as fairly constant (e.g., Shackleton 1977; Duplessy et al. 1984).

Interspecific differences in $\delta^{13}C$ values, however, may be due to microhabitat differences, as postulated by Grossman (1984) and Berger and Vincent (1986). This is demonstrated in the recent data of Zahn et al. (1986), regarding the key species *C. wuellerstorfi* and *U. peregrina* (group). The clue is in the fact that the difference in $\delta^{13}C$ has changed in predictable ways through glacial and interglacial times. This conclusion is corroborated by the fact that the interspecific $\delta^{13}C$ difference increases in a nonlinear mode with the enhanced accumulation of organic carbon (Fig. 4). This carbon preferentially affects the ambient pore water environment of the infaunal species *U. peregrina* and not the markedly epifaunal *C. wuellerstorfi* (Lutze and Thiel 1987). Indeed, the $\delta^{13}C$ difference between the two species attains a maximum of about 2.0‰, which comes very close to the ^{13}C depletion by 2.7‰ reported by Grossman (1984), for the isotopic composition of inorganic dissolved carbon in the pore water of the 0–2 cm interval as compared to free bottom water.

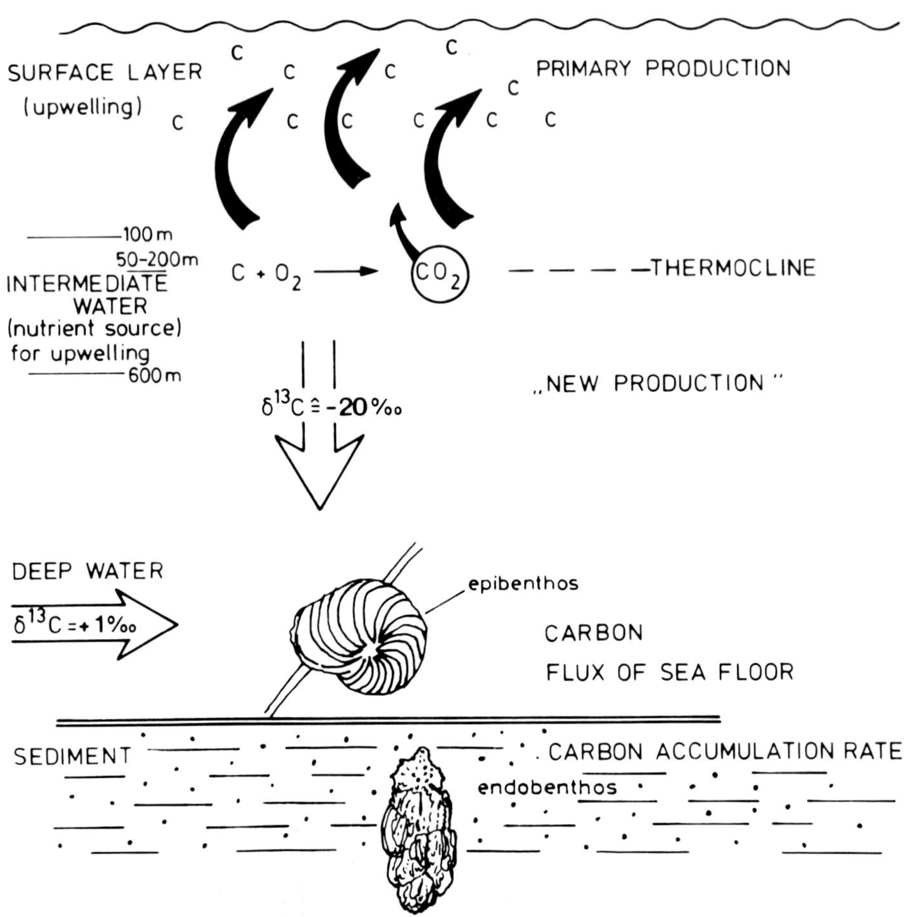

Fig. 3—Basic pattern of organic C reservoirs in the ocean and their effect on the $\delta^{13}C$ composition of benthic Foraminifera.

In summary, carbon isotope values of benthic Foraminifera reflect both a global signal and a locally derived signal. The $\delta^{13}C$ signal of infaunal species shows a stronger but nonlinear imprint from the local carbon flux and sea surface productivity, whereas the signal of epibenthic species such as *C. wuellerstorfi* is largely dominated by global changes in the geochemical cycling of carbon in the deep ocean (e.g., Duplessy et al. 1984). Nevertheless, local variations in sea surface productivity were shown to influence also the epibenthic carbon isotope record of *C. wuellerstorfi* (Woodruff and Savin 1985; Sarnthein et al. 1988). During the Last Glacial Maximum, the bottom water of the world ocean (below 2000 m w.d.) was generally ^{13}C depleted by 0.45‰, but *C. wuellerstorfi* was ^{13}C depleted by up to 0.9‰ below

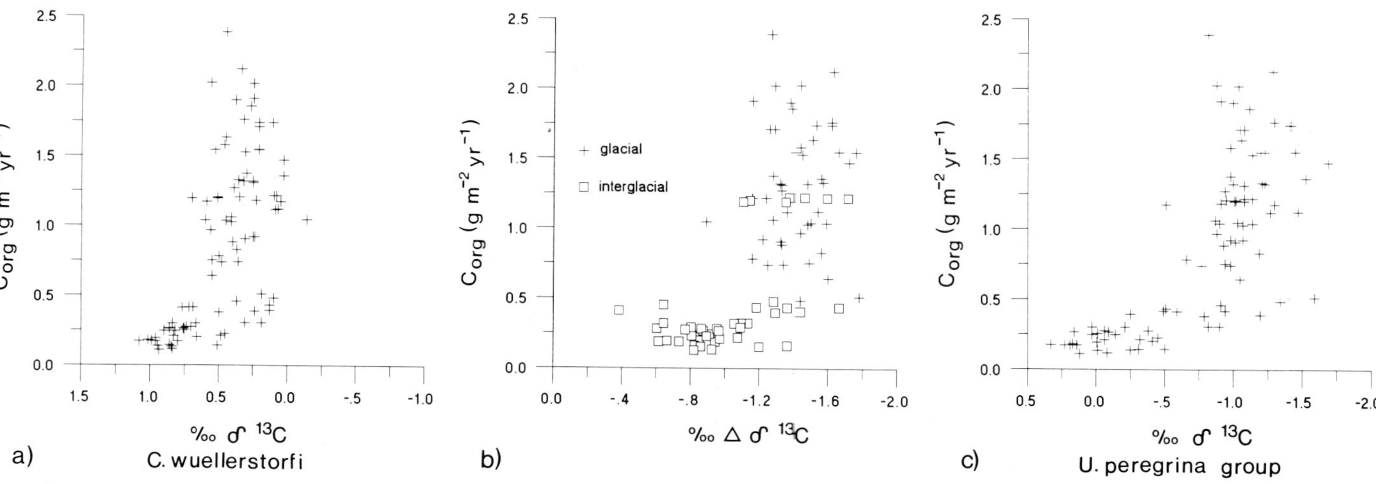

Fig. 4—Relation between carbon accumulation rates in METEOR core 12392 and (*a*) $\delta^{13}C$ values of *Cibicidoides wuellerstorfi*; (*b*) the $\delta^{13}C$ departure between (*a*) and (*c*); and (*c*) $\delta^{13}C$ values of *Uvigerina peregina* group (data modified from Zahn et al. 1986).

pronounced synglacial high-productivity belts. The excess maximum of 0.45‰ precisely corresponds to that observed by Kroopnick (1971) for the modern nepheloid layer near the seafloor below equatorial upwelling. Accordingly, it still remains unsolved as to how to quantify the imprint of local productivity and carbon flux on the *epi*benthic $\delta^{13}C$ record and how one can isolate it from that of global deep-water ventilation (Berger and Vincent 1986). Woodruff and Savin (1985) suggested comparison of coeval differences in $\delta^{13}C$ between different sites with local differences in the sediment organic carbon concentration. They found a 1‰ depletion in ^{13}C approximately corresponding to an absolute increase of organic carbon by 3‰. However, the carbon concentrations are only partly and nonlinearly controlled by paleoproductivity (Sarnthein et al. 1988). Also, the approach of Zahn et al. (1986), which shows that the $\delta^{13}C$ difference values of coeval epi- and endobenthic Foraminifera are plainly linked to carbon flux rates, escapes clear quantification since a large proportion of the data in the regression attain a constant maximum ^{13}C depletion near 2‰ (Fig. 4). Moreover, the influence of primary productivity on epibenthic $\delta^{13}C$ can be estimated at any single site by comparing a high-resolution epibenthic $\delta^{13}C$ record with a $\delta^{13}C$ record from the thermocline as obtained from deep-living planktonic foraminiferal species such as *G. inflata*. Recent studies have shown that in cases where the planktonic and benthic curves run strictly parallel and the planktonic forms are more ^{13}C depleted than the benthic forms, one may assume that the bottom water $\delta^{13}C$ and CO_2 are largely controlled by the vertical particulate carbon flux (Sarnthein and Tiedemann 1988). Conversely, if the two $\delta^{13}C$ curves vary independently and if the benthic signal is markedly more ^{13}C depleted, it can be expected to depend largely on the lateral advection of more or less oxygenated deep water, independent of local surface ocean productivity.

In summary, epibenthic $\delta^{13}C$ values can provide some semiquantitative clues to paleoproductivity, but quantitative estimates still remain elusive.

Acknowledgements. We gratefully acknowledge the careful reviews of M.V. Angel, W.H. Berger, P. Jumars, and E. Suess. This study was supported by the Sonderforschungsbereich 313 at Kiel University (Deutsche Forschungsgemeinschaft) and by the German National Climate Project (Grant KF 2004/1).

REFERENCES

Altenbach, A.V. 1985. Die Biomasse der benthischen Foraminiferen. Auswertungen von "Meteor"-Expeditionen im östlichen Nordatlantik. Ph.D. diss., Univ. of Kiel.

Altenbach, A.V. 1988. Deep sea benthic foraminifera and flux rates of organic carbon. *Rev. Paléobiol.* (vol. spéc.) **2**: 719–720.

Basov, I.A. 1974. Biomass of benthic foraminifers in the region of the South Sandwich Trench and Falkland Islands. *Oceanology* **14**: 277–279 (Acad. Sci. USSR, trans. by Amer. Geophys. Union).

Basov, I.A., and Khusid, T.A. 1983. Biomass of benthic foraminifera in sediments of the Sea of Okhotsk. *Oceanology* **23 (4)**: 489–495 (Acad. Sci. USSR, trans. by Amer. Geophys. Union).

Berger, W.H.; Fischer, K.; Lai, C.; and Wu, G. 1987. Ocean productivity and organic carbon flux. Part I. Overview and maps of primary production and export production. SIO Reference Series 87–30. La Jolla, CA: Scripps Inst. of Oceanography.

Berger, W.H., and Vincent, E. 1986. Deep-sea carbonates: reading the carbon-isotope signal. *Geol. Rundschau* **75**: 249–270.

Bernstein, B.B.; Hessler, R.R.; Smith, R.; and Jumars, P.A. 1978. Spatial dispersion of benthic foraminifera in the abyssal central North Pacific. *Limnol. Ocean.* **23 (3)**: 401–416.

Boersma, A. 1986. Biostratigraphy and biogeography of Tertiary bathyal benthic foraminifers: Tasman Sea, Coral Sea, and on the Chatham Rise. In: Initial Reports of the Deep Sea Drilling Project, eds. J.P. Kennet, C.C. von der Borch, et al., vol. 90, pp. 961–1035. Washington, D.C.: U.S. Govt. Printing Office.

Boltovskoy, E., and Lena, H. 1969. Seasonal occurrences, standing crop and production in benthic foraminifera of Puerto Deseado. *Contr. Cushman Found. Foram. Res.* **20 (3)**: 87–95.

Caralp, M.-H. 1984. Impact de la matière organique dans des zones de forte productivité sur certains foraminiferes benthiques. *Oceanol. Acta* **7 (4)**: 509–515.

Corliss, B. 1985. Microhabitats of benthic foraminifera within deep-sea sediments. *Nature* **314**: 435–438.

DeLaca, T.E.; Karl, D.M.; and Lipps, J.H. 1981. Direct use of dissolved organic carbon by agglutinated benthic foraminifera. *Nature* **289**: 287–289.

Duplessy, J.-C.; Moyes, J.; and Pujol, C. 1980. Deep water formation in the North Atlantic Ocean during the last ice age. *Nature* **286**: 479–482.

Duplessy, J.-C.; Shackleton, N.J.; Matthews, R.K.; Prell, W.; Ruddiman, W.F.; Caralp, M.; and Hendy, C.H. 1984. ^{13}C record of the benthic foraminifera in the last Interglacial ocean: implications for the carbon cycle and the global deep water circulation. *Quatern. Res.* **21 (2)**: 244–263.

Fetter, C. 1973. Recent deep-sea benthic foraminifera from the Alpha Ridge Province of the Arctic Ocean. In: Benthic Ecology of the High Arctic Deep-sea, eds. A.Z. Paul and R.L. Menzies, pp. 296–337. Report of the Dept. of Oceanography, Florida State Univ.

Fontugne, M.R., and Duplessy, J.-C. 1986. Variations of the monsoon regime during the upper Quaternary: evidence from carbon isotopic record of organic matter in north Indian Ocean sediment cores. *Paleogeog. Pal. Pal.* **56**: 69–88.

Gerlach, S.A.; Hahn, A.E.; and Schrage, M. 1985. Size spectra of benthic biomass and metabolism. *Mar. Ecol. Prog. Ser.* **26**: 161–173.

Gooday, A.J. 1986. Meiofaunal foraminiferans from the bathyal Porcupine Seabight (northeast Atlantic): size structure, standing stock, taxonomic composition, species diversity and vertical distribution in the sediment. *Deep-Sea Res.* **33 (10)**: 1345–1373.

Gooday, A.J. 1988. A response by benthic foraminifera to the deposition of phytodetritus in the deep sea. *Nature* **332**: 70–73.

Grossman, E.L. 1984. Stable isotope fractionation in live benthic foraminifera from the southern California borderland. *Paleogeog. Pal. Pal.* **47**: 301–327.

Khusid, T.A. 1974. Distribution of benthic foraminifers off the West Coast of South

America. *Oceanology* **14**: 900–904 (Acad. Sci. USSR, trans. by Amer. Geophys. Union).

Koopman, B. 1981. Sedimentation von Saharastaub im subtropischen Atlantik während der letzten 25,000 Jahre. *"Meteor" Forsch. Ergeb. C* **35**: 23–59.

Kroopnick, P. 1971. Oxygen and carbon in the oceans and atmosphere: stable isotopes as tracers for consumption, production, and circulation models. Ph.D. diss., Univ. of California at San Diego.

Labracherie, M. 1980. Les Radiolaires temoins de l'évolution hydrologique depuis le dernier maximum glaciaire au large du Cap Blanc (Afrique du Nord Ouest). Paleogeog. Pal. Pal. **32**: 163–184.

Lee, J.J. 1974. Towards understanding the niche of foraminifera. In: Foraminifera, eds. R.H. Hedley and C.G. Adams, vol. 1, pp. 207–260. London: Academic Press.

Lipps, J.H. 1983. Biotic interactions in benthic foraminifera. In: Biotic Interactions in Recent and Fossil Benthic Communities, eds. M.J.S. Fevesz and D.L. McCall, pp. 331–375. London: Plenum.

Lutze, G.-F. 1986. Uvigerina species of the eastern North Atlantic. *Utrecht Micropal. Bull.* **35**: 21–46.

Lutze, G.-F., and Coulbourn, W.T. 1984. Recent benthic foraminifera from the continental margin of Northwest Africa: community structure and distribution. *Mar. Micropal.* **8**: 361–401.

Lutze, G.-F.; Pflaumann, U.; and Weinholz, P. 1986. Jungquartäre Fluktuationen der benthischen Foraminiferenfaunen in Tiefsee-Sedimenten vor NW-Afrika: eine Reaktion auf Produktivitätsänderungen im Oberflächenwasser. *"Meteor" Forsch. Ergeb. C* **40**: 163–180.

Lutze, G.-F., and Thiel, H. 1987. *Cibicidoides wuellerstorfi* and *Planulina ariminensis*, elevated epibenthic foraminifera. Univ. of Kiel, *Ber. Sonderforschungsbereich 313* **6**: 17–30.

Murray, W.M. 1973. Distribution and Ecology of Living Benthic Foraminiferids. London: Heinemann.

Paul, A.Z., and Menzies, R.J. 1974. Benthic ecology of the high Arctic deep sea. *Marine Biol.* **27**: 251–262.

Saidova, Ch.M. 1967. Biomassa i kolichestvennoe raspredelenie zhivykh foraminifer v rayone Kurilo-Kamchatskogo zheloba. *Doklady Akad. Nauk SSSR* **174 (1)**: 207–209.

Sarnthein, M., and Tiedemann, R. 1988. Towards a high-resolution stable isotope stratigraphy of the last 3.4 million years, ODP sites 658 and 659 of northwest Africa. In: Proc. ODP, eds. M. Sarnthein et al., vol. 108B. Washington, D.C.: US Govt. Printing Office, in press.

Sarnthein, M.; Winn, K.; Duplessy, J.-C.; and Fontugne, M.R. 1988. Global carbon variations of surface ocean productivity in low and mid latitudes: influence on CO_2 reservoirs of the deep ocean and atmosphere during the last 21,000 years. *Paleocean.* **3** (3): 361–399.

Sen Gupta, B.K.; Lee, R.F.; and Mallory, S.M. 1981. Upwelling and an unusual assemblage of benthic foraminifera on the northern Florida continent. *J. Paleontol.* **55 (4)**: 853–857.

Shackleton, N.J. 1977. Carbon-13 in Uvigerina: tropical rainforest history and the equatorial Pacific carbonate dissolution cycles. In: The Fate of Fossil Fuel in the Oceans, eds. N.R. Anderson and A. Malahoff, pp. 401–427. New York: Plenum.

Smith, K.L., Jr.; Clifford, C.H.; Eliason, A.H.; Walden, B.; Rowe, G.T.; and Teal, J.M. 1976. A free vehicle for measuring benthic community metabolism. *Limnol. Oceano.* **21**: 164–170.

Snider, L.J.; Burnett, B.R.; and Hessler, R.R. 1984. The composition and distribution of meiofauna and nanobiota in a central North Pacific deep-sea area. *Deep-Sea Res.* **31**: 1225–1249.

Thiede, J. 1988. Scientific cruise report of Arctic Expedition ARK IV/3. Bremerhaven, FRG: Reports on Polar Research.

Van Der Zwaan, G.J. 1982. Paleoecology of late Miocene Mediterranean foraminifera. *Utrecht Micropal. Bull.* **25**.

Wefer, G., and Lutze, G.-F. 1976. Benthic foraminifera biomass production in the western Baltic. *Kieler Meeresforsch. Sonderheft* **3**: 76–81.

Woodruff, F. 1985. Changes in Miocene deep-sea benthic foraminiferal distribution in the Pacific Ocean: relationship to paleoceanography. *Geol. Soc. Amer. Mem.* **163**: 131–175.

Woodruff, F., and Savin, S.M. 1985. $\delta^{13}C$ Values of Miocene Pacific benthic foraminifera: correlations with sea level and biological productivity. *Geology* **13**: 119–122.

Woodruff, F.; Savin, S.M.; and Douglas, R.G. 1980. Biological fractionation of oxygen and carbon isotopes by recent benthic foraminifera. *Mar. Micropal.* **5**: 3–11.

Zahn, R.; Winn, K.; and Sarnthein, M. 1986. Benthic foraminiferal $\delta^{13}C$ and accumulation rates of organic carbon (*Uvigerina peregrina* group and *Cibicidoides wuellerstorfi*). *Paleocean.* **1 (1)**: 27–42.

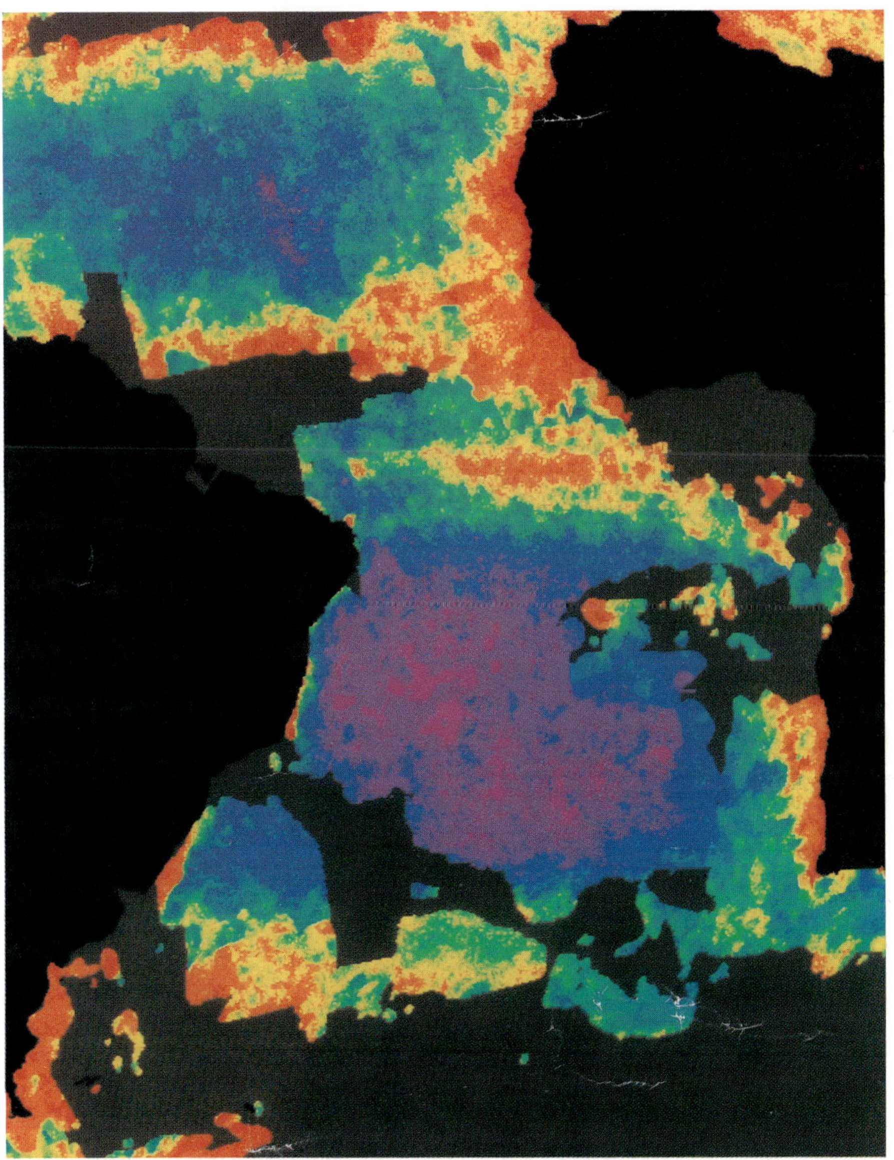

Plate 1—Portion of the first satellite-derived image estimating global distribution of phytoplankton, prepared by the NSF/NASA-sponsored U.S. Global Ocean Flux Study Planning Office, Woods Hole Oceanographic Institution, Woods Hole, MA 02543, with contribution from the Goddard Space Flight Center, University of Miami, and University of Rhode Island. Reproduced by permission. Observations are from the Coastal Zone Color Scanner, taken in December 1981. Red, orange, and yellow: high chlorophyll concentrations. Purple and blue: low chlorophyll values. Grey: no data (cloud).

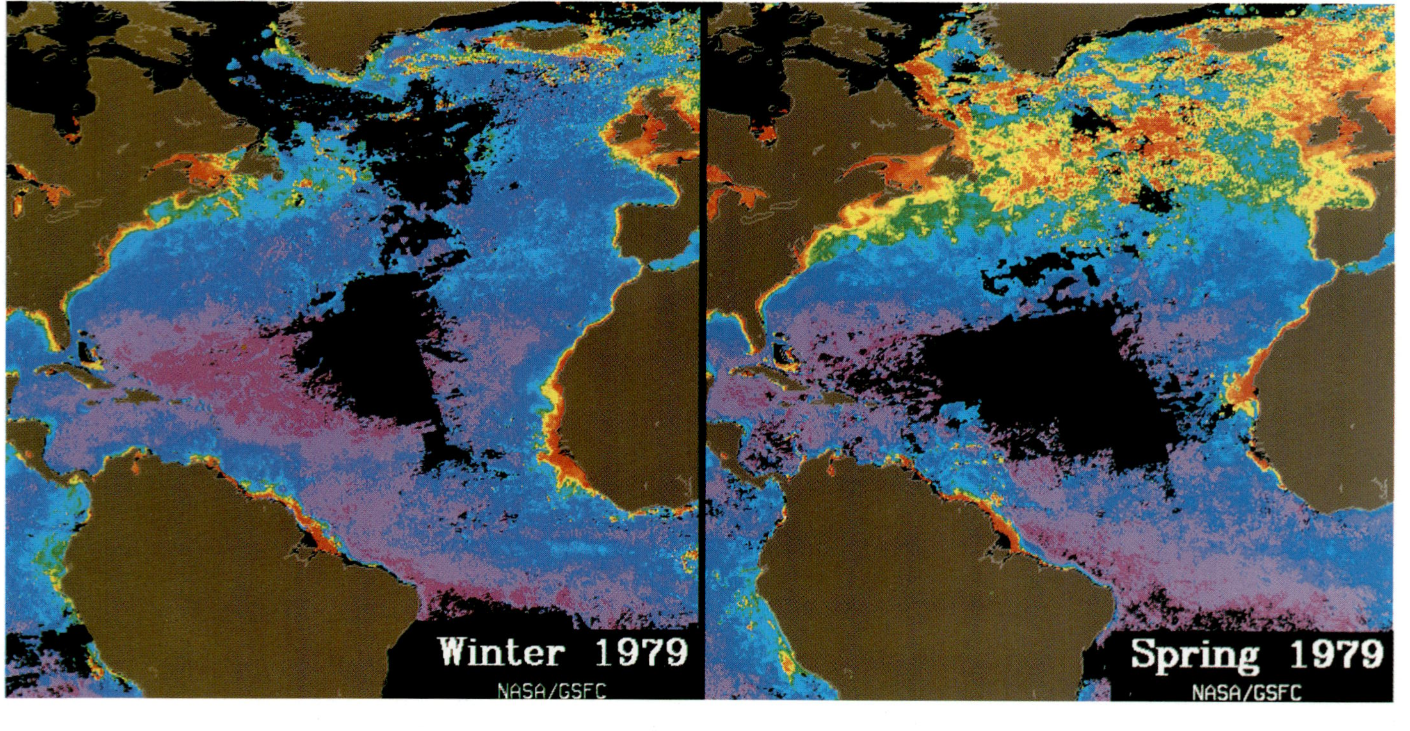
Winter 1979
NASA/GSFC
Spring 1979
NASA/GSFC

Plates 2 and 3— Phytoplankton, the single-celled microscopic plants that form the base of the marine food web, require light and nutrients to grow. As phytoplankton and their associated chlorophyll pigments become more abundant, ocean color shifts from blue to green. Taking advantage of this fact, NASA developed the Coastal Zone Color Scanner (CZCS) which was launched on the Nimbus-7 satellite in 1978. The satellite ocean color measurements derived from the CZCS have made possible our first synoptic view of the distribution and abundance of phytoplankton over the global oceans. More importantly, it has provided us with the ability to monitor how these patterns change both in time and space.

From the temperate regions to the poles, the nutrient supply of the upper ocean is replenished through the deep vertical mixing driven by surface cooling and the strong winter winds. As the seasons progress from winter to spring, the amount of sunlight available for phytoplankton growth increases. This period of rapid growth and accumulation of phytoplankton biomass in the upper waters is referred to as the "Spring Bloom." Its development and intensity are shown in these two seasonal composites of CZCS data acquired during the winter (January-March) and spring (April-June) of 1979. Areas rich in phytoplankton are shown as orange-red (greater than 1 mg/m^3) progressing through yellow, green, and blue with purple (less than 0.1 mg/m^3) representing areas with the lowest chlorophyll concentrations.

While the amount of data shown here is several hundred times more than oceanographers have been able to gather from ships, large regions of the Central Atlantic are still blank since the CZCS could operate only intermittently. Still, the Bloom's intensification with time and its broad spatial extent are clearly evident. More continuous high productivity of the coastal zones and major fisheries' grounds is also seen. The Spring Bloom of phytoplankton in the North Atlantic converts a significant fraction of atmospheric carbon dioxide (CO_2) into particulate matter through photosynthesis. A major focus for oceanographers in the next decade is to understand the dynamics responsible for this global phenomenon and to place it in perspective with the atmospheric CO_2 increase and projected effects on climate.

Credits: NASA/Goddard Space Flight Center (G. C. Feldman, W. E. Esaias, C. R. McClain) and University of Miami/RSMAS (R. H. Evans, J. W. Brown).

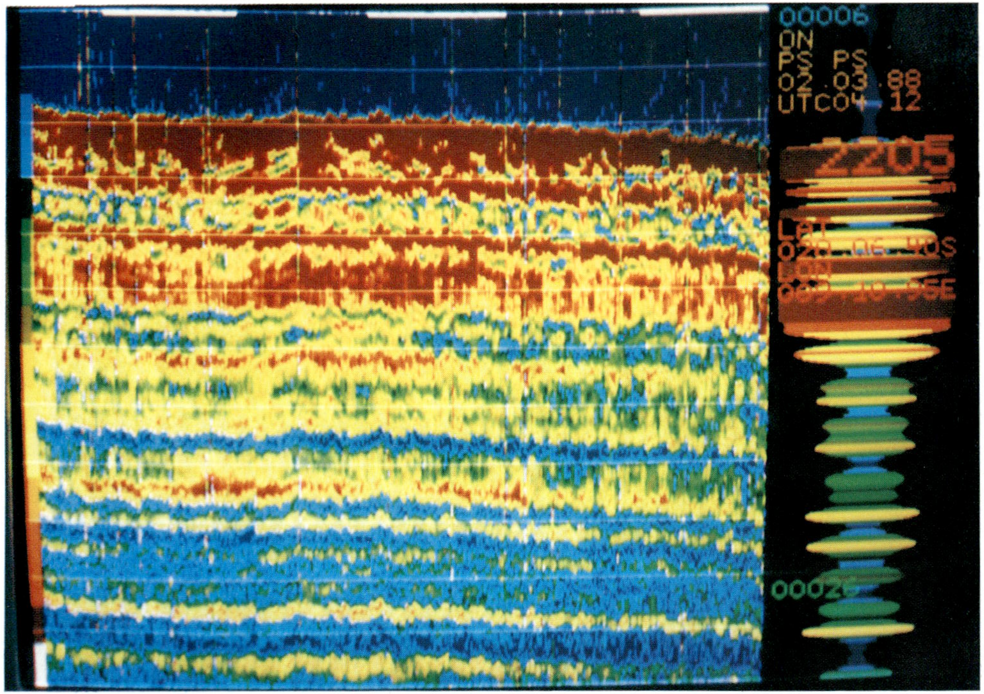

Plate 4—Left: high resolution seismic profile from Walvis Ridge, 20°6′S, 9°11′E, generated by Parasound sonar system of RV METEOR. Frequency ca. 3.5 kHz. Penetration is approximately 100 m. Echo structure shows cyclic deposition in an area of high pelagic productivity; the sediment cycles are presumably generated by glacial/interglacial fluctuations and are associated with variations in supply of biogenous materials.

Right: vertical cut of hydraulic piston core section from Leg 112 (Site 686) from upwelling area off Peru (water depth: 450 m) showing finely stratified sediments rich in organic matter and other biogenous materials. Note alternation between light and dark laminae. Light layers consist of either diatoms or coccoliths, in many cases strongly dominated by a few species.

Productivity of the Ocean: Present and Past
eds. W.H. Berger, V.S. Smetacek and G. Wefer, pp. 271–289
John Wiley & Sons Limited

Lipid Biomarkers as Geochemical Tools for Paleoceanographic Study

F.G. Prahl and L.A. Muehlhausen

College of Oceanography
Oregon State University
Corvallis, OR 97331, U.S.A.

Abstract. This article reviews the geochemistry of two types of source-specific organic molecules ("biomarkers") present in marine sediments throughout the world. The first of these lipid biomarkers is a series of long-chain (C_{37},C_{38},C_{39}) alkenones derived from a particular class of marine phytoplankton. The second is a series of long-chain saturated hydrocarbons and fatty acids derived from the surface wax of vascular plants and introduced to the marine environment from the land via atmospheric and riverine erosional processes. It is shown that (*a*) both types of biomarkers can be calibrated empirically and that (*b*) these calibrations can be employed as effective geochemical tools to gain insight into various paleoceanographic problems. Unsaturation patterns in the marine biomarker series and the total abundance of these compounds measured with depth in dated sediment cores potentially provide an accurate, unambiguous record of how surface water temperature and an important component of marine primary productivity, respectively, have changed in the ocean over glacial/interglacial time scales. Results from mixing model analyses using data for the terrestrial biomarker series suggest that sediments accumulating along continental margins and in the deep sea contain significantly more land-derived organic carbon ($\geq$ 20%) than has been accepted from conventional interpretation of stable isotope data. A reinterpretation of the isotopic data suggests that the lipid biomarker assessment of the terrestrial organic carbon component contained in these sediments, distal from the land, is completely reasonable.

INTRODUCTION

The organic matter buried in ocean sediments is a composite of residual materials derived from marine and terrestrial sources. The rate of transfer of organic matter from these sources to sediments is small, generally 0.1 to 1% of the rate of production (Romankevich 1984). The chemical composition of this small fraction of residual organic matter that survives the process of sedimentation is an extensively modified version of the original whole and bears at best only broad similarity to it. Nonetheless, the residual fraction may contain important clues illuminating the paleoceanographic conditions extant at the time of sediment deposition. Traditionally, the relative proportion of marine and terrestrial organic matter contained in a given sediment has been ascertained through analyses of the stable carbon isotope composition of bulk organic matter. As a general rule, $\delta^{13}C$ values measured on the total sedimentary organic carbon are assumed to reflect the simple binary mixture of organic carbon derived from two isotopically definable and distinct sources: marine plankton and C_3 vascular plants (Degens 1969). Although the isotopic values of the organic carbon biosynthesized by these biological sources can be quite variable (Galimov 1985 and references therein), average values in the range of -18 to $-21‰$ and -25 to $-27‰$ are typically chosen as the marine and terrestrial endmembers, respectively, for use in the mixing models (Emerson et al. 1987; Hedges and Mann 1979 and references therein). The results from such mixing model analyses have suggested that terrestrial organic carbon comprises a negligible proportion of the total organic carbon contained in all ocean sediments except those deposited on continental shelves (Romankevich 1984) and major river deltas (Ittekkot 1988). Sediments accumulating along continental margins and particularly in the deep sea, regions furthest removed from land, are generally assumed to contain organic carbon derived predominantly from marine productivity (Emerson et al. 1987).

Marine and terrestrial autotrophs, which produce organic matter from inorganic forms of carbon, biosynthesize molecules in some cases that are highly source-specific. These organic compounds are often referred to as molecular biomarkers. Upon the death of the source organism, some of these molecules resist remineralization and may be transferred and preserved as fossils in various sedimentary sinks. The persistence of these molecules in the environment reflects some combination of their intrinsic chemical stabilities and the physicochemical protection afforded them by their specific association with detrital matrices. Systematic study of molecular biomarker information contained within sedimentary records provides an opportunity (*a*) to decode the marine organic carbon component of ocean sediments and apportion it into more source-specific compartments, (*b*) to reevaluate by an independent geochemical means whether the terrestrial input of organic carbon to sediments deposited beyond the immediate vicinity of major land masses is, in fact, quantitatively insignificant, and (*c*) to determine through analyses of stratigraphically related sediment horizons how the different identified, contributing sources of sedimentary organic carbon have varied at a given geographic location over geological time and, hence, over changing climatic conditions.

In a recent review article, de Leeuw (1986) enumerated the wide variety of molecular biomarkers currently identified in various sedimentary deposits and discussed how such organic geochemical information can be used in a qualitative sense to decipher the origin of sedimentary organic carbon. In this paper, we will focus on two types of lipid molecular biomarkers, one specific to an important class of marine phytoplankton and one specific to vascular plants. Our goal is to demonstrate how, with proper calibration, systematic study of these source-specific compounds preserved in the marine sedimentary record can lead to new insight and a refined overall understanding of various paleoceanographic problems.

MARINE ORGANIC CARBON

Prymnesiophyte Productivity

Currently, the most well-defined lipid biomarkers of the marine organic carbon component deposited in ocean sediments are a series of long-chain, unsaturated methyl and ethyl ketones, collectively known as alkenones. The components of this series have C_{37}, C_{38}, and C_{39} chain lengths and typically contain two or three double bonds located in biosynthetically unusual positions within the chain (de Leeuw et al. 1980). The alkenone series is accompanied by a methyl and ethyl ester of a di-unsaturated C_{36} fatty acid (Marlowe et al. 1984).

These compounds were first reported in recent sediments from the Black Sea and Walvis Ridge off SW Africa and have subsequently been found almost universally in ocean sediments around the world (Brassell et al. 1986). The various sediments containing these compounds range in age from contemporary to Eocene (Brassell et al. 1986) and, in the most recent case, Cretaceous (Farrimond et al. 1987). Fortuitously, an algal source of these compounds was discovered, the marine coccolithophorid *Emiliania huxleyi* (Volkman et al. 1980). Marlowe et al. (1984) systematically investigated the occurrence of this series in other types of marine algae and established that it is taxonomically restricted to a narrow range of algae belonging to the class Prymnesiophyceae, of which *E. huxleyi* is an important contemporary member. Although much is yet to be learned, currently known facts about the geochemistry of the long-chain alkenones suggest that these molecular biomarkers deposited within sediment represent a well-preserved fossil of the original source and contain valuable clues about sea-surface water temperature (SST) and marine primary productivity in the past.

Paleotemperature Indicators

Laboratory experiments with cultures of *E. huxleyi* grown under controlled conditions have demonstrated that the overall degree of unsaturation in the long-chain alkenone series biosynthesized by this plant varies inversely with water temperature (Brassell et al. 1986). In view of this feature of the plant's physiology and a general observation that unsaturation in the series systematically increases in sediments deposited at higher latitudes, Brassell et al. (1986) proposed that these molecular biomarkers geochemically sequester a record of the overlying surface water temperature present at the time of deposition.

A detailed stratigraphic record now exists defining how unsaturation patterns in the long-chain alkenone series vary downcore in sediments deposited over the past 1 My at a tropical location in the Northeast Atlantic Ocean (Brassell et al. 1986). Over the past 120 ky of depositional history at this oceanic location, a striking correlation ($r = 0.932$, $n = 15$) was noted between values of a parameter which quantifies overall unsaturation in the alkenone series (U^K_{37}) and $\delta^{18}O$ values measured in planktonic foraminiferal tests from corresponding depth horizons. A more general temporal correspondence in values of these two geochemical parameters continued further back in geological time to 550 ky B.P. ($r = 0.676$, $n = 30$). The observed correlations led Brassell et al. (1986) to conclude that the long-chain alkenone series preserved in deep-sea sediments furnishes a direct organic geochemical means of estimating relative changes in overlying SST at a given oceanic location.

Using laboratory cultures of *E. huxleyi*, Prahl and Wakeham (1987) more recently established that values of the alkenone unsaturation index (U^K_{37}) vary linearly with growth temperature over a studied range of 8° to 25°C. Subsequent analyses of long-chain alkenone compositions in suspended particulate materials collected in ocean waters of known temperatures have confirmed that the laboratory calibration curve relating U^K_{37} values to growth temperature is valid for field study. Initial use of this calibration curve as a paleoceanographic tool indicates that SST predicted from values of U^K_{37} measured in sediment horizons of known age are consistent with estimates based on analyses of microfossil assemblages at least back to the Last Glacial Maximum, 18 ky B.P. (Prahl and Wakeham 1987). The concordance between these two independent assessments of SST would suggest the lipid biomarker composition biosynthesized by the living phytoplankton source is not appreciably altered by diagenetic processes after the death of the organism and entry into the geological record. Prahl et al. (1988) have recently identified a means by which intensive properties of the overall biomarker series can be examined to discern periods in the geological past when "SST" estimates made from an analysis of U^K_{37} values recorded in sediments might be inaccurate as a consequence of diagenetic effects.

Assuming diagenetic effects are negligible, unsaturation patterns in the long-chain alkenone series preserved in sediments are influenced presumably only by the temperature of the water in which the phytoplankton source once grew (Brassell et al. 1986; Prahl and Wakeham 1987). On the other hand, $\delta^{18}O$ values measured in the tests of planktonic foraminifera preserved in sediments record not only the temperature but also the oxygen isotopic composition of the water environment in which these protozoans lived. The latter factor, influenced by changes in global ice volume, is considered to have primary control on stratigraphic variations in values of $\delta^{18}O$ noted with depth in deep-sea sediment cores. To explain the strong correlation between U^K_{37} and $\delta^{18}O$ measured in sediments deposited over the last 120 ky at a location in the northeastern tropical Atlantic, Brassell et al. (1986) suggested that changes in SST may have had greater influence on variations in the oxygen isotopic composition of planktonic foraminiferal tests than has commonly been thought by isotope stratigraphers. Future study of long-chain alkenone compositions in concert with isotopic stratigraphy offers a promising means of decoupling the effects that SST and global ice volume have had on values of $\delta^{18}O$ recorded in planktonic foraminiferal tests deposited over glacial/interglacial time scales in particular oceanic locations. However, before this geochemical approach can be used with confidence, considerably more must be learned about the relative life histories and distributional patterns of planktonic foraminifera and the phytoplankton that produce the long-chain alkenones. It cannot be assumed that these organisms live at the same water depths, produce at the same time during the year and, therefore, record as geochemical signals in sediments an equivalent average surface water temperature.

Paleoproductivity Indicators

In calibrating the response of unsaturation to growth temperature, the cellular abundance of the long-chain alkenones was also established (Prahl et al. 1988). Although unsaturation patterns in this biomarker series change dramatically as a function of growth temperature, total alkenone concentrations per cell are essentially constant and independent of this environmental variable. The long-chain alkenones constitute $\approx$8% of the total organic carbon in living cells of *E. huxleyi* and presumably other prymnesiophytes also capable of this biochemical synthesis. By comparison, the photosynthetic pigment chlorophyll *a* accounts for $\leq$ 1% of the total organic carbon content in the cell. At least for the living algae, total alkenone concentration provides a direct measure of biomass.

Based on these observations, one can suggest that downcore profiles of long-chain alkenone abundance in sediments would provide, for a given oceanic location, a record of paleoproductivity for an important class of marine algae. However, before this tool can be used reliably, it will be necessary to demonstrate that stratigraphic variations in the biomarker concentration reflect only changes in productivity and not also changes in preservation efficiency within the water column and at the water-sediment interface over time. At present, few studies have formally addressed the issue of preservation efficiency for the long-chain alkenones. Thus, this aspect of the geochemistry of these compounds is currently left to conjecture. Nonetheless, some promising background information exists that justifies excitement for such a future application.

Results of preliminary experiments with marine copepods fed in the laboratory a diet of *E. huxleyi* suggest these biomarkers are poorly assimilated by planktonic herbivores. Rather than assimilation and metabolic degradation, the primary fate of these lipids in the food chain appears to be release within fecal material, an important agent of vertical transport for these biomarkers from surface waters to the geological record (Corner et al. 1986 and references therein). Using sediment traps deployed over very short time periods in the Peru Upwelling Zone, Volkman et al. (1983) measured the vertical flux of long-chain alkenones and other types of planktonic lipids. Comparison of these values with accumulation rates measured for corresponding compounds in underlying sediments indicated that long-chain alkenones are considerably more resistant to biodegradation in the sedimentation process than are most other lipids of planktonic origin.

Nonetheless, despite their apparent resistance to biodegradation, limited evidence suggests these biomarkers are not transferred conservatively from the water column to the ocean floor. Prahl and colleagues (unpublished data) recently measured the total long-chain alkenone concentration in a single sediment trap sample (3.4% C_{org}) from 3900 m water depth at MANOP Site C (1°N 138°W) in the tropical North Pacific Ocean. The sample represented an average of more than one year (434 days) of sedimentation at this location. Comparison of this single measurement (300 $\mu g/gC_{org}$) with corresponding concentrations (115 $\mu g/gC_{org}$) measured in underlying surface sediment (0.3% C_{org}, 4400 m water depth) showed that these biomarkers relative to total organic carbon are preferentially "lost" rather than concentrated in the sedimentation process. Although biodegradation would provide the simplest explanation for the apparent loss, further field study is necessary to confirm that there is no other explanation for the data. However, whatever the cause for the threefold decrease in alkenone concentration ($/gC_{org}$) in the bottom 500 m of the MANOP Site C water column, the "removal" process does not selectively attack and alter particular components of the alkenone series to any major extent. The intensive properties of the alkenone series preserved in bottom sediments from MANOP Site C and elsewhere in the ocean are quite similar to those observed in the sediment trap particles and those expected for the living phytoplankton source (Prahl et al. 1988).

Jasper (1988) measured long-chain alkenone concentrations and stable isotope compositions for total organic carbon in sediments from 35 depth horizons of a 105 ky DSDP record from the Pigmy Basin in the Gulf of Mexico. In addition to marine inputs, this sedimentary environment has received, over this geological period of time, varying amounts of terrestrial organic carbon from discharge by the Mississippi River. A scatter plot of long-chain alkenone concentrations ($/gC_{org}$) versus corresponding $\delta^{13}C_{org}$ values measured on total organic carbon in the sediments showed that these two parameters are linearly well correlated ($r = 0.85$; Jasper, personal communication). An obvious interpretation of this linear relationship is that the quantity of these molecular biomarkers preserved at any given sediment depth is directly proportional to the quantity of total marine organic carbon preserved in that horizon. This interpretation assumes that downcore variations in $\delta^{13}C_{org}$ values sensitively reflect relative changes in the admixture of terrestrial and marine organic carbon over time. If the geochemical relationship documented in Jasper's study is, in fact, a phenomenon demonstrable in other oceanic settings, it will be intriguing because the long-chain alkenones are not universal indicators of marine primary producers. Rather, these biomarkers derive from a very specific class of phytoplankton (Marlowe et al. 1984 and references therein). Therefore, synchrony between an indicator of prymnesiophyte productivity (long-chain alkenones) and an indicator of total primary productivity ($\delta^{13}C$ of total marine organic carbon) would not necessarily be expected. Nonetheless, in view of Jasper's observation, it is possible that vertical profiles of total alkenone abundance with depth in sediment cores provide a record of how long-term climatic changes have influenced not only a specific type of algal productivity but also total primary productivity in different regions of the ocean.

TERRESTRIAL ORGANIC CARBON

Almost universally, vascular plants biosynthesize as surface waxes long-chain ($\geq C_{20}$) series of *n*-alkanes, *n*-alcohols, and *n*-acids (Kollatukudy 1976 and references therein). These series display source-specific features, not only in terms of long carbon chain length but also in terms of carbon number preference. Major components of the *n*-alkane series are predominantly odd carbon-numbered while components of the other two series are predominantly even carbon-numbered.

Through erosional processes involving atmospheric or aqueous media, these lipid biomarkers, associated with bulk terrestrial organic carbon, are transported within particulate material from the land to various marine depositional environments. Once contained in marine sediments, these compounds display remarkable stability stemming more likely from protection afforded by their particulate form rather than intrinsic chemical properties

(Cranwell 1982 and references therein). Sufficient background information is now available to show that these lipid biomarkers of vascular plants provide not only a qualitative indication of the presence of terrestrial organic carbon in marine sediments but also, with proper calibration, can be used to make reasonable quantitative estimates of how much terrestrial organic carbon is present. The following discussion reviews the current state of knowledge justifying the latter application of terrestrial lipid biomarkers in environmental study.

Riverine Inputs

Prahl and Carpenter (1984) examined plantwax *n*-alkane distributions in continental shelf and slope sediments off the southern coast of Washington state, U.S.A. This coastal region receives its predominant detrital input of organic material essentially from a point source, the mouth of the Columbia River. Plantwax *n*-alkane concentrations normalized to total organic carbon showed a general decrease with distance offshore in this coastal region. The cause of this trend was ascribed to dilution of the biomarker series with an increasing proportion of marine relative to terrestrial organic carbon in sediments furthest removed from the source of the land-derived detrital material.

The plantwax *n*-alkane content of sediments from a variety of locations throughout the Columbia River basin was also examined in the study (Prahl and Carpenter 1984). Although the riverine sediments displayed a wide range of textures and organic carbon content ($\leq$ 0.3 to 3% by weight), plantwax *n*-alkane concentrations ($/gC_{org}$) were surprisingly constant (277 ± 87 $\mu g/gC_{org}$; n = 14) throughout the basin. Using a simple binary mixing model based on this observation, the terrestrial component of total organic carbon was determined for a variety of surface sediments accumulating on the Washington continental shelf and slope. The model assumed (*a*) plantwax *n*-alkanes and bulk terrestrial organic carbon are exported at the mouth of the Columbia River to the Washington coastal region in a fixed proportion (277 ± 87 $\mu g/gC_{org}$) and (*b*) the fundamental process that causes a change in the value of this proportion measured in coastal sediments is dilution with marine organic carbon containing none of the terrestrial *n*-alkane series.

The results of this exercise are displayed on the map in Fig. 1. Interestingly, this analysis suggests that the organic-rich ($\approx$3% C_{org}), hemipelagic muds accumulating on the continental slope of this coastal region contain a significant fraction of terrestrial organic carbon, i.e., 20–30%. This estimate, if correct, is certainly notable and controversial as it is 5–10 times greater than previous estimates of the terrestrial component of total organic carbon in this region. The previous estimates were also made using a mixing model

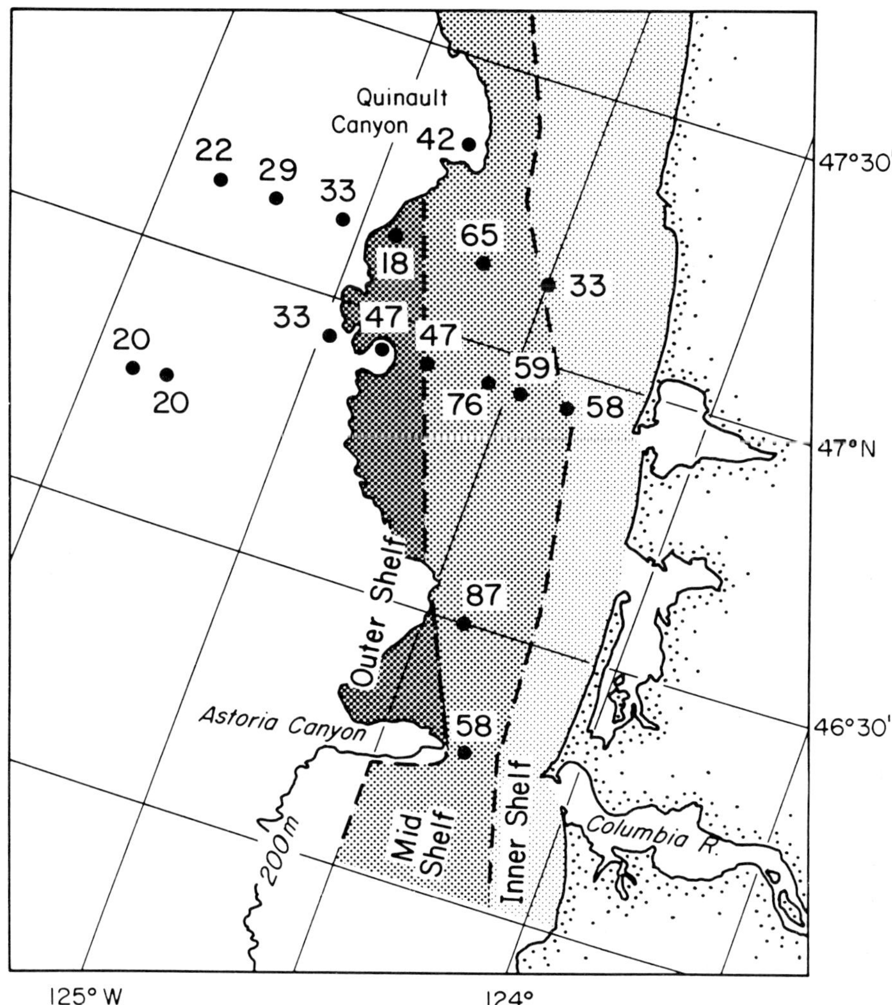

Fig. 1—Map displaying how the terrestrial component of total organic carbon varies spatially in Washington coastal sediments. The percentage terrestrial organic carbon is calculated from the concentration ($/gC_{org}$) of plantwax *n*-alkanes measured in sediments from each sampling location normalized to the concentration of plantwax *n*-alkanes exported by the Columbia River to this coastal region, i.e., 277 ± 87 $\mu g/gC_{org}$ (Prahl and Carpenter 1984).

calculation based on lignin and stable carbon isotope data (Hedges and Mann 1979) rather than lipid data. The apparent discrepancy in the predictions made from these two independent geochemical data sets can only be resolved through careful reexamination of the values assigned in respective mixing models to the pure terrestrial endmember for plantwax *n*-alkanes and lignin. An average value of $\delta^{13}C$ measured for bulk organic matter in Holocene sediments from the Washington continental slope is $-21.9 \pm 0.3‰$ (n = 6; Hedges and Mann 1979; Hedges and van Geen 1982). This value could easily accommodate either estimate of the terrestrial organic carbon fraction (20–30% or $\leq$ 3%) depending upon how one chooses to interpret it.

Hedges and Mann (1979) based their estimate of a 3–5% terrestrial component in slope sediments on $\delta^{13}C$ values of -25.5 and $-21.6‰$ for the pure terrestrial and marine organic carbon endmembers, respectively. However, if the marine endmember were only 1‰ isotopically heavier (i.e., $-20.6‰$) and the terrestrial endmember remained the same, a simple mixing model analysis of stable carbon isotope data alone would predict that the organic carbon accumulating in Washington continental slope sediments is $\approx$25% terrestrial. Although significantly different from that of Hedges and Mann (1979), the latter estimate, based exclusively on a small change in the $\delta^{13}C$ value chosen for the marine endmember, would be quite consistent with predictions made from data for plantwax *n*-alkanes. Without compelling support from independent geochemical measurements, either choice for the isotopic value of the marine endmember (-20.6 or $-21.6‰$) would seem reasonable considering that $\delta^{13}C$ values measured in bulk plankton, the ultimate source of the marine component of organic carbon in these sediments, span a range of values typically lying between -18 and $-21‰$ (Degens 1969).

Detrital organic carbon exported at the mouth of the Columbia River is hydrodynamically sorted and, as a consequence, chemically fractionated as it is dispersed throughout the coastal region off southern Washington state (Prahl 1985). Relatively fresh, coarse-grained vascular plant debris accumulates within a silt deposit situated along the continental shelf, while more highly degraded, fine-grained terrestrial organic materials are preferentially transported further offshore and deposited as hemipelagic muds on the continental slope. The effect of spatial variations in the quality of terrestrial organic carbon deposited in sediments from this region was not considered in either of the two biomarker models described above. Rather, both geochemical approaches assumed implicitly that the detrital organic carbon component of shelf and slope sediments is described quantitatively by the same value for the terrestrial endmember. Perhaps the fundamental cause for the large discrepancy in the current model predictions, based on lipid and lignin biomarker data, lies in making this assumption. If biomarkers are to provide reliable estimates of the terrestrial component of total organic carbon in coastal environments, experiments must be designed that either confirm the validity of this assumption for a given biomarker or correct for the effect of chemical fractionation caused by hydrodynamic sorting.

Aeolian Inputs

Long-range atmospheric transport of dust represents the primary pathway by which detrital organic carbon is transferred from the land to remote areas of the ocean. As a consequence of the Sea-Air Exchange (SEAREX) Program, the organic compositions of atmospheric particulate materials collected at various remote locations in the ocean have been characterized in some detail. The results of this work suggest that the organic carbon contained in atmospheric particulate materials from these regions is largely of terrestrial origin despite the great distance of the sampling locations from major land masses. Based on an analysis of stable carbon isotope data, Chesselet et al. (1981) estimated that ≥ 80% of the particulate organic carbon contained in marine aerosols collected at distal locations in the Atlantic (Sargasso Sea) and Pacific (Enewetak Atoll) Ocean is land-derived. The major abundance of long-chain series of vascular plantwax lipids (*n*-acids, *n*-alcohols, *n*-alkanes) measured in these same aerosols (Gagosian and Peltzer 1987) corroborates the stable carbon isotope assessment that these aerosols contain a significant component of terrestrial organic carbon.

From time series data for Enewetak and three different sets of model assumptions, Zafiriou et al. (1985) estimated the annual flux of organic carbon from the atmosphere to this remote part of the Pacific Ocean to be 0.50 ± 0.14 gC/m^2-yr. They noted that this estimate, if correct, represents approximately 1% of the annual primary productivity in such oligotrophic regions of the ocean. Therefore, if the particulate organic carbon introduced atmospherically to this deep sea environment is largely refractory and its primary fate is settling to the seafloor, this input would make a significant contribution to the total organic carbon inventory in underlying sediments. Typically, ≤1% of the primary productivity in surface waters of the open ocean escapes recycling in the water column or at the water-sediment interface and is incorporated as a relatively stable component of the sedimentary record (Emerson et al. 1987). Thus, their rough estimate, based solely on generalized data for total organic carbon, suggests terrestrial organic carbon introduced via the deposition of aeolian dusts could account for as much as 50% of the total organic carbon preserved in deep-sea sediments (Zafiriou et al. 1985). Specific organic geochemical measurements of underlying sediment deposits were not made to validate this view, however.

Recently, detailed analyses of lipid compositions in eleven depth horizons of a deep sea core from MANOP Site C (W8402A-14GC) have been made (Muehlhausen, unpublished data). The time spanned by these samples covers the past 26 ky of sediment deposition at this location. Interestingly, a series of *n*-fatty acids (C_{20} to C_{30}) characteristic of surface waxes from terrestrial vascular plants is one of the most prominent features of the total lipid composition in these sediments. Fig. 2 illustrates that these biomarkers of organic carbon derived from the land occur in similar abundance to the long-chain alkenones originating from a specific type of marine primary productivity in overlying waters. The combined concentration of all components in this *n*-fatty acid series, C_{20} to C_{30}, averaged 224 ± 73 μg/gC_{org} (n = 11) over the past 26 ky of sediment deposition.

In addition to estimates for total organic carbon, Zafiriou et al. (1985) also estimated from their atmospheric time series data the annual aeolian input of terrestrial *n*-fatty acids to the sea near Enewetak. Influxes for these biomarkers, predicted from the same three sets of model assumptions used for the total organic carbon assessment, ranged from 160 to 760 μg/m^2-yr, averaging 410 ± 310 μg/m^2-yr. Using the aeolian flux estimates, the relative concentration of total terrestrial *n*-fatty acids to total organic carbon in the average dust particle can be calculated. The value is 820 ± 660 μg/gC_{org}. With two fundamental assumptions, this value can be used to determine the fraction of terrestrial organic carbon in sediments deposited at MANOP Site C. The estimate is given from a simple proportion between the average concentration of terrestrial *n*-fatty acids measured in MANOP Site C sediments (224 ± 73 μg/gC_{org}) and the concentration of these biomarkers predicted for the average aeolian dust particle (820 ± 660 μg/gC_{org}). The result of this calculation, based in this case on a comparison of chemical composition and not flux measurements (Zafiriou et al. 1985), again indicates that the terrestrial fraction of total organic carbon in these remote ocean sediments is quite significant. Terrestrial organic carbon could constitute 27 ± 23% of the total organic carbon.

The previous analysis has required the following two assumptions. First, the concentration of terrestrial *n*-fatty acids relative to total organic carbon in atmospheric particulate materials collected at Enewetak represents an "average" value for the land-derived aeolian dusts introduced to surface waters at MANOP Site C. This assumption is considered reasonable because both locations lie beneath the northeast trade winds and thus, very likely receive inputs of aeolian dusts derived from the same source regions (Gagosian and Peltzer 1987). Second, the aeolian dusts accumulating in underlying sediments are described by the same relative chemical composition. That is, the quantitative relationship between the biomarker and total organic carbon is not significantly altered as a consequence of the settling process. Notably, the estimation of the terrestrial organic carbon

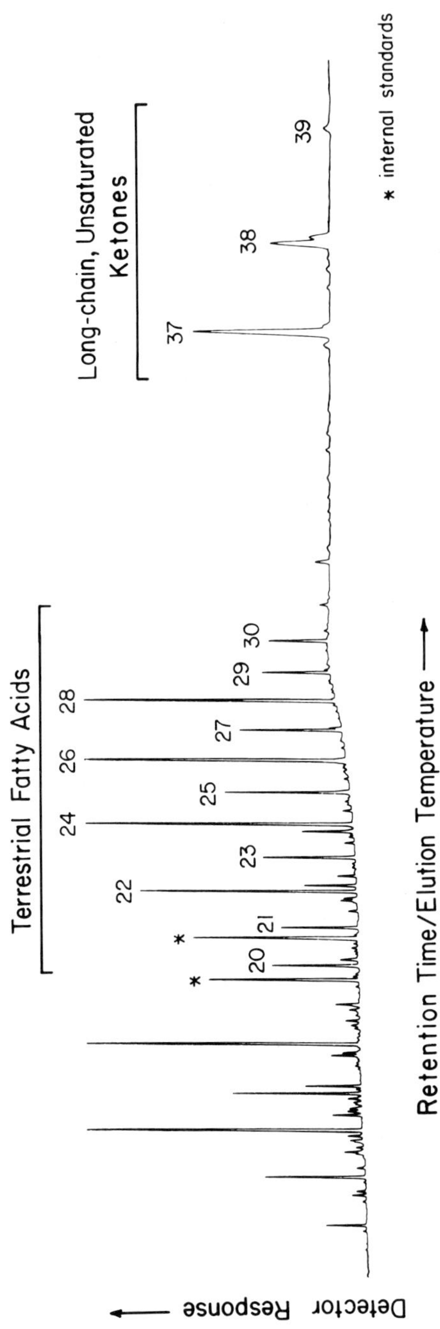

Fig. 2—Capillary gas chromatogram of the lipid fraction extracted and isolated from sediments deposited at MANOP Site C (4400 m water depth) located in the tropical North Pacific (1°N 138°W) illustrating (*a*) the fatty acid series (C_{20} through C_{30}) derived from the surface waxes of vascular plants and introduced as a chemical component of land-derived aeolian dusts and (*b*) the long-chain (C_{37}, C_{38}, C_{39}) alkenone series derived from prymnesiophyte productivity in overlying surface waters.

fraction in sediments from the Washington continental slope (see earlier discussion) was formulated on an analogous set of assumptions.

The stable carbon isotope composition of bulk organic matter in four near-surface horizons of a core from MANOP Site C has been measured (McCorkle 1987). $\delta^{13}C_{org}$ values averaged $-19 \pm 0.5‰$ in these sediments of low total organic carbon content ($\leq 0.35\%$). This average lies toward the isotopically heaviest edge of the range of values typifying bulk organic matter produced by marine plankton (i.e., -18 to $-21‰$). Therefore, purely on the basis of comparison with the isotopic composition of bulk organic carbon in whole marine plankton, it would seem that sediments at MANOP Site C could not contain a 27% component of isotopically light ($\approx -25.5‰$) terrestrial organic carbon (Emerson et al. 1987). Of course, this conclusion assumes that the isotopic composition of the marine organic carbon preserved in these sediments falls within the range characteristic of bulk plankton. The average isotopic composition of a 73% (X_m) marine component ($\delta^{13}C_m$) required to disguise a 27% (X_t) terrestrial component ($\delta^{13}C_t = -25.5‰$) of total organic carbon can be calculated from the value of $\delta^{13}C$ measured for bulk organic matter in these sediments ($-19‰$) and a simple mass balance equation $\{\delta^{13}C \text{ (measured)} = (X_t\delta^{13}C_t + X_m\delta^{13}C_m)/100\%\}$. The value of $\delta^{13}C_m$ calculated by this expression is $-16.6‰$.

Three major classes of biochemicals (amino acids, lipids, and sugars) comprise the bulk of the organic carbon synthesized by marine plankton. Earlier studies (Degens 1969) have shown that the stable carbon isotopic composition of these different biochemical classes is systematically offset from that of the total organic carbon in these organisms (Table 1). Relative to the total organic carbon in plankton, amino acids, and sugars, viewed as whole biochemical classes, are enriched in ^{13}C by $+2.7$ and $+1.1‰$, respectively, while lipids are depleted in ^{13}C by as much as $-9.5‰$. If a proteinaceous component of plankton were selectively preserved in the MANOP Site C sediments, the value of $\delta^{13}C_m$ required to appropriately model the bulk marine organic carbon

TABLE 1 Average isotopic composition of organic carbon in major biochemical fractions of marine plankton[1].

Biochemical Fraction	*$\delta^{13}C$ (vs. PDB)*	*^{13}C Enrichment[2]*
Total Organic Carbon	$-20.3 \pm 0.5‰$	$0.0‰$
Total Amino Acids	$-17.6 \pm 1.1‰$	$+2.7‰$
Total Lipids	$-29.8 \pm 5.3‰$	$-9.5‰$
Total Sugars	$-19.2 \pm 1.1‰$	$+1.1‰$

[1] data abstracted from Degens (1969) and references therein
[2] difference between isotopic composition measured in biochemical fraction and total organic carbon

contained in the sediments would need to be isotopically heavier than the value measured for total organic carbon in unaltered marine plankton.

A physicochemical mechanism can be suggested to account for selective preservation of ^{13}C enriched planktonic amino acids in high carbonate, low organic carbon content sediments like those depositing at MANOP Site C. Biological carbonates, primarily foraminiferal tests, comprise 80 to 90% of the dry mass of these sediments (Emerson et al. 1987). Studies have shown that a characteristic organic matrix composed of glycoprotein is associated with the inorganic skeleton of calcareous plankton (Degens 1976 and references therein). The mechanism for selective preservation assumes that calcified tissues, sequestered within and physically protected by a solid matrix, demonstrate greater resistance to biological degradation than do unprotected organics derived from the soft-body parts of plankton. Hence, as an indirect consequence of remineralization processes, amino acid-rich, calcified tissue, being less available to biodegradation, is concentrated as an increasingly important component of the residual organic carbon pool in the sedimentary particulate materials.

In Table 2, the amino acid composition of the "average" protein in mixed marine plankton is compared with that measured in calcified tissue from

TABLE 2 Average amino acid composition (# residues/1000 residues) of mixed marine plankton and of calcified tissue in foraminiferal tests.

Amino Acid	*Plankton*[1]	*Forams*[2]	*% Enrichment*[3]
Aspartic Acid	103	212	+106
Threonine	58	82	+41
Serine	53	84	+58
Glutamic Acid	121	114	−6
Proline	43	–	–
Glycine	140	138	−1
Alanine	99	100	+1
Cysteine	14	–	–
Valine	56	60	+7
Methionine	23	9	−61
Isoleucine	38	36	−5
Leucine	71	44	−38
Tyrosine	31	23	−26
Phenylalanine	34	35	+3
Lysine	63	20	−68
Histidine	16	22	+38
Arginine	45	21	−53

[1] data from Degens (1969) and references therein
[2] data from King (1977)
[3] enrichment of individual amino acid content in calcified tissue from foram tests relative to the "average" protein in mixed marine plankton: % Enrichment = (Forams − Plankton)/Plankton × 100.

foraminiferal tests. A large percentage enrichment of aspartic acid in the calcified tissue (+106%) is the most obvious difference noted in this comparison. Other compositional differences can also be observed, however. In the context of the present discussion, consideration of the overall difference in these amino acid compositions is particularly relevant to the proper assignment of a value for $\delta^{13}C_m$.

Although the isotopic composition of the total amino acid pool in mixed marine plankton (Table 2) averages +2.7‰ heavier than the total organic carbon in these organisms (Table 1), the internal $\delta^{13}C$ variation for the individual amino acids comprising this pool spans a range of 17‰ (Degens 1969 and references therein). The amino acids most enriched in ^{13}C are aspartic acid, glycine, serine, and threonine; those most depleted in ^{13}C are leucine, isoleucine, and the two aromatic amino acids, phenylalanine and tyrosine (Galimov 1985 and references therein). Interestingly, relative to bulk plankton, the isotopically heavy amino acid residues are significantly more abundant in the protein of calcified tissue while the isotopically light residues are less abundant (see % Enrichment in Table 2). On this basis, the offset noted between the $\delta^{13}C$ values of the total amino acid pool in calcified tissue of foram tests and of total organic carbon in mixed marine plankton (Table 1) could be considerably greater than +2.7‰.

Macko et al. (1987) recently reported data defining the stable carbon isotopic fractionation in individual amino acid residues isolated from cultures of a blue-green algae (*Anabaena sp.*) and a heterotrophic bacteria (*Vibrio harveyi*). From these data (see Fig. 5 in Macko et al. 1987), the data for the average amino acid composition in mixed plankton and foraminiferal tests (Table 2) and the number of carbon atoms per individual amino acid residue, it is possible to estimate ^{13}C enrichment of the protein in the shells of planktonic foraminifera relative to that in the whole plankton. This mass balance calculation predicts that the average carbon atom in protein from foraminiferal shells is ≈2.1‰ heavier than that in the protein of whole plankton. This estimate assumes the individual amino acid residues in mixed plankton and foraminiferal shells have the same isotopic composition as those reported for the cultures of microorganisms studied by Macko et al. (1987). Given that the data in Table 1 and the result of the latter calculation are correct, it would appear the organic carbon in the protein fraction of calcified tissue could be perhaps 5‰ (2.1‰ plus 2.7‰) heavier than the total organic carbon in mixed marine plankton. If this is so and the protein fraction of calcified tissue represented one-quarter or more of the total organic carbon deposited at MANOP Site C, a 27% component of terrestrial organic carbon in these sediments would not be easily recognized from measurement of bulk $\delta^{13}C_{org}$ values.

At present, the proposed explanation to isotopically disguise a significant terrestrial component of total organic carbon in these sediments must remain hypothetical as its two basic assumptions can not be adequately tested from existing data in the literature. No data are yet published to formally substantiate whether the predicted 5‰ enrichment of ^{13}C in calcified protein relative to the total organic carbon in bulk marine plankton is reasonable or unreasonable. And, few studies have evaluated how much of the total organic carbon content in calcareous oozes like those accumulating at MANOP Site C is attributed to amino acids. To date, most work on calcareous oozes has been done from a micropaleontology perspective and has only examined the amino acid composition and content of a minor, coarse particulate fraction from such sediments (Müller 1984 and references therein). Müller and Suess (1977), however, have reported data for amino acids in whole calcareous sediments from a productive, shallow water lagoon in the Central Pacific Ocean. These sediments, largely composed of aragonite derived from coral, displayed an overall amino acid composition intermediate between that of the calcified tissue of coral and dissolved organic matter in seawater. The latter composition looks somewhat like that of mixed marine plankton (Table 2). Total hydrolyzable amino acids accounted for roughly 25% of the total organic carbon in the lagoonal sediments. Although the calcareous ooze depositing at MANOP Site C is largely calcite derived from foraminifera, the results of Müller and Suess (1977) are very likely quite similar to what one would find from an analysis of amino acids in such deep-sea environments. If so, the presence of a significant terrestrial component of total organic carbon in remote, deep-sea sediments can not be easily discounted on the basis of stable carbon isotopic measurements of the bulk sedimentary organic matter alone. A more detailed organic geochemical investigation is warranted to formally establish the magnitude that terrestrial sources actually contribute to the total organic carbon record accumulating in sediments from the deep sea.

CONCLUSIONS

Organic geochemists have now identified a variety of molecular biomarkers not only suitable for determining past hydrographic conditions in the surface ocean and the assortment of sources that contribute organic carbon to sediments but also potentially suitable for characterizing how diagenetic processes specifically act to alter and degrade sedimentary organic matter within the water column and at the water-sediment interface (de Leeuw 1986 and references therein). In order to utilize the full value of these organic geochemical tools in paleoceanographic study, considerable effort must be expended to calibrate them individually. The biomarkers can not be calibrated by theoretical means, they must be calibrated empirically through systematic laboratory and field study. As shown in this article through the use of specific examples, the effort spent in making such calibrations may have considerable pay-off by providing a new set of clues with which to refine an overall understanding of particular paleoceanographic problems. Future usage of biomarker techniques in tandem with stable isotope techniques very likely will provide for the geochemist a more reliable means to obtain answers sought for important environmental questions.

Acknowledgements. The quality of this article has benefited greatly from the constructive comments and criticisms of the following people: S. Emerson, P. Müller, C.E. Reimers, and E. Suess. We are grateful to the ACS Petroleum Research Fund (PRF #16735–AC2) and the National Science Foundation (OCE-8419861, 8600418, and 8716244) for providing research support that has made it possible in a variety of ways to write this review article.

REFERENCES

Brassell, S.C.; Eglinton, G.; Marlowe, I.T.; Pflaumann, U.; and Sarnthein, M. 1986. Molecular stratigraphy: a new tool for climatic assessment. *Nature* **320**: 129–133.

Chesselet, R.; Fontugne, M.; Buat-Ménard, P.; Ezat, U.; and Lambert, C.E. 1981. The origin of particulate organic carbon in the marine atmosphere as indicated by its stable carbon isotopic composition. *Geophys. Res. Lett.* **8**: 345–348.

Corner, E.D.S.; O'Hara, S.C.M.; Neal, A.C.; and Eglinton, G. 1986. Copepod faecal pellets and the vertical flux of biolipids. In: The Biological Chemistry of Marine Copepods, eds. E.D.S. Corner and S.C.M. O'Hara, pp. 260–321. Oxford: Oxford Science Publications.

Cranwell, P.A. 1982. Lipids of aquatic sediments and sedimenting particulates. *Prog. Lipid Res.* **21**: 271–308.

Degens, E.T. 1969. Biogeochemistry of stable carbon isotopes. In: Organic Geochemistry, eds. G. Eglinton and E.T.J. Murphy, pp. 304–329. Berlin: Springer.

Degens, E.T. 1976. Molecular mechanisms on carbonate, phosphate and silica deposition in the living cell. *Topics Curr. Chem.* **64**: 1–112.

de Leeuw, J.W. 1986. Sedimentary lipids and polysaccharides. In: Organic Marine Chemistry, ed. M.L. Sohn, pp. 33–61. ACS Publication.

de Leeuw, J.W.; van der Meer, F.W.; Rijpstra, W.I.C.; and Schenck, P.A. 1980. On the occurrence and structural identification of long chain alkenones and hydrocarbons in sediments. In: Advances in Organic Geochemistry 1979, eds. A.G. Douglas and J.R. Maxwell, pp. 211–217. New York: Pergamon.

Emerson, S.; Stump, C.; Grootes, P.M.; Stuiver, M.; Farwell, G.W.; and Schmidt, F.G. 1987. Estimates of degradable organic carbon in deep-sea surface sediments from ^{14}C concentrations. *Nature* **329**: 51–53.

Farrimond, P.; Eglinton, G.; and Brassell, S.C. 1987. Alkenones in Cretaceous black shales, Blake-Bahama basin, western North Atlantic. *Org. Geochem.* **10**: 897–903.

Gagosian, R.B., and Peltzer, E.T. 1987. The importance of atmospheric input of terrestrial organic material to deep sea sediments. *Org. Geochem.* **10**: 661–669.

Galimov, E.M. 1985. The Biological Fractionation of Isotopes, p. 261. Orlando, FL: Academic Press.

Hedges, J.I., and Mann, D.C. 1979. The lignin geochemistry of marine sediments from the southern Washington coast. *Geochim. Cosmochim. Acta* **43**: 1809–1818.

Hedges, J.I., and van Geen, A. 1982. A comparison of lignin and stable carbon isotope compositions in Quaternary marine sediments. *Marine Chem.* **11**: 43–54.

Ittekkot, V. 1988. Global trends in the nature of organic matter in river suspensions. *Nature* **332**: 436–438.

Jasper, J. 1988. An organic geochemical approach to problems of glacial-interglacial climatic variability. Ph.D. diss. Woods Hole Oceanographic Institution, p. 312.

King, K., Jr. 1977. Amino acid survey of Recent calcareous and siliceous deep-sea microfossils. *Micropaleon.* **223**: 180–193.

Kollatukudy, P.E. 1976. Chemistry and Biochemistry of Natural Waxes, p. 459. Amsterdam: Elsevier.

Macko, S.A.; Fogel (Estep), M.L.; Hare, P.E.; and Hoering, T.C. 1987. Isotopic fractionation of nitrogen and carbon in the synthesis of amino acids by microorganisms. *Chem. Geol. (Isotope Geosci. Sec.)* **65**: 79–92.

Marlowe, I.T.; Green, J.C.; Neal, A.C.; Brassell, S.C.; Eglinton, G.; and Course, P.A. 1984. Long chain alkenones in the Prymnesiophyceae. Distribution of alkenones and other lipids and their taxonomic significance. *Br. J. Phycol.* **19**: 203–216.

McCorkle, D.C. 1987. Stable carbon isotopes in deep sea pore waters: modern geochemistry and paleoceanographic applications. Ph.D. diss., p. 209. Univ. Washington, Seattle.

Müller, P.J. 1984. Isoleucine epimerization in Quaternary planktonic foraminifera: effects of diagenetic hydrolysis and leaching, and Atlantic-Pacific intercore correlations. *Meteor Forsch.-Ergeb. C* **38**: 25–47.

Müller, P.J., and Suess, E. 1977. Interactions of organic compounds with calcium carbonate-III. Amino acid composition of sorbed layers. *Geochim. Cosmochim. Acta* **41**: 941–949.

Prahl, F.G. 1985. Chemical evidence of differential particle dispersal in the southern Washington coastal environment. *Geochim. Cosmochim. Acta* **49**: 2533–2539.

Prahl, F.G., and Carpenter, R. 1984. Hydrocarbons in Washington coastal sediments. *Est. Coast. Shelf Sci.* **18**: 703–720.

Prahl, F.G.; Muehlhausen, L.A.; and Zahnle, D.L. 1988. Further evaluation of long-chain alkenones as indicators of paleoceanographic conditions. *Geochim. Cosmochim. Acta.* **52**: 2303–2310.

Prahl, F.G., and Wakeham, S.G. 1987. Calibration of unsaturation patterns in long-chain ketone compositions for paleotemperature assessment. *Nature* **330**: 367–369.

Romankevich, E.A. 1984. Geochemistry of Organic Matter in the Ocean, p. 334. Berlin: Springer.

Volkman, J.K.; Eglinton, G.; Corner, E.D.S.; and O'Hara, S.C.M. 1980. Novel unsaturated straight chain C37–C39 methyl and ethyl ketones in marine sediments and a coccolithophore *Emiliania huxleyi*. In: Advances in Organic Geochemistry 1979, eds. A.G. Douglas and J.R. Maxwell, pp. 219–227. New York: Pergamon.

Volkman, J.K.; Farrington, J.W.; Gagosian, R.B.; and Wakeham, S.G. 1983. Lipid composition of coastal marine sediments from the Peru Upwelling Region. In: Advances in Organic Geochemistry 1981, ed. M. Bjoroy, pp. 228–240. Chichester: Wiley.

Zafiriou, O.C.; Gagosian, R.B.; Peltzer, E.T.; Alford, J.B.; and Loder, T. 1985. Air-to-sea fluxes of lipids at Enewetak Atoll. *J. Geophys. Res.* **90 (D1)**: 2409–2423.

Standing, left to right:
Gert De Lange, Alexander Altenbach, Peter Müller, Thorsten Steiger, Steve Emerson

Seated, left to right:
Clare Reimers, Frederick Prahl, Barry Hargrave, Erwin Suess, Peter Jumars

Productivity of the Ocean: Present and Past
eds. W.H. Berger, V.S. Smetacek and G. Wefer, pp. 291–311
John Wiley & Sons Limited

Group Report
Transformation of Seafloor-arriving Fluxes into the Sedimentary Record

P.A. Jumars, Rapporteur
A.V. Altenbach
G.J. De Lange
S.R. Emerson
B.T. Hargrave
P.J. Müller
F.G. Prahl
C.E. Reimers
T. Steiger
E. Suess

INTRODUCTION

Seafloor processes control the transfer of biogenous detritus into the sediment by burial or into the overlying water by nutrient reflux and resuspension. Either transfer is in part biologically mediated and can involve complex interactions among micro-, meio-, and macrofauna. Recent reports show that phytodetritus that reaches the seafloor hosts vigorous bacterial and protozoan activity (Lochte and Turley 1988; Gooday 1988), so initial breakdown may be surprisingly rapid and direct. This finding at first seems to contradict observations on microbes attached to sinking particles intercepted in the water column (Karl et al. 1988; Cho and Azam 1988). These latter authors showed that biomass as well as microbial activity on sinking particles diminishes with increasing water depth. How then is it possible that the microbial population attached to particles recovered from the seafloor is so unusually active? Could it be that only after the phytodetritus reaches the seafloor is it populated by specialized, barophilic heterotrophs that would metabolize the most labile material first (Deming and Colwell 1985; Suess 1988)? The most labile material arriving on the bottom comes in the form of rapidly sinking particles, whose residence time in the water column is short and exposes attached organisms to rapidly increasing pressure. Clearly, the mechanism of turnover of incoming flux at the seafloor is not merely an extension of that in the water column: the biological species that mediate that turnover are different from those in the water column; a suite of oxidants rarely used in organic matter degradation in the water column often comes into play at the seafloor; solute transport is dominated by molecular diffusion and animal pumping from the sediment-water interface downward; and, unlike the case in the water column, virtually all benthic bacteria live attached to particles.

Our specific focus thus was to understand (well enough to identify and evaluate indicators of past surface-water productivities and bottom-arriving fluxes) transformation into the sedimentary record of the flux of matter arriving at the bottom. As addressed by Groups 1 and 2 (see Williams et al. and Bruland et al., both this volume), there is no simple proportionality between surface production and the bottom-arriving flux. New production arriving at the bottom below 1000 m water depth is significant in the carbon cycle because even the major fraction that is oxidized to CO_2 is kept out of contact with the atmosphere by circulation patterns for roughly 10^3 yr. We emphasize that there are differences in benthic processes between continental margins and "open-ocean," depositional environments. The former contain the bulk of buried organic carbon and are sites of injection from the benthos of remineralized and resuspended material into midwater circulation systems (Romankevich 1984; Jahnke and Jackson 1987). Open-ocean sediments appear stratigraphically and diagenetically less complicated, however, and may thus allow easier reconstruction of the paleoenvironment.

We selected a limited goal, avoiding in general environments of turbidite and slump deposition, contourites and reworked facies that produce records difficult or impossible to interpret in terms of paleoproductivity. Thus for the most part we do not treat deep-sea regions of episodic erosion, although they may be quite widespread (Hollister et al. 1984). We also made no attempt to include in our goal estimates of past contributions to seafloor food supplies from chemolithoautotrophy at hydrothermal or cold-seep sites, except insofar as this chemoautotrophic signal might be confused with the rain of material from the surface ocean. Chemolithotrophic production and its effect on the sedimentary record is just emerging as a field of research.

A VIEW OF THE SEABED UNDER DIFFERING CARBON FLUX REGIMES

The key in the generation of vertical geochemical structure is the ratio in fluxes of organic carbon versus oxygen into the seabed. The succession of dissolved oxidants is the focus of the classic approach to defining geochemical structure. It varies strongly with sedimentation rate. Oxygen flux is usually calculable from pore water gradients (Fig. 1) and is generally proportional to bottom-water O_2 concentration if organic carbon is supplied in excess. Even complete knowledge of organic carbon flux and bottom-water O_2 concentration, however, is insufficient to allow accurate prediction of the proportion of organic carbon that will survive diagenesis, for organic matter arriving at the seabed is a chemically heterogeneous mixture. This

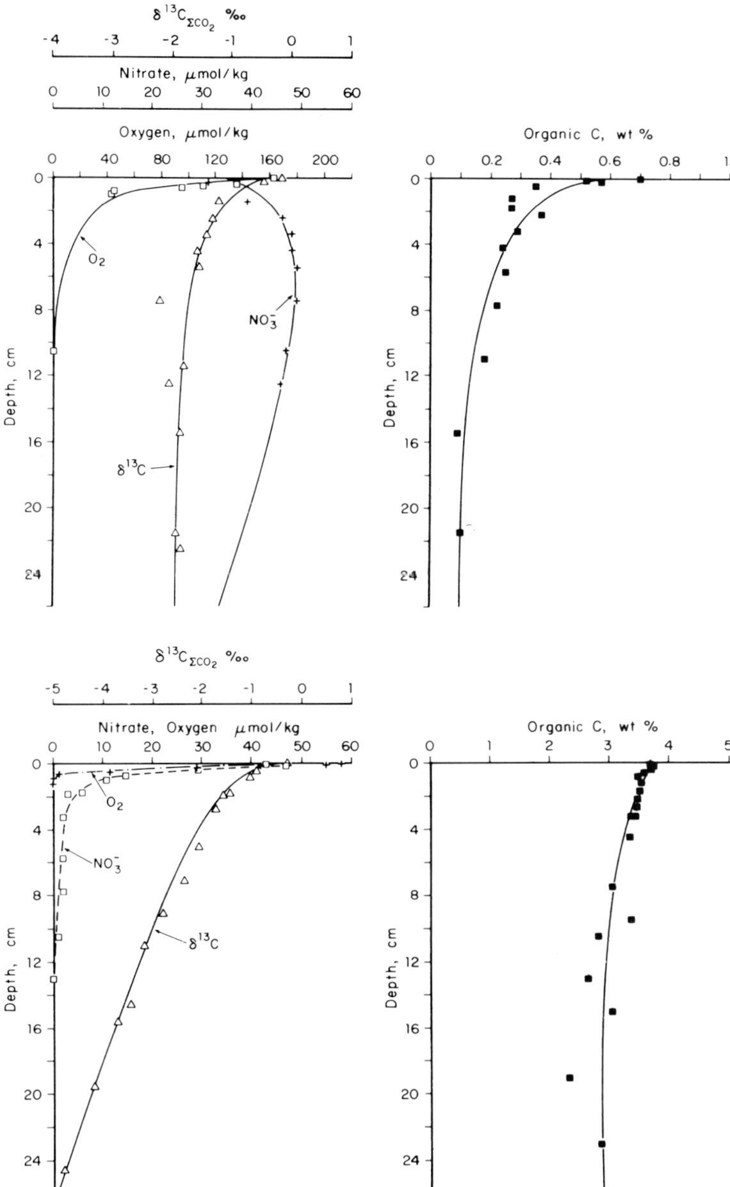

Fig. 1—Depth distributions of chemical properties at an open-ocean site (1°N, 139°W; MANOP site C; top panels) and a hemipelagic site (33°N, 118°W; San Clemente Basin; bottom panels), comparing pore-water O_2, NO_3, and $\delta^{13}C$ and solid organic carbon contents. Data are compiled from Reimers et al. (1984), Reimers (1987), McCorkle (1987), and Shaw (1988). (Figure prepared by C. Reimers and S. Emerson).

heterogeneity is most often represented via multicomponent models that assume first-order behavior for each component (Reimers, this volume, Eq. 1).

There is a wide and no doubt very complex spectrum of reactivities among the organic carbon molecules arriving at the seabed. Although carrying the idea of a full spectrum is clearly more accurate, for the sake of convenience we will adopt a less accurate three-component, qualitative model (cf. Reimers, this volume). Seasonality of benthic flux measurements (Smith and Baldwin 1984) requires that a major fraction of the bottom-arriving flux of organic matter have a lifetime of several months or less and that it not be mixed extensively into the sediments by bioturbation. Recent ^{14}C measurements of organic carbon in the bioturbated layer (Emerson et al. 1987) imply that a substantial component (30–40%) of this layer's organic matter has a degradation residence time of less than 1000 yr, with the remaining refractory portion degraded on time scales much longer than the bioturbated layer residence time ($> 10^4$ yr). The most refractory material appears to be degraded more readily by aerobic than anaerobic processes (Reimers, this volume; Emerson and Hedges 1988). At present, a glaring lack exists in our knowledge of the chemical identities of these classes of organic matter. These patterns are consistent, however, with the existence of organic compound classes that yield too few kcal mol^{-1} when oxidized with SO_4^{2-} to repay the capital and operating costs of the metabolic machinery that would be required to carry out the oxidation.

Another pair of related problems that would benefit from more precise knowledge of chemical structure and classification at the level of chemical compounds are those of the terrestrial contribution and of the contribution of marine material originally deposited elsewhere to the sedimentary record at a given site (e.g., Prahl and Muehlhausen, and Walsh, both this volume). Conventional wisdom formulated primarily from analysis of $\delta^{13}C$ holds that sedimentary organic carbon in the open ocean is predominantly marine derived. Illumination of aeolian fluxes, geochemical evidence of horizontal midwater transport from continental margins to the open ocean (Martin et al. 1985) and uncertainty in $\delta^{13}C$ values previously assigned to the marine endmember all have led to skepticism about earlier assessments. That 20% of sedimentary organic matter in the open ocean is terrestrially produced is now within reasonable model estimates based on biomarker analyses (Prahl, this volume, Fig. 1). If this terrestrial material has slow reaction kinetics due either to decomposition during its extensive horizontal transit, to its initially highly degraded chemical form on entry into the ocean, or to protective incorporation within detrital particles, it may make a major contribution to the sedimentary organic carbon preserved in the geological record. The problem of horizontal redistribution is a most thorny one. By working only at sites of apparently continuous sedimentary accretion, one must be biased towards sites that are the ultimate resting places for material resuspended after initial deposition elsewhere. Perhaps this bias of focusing on regions from which present-day erosion is unlikely helps, together with more traditional explanations, to account for anomalies in ^{14}C ages of core-top sediments.

Provided reliable calibration of terrestrial biomarkers can be achieved, geographic maps of terrestrial contributions to total sedimentary organic carbon could be constructed. This information would provide, albeit indirectly, an assessment of marine contributions to the organic carbon record preserved in sediment cores, thus facilitating the use of such records in paleoenvironmental reconstructions. Lipid biomarkers of terrestrial organic carbon occur as homologous series of compounds (Prahl and Muehlhausen, this volume, and references therein). Compositional characteristics of those series can vary with source region (Gagosian et al. 1987). Variations in the intensive properties of these series deposited with depth in marine sediments could provide indications of how source regions or mechanisms of delivery (aeolian transport versus riverine transport and tubidity current dispersal) of terrestrial organic carbon to a given oceanic location changed with long-term climatic variations.

A commonly used method to estimate the terrigenous organic matter fraction in marine sediments is Rock-Eval pyrolysis. This method, originally developed to screen the quality and maturity of hydrocarbon source rocks (Espitalie et al. 1977), has also become one of the standard methods for typing organic matter in paleoceanographic studies, e.g., in the ocean drilling program (Peters and Simoneit 1982). The hydrogen index obtained by Rock-Eval pyrolysis is the amount of pyrolyzable hydrocarbons normalized to total organic carbon ($mgHC/gC_{org}$) and equivalent to the H/C ratio of the organic material. The proportions of marine and terrigenous organic matter in a sediment sample, in principle, can be partitioned by its hydrogen index because the material derived from marine plants is richer in hydrogen than the organic remains of terrestrial vegetation. A problem associated with the Rock-Eval technique, however, is that bulk sediment samples may yield spuriously low hydrogen index values due to the adsorption of hydrocarbons onto clay mineral surfaces ("mineral matrix effect," e.g., Espitalie et al. 1984). This effect is particularly important in sediments with low organic carbon contents ($<1\%$ C_{org}) which dominate the geological record. Hydrogen indices based on whole-rock samples therefore tend to overestimate the terrestrial component in marine sediments. Low hydrogen index values, in particular, should therefore be verified by some other means, e.g., by pyrolyzing organic matter concentrated via chemical extraction.

Both inherent organic matter properties and the agents acting on it determine the fraction buried. Much organic matter consumption is thought to be bacterial but to be strongly modulated by the numerous kinds of animals that graze bacteria. The water-column view has shifted from one of bacteria as the overwhelmingly dominant remineralizers of labile organics to one of the predators of bacteria as remineralizers and regenerators of DOM (Blackburn and Fenchel 1979), and there is no reason to expect the benthos to differ in this regard. Feeders on bacteria (e.g., foraminiferans, many meiofauna, and deposit feeders) increase mineralization by both digesting and respiring assimilated bacterial organic matter and also by making room for younger, more rapidly growing and respiring bacterial individuals on fresh surfaces. As organic material becomes more refractory, greater inefficiency in bacterial assimilation becomes evident (Benner et al. 1988), and bacteria relative to bacterivores assume a correspondingly greater direct role in respiration and mineralization (e.g., Cho and Azam 1988). Studies under simulated *in situ* conditions of organic carbon cycling in sediment-trap and box-core samples from the Biscay and Demerara abyssal plains showed that <10% of the flux was buried and 13–30% of biological consumption was due to bacteria (Rowe and Deming 1985).

This line of argument emphasizes the paucity of direct information on deep-sea organisms. There is little information on standing stocks and identities of species under the open ocean, and there are almost no direct measurements of activities or rates of population change (Rowe 1983). Surprises in animal abundance patterns continue to arise. There is now good reason to suspect that foraminiferans dominate many and perhaps most deep-sea biotas in biomass (Altenbach and Sarnthein, this volume; Fig. 2). A carbon budget partitioning benthic boundary layer respiration exists for but a single deep-sea site, Santa Catalina Basin (Smith et al. 1987). It lies at about 1300 m and has bottom-water oxygen concentrations of only about 15 μM/liter because the basin's sill lies within the oxygen minimum zone of the eastern North Pacific. Measurements suggest that 38–68% of benthic carbon mineralization occurs *above* surface sediments (but within the bottom boundary layer), with 18–55% being due to free-living bacterioplankton. This study site, whose surface sediments contain 5–7% organic carbon, cannot be considered representative of abyssal plains far from land, or even of slope depths outside semi-enclosed basins. The fact that the overall budget does not balance, i.e., that vertical sedimentation into sediment traps can account for only 17–43% of benthic respiration emphasizes the need to evaluate the role of horizontal transport in the deep sea.

Although Smith et al. (1987) did not attempt to partition community respiration on the basis of animal size, some speculation on partitioning of roles at their and other deep-sea sites seems safe based on shallow water information. Larger organisms dominate the process of bioturbation (Jumars and Wheatcroft, this volume), but small organisms almost always dominate the process of POC degradation in terms of any overall rate measure (because metabolic rate scales as body volume or weight to some exponent

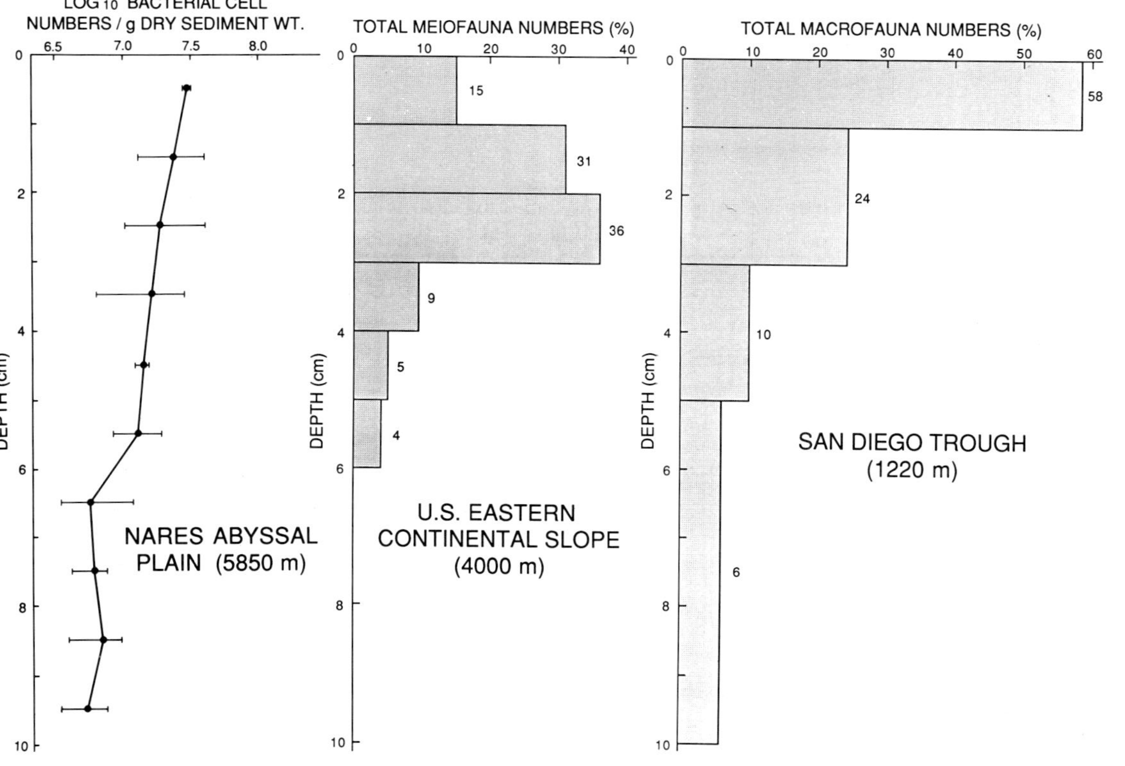

Fig. 2—Depth distributions of biotic components at several deep-sea sites. Bacterial data are from J. Deming (unpublished); the bars represent ranges for three replicates at each depth. Data for meiofauna come from Coull et al. (1977) and data for macrofauna from Jumars (1978). (Figure prepared by B. Hargrave).

between 2/3 and 3/4). It is important to remember, however, that this overall rate would fall (at least for all but the most labile and initially poorly colonized organics) if the bacteria (and their immediate predators in turn) were not being eaten. This consideration underscores the fallacy—demonstrated long ago in studies of forest litter decomposition (e.g., Bocock 1964)—of concluding from work with an isolated animal component or an isolated bacterial component that degradation rates in the deep sea are slow. Removal of the influence of metazoans may be one of the reasons for enhanced efficiency of preservation of organic matter in anaerobic environments (Kristensen and Blackburn 1987).

Interaction between sediment- and water-moving animals and organic matter-digesting bacteria helps to explain the seeming paradox that a greater fraction of organic matter is buried where particulate bioturbation is deeper and more intense (Aller and Cochran 1976; Emerson et al. 1985). The same phenomenon can be seen in radioisotopes whose decay constants roughly match those of the organic matter in question. The effects of this mixing downward by animals are compounded in organic matter, however, by shifting material of intermediate reactivity to anaerobic strata where its decay rate may drop dramatically. An understanding of what chemical species and what proportions of them succeed in running the interfacial gantlet is thus unlikely to be achieved by contemplation of mean vertical snapshots such as those of Fig. 1. Spatial and temporal variability on the scales of organisms is the rule. Aggregations of a few bacteria can make the space among them anoxic. Tube and burrow environments differ radically from the horizontal mean condition (Reimers, this volume).

A nearly completely unexplored environment is that of animal guts. Every particle in oxic deposits passes through them more than once judging by analogy with shallow water. We know deep-sea animal guts are sites of elevated microbial abundance and probably elevated activity. Some of the fastest growing bacterial species isolated from deep-sea sedimentary environments have come from animal guts (Deming and Colwell 1982). Animal guts may have either high or low pH and may be aerobic or anaerobic, but we know of no direct measurements on a single deep-sea species. We suspect that degradation within animal guts or in fecal pellets supporting high bacterial growth rates may be a major cause for the lack of entry of the labile component into even the first centimeter of the sedimentary record. Furlong and Carpenter (1988) found, for example, that $< 1\%$ of the pheopigment flux out of the water column of a coastal fjord accumulated in the upper 2 cm of sediment. A corollary is that early degradation is likely to be poorly modeled by incubation experiments with particles that are not continuously transiting among these heterogeneous gut and external microenvironments. Particularly steep redox gradients in the horizontal are likely to result at high organic fluxes; in order to fulfill its respiratory function, animal ventilation of burrows will have to be most intense when animals are large and oxygen demand of the surrounding sediments is great.

Upward pore-water flow due to pressure differentials in geological formations—created by tectonics or sediment loading—adds further to the diversity of benthic interfacial environments. Pore fluids venting along active continental margins and seepage along passive margins provide nutrients and dissolved gases to the benthos and might, in places, mask those benthic signals usually controlled by supply from the sea surface. Particularly severe interference with the paleoproductivity reconstructions using $\delta^{13}C$ contents of benthic foraminiferan tests is expected from the injection of ^{13}C-depleted ΣCO_2. This depletion is generated by methane-based chemoautotrophic benthic communities (Kulm et al. 1986; Childress et al. 1986) and mimics, even at slow venting rates, enormous fluxes of phytodetritus to the seafloor. At present it is unclear how much of the ocean floor sediments are affected by upward expulsion of pore fluids, but any stable carbon isotope signal of benthic Foraminifera that indicates abnormally ^{13}C-depleted metabolic ΣCO_2 in the benthic environment should be examined for evidence of venting before it is interpreted in terms of apparent oxygen utilization of normal oceanic bottom waters. The degree of ^{13}C depletion in pore waters from the Oregon subduction zone is between $-5‰$ PDB at the sediment-water interface and $-34‰$ PDB at about 100 cm depth below (Suess and Whiticar 1988), a substantial change over normal profiles (Fig. 1). Benthic forams are present at these vent environments but have not been examined in as much detail as the macrofauna. If this problem proves severe, there may be help in the form of biomarkers. Chemoautotrophic bacteria are rich in otherwise exotic structural compounds (Langworthy 1985; Taylor 1984). In conjunction with depleted $\delta^{13}C$, they might be used to diagnose vent environments.

SEABED RESPONSE TO TEMPORAL VARIATION IN FLUXES

Among the most far-reaching conclusions of recent modeling is the rapid utilization, on time scales of < 1 yr, of rapidly deposited pulses of organic matter (Reimers, this volume). Equally supportive is the observed speed of population response to such pulses (Gooday 1988; Lochte and Turley 1988). Data are too sparse, however, to know whether such inputs and responses are seasonally predictable or more episodic both within and between years. Among the many implications of such rapidity in benthic response is that *in situ* experiments of relatively short duration might be very informative.

Perhaps the most informative experimental manipulations, however, have been carried out by nature herself on time scales normally inaccessible to experimenters. Sediments initially deposited under high fluxes of organic matter have been redeposited by turbidity currents to regions of presently low organic carbon rain rates. Organic matter of distally deposited and ungraded silty clays, now buried under a thin and well oxidized pelagic sediment cover of the Madeira abyssal plain, is presently being remineralized with oxygen from bottom water (Colley et al. 1984; Wilson et al. 1985; de Lange et al. 1987). Total organic content of the turbidite is much lower above than below the oxygen front, illustrating the slow kinetics of anaerobic processes when operating on organic material of intermediate reactivity. Furthermore, to achieve such penetration, oxygen must diffuse past less reactive organic matter in sediments now overlying the turbidite to consume the more oxidizable material below. Similar "burn down" phenomena may operate at a given locality, without redeposition by turbidity currents, following a decrease in organic carbon flux or increase in bottom water oxygen concentration. Such changes occurred after the last glacial period and may be continuing today (Wilson et al. 1986). This phenomenon dramatically illustrates the importance of oxygen to organic matter degradation in deep-ocean marine sediments.

A similar "carbon burn-down" phenomenon may have affected Mediterranean sapropels, organic carbon-rich sediments originally deposited under anoxic conditions, whose upper portions are now peculiarly depleted in organic carbon (Emeis et al. 1988; de Lange et al., submitted). Reestablishment of oxic bottom waters may have mineralized this reactive organic matter and produced a redox front similar to that seen in the Madeira abyssal plain. Conversely, increased organic sedimentation effectively caps off organic matter that would otherwise decompose.

A major under-explored arena in diagenesis comprises regions in which erosion, physical reworking, and redeposition occur sporadically (e.g., Nowell and Hollister 1985). Such regions are difficult operating environments for equipment as well as steady-state chemical models, and neither flux chambers nor pore water gradients in solute concentration may indicate the quantitatively important fluxes during sediment transport episodes. We know that floras and faunas of such regions differ from those where sediment transport is infrequent (Jumars and Wheatcroft, this volume), but we do not know whether mean remineralization rates and net long-term burial rates of organic carbon are radically out of line with those found at comparable depths in the presence of steady accretion. Similarly, we do not know what fraction of the organic carbon being buried below animal reworking depths is contained in major turbidity flows and slumps.

ESTIMATION OF PAST PRODUCTION AND FLUXES

Based on the preceding background, which details recent insights into early diagenesis and seafloor processes, we again emphasize that most buried signals record bottom-arriving flux and not surface-ocean paleoproduction. There are no biogenic particles or organic chemicals known to be completely immune to degradation, whose accumulation rate in the sediments could be taken as a direct measure of their production rate at the sea surface. The problem of estimating paleoproduction is conceptually divided into quantifying the relationship between *(a)* total production and export flux (new production), *(b)* export flux and bottom-arriving flux, *(c)* bottom-arriving flux and shallow burial flux, and *(d)* shallow burial flux and deep burial flux. We examined primarily the third relationship, with some consideration of the fourth, whereas one would have to combine the recommendations of Groups 1 and 2 (see Williams et al. and Bruland et al., both this volume) with ours in order to evaluate the prospects for recording surface production in bulk sedimentary variables.

The first attempt to estimate surface ocean production short-circuited these subdivisions and predicted surface production directly from the deeply buried sedimentary record (Müller and Suess 1979). In so doing the immense complexities of all the intervening transformations were reduced to one empirical function. Recently, Sarnthein et al. (1988) have improved this approach by greatly expanding the data base for calibration and by incorporating relationships between surface production and export flux and between export flux and water depth, the latter to take into account degradation in the water column. Both approaches contain as a variable the bulk sedimentation rate, ostensibly to account for seafloor processes. In essence the loss of organic matter by benthic degradation is higher when the sedimentation rate is slower, but this control of carbon preservation is not without controversy. Emerson (1985) and Emerson et al. (1985) stress instead the rate of bioturbation as another of the controlling factors. In stressing bioturbation they introduce another subdivision in the carbon burial pathway, i.e., the relationship between shallow burial and deep burial (the depth below which little additional decrease in sedimentary organic carbon is observed). According to this view, the loss at the interface (as envisioned by Müller and Suess 1979; Reimers and Suess 1983; and Sarnthein et al. 1988), is a two-step process whereby higher bioturbation rate increases carbon preservation. The mechanism can perhaps best be envisioned as a feedback in which more intense bioturbation more rapidly buries more reactive material and the degradation of that material in turn limits further penetration of oxygen into the sediments. The issue is far from being resolved, but we suspect that explicit models exploring more than two types of reactive organic matter, different magnitudes of carbon residence times within the bioturbated layer, fast and slow sedimentation rate regimes, and high and low bottom-arriving fluxes of organic matter may succeed in pinpointing the mechanism(s) behind the empirical observation that a greater fraction of organic carbon is preserved under greater absolute flux rates.

At present the sole means, independent of sedimentary organic matter, to estimate surface production on the basis of sedimentary evidence derives from the relation between qualitative structure of the planktonic death assemblage and surface-ocean productivity (Mix, this volume). Interestingly, the approach gives results divergent from those based on bulk organic carbon preservation. Nevertheless, so long as they are relatively insensitive to taphonomic (preservational) artifacts, species compositions of buried planktonic microfossils can provide useful estimates of surface production.

For hindcasting of bottom-arriving fluxes, several direct and indirect means are promising enough to warrant further investigation. Interpretations of any of them are still aided by correlative measurements of TOC, total carbonate, and biogenic opal in the same stratum. Especially exciting is the recent finding of species of benthic forams that respond opportunistically to pulsed inputs of organic matter (Gooday 1988). Not only may analysis of benthic foram community structure allow hindcasting of the magnitude of flux (Altenbach and Sarnthein, this volume), but the presence or abundance of these "event opportunists" might also provide an indication of the short-term variation in the flux.

Among the refractory materials whose abundances may yield estimates of paleofluxes, the long-chained alkenones of coccolithophorids discussed by Prahl and Muehlhausen (this volume) are particularly promising biomarkers. They are relatively well characterized chemically and appear to be highly refractory. Their decomposition kinetics require additional study, however, in order to test their adequacy as paleoflux estimators. If the kinetics are genuinely first order and the same is true of the major constituents of TOC, then it is not clear that these alkenones provide a better estimate of bottom-arriving organic flux due to coccolithophorids than TOC itself provides of the bulk arriving organic flux. To be superior would require that the alkenone degradation rate is both slower and more constant than the rate for TOC decomposition at the site. Nevertheless, biomarkers such as those produced by coccolithophorids exist as a series of molecules both structurally and functionally different from each other. Hence, the series potentially displays differential reactivities under variable environmental conditions (Prahl et al. 1988). Such differences can be exploited by studying stratigraphic changes in composition of the chemical series. Examinations of this sort could resolve to what extent preservational factors for the biomarkers vary with time and location. Parallel information cannot be obtained from the single measurement of a bulk property such as TOC.

A very promising indirect method is to measure a fossil signature of the gradients in bottom-water oxygen concentration as derived from the metabolic ΣCO_2 input (Fig. 1). Required is a pair of preserved, calcium carbonate-containing species that record, at two distinct and known levels within (or one level within and one above) the sediment, at least one of the parameters whose gradients correlate with magnitude of the organic carbon flux. The most useful combination of species and geochemical parameters found to date appears to be *Cibicidoides wuellerstorfi* and *Uvigerina peregrina*, with $\delta^{13}C$ measurements of the tests (Zahn et al. 1986; Altenbach and Sarnthein, this volume). *C. wuellerstorfi* is not found living within a diffusive sublayer; it lives attached to worm tubes and other objects protruding above the bed, usually at heights of several mm above the bed. Thus it is clearly above any concentration gradient and is bathed in water containing bottom-water characteristics of ΣCO_2 and $\delta^{13}C$. The sole exception would be those rare individuals poised in a secondary circulation (e.g., in the lee of a tube or nodule, Nowell and Jumars 1984) that entrains water from the diffusive sublayer. *U. peregrina* lives infaunally and is bathed in pore waters enriched in the lighter isotope via organic matter degradation. This example clearly demonstrates the need for, and value of, ecological information at the level of species, individuals, and microenvironments. In principle, Cd/Ca ratio in the same foram tests should afford the same opportunity as ^{13}C, but the method remains to be explored.

RECOMMENDATIONS FOR FUTURE RESEARCH

Source and Chemical Identification of Organic Carbon Fractions

A fundamental block to further understanding of transformation of bottom-arriving organic carbon fluxes into the sedimentary record is the lack of knowledge concerning the identities of chemical compounds in those fluxes and the specific reactions in which they engage. One of the most troublesome unknowns in hindcasting bottom-arriving fluxes is the lack of a sufficiently precise means of estimating the terrigenous contribution to the total organic carbon preserved. The bulk of the terrigenous contribution remains chemically unidentified beyond crude operational catch-alls such as "humic acids." Acute problems with this imprecise identification are that the category includes a broad spectrum of reactivities and that numerous chemical transformations of a substance can occur without removing it from the category. Casual use of the analysis and the term lull one into believing that one knows more about a sample's chemistry than is true. Until estimates of the terrigenous contribution converge, we recommend a two-pronged attack combining a search for additional biomarkers with finer chemical subdivision of the bulk terrigenous component.

Now that it has become clear that a two-component, first-order degradation model is inadequate to treat the deep-sea sedimentary record and we have been forced to a three-component model for even a qualitative understanding, it seems appropriate to ask whether the best approach is to continue to treat degradation as a first-order, multi-component process or whether more complex kinetics should be explored. Most degradation is biochemical, and even the simplest degradation of one substrate by one enzyme is poorly treated as simply first order. Resolution of the best kinetic approach again seems unlikely, however, until the substances whose kinetics are being studied are better identified. The problem also is acute in benthic biological oceanography, where the specific identities of organic substrates digested and assimilated in nature by benthic organisms, in particular by heterotrophic bacteria and deposit feeders, are unknown.

Use should be made of the turbidite burn-down effect as a natural "experiment" on very long time scales. Chemical analyses should be done on both sides of the oxidation front to examine the sensitivity of specific biomarkers to long-term differences in redox conditions. The ideal indicator of terrigenous source, paleotemperature or paleoflux (although for the turbidite no such flux determination would be attempted), should not change in its intensive or extensive properties across the front. Those chemical species that do fall in concentration on the oxic side of the front are less resistant to oxic degradation than to anaerobic degradation on time scales otherwise beyond experimental reach.

Calibration of Paleoflux Estimators

All the paleoflux estimators require calibrations, which should take the form of field programs that measure fluxes arriving at the seabed and compare model-estimated and directly measured fluxes of organic matter. At present, two methods appear to provide the most reliable organic carbon flux measurements. The first involves sediment trapping above the bottom boundary layer. Within the bottom boundary layer material can be resuspended or at least be held turbulently in resuspension to heights of 50–100 m. The second method is to *(a)* estimate burial rates and add them to chamber measurements of total benthic respiration of those organisms living on and in the seabed and *(b)* to estimate respiration of organisms not caught in such chambers (Smith et al. 1987). Calibration should be over the entire range of fluxes for which hindcasts are to be made and should cover a variety of sediment types (oozes, clays, and hemipelagic sediments) and a range of bottom boundary layer flow regimes and oxygen concentrations. Ideally, several promising methods would be intercalibrated to assess relative precision and accuracy. Furthermore, such a set of observations would allow multivariate statistical analysis (e.g., multiple regression) to assess the added value of using several independent measurements. In addition to alkenone biomarker concentrations in bulk sediments and $\delta^{13}C$ and Cd/Ca ratios in foraminiferal tests, more traditional measurements (TOC, total carbonate, reactive silica) should be taken for objective multivariate evaluation of their utility.

Development of Additional Paleoflux Estimators

The demonstrated utility of using benthic foraminiferans suggests that additional species indicators of fluxes and seasonal variation in fluxes should be sought. The basic natural history of foraminiferans is poorly understood. The exciting possibility exists, however, that horizontal as well as vertical fluxes can be bracketed by species composition. It is likely, for example, that *C. wuellerstorfi* is responding to horizontal fluxes or turbulent vertical fluxes since it resides above the diffusive sublayer (Fig. 3). The limited studies of living forams (Altenbach et al. 1987 and references therein) already indicate that some combination of paleo-flow and paleo-carbon rain rate reconstruction for the bottom boundary layer is feasible.

Xenophyophorans should also be examined. This phylum of protozoans inhabits the deep-sea floor and secretes barite that may be fossilizable (Schulze 1905; Tendal and Gooday 1981). This biological production of benthic barite and its effect on the sedimenting Ba flux (see Bruland et al., this volume) needs to be evaluated. The exciting possibility exists that xenophyophorans may preserve or even amplify the seabed-arriving Ba

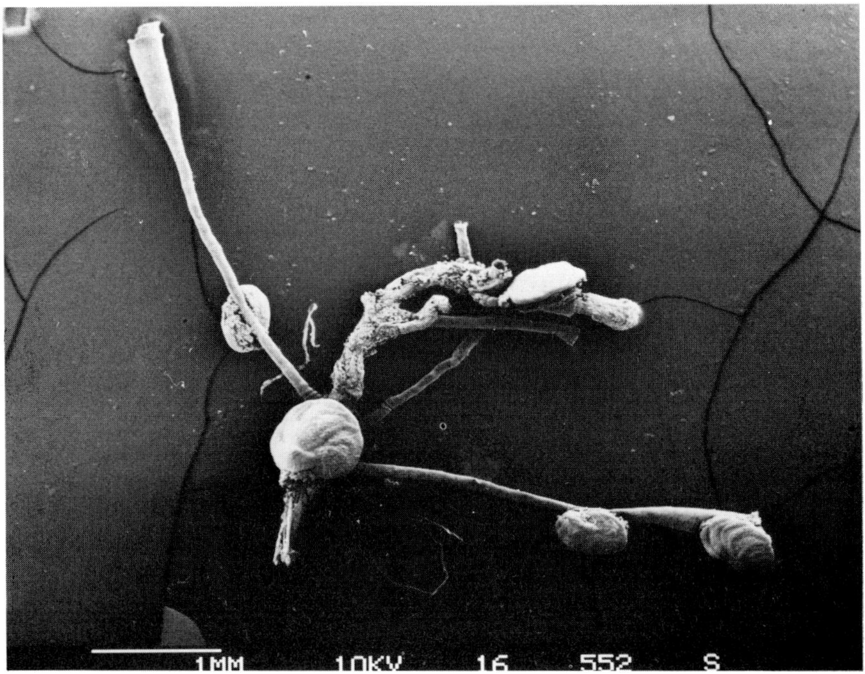

Fig. 3—A scanning electron micrograph of five individuals of *Cibicidoides wuellerstorfi* attached to a hydrozoan (photograph provided by A. Altenbach).

signal. The long-known association of barite, organic matter, and opaline silica has now been identified as largely due to syngenetic formation of $BaSO_4$ precipitates during aggregation, aging, and settling of siliceous phytodetritus (Bishop 1988). Previous interpretation favored direct biological production by siliceous phytoplankton, but culturing experiments consistently have failed to show it. Marine haptophytes (*Pavlova lutherii*) are known to secrete barite along with energy storage products in their intracellular vacuoles (Fresnel et al. 1979). Little is known, however, about the contribution of microflagellates to total primary production. Syngenetic formation within decomposing particles explains several of the characteristic features of dissolved and particulate Ba patterns in the water column as well as an increase in Ba flux with water depth. The mechanism proposed by Bishop (1988) also encourages, albeit circuitously, application of Ba accumulation as a paleoproductivity indicator (Schmitz 1987).

Tentatively, the accumulation of barite in the sediment would be some function, as yet unknown, of total supply of siliceous phytodetritus and residence time in the water column. Bio-barite produced by haptophytes would modulate such a function but would not be expected to be an independent source of significant magnitude. Excretion of bio-barite and its incorporation into tests by Xenophyophora might affect accumulation of barite in those areas where increased supply of detritus also stimulates the growth of xenophyophore populations. It is intriguing to pursue the utility of bio-barite as a proxy for productivity because, unlike any of the other indicators (C_{org}, opal, $CaCO_3$) which suffer extensive loss during transfer from the sea surface to the ocean floor, barite abundance actually increases during transfer.

Systematic means of identifying additional biomarkers for studies of high- and low-productivity regions of the ocean are beginning to become evident. Lipids are generally the most structurally diverse biochemicals and hence display the greatest source specificity to the organisms that produce them, allowing in the ideal situation unique identification of the species that produce them. Lipid components traceable to protective coatings or cell walls of autotrophs are typically the most refractory biochemicals in organisms and as such are found preserved in sediments. Although most reported biomarkers are for organisms at the base of the food web (de Leeuw 1986 and references therein), too few studies have been conducted to generalize that distinctive lipids are restricted to organisms of low trophic position. In addition to the alkenones of coccolithophorids, compounds such as 4-methylsterols of dinoflagellates (de Leeuw 1986) and the hopanoids of cyanobacteria (Rohmer et al. 1984) should be investigated as biomarkers of paleoflux in the geological record. Given the above patterns and their physical presence in the fossil record, the resting cysts of dinoflagellates and spores of diatoms would seem to be promising candidates as sources for new paleoenvironmental biomarkers. Nearly unexplored is the potential of bottom-dwelling bacterial or animal species to produce diagnostic biomarkers that persist in the sedimentary record. Functional groups of bacteria produce distinctive lipids (Goosens et al. 1987), but refractory components remain to be found, with the exception of ether lipids of Archaebacteria in ancient sediments and Recent hydrothermal environments (Comita and Gagosian 1983). Animals produce a range of antibiotics to prevent bacterial attack of their bodies and structures (tubes and burrows). Despite these ideas on potentially good candidates for sources of biomarkers, the search is still largely empirical, and new biomarkers are found primarily through serendipity.

Use of the carbon isotopic ratio in benthic foraminiferans for estimating past gradients in $\delta^{13}C$ of marine pore waters would benefit if additional species (beyond *U. peregrina*) could be identified that cover a greater depth range within diagenetic chemical gradients. Species that appear to occupy the stratum where oxygen falls to undetectable levels (Corliss 1985) may be useful for estimating paleoceanographic bottom water oxygen levels (McCorkle and Emerson 1988) if they are abundant enough geographically and can be calibrated using recent individuals and present O_2 levels. If pairs of species that are widely separated vertically in life positions are used, however, care must be taken to determine that they are contemporaneous in the fossil record.

Assessment of Biological Responses and Effects

Rapid population response of benthic foraminiferans to seasonally enhanced flux in the deep sea (Gooday 1988) came as a surprise but may well be widespread. Metazoan meiofauna have shown similar seasonal responses at shelf or slope depths off both U.S. coasts (Cammen, in preparation; Fleeger, in preparation). Seasonal changes in sedimentation flux need to be linked to biological and geochemical changes occurring at the sediment-water interface. No deliberately seasonal studies have been initiated to correlate biological responses (changes in numbers, reproductive conditions, vertical distributions of organisms) or geochemical changes in particulate matter and pore water profiles to variable supply rates of particulate organic matter deposition (Reimers, this volume). In general geochemical and biological studies have been poorly coupled.

Experiments under actual or simulated *in situ* deep-sea conditions are needed to investigate the impact of invertebrate feeding and bioturbation on preservation of inorganic and organic material. The unique microenvironment (physically mixed and anaerobic) within some animal guts can create conditions for dissolution, precipitation, or degradation that do not occur in ambient sediments and are not yet reproducible in the laboratory. Animal feeding should also stimulate microbial activity and thus increase the rate of organic matter degradation inferred from simple microbial incubations alone.

REFERENCES

Aller, R.C., and Cochran, J.K. 1976. ^{234}Th/^{238}U disequilibrium in near-shore sediment: particle reworking and diagenetic time scales. *Earth Planet. Sci. Lett.* **29**: 37–50.

Altenbach, A.V.; Lutze, G.F.; and Weinholz, P. 1987. Beobachtungen an Benthos-Foraminiferen (Teilprojekt A3). Univ. Kiel, *Ber. Sonderforschungsbereich 313* **6**: 1–86.

Benner, R.; Lay, J.; K'nees, E.; and Hodson, R.E. 1988. Carbon conversion efficiency for bacterial growth on lignocellulose: implications for detritus-based food webs. *Limnol. Oceanogr.* **33**(5): in press.

Bishop, J.K.B. 1988. The barite-opal-organic carbon association in oceanic particulate matter. *Nature* **332**: 341–343.

Blackburn, T.H., and Fenchel, T. 1979. Bacteria and Mineral Cycling. London: Academic Press.

Bocock, K.L. 1964. Changes in the amounts of dry matter, nitrogen, carbon and energy in decomposing woodland leaf litter in relation to the activities of the soil fauna. *J. Ecology* **52**: 272–284.

Childress, J.J.; Fischer, C.R.; Brooks, J.M.; Kennicutt, C.M., II; Bidigare, R.; and Anderson, A.E. 1986. A methanotrophic marine molluscan (Bivalvia, Mytilidae) symbiosis: mussels fueled by gas. *Science* **233**: 1306–1308.

Cho, B.C., and Azam, F. 1988. Major role of bacteria in biogeochemical fluxes in the ocean's interior. *Nature* **332**: 441–443.

Colley, S.; Thompson, J.; Wilson, T.R.S.; and Higgins, N.C. 1984. Post-depositional migration of elements during diagenesis in brown clay and turbidite sequences in the North East Atlantic. *Geochim. Cosmochim. Acta* **48**: 1223–1235.

Comita, P.B., and Gagosian, R.B. 1983. Membrane lipid from deep-sea hydrothermal vent methanogen: a new macrocyclic glycerol diether. *Science* **222**: 1329–1331.

Corliss, B.H. 1985. Microhabitats of benthic Foraminifera within deep-sea sediments. *Nature* **314**: 435–438.

Coull, B.C.; Ellison, R.L.; Fleeger, J.W.; Higgins, R.P.; Hope, W.D.; Hummon, W.D.; Rieger, R.M.; Sterrer, W.E.; Thiel, H.; and Tietjen, J.H. 1988. Quantitative estimates of meiofauna from the deep sea off North Carolina. *Mar. Biol.* **39**: 233–240.

de Lange, G.; Jarvis, I.; and Kuijpers, A. 1987. Geochemical characteristics and provenance of late Quaternary sediments from the Madeira Abyssal Plain, N. Atlantic. In: Geology and Geochemistry of Abyssal Plains, ed. P.P.E. Weaver and J. Thomson, pp.179–313. Blackwell Scientific Publications, London. *Spec. Publ. Geol. Soc. Lond.* **31**.

de Lange, G.; Middelburg. J.J.; and Pruysers, P.A. Middle and Late Quaternary deposition sequences and cycles in the eastern Mediterranean. *Sedimentology*, submitted.

de Leeuw, J.W. 1986. Sedimentary lipids and polysaccharides. In: Organic Marine Chemistry, ed. M.K. Sohn, pp.31–61. Washington, D.C.: Am. Chem. Soc.

Deming, J.W., and Colwell, R.R. 1982. Barophilic bacteria associated with digestive tracts of abyssal holothurians. *Appl. Environ. Microbiol.* **44**: 1222–1230.

Deming, J.W., and Colwell, R.R. 1985. Observations of barophilic microbial activity in samples of sediment and intercepted particulates from the Demerara Abyssal Plain. *Appl. Env. Microbiol.* **50**: 1002–1005.
Emeis, K.-C.; Camerlenghi, A.; McKenzie, J.A.; Rico, D.; and Sprovieri. 1988. Pleistocene/upper Pliocene sapropels in the Tyrrhenian Sea recovered during ODP Leg 107. In: Proc. Initial Repts. (Pt.B) ODP 1, ed, K.A. Kastens et al., vol. 107. College Station, Texas, in press.
Emerson, S. 1985. Organic carbon preservation in marine sediments. In: The Carbon Cycle and Atmospheric CO_2: Natural Variations Archean to Present, eds. E.T. Sundquist and W.S. Broecker. *Geophys. Monog.* **32**: 78–87. Washington, D.C.: Amer. Geophys. Union.
Emerson, S.; Fischer, K.; Reimers, C.E.; and Heggie, D. 1985. Organic carbon dynamics and preservation in deep-sea sediments. *Deep-Sea Res.* **32**: 1–21.
Emerson, S., and Hedges, J.I. 1988. Processes controlling the organic carbon content of open ocean sediments. *Paleocean.*, in press.
Emerson, S.; Stump, C.; Grootes, P.M.; Stuiver, M.; Farwell, G.W.; and Schmidt, F.H. 1987. Organic carbon in surface deep-sea sediments: C-14 concentration. *Nature* **329**: 51–54.
Espitalie, J.; Laporte, J.L.; Madec, M.; Marquis, F.; Leplat, P.; Paulet, J.; and Boutefeu, A. 1977. Méthode rapide de caractérisation des roches mères de leur potentiel pétrolier et de leur degré d'évolution. *Rev. Inst. Fr. Petr.* **32**: 23–42.
Espitalie, J.; Nakadi, K.S.; and Trichet, J. 1984. Role of the mineral matrix during kerogen pyrolysis. *Org. Geochem.* **6**: 365–382.
Fresnel, J.; Galle, P.; and Gayral, P. 1979. Micro-analysis of the vacuolar crystals of two uni-cellular marine chromophytes *Exanthemachrysis gayralliae, Pavlova* sp., (Pyrmnesiophyceae, Pavlovaceae). *C.R. Acad. Sci. Ser. D Sci. Nat.* **288**: 823–826.
Furlong, E.T., and Carpenter, R. 1988. Pigment preservation and remineralization in oxic coastal marine sediments. *Geochim. Cosmochim. Acta* **52**: 87–99.
Gagosian, R.B.; Peltzer, E.T.; and Merrill, J.T. 1987. Long-range transport of terrestrially derived lipids in aerosols from the South Pacific. *Nature* **325**: 800–903.
Gooday, A.J. 1988. A response by benthic Foraminifera to the deposition of phytodetritus in the deep sea. *Nature* **332**: 70–73.
Goosens, H.; Rijpstra, W.I.C.; Duren, R.R.; de Leeuw, J.W.; and Schenck, P.A. 1987. Bacterial contribution to sedimentary organic matter: a comparative study of lipid moieties in bacteria and Recent sediments. *Org. Geochem.* **10**: 683–696.
Hollister, C.D.; Nowell, A.R.M.; and Jumars, P.A. 1984. Dynamic abyss. *Sci. Am.* **250**(3): 42–53.
Jahnke, R.A., and Jackson, G.A. 1987. Role of sea floor organisms in oxygen consumption in the deep North Pacific Ocean. *Nature* **329**: 621–623.
Jumars, P.A. 1978. Spatial autocorrelation with RUM (Remote Underwater Manipulator): vertical and horizontal structure of a bathyal benthic community. *Deep-Sea Res.* **25**: 589–604.
Karl, D.M.; Knauer, G.A.; and Martin, J.H. 1988. Downward flux of particulate organic matter in the ocean: a particle decomposition paradox. *Nature* **332**: 438–441.
Kristensen, E., and Blackburn, T.H. 1987. The fate of organic carbon and nitrogen in experimental marine sediment systems: influence of bioturbation and anoxia. *J. Marine Res.* **45**: 231–257.
Kulm, L.D.; Suess, E.; Carson, B.; Lewis, B.T.; Ritger, S.D.; Kadko, D.; Thornburg, T.M.; Embley, R.; Rugh, W.; Massoth, G.; Langseth, M.; and

Cochrane, G. 1986. Oregon margin subduction zone: venting, fauna and carbonates. *Science* **231**: 561–566.

Langworthy, T.A. 1985. Lipids of Archaebacteria. In: The Bacteria, ed. I.C. Gunsalus, vol. VIII, pp. 459–497. New York: Academic Press.

Lochte, K., and Turley, T.M. 1988. Bacteria and cyanobacteria associated with phytodetritus in the deep sea. *Nature* **333**: 67–69.

Martin, J.H.; Knauer, G.A.; and Broenkow, W.W. 1985. VERTEX: the lateral transport of manganese in the northeast Pacific. *Deep-Sea Res.* **32**: 1405–1427.

McCorkle, D.C. 1987. Stable carbon isotopes in deep-sea pore waters: modern geochemistry and paleoceanographic applications. PhD. diss., Univ. of Washington, Seattle.

McCorkle, D., and Emerson, S. 1988. The relationship between pore water carbon isotopic composition and bottom water oxygen concentration. *Geochim. Cosmochim. Acta*, in press.

Müller, P.J., and Suess, E. 1979. Productivity, sedimentation rate and sedimentary organic matter in the oceans. I. Organic carbon preservation. *Deep-Sea Res.* **26**: 1346–1362.

Nowell, A.R.M., and Hollister, C.D., eds. 1985. Deep Ocean Sediment Transport. *Marine Geol.* **66**: 1–420.

Nowell, A.R.M., and Jumars, P.A. 1984. Flow environments of aquatic benthos. *Ann. Rev. Ecol. Syst.* **15**: 303–328.

Peters, K.E., and Simoneit, B.R.T. 1982. Rock-Eval pyrolysis of Quaternary sediments from Leg 64, Sites 479 and 480, Gulf of California. In: Initial Reports of the Deep-Sea Drilling Program, 64, Pt. 2, eds. J.R. Curray and D.G. Moore, pp. 925–931. Washington, D.C.: US Govt. Printing Office.

Prahl, F.G.; Muehlhausen, L.A.; and Zahnle, D.L. 1988. Further evaluation of long-chain alkenones as indicators of paleoceanographic conditions. *Geochim. Cosmochim. Acta*, in press.

Reimers, C. 1987. An *in situ* microprofiler instrument for measuring interfacial pore water gradients: methods and O_2 profiles from the North Pacific Ocean. *Deep-Sea Res.* **34**: 2019–2035.

Reimers, C.; Kalhorn, S.; Emerson, S.; and Nealson, K. 1984. Oxygen consumption rates in pelagic sediments from the central Pacific: first estimates from microelectrode profiles. *Geochim. Cosmochim. Acta* **48**: 903–910.

Reimers, C.E., and Suess, E. 1983. Partitioning of organic fluxes and sedimentary organic decomposition rates in the oceans. *Marine Chem.* **13**: 141–168.

Rohmer, M.; Bouvier-Nave, P.; and Ourisson, G. 1984. Distribution of hopanoid triterpenes in prokaryotes. *J. Gen. Microbiol.* **130**: 1137–1150.

Romankevich, E.A. 1984. Geochemistry of Organic Matter in the Ocean. New York: Springer-Verlag.

Rowe, G.T., ed. 1983. Deep-sea Biology. The Sea, vol. 8. New York: Wiley.

Rowe, G.T., and Deming, J.W. 1985. The role of bacteria in the turnover of organic carbon in deep-sea sediments. *J. Marine Res.* **43**: 925–950.

Sarnthein, M.; Winn, K.; Duplessy, J.-C.; and Fontugne, M.R. 1988. Global variations in surface productivity in low and mid latitudes: influence on CO_2 reservoirs of the deep ocean and atmosphere during the last 21,000 years. *Paleocean.* **3**: in press.

Schmitz, B. 1987. Barium, equatorial high productivity, and the northward wandering of the Indian continent. *Paleocean.* **2**: 63–77.

Schulze, F.E. 1905. Die Xenophyophoren, eine besondere Gruppe der Rhizopoden. Wissenschaftliche Ergebnisse der *Deutschen Tiefsee-Expedition 1898–1899 mit dem*

Dampfer VALDIVIA 1889–1899, ed. C. Chun. Band 11, Heft 1. Jena: Verlag G. Fischer.

Shaw, T. 1988. The early diagenesis of trace metals in nearshore sediments. Ph.D. diss., Univ. of California, San Diego.

Smith, K.L., Jr., and Baldwin, J.R. 1984. Seasonal fluctuations in deep-sea sediment community oxygen consumption: central and eastern North Pacific. *Nature* **307**: 624–626.

Smith, K.L., Jr.; Carlucci, A.F.; Jahnke, R.A.; and Craven, D.B. 1987. Organic carbon mineralization in the Santa Catalina Basin: benthic boundary layer metabolism. *Deep-Sea Res*. **34**: 185–211.

Suess, E. 1988. Effects of microbe activity. *Nature* **333**: 17–18.

Suess, E., and Whiticar, M.J. 1988. Methane-derived CO-2 in pore fluids expelled from the Oregon subduction zone. *Paleogeog. Pal. Pal.* (Special Issue). Kaiko Symposium: Geology, Geochemistry and Biology of the Trench Subduction Zone, ed. H. Okada and J.P. Cadet, in press.

Taylor, R.F. 1984. Bacterial triterpenoids. *Microbiol. Rev*. **48**: 181–198.

Tendal, O.S., and Gooday, A.J. 1981. Xenophyophoria in bottom photographs from the bathyal and abyssal NE Atlantic. *Oceanol. Acta* **4**: 415–422.

Westrich, J.T., and Berner, R.A. 1984. The role of sedimentary organic matter in bacterial sulfate reduction: the G model tested. *Limnol. Ocean*. **29**: 236–249.

Wilson, T.R.S.; Thomson, J.; Colley, S.; Hydes, D.J.; Higgs, N.C.; and Sorensen, J. 1985. Early organic diagenesis: the significance of progressive subsurface oxidation fronts in pelagic sediments. *Geochim. Cosmochim. Acta* **49**: 811–822.

Wilson, T.R.S.; Thomson, J.; Hydes, D.J.; Colley, S.; Culkin, F.; and Sorensen, J. 1986. Oxidation fronts in pelagic sediments: diagenetic formation of metal-rich layers. *Science* **232**: 972–975.

Zahn, R.; Winn, K.; and Sarnthein, M. 1986. Benthic foraminiferal $\delta^{13}C$ and accumulation rates of organic carbon: *Uvigerina peregrina* group and *Cibicidoides wuellerstorfi*. *Paleocean*. **1**: 27–42.

Productivity of the Ocean: Present and Past
eds. W.H. Berger, V.S. Smetacek and G. Wefer, pp. 313–340
John Wiley & Sons Limited

Pleistocene Paleoproductivity: Evidence from Organic Carbon and Foraminiferal Species

A.C. Mix

College of Oceanography
Oregon State University
Corvallis, OR 97331, U.S.A.

Abstract. One possible control on long-term changes in atmospheric carbon dioxide concentration is oceanic productivity. Evidence from organic carbon and foraminiferal transfer functions suggests that productivity was higher during glacial episodes than at present, especially under the equatorial upwelling regime. The different techniques disagree in detail, so the conclusions must be limited. Reconstructions based on organic carbon indicate larger changes than those based on foraminifera, and in low productivity areas the two different methods yield poorly correlated estimates. Future research must resolve these differences, work toward global reconstructions, and clarify the relationship between primary productivity and the export flux of carbon from the sea surface.

INTRODUCTION

During the late Quaternary the earth experienced major changes in atmospheric CO_2 concentration (Delmas et al. 1980; Neftel et al. 1982; Barnola et al. 1987). This discovery, along with the analysis of carbon isotopes in foraminifera (Shackleton 1977; Broecker 1982; Shackleton and Pisias 1985), has brought the oceanic carbon cycle to the forefront of paleoceanographic research. It is now clear that redistribution of carbon within the ocean must have caused atmospheric CO_2 changes during the ice ages (Broecker and Peng 1986). Biological productivity is one of the mechanisms responsible for partitioning carbon within the ocean. Thus, to constrain global models of the carbon cycle we must reconstruct the history and mechanisms of variations in paleoproductivity.

$\Delta\delta^{13}C$ AND THE OCEAN'S BIOLOGICAL PUMP

Most of the exchangeable carbon related to glacial-interglacial CO_2 change is oceanic. Nutrient-limited biological production moves ^{13}C-depleted organic carbon from surface to deep waters. Sinking of organic matter leaves surface waters depleted in CO_2, but enriched in ^{13}C (Kroopnick 1985). This effect is variously called the "biological pump" and the "carbon pump" (Berger and Vincent 1986). A stronger biological pump, along with feedback from the carbonate system (Broecker and Peng 1987), draws down pCO_2 in surface waters and the atmosphere. In the process, it increases surface water $\delta^{13}C$ and steepens the surface-to-deep $\delta^{13}C$ gradient ($\Delta\delta^{13}C$). To understand the mechanisms of paleo-CO_2 change and carbon isotope distributions in the ocean, we must document the time history and understand the cause of variations in the biological pump.

The longest and most detailed record of biological pump variations comes from core V19-30 in the eastern equatorial Pacific. Shackleton and Pisias (1985) use the $\delta^{13}C$ difference between the benthic foraminifera *Uvigerina sp.* and the planktonic foraminifera *Neogloboquadrina dutertrei* in this core to estimate the effect of the global biological pump on the past history of atmospheric CO_2. They assume here that (*a*) the benthic record in V19-30 reflects the global average deep water signal, (*b*) the planktonic record in this core gives a global average signal for nutrient-free surface waters, and (*c*) other mechanisms do not contribute significantly to CO_2 variations.

With these assumptions, Shackleton and Pisias (1985) interpret CO_2 variations over the last 340,000 years. Cyclic changes coinciding with the Milankovitch orbital periods of 100 ky (eccentricity), 41 ky (obliquity), and 23 ky (precession) dominate this record. Further, $\Delta\delta^{13}C$ changes in this core occur before $\delta^{18}O$ (approximate ice volume) changes in all of these frequency bands. If correct, this demonstrates that productivity and CO_2 changes are not a response to glaciation. Instead, they may be part of the forcing of the ice ages (Pisias and Shackleton 1984).

Several problems exist with reading the $\delta^{13}C$ record. For example, Zahn et al. (1986) demonstrate that the benthic foraminifera *Uvigerina* does not reliably record deep ocean $\delta^{13}C$. This may be because it lives within the sediment. If so, shell formation would reflect chemistry of pore water rather than that of bottom water (McCorkle et al. 1985). These authors recommend analysis of species living at the sediment-water interface such as *Cibicides wuellerstorfi* (also see Altenbach and Sarnthein, this volume). Recent analyses of *C. wuellerstorfi* in eastern Pacific sediments (Pisias et al. 1988; Shackleton, personal communication, 1988) support Shackleton's original assertion that the benthic isotope record in core V19-30 reflects deep water variations at the site. To what extent this is representative of the global average deep water signal remains uncertain.

Another possible problem is that V19-30 is in a high productivity, shallow thermocline area in the eastern equatorial Pacific. The planktonic species used, *N. dutertrei*, does not live at the sea surface here (Fairbanks et al. 1982). Further complicating matters, *N. dutertrei* may record large variations in their isotope content between upwelling and non-upwelling events. They seem to live at the sea surface and produce smaller specimens during cold upwelling events (Wefer et al. 1983). The isotope effects associated with this may be highly size dependent. The site of core V19-30 may have experienced large changes in paleoproductivity during the late Quaternary (Pedersen 1983; Lyle et al. 1988), so it is reasonable to suspect that $\delta^{13}C$ of *N. dutertrei* in V19-30 does not reflect the global nutrient-free signal without local contaminating effects.

Mix and Shackleton (1986) and Curry and Crowley (1987) provide additional data sets from different oceanographic settings to compare with the $\Delta\delta^{13}C$ record from V19-30. Both studies analyze $\delta^{13}C$ in surface dwelling planktonic foraminifera preserved in cores from the tropical Atlantic, in areas with a deep thermocline. To calculate the $\Delta\delta^{13}C$ index, both continue to use the V19-30 benthic record as the global deep water signal. They do not use the Atlantic benthic records because these include local watermass effects related to North Atlantic deep water variations. There are some subtle differences in the $\Delta\delta^{13}C$ record derived from these studies when compared to the Pacific record of Shackleton and Pisias (1985). For example, the $\Delta\delta^{13}C$ index generated using the Atlantic planktonic foraminifera, unlike that from the Pacific, does not contain a coherent 23 ky precessional cycle. In general, however, these other data sets support the V19-30 $\Delta\delta^{13}C$ record as an approximate index of the global biological pump. More importantly, they confirm that changes in $\Delta\delta^{13}C$ occur before changes in $\delta^{18}O$. As stated by Shackleton and Pisias (1985), changes in the efficiency of the ocean's carbon pump are not a simple response to the ice ages.

The last assumption of Shackleton and Pisias (1985), that the biological pump is the dominant mechanism controlling atmospheric CO_2, can be assessed by comparing the $\Delta\delta^{13}C$ record with the paleo-CO_2 record from the Vostok ice core (Barnola et al. 1987). Figure 1 illustrates these records. Both records were interpolated at 2.5 ky, the mean sampling interval of the Vostok record, and smoothed to eliminate fluctuations of less than 6 ky period. The visual agreement between these two records is good, although with current time scales the correlation is relatively low ($r=-.54$, $n=63$). Considering uncertainties in the time scales in both records, and the analytical errors of measurement, it is difficult at present to say that the two records are significantly different. As these records are refined, our ability to interpret differences between the two records will improve. Eventually this will yield insight into feedback mechanisms related to the biological pump or to other unrelated mechanisms that contribute to CO_2

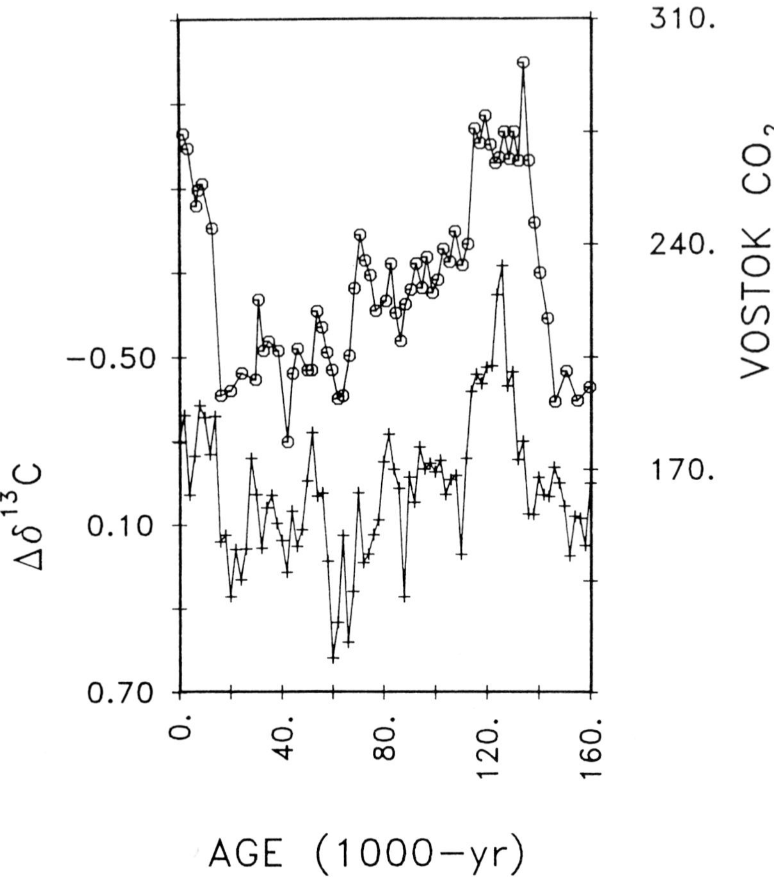

Fig. 1—Comparison of the $\Delta\delta^{13}C$ estimate of strength of the ocean's biological carbon pump in ‰(+) from Shackleton and Pisias (1985) with ice core CO_2 data in ppm (○) from Barnola et al. (1987). The strong similarity confirms the importance of biological pump variations to atmospheric CO_2.

change. For now, however, the major conclusion is that the overall similarities support the importance of the ocean's carbon pump to atmospheric CO_2 change.

Changing paleoproductivity is a mechanism that may drive variations in the ocean's vertical carbon isotope gradients (Broecker 1982; Boyle 1986), but it is not the only one. Circulation changes without any changes in productivity could be invoked to explain the carbon isotope gradients (e.g., Toggweiler and Sarmiento 1985). While not denying the other mechanisms, this paper concentrates on the Pleistocene record of paleoproductivity, to assess the possible role of productivity in explaining past variations in the oceanic carbon pump.

RECONSTRUCTIONS OF PALEOPRODUCTIVITY

Attempts to estimate paleoproductivity have employed many different methods. This paper emphasizes recent work on foraminifera and organic carbon in Pleistocene sediments.

As the dominant biological component, measurement of organic carbon content of sediment seems a logical choice for reconstructing biological carbon production. Unfortunately, deep-sea sediments preserve only a small fraction (typically less than one per cent) of the organic carbon produced at the sea surface (Romankevich 1984). Müller and Suess (1979) first attempted to correct for the observed effect of lower carbon concentrations at lower sedimentation rates. Their paleoproductivity equation reconstructs original production, R, from percent organic carbon in sediments. The equation is

$$R = (\%C \cdot BD)/(.003 \cdot S^{.3}) \tag{1}$$

where %C is the weight percent organic carbon in deep-sea sediments, BD is the dry bulk density (dry weight/wet volume) in g cm^{-3}, and S is the sedimentation rate in cm ky^{-1}.

Emerson (1985) argues against this approach. He suggests that the major factors controlling the preserved record of organic carbon are the carbon flux to the seafloor and deep-water oxygen contents. The paleoproductivity equation does not consider the oxygen effect. Thus, in Emerson's view, down-core variations could reflect changing deep water oxygen content, in addition to productivity.

Sarnthein et al. (1987) revise the earlier equation, finding

$$R = 15.9\,(\%C \cdot S \cdot BD)^{.66}\,(S(1-\%C/100))^{-.71}\,Z^{.32}\,. \tag{2}$$

Here, %C is the weight percent organic carbon, BD is the dry bulk density (dry weight/wet volume) in g cm^{-3}, and S is the sedimentation rate in cm ky^{-1}. Z is the water depth in meters. As with the Müller and Suess (1979) equation, this one does not explicitly address the effects of deep water oxygen on organic carbon preservation. Based on application of this equation to core-top sediments under different oxygen regimes, however, Sarnthein et al. (1987) suggest that deep water oxygen does not significantly affect the paleoproductivity estimate.

A third version of the organic carbon paleoproductivity equation is given by Sarnthein et al. (1988). Here, new (export) productivity is estimated as well as primary productivity. The two sets of equations are:

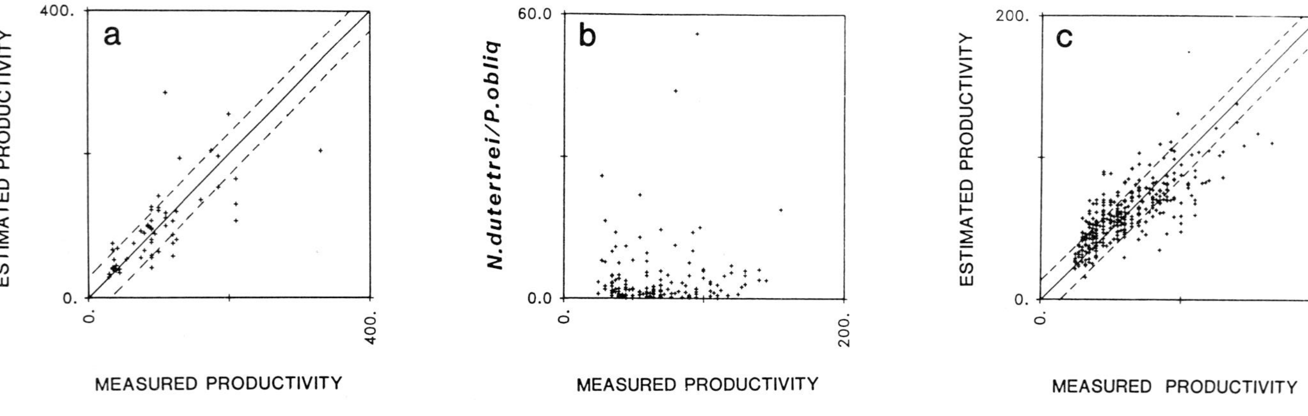

Fig. 2—(*a*) Calibration data for the organic paleoproductivity equation (from Sarnthein et al. 1988). The dashed error envelope (mean absolute value of residuals) is ±30 gC m^{-2} yr^{-1}. (*b*) A test of the foraminiferal fertility index of Berger and Killingley (1977), *N. dutertrei*/*P. obliquiloculata*, in Atlantic core tops. Lack of correlation to modern productivity suggests that this is not a generally applicable index in this area. (*c*) Calibration data for the foraminiferal transfer function (equation FAP-6) for paleoproductivity. The error envelope (mean absolute value of residuals) is ±12 gC m^{-2} yr^{-1}.

$$R_{new} = 0.0238 \cdot \%C^{.6429} \cdot S^{.8575} \cdot BD^{.5364} \cdot Z^{.8292} (S(1-\%C/100))^{-.2392} \quad (3)$$

and

$$R_{total} = 2 \cdot R_{new} \qquad (\text{for } R_{new} > 100) \quad (4)$$

or

$$R_{total} = 20 \cdot \sqrt{R_{new}} \qquad (\text{for } R_{new} \leq 100) . \quad (5)$$

As the earlier indices (Eqs. 1 and 2) are written for total primary productivity, here we will make comparisons based on primary productivity rather than export productivity. Figure 2a illustrates the organic carbon calibration data set. This estimator of primary paleoproductivity has an error envelope (average absolute value of residuals) of ± 30 gC m^{-2} yr^{-1}. The estimates of new productivity are more relevant to large-scale carbon cycling, and ultimately, expressing estimates as new productivity will be important. To do this properly we will need good calibration data for global patterns of new productivity, which are not available at present.

The preservation question is not yet settled. Emerson and Hedges (1988) again emphasize that the observed positive correlation between organic carbon percent and sedimentation rate is not well understood. In their view, the existence of a "sealing effect" associated with more rapid burial would require time constants for organic degradation on the order of 10,000 years (i.e., larger than residence times in the sedimentary mixed layer). Emerson et al. (1985) suggest, based on pore water data and sediment trap fluxes, that most of the organic carbon reaching the deep sea is rapidly degraded with a time constant on the order of 10 to 100 years. Emerson et al. (1987) argue from organic carbon ^{14}C data that 20–40% of the carbon in the sedimentary mixed layer is labile with a degradation time of less than 1000 years. Thus, for this fraction of the organic matter, a sealing effect related to sedimentation rate probably does not exist.

The small fraction of organic carbon remaining in the sediment column may take much longer to degrade (Grundmanis and Murray 1982; Müller and Mangini 1980). This fraction of the organic flux perhaps could record a sealing effect due to sedimentation rate. The problem becomes more complex if various types of organic matter have a broad spectrum of sensitivities to degradation (see Reimers, this volume). In this more realistic case, reconstructing paleoproductivity from total organic carbon content would require a predictable relationship between the various organic components produced and total primary productivity. A further complication is that some fraction of organic matter may be protected from degradation by adsorption on clays or enclosure in a matrix of shell or protein (Suess and Müller 1980). If so, the reactivity of the organic matter may not be related in a simple way to compound type.

The argument about deep water oxygen relies on whether diagenesis is faster under oxic conditions than under suboxic or anoxic conditions. Although this apparently is not true for fresh organic matter, Emerson and Hedges (1988) believe that it is true for the relatively refractory organic matter in deep-sea sediments. They admit that the data supporting this are circumstantial. If this rate difference exists, bioturbation adds a further complication. Faster benthic mixing would stir organic materials down into anoxic or suboxic sites, and thus enhance preservation of organic carbon.

Evidence for the oxygen effect comes from Curry and Lohmann (1985), who show increasing organic carbon content below 3800 m in the tropical Atlantic during glaciation. They interpret low $\delta^{13}C$ of benthic foraminifera here as the presence of a nutrient-rich, oxygen-poor water mass at the deeper sites. Shallower sites in the same area were bathed with more oxygenated water and had lower organic carbon contents despite similar surface-water productivity.

Evidence also exists favoring dominance of the productivity effect. Pedersen et al. (1988) show that organic carbon in the eastern equatorial Pacific does not covary with molybdenum concentration, contrary to predictions for sedimentary anoxia (Bertine and Turekian 1973). Note, however, that this constraint excludes only complete anoxia and does not preclude low but non-zero oxygen content of the deep water. In the same area, Lyle et al. (1988) demonstrate linkage between mass accumulation rates of organic carbon, calcite, and biogenic opal. As oxygen does not affect preservation of calcite and opal, they conclude that productivity must drive the bulk of the signal in this area, and that the oxygen effect, if real, is negligible in this area.

Organic carbon is not the only available index of paleoproductivity. Planktonic foraminifera may also provide useful information. Single species abundances point to areas of upwelling. For example, Thiede (1983) infers that *Neogloboquadrina dutertrei* is the species most clearly associated with coastal upwelling in the Atlantic and Pacific Oceans. Kipp (1976) and Bé (1977) note the prevalence of this species in open ocean upwelling regimes (offshore and equatorial) as well. Prell and Curry (1981) note the abundance of *Globigerina bulloides* in the cool coastal upwelling system of the Arabian Sea. The presence of *G. bulloides* here cannot be attributed to advection of cooler water from the high latitudes, so the link to upwelling is certain. Oxygen isotope data from Ganssen and Sarnthein (1983) support this interpretation, as $\delta^{18}O$ values from *G. bulloides* in surface sediments off N.W. Africa are consistent with temperatures in the upwelling season, whether it is summer or winter. Prell (1984) uses down-core abundances of this species as an index of monsoonal upwelling. With these data, Prell and Kutzbach (1987) link orbital modulation of insolation, the strength of the Asian monsoon, and upwelling in the Arabian Sea. These studies, like most using foraminiferal species, emphasize the physical process of upwelling rather than biological production.

Berger and Killingley (1977) approach a biological reconstruction, using the ratio of the planktonic foraminifera *Neogloboquadrina dutertrei* to *Pulleniatina obliquiloculata*, as a qualitative "fertility index," *N. dutertrei* is more closely linked to equatorial upwelling than *P. obliquiloculata*. We test this idea in Atlantic core-top sediments by comparing the ratio *N. dutertrei*/*P. obliquiloculata* to modern primary productivity (as summarized by Berger et al. 1987). Figure 2b shows this ratio for all Atlantic sites containing both species and demonstrates no correlation between the *N. dutertrei*/*P. obliquiloculata* ratio and productivity. Thus, while the idea of using species ratios as indices of fertility is good, this particular index does not yield a reliable measure of modern productivity in the Atlantic Ocean.

We extend Berger and Killingley's idea of a species-based index of productivity here by using all of the foraminiferal species to build a quantitative estimate of productivity using standard transfer function techniques (Imbrie and Kipp 1971). Estimates of modern primary productivity are from Berger et al. (1987). Although some significant assumptions go into this summary, it is used here because it is the only available map with global coverage. A transfer function of the form:

$$R = \sum_{i=1}^{n} (a_i \cdot X_i) + c \tag{6}$$

relates the modern seafloor foraminiferal assemblages (the CLIMAP 1981 Atlantic core tops) to productivity. Here X_i is the foraminiferal assemblages and their nonlinear combinations. Q-mode factor analysis defines these assemblages (Table 1). The equation coefficients a_i and c are constants, found by multiple regression of the core-top foraminiferal assemblages on the modern productivity values from each site (Table 2). We use this equation to generate paleoceanographic maps and time series. First we apply the core-top factors to down-core species abundances. We then insert the down-core factors into the paleoproductivity equation to obtain estimates of past productivity.

This quantitative index of productivity based on foraminiferal species (equation FAP-6) reconstructs better than 60% of the modern variance in productivity (Fig. 2c). The equation has an error envelope of $\pm$ 12 gC m^{-2} yr^{-1} (the average absolute value of residuals). This compares with the error envelope of $\pm$ 30 gC m^{-2} yr^{-1} for the organic carbon paleoproductivity estimates of Sarnthein et al. (1988). The foraminiferal estimates are frequently too low at high productivities ($>$120 gC m^{-2} yr^{-1}). This suggests limits to the response of foraminifera in high-productivity regimes. That is,

TABLE 1 Factor description matrix: Atlantic core-top foraminifera.

	Species	1 Tropical	2 Transi-tional	3 Polar	4 Gyre	5 Sub-polar	6 Sub-tropical
		Factor Scores					
1	*O. universa*	.022	.029	−.002	.039	.011	−.052
2	*G. conglobatus*	.023	.006	.001	.019	−.008	.003
3	*G. ruber* (pink)	.112	−.039	.004	.073	.018	−.068
4	*G. ruber* (white)	.932	.028	.023	−.124	−.046	.007
5	*G. tenellus*	.045	.008	−.001	−.030	−.001	.011
6	*G. sacc.* (no sac)	.206	−.048	.006	.328	.009	−.088
7	*G. sacc.* (sac)	.102	−.028	.003	.235	.003	−.048
8	*S. dehiscens*	−.001	.002	.000	.042	−.005	.006
9	*G. aequilateralis*	.100	.009	−.001	.036	−.001	.009
10	*G. calida*	.029	.016	−.003	−.001	.001	.006
11	*G. buloides*	−.033	.029	.074	.083	.615	.646
12	*G. falconensis*	.056	.315	−.034	−.118	−.083	.409
13	*G. digitata*	.007	.009	.000	.021	−.003	−.002
14	*G. rubescens*	.034	−.002	.000	−.015	.003	.002
15	*G. quinquiloba*	−.002	−.064	.120	−.005	.166	.059
16	*G. pachyderma* (L)	−.012	.007	.988	−.011	−.051	−.042
17	*G. pachyderma* (R)	−.010	.109	−.007	−.062	.512	−.393
18	*N. dutertrei*	.004	.045	−.004	.600	−.014	.017
19	*P. obliquiloculata*	.013	−.007	−.001	.263	−.003	.024
20	*G. inflata*	−.027	.903	.013	.062	−.074	−.091
21	*G. truncat.* (L)	.027	.134	−.002	−.023	−.033	.174
22	*G. truncat.* (R)	.046	.073	−.007	−.040	−.012	.032
23	*G. crassaformis*	.008	.009	.000	.068	−.008	.001
24	P-D intergrade	−.005	.144	−.007	−.007	.428	−.425
25	*G. hirsuta*	.015	.098	−.008	−.022	−.022	.082
26	*G. scitula*	.014	.048	−.010	−.021	.017	.057
27	*G. menardii*	.057	−.002	.002	.457	−.023	−.028
28	*G. tumida*	−.010	−.009	.000	.369	−.019	.068
29	*G. glutinata*	.198	−.061	−.029	−.015	.359	.033

after a certain point further increases of productivity do not appear to modify foraminiferal assemblages.

We test for artifacts in this productivity transfer function by comparing the estimates to water depth and mean sea-surface temperature (Fig. 3). As foraminifera in deeper sites are more dissolved, any bias due to carbonate dissolution would show up as a correlation between the productivity estimate and water depth. Figure 3a demonstrates lack of correlation between the modern productivity and water depths at the core-top sites ($r=-.11$, $n=356$).

TABLE 2 Foraminiferal transfer function for primary productivity.

Term (i)	X_i	a_i
1	ONE-SQ	72.6576
2	TWO-SQ	−145.8828
3	THREE-SQ	−69.8703
4	FOUR-SQ	−38.2067
5	FIVE-SQ	0.0000
6	SIX-SQ	−91.2309
7	1×2	0.0000
8	1×3	−223.3566
9	1×4	71.9344
10	1×5	302.8880
11	1×6	−64.5883
12	2×3	−48.2915
13	2×4	89.4054
14	2×5	0.0000
15	2×6	0.0000
16	3×4	0.0000
17	3×5	69.4658
18	3×6	81.1349
19	4×5	0.0000
20	4×6	−84.9721
21	5×6	−43.3413
22	ONE	−173.1118
23	TWO	75.3244
24	THREE	0.0000
25	FOUR	0.0000
26	FIVE	−106.2444
27	SIX	0.0000
INTERCEPT	c =	136.9832

Similarly, Fig. 3b shows that there is no correlation between errors in the productivity estimates (i.e., the residuals $R_{estimate} - R_{observed}$) and water depth (r=−.20, n=356). This is true over a very broad depth range (~200 m to ~5800 m). Thus, there is no apparent preservation effect on this estimate of paleoproductivity.

There is also no correlation between modern productivity and mean sea-surface temperature (r=.08, n=356), as shown in Fig. 3c. Although locally in the tropics high primary productivity may be linked to cool upwelling

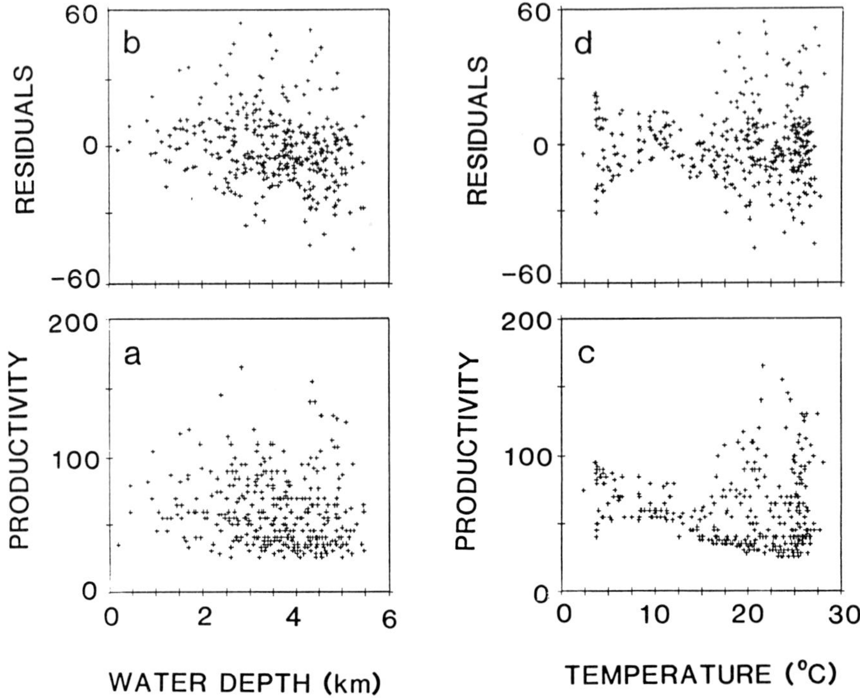

Fig. 3—Tests for artifacts in the foraminiferal transfer function for paleoproductivity (FAP-6). (*a*) Lack of correlation between productivity and water depth indicates independence in the calibration data set. (*b*) No correlation between estimate residuals (core-top estimated productivity minus modern observed) and water depth indicates no significant bias induced by moderate amounts of carbonate dissolution. (*c*) Lack of correlation between productivity and mean annual sea-surface temperature indicates independence in the calibration data set. (*d*) No correlation between estimate residuals and mean annual temperature indicates no significant bias related to sea-surface temperatures.

systems, on a large scale, sea-surface temperature and primary productivity are independent. This is a necessary condition for transfer functions designed to reconstruct temperature and productivity from the same multivariate data set of foraminiferal species. A second necessary condition is the lack of correlation of the errors in core-top productivity estimates with sea-surface temperatures. This is demonstrated in Fig. 3d (r=0.00, n=356). It does appear, however, that the mean absolute value of the residuals is larger at warmer sites. This may reflect the larger range of modern productivity values in tropical sites within our calibration data set (Fig. 3c).

Absence of biases in the paleoproductivity estimates related to either moderate carbonate dissolution or sea-surface temperature changes gives confidence that the index can be used down core. It appears that relationships (if any) between down-core estimates of temperature and productivity based on foraminifera are not artifacts of the methodology. Even though they are based on the same foraminiferal data, the temperature and productivity transfer functions are statistically independent.

THE PLEISTOCENE RECORD

Last Glacial Maximum

With this quantitative foraminiferal index, we reconstruct primary productivity of the whole Atlantic at the last glacial maximum (18,000 years ago). The foraminiferal species data are those of CLIMAP (1981). In Fig. 4 we compare modern productivity, as summarized by Berger et al. (1987), with transfer function estimates made for core tops (the calibration data set) and for the last glacial maximum. In the modern world (Fig. 4a), productivity is low in the central subtropical gyres with values of about 30 gC m^{-2} yr^{-1}. Productivity is highest in the eastern boundary currents, especially under the Canary Current off N.W. Africa and the Benguela Current off Angola. Values here are greater than 100 gC m^{-2} yr^{-1}. Transfer-function estimates made with the core-top foraminifera (Fig. 4b) represent the spatial distribution and absolute value of these features well, as expected from Fig. 2c. At the last glacial maximum (Fig. 4c), the pattern of lower productivity in the central gyres remains. Values here are slightly higher than at present. The minimum is near 45 gC m^{-2} yr^{-1}. Highest values, about 120 gC m^{-2} yr^{-1}, occur along the equator instead of in the eastern boundary currents.

The largest increases in glacial productivity occur along the equator. The general increase in productivity in glacial time, especially along the equator, is consistent with previous reconstructions based on organic carbon data (Müller and Suess 1979; Müller et al. 1983; Sarnthein et al. 1987). In the foraminiferal reconstruction, however, glacial productivity is slightly lower than at present off N.W. Africa and along the North American margin. The result off N.W. Africa conflicts with the reconstruction of Sarnthein et al. (1987), which infers a large productivity increase in this area at the last glacial maximum. Note, however, that the sites of N.W. Africa considered by Sarnthein et al. (1987) are mostly under coastal upwelling regimes. The cores in the glacial maximum reconstruction here are mostly offshore. Thus, the two data sets are not easily comparable. It is conceivable that the histories of coastal and offshore paleoproductivity are different. For example, Sarnthein et al. (1988) infer from organic carbon data a decrease in productivity in one central North Atlantic core and a few off N.W. Africa, in qualitative agreement with the foraminiferal reconstruction presented here.

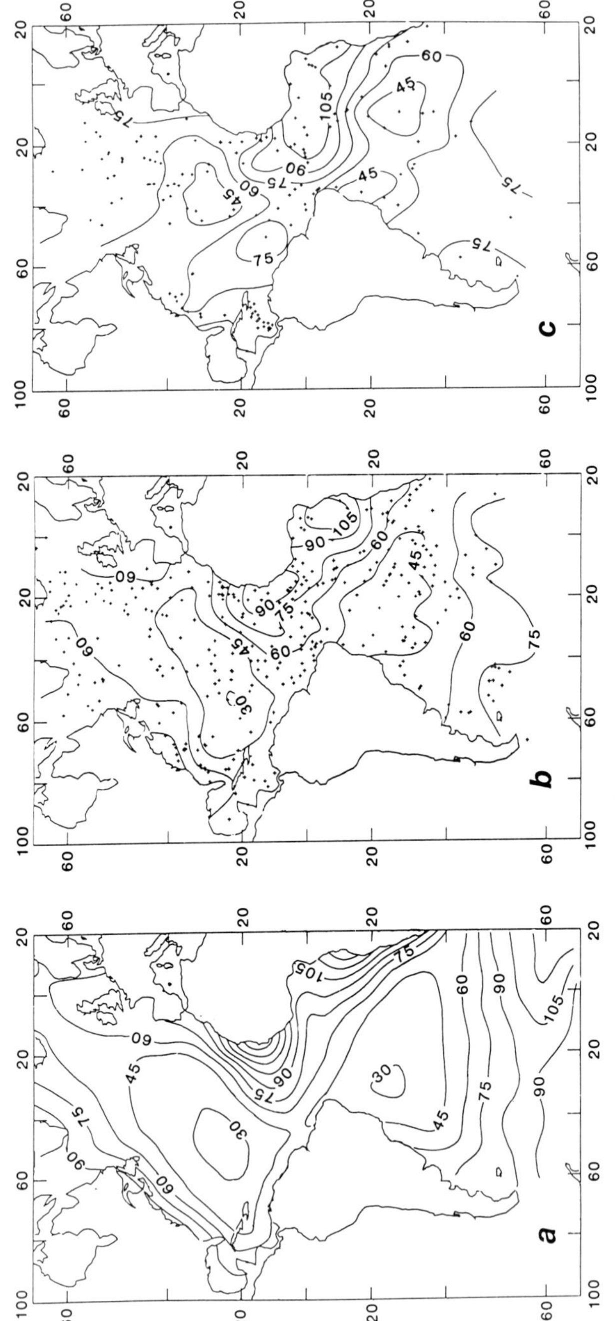

Fig. 4—Atlantic primary productivity, in gC m^{-2} yr^{-1}. (*a*) Modern primary productivity, as summarized by Berger et al. (1987). (*b*) Core-top productivity estimates using foraminiferal transfer function FAP-6. (*c*) Productivity at the last glacial maximum (18,000 years ago), using equation FAP-6 and the CLIMAP (1981) foraminiferal species data.

Increased equatorial productivity is consistent with equatorial Atlantic cooling at the glacial maximum, which CLIMAP (1981) and Mix et al. (1986) suggest was due to stronger winter upwelling. These studies, however, also estimate significant cooling along the eastern boundary currents, especially off N.W. Africa. If correct, the foraminiferal reconstructions may suggest that cooling in these boundary currents was due to advection of nutrient-depleted surface waters from higher latitudes, rather than local upwelling of cold nutrient-rich waters.

On average, Atlantic primary productivity reconstructed by the foraminiferal species transfer-function increases at the glacial maximum by 18±5%, from 57±1.0 gC m^{-2} yr^{-1} at present to 67±1.6 gC m^{-2} yr^{-1} in glacial time. These values are based on area weighted averages of the glacial and interglacial reconstructions between 60°N and ~40°S, gridded with a 4° mesh. The southern boundary is diagonal between southern South America and South Africa, as no foraminiferal data exist south of this line. Error estimates are the standard deviations of the means, calculated using the actual number of samples analyzed; 356 modern core tops and 151 samples of the last glacial maximum. The location of the largest productivity change, in the equatorial upwelling system, is consistent with stronger and more zonal glacial trade winds, which would drive stronger equatorial divergence. High rates of glacial intermediate-water formation (Boyle and Keigwin 1988) could also contribute to the increased exchange between intermediate and surface waters implied by this reconstruction. Thus, both low- and high-latitude processes could be involved as causal mechanisms.

This reconstruction of productivity at the last glacial maximum, if correct, has significant implications for models of atmospheric carbon dioxide. The 17% increase in Atlantic primary productivity at the glacial maximum compares to a predicted increase in export of carbon to the deep sea of 25% in the model of Boyle (1988). This model reproduces both the observed CO_2 changes and the observed change in $\Delta\delta^{13}C$. Caution is appropriate in comparing these numbers, however. Biological pumping of carbon reflects "new productivity" rather than primary productivity (Eppley and Peterson 1979). As yet we have no reliable way of quantifying the recycling of carbon within near-surface waters in the past. It does appear, however, that new and primary productivity are related in the modern ocean, at least at lower productivities (Eppley and Peterson 1979; Martin et al. 1987). If we use the relationship between new and primary productivity proposed by Eppley and Peterson (1979), an 18% increase in glacial primary productivity is amplified to a 38% increase in new productivity. This transform is highly uncertain for the modern world, so application to the past must be done cautiously. All we can say at present is that the inferred increase in glacial productivity is roughly consistent with CO_2 models driven by low-latitude productivity variations

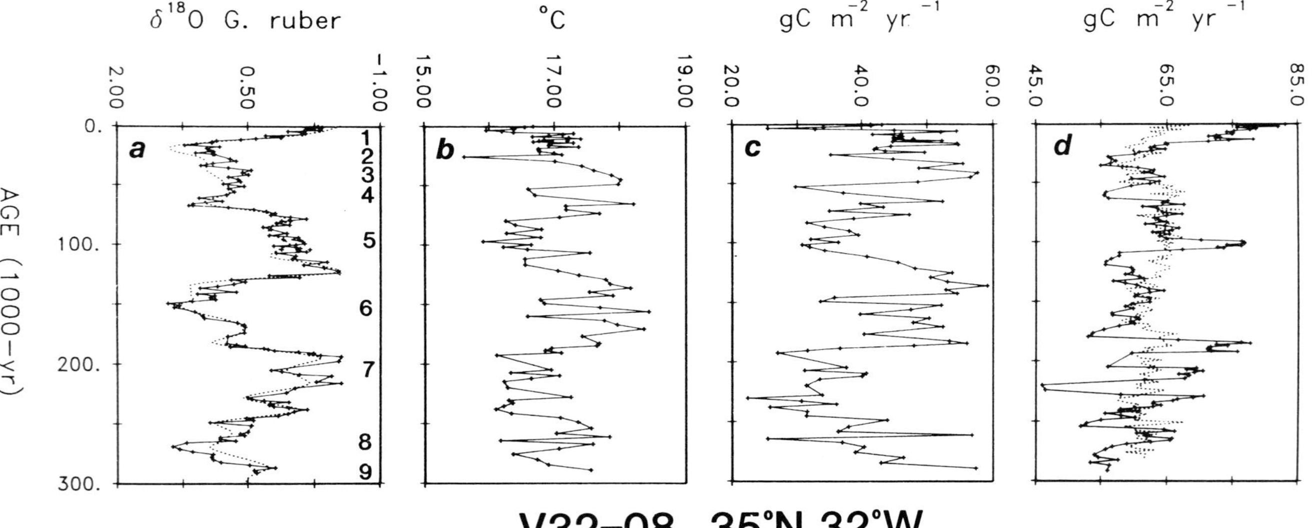

Fig. 5—Down-core records in core V32-08. (*a*) Oxygen isotope stratigraphy, based on analysis of white *Globigerinoides ruber* (solid line). The dashed line is the reference $\delta^{18}O$ time scale of SPECMAP (Imbrie et al. 1984). Traditional oxygen isotope stages are numbered. (*b*) Foraminiferal sea-surface temperature estimates for winter. (*c*) Foraminiferal transfer-function estimates of primary productivity. (*d*) Primary productivity estimated from organic carbon data, using the equations of Sarnthein et al. (1988). Solid line uses sedimentation rates as defined by the $\delta^{18}O$ stratigraphy, while the dashed line assumes a constant sedimentation rate of 2.4 cm kyr^{-1}.

An additional caution is that present reconstructions are not of global extent, and it is the global budget of paleoproductivity that would drive the partitioning of carbon within the oceans. The Pacific and Indian Oceans may have different histories than the Atlantic (Sarnthein et al. 1988). Thus, any application of these results to the global budget of productivity must be tentative. The present reconstruction does not preclude productivity variations in the high latitudes, and their role in the history of CO_2, as it does not extend into the Southern Oceans.

Time Series: The Last 300,000 Years

The transfer-function for paleoproductivity can also be applied to time series data sets. This provides an opportunity to compare the paleoproductivity reconstructions based on this foraminiferal transfer-function with those based on organic carbon. Here we illustrate two time series: Core V32-08 is in the central subtropical Atlantic (34°47′N, 32°25′W, 3252 m depth) and core V30-36 is under the N. Equatorial Current (5°21′N, 27°19′W, 4245 m depth). Sedimentation rates average 2.4 cm ky^{-1} in V32-08, and 1.8 cm ky^{-1} in V30-36. Time control for these cores comes from correlation of oxygen isotope ratios ($\delta^{18}O$) to the SPECMAP time scale (Imbrie et al. 1984) as shown in Figs. 5a and 6a.

Temperature changes (based on the foraminiferal transfer-function equation FA-20 of Molfino et al. 1982) at the sites are quite different. Core V32-08 records warmer sea-surface temperatures during glacial maxima than at present (Fig. 5b). This is consistent with its position in the subtropical warm pool of CLIMAP (1981). In contrast, core V30-36 records cooler temperatures at glacial maxima than at present (Fig. 6b).

Foraminiferal estimates of primary productivity from both cores (Figs. 5c, 6c) are highest during glacial episodes (stages 2–4,6,8, indicated by more positive $\delta^{18}O$). Unlike the temperature estimates, productivity variations based on the same foraminiferal species data from the two sites covary ($r=0.46$, $n=145$). Core V30-36 always gives higher estimates, consistent with its position under the North Equatorial Current. Paleoproductivity maxima are positively correlated ($r=0.63$, $n=145$) with warm temperatures at V32-08 (Fig. 5b,c). In contrast, the correlation of productivity to temperature in core V30-36 (Fig. 6b,c) is weak ($r=.27$, $n=145$). Although the overall correlation is positive, some of the extreme high-productivity events correspond to cold episodes (~130, 180, 260 ky B.P.) at this site.

The parallel productivity signals in the two sites are not artifacts of similar foraminiferal faunal variations. Indeed, the faunal data are quite different. Core V32-08 is dominated by the subtropical and transitional assemblages (*G. ruber, G. inflata,* and *G. bulloides*). Core V30-36 contains mostly tropical and gyre-margin species (*G. sacculifer, N. dutertrei, G. ruber*, and *G. menardii*).

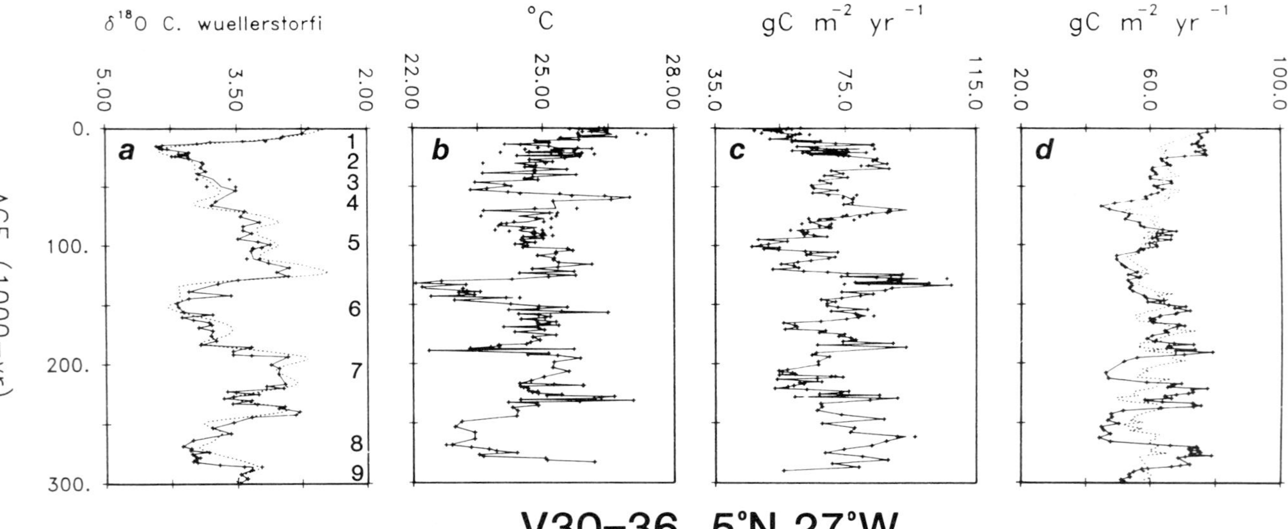

Fig. 6—Down-core records in core V30-36. (*a*) Oxygen isotope stratigraphy, based on analysis of *Cibicides wuellerstorfi* (solid line). The dashed line is the reference $\delta^{18}O$ time scale of SPECMAP (Imbrie et al. 1984). Traditional oxygen isotope stages are numbered. (*b*) Foraminiferal sea-surface temperature estimates for winter. (*c*) Foraminiferal transfer-function estimates of primary productivity. (*d*) Primary productivity estimated from organic carbon data, using the equations of Sarnthein et al. (1988). Solid line uses sedimentation rates as defined by the $\delta^{18}O$ stratigraphy, while the dashed line assumes a constant sedimentation rate of 1.75 cm kyr^{-1}.

Are these reconstructions of productivity consistent with those made using organic carbon data? To test this, we analyzed calcite and organic carbon content in cores V32-08 and V30-36 (data in appendix), using the techniques of Weliky et al. (1983). Figures 5d and 6d illustrate paleoproductivity values calculated with the equations 3–5 above. Bulk densities used in these estimates come from $\%CaCO_3$ data, using the equation of Curry and Lohmann (1985).

Because of the sedimentation rate term in the organic carbon paleoproductivity equations, two estimates are given. The solid lines in figures 5d and 6d use sedimentation rates as defined by correlation of $\delta^{18}O$ to the SPECMAP time scale of Imbrie et al. (1984), while the dashed lines in the same figures assume constant sedimentation rates of 2.40 cm ky^{-1} for core V32-08 and 1.75 cm ky^{-1} for core V30-36. The productivity estimates made in these different ways are not similar. Given this sensitivity to small changes in sedimentation rates, it is clear that the application of organic carbon data to paleoproductivity requires extremely accurate chronologies. This potential for error restricts the interpretation of organic carbon data to large changes where the influence of errors in determining sedimentation rates can be excluded.

The productivity variations estimated from organic carbon do not give consistent glacial-interglacial patterns and are quite different from those based on the foraminiferal transfer-function. In V30-36, the correlation between foraminiferal and organic carbon estimates is essentially zero, with $r=0.06$ ($n=145$). The correlation between the organic carbon and foraminiferal paleoproductivity estimates in V32-08 is better, but still low with $r=.20$ ($n=145$). Thus, the two methods of reconstructing Pleistocene paleoproductivity discussed here, one based on organic carbon and another on foraminiferal species, do not agree, at least in the relatively low productivity sites considered here.

DISCUSSION AND CONCLUSIONS

If the foraminiferal reconstruction of the last glacial maximum is correct, a link between productivity, the carbon isotope record of biological pump strength, and the paleo-CO_2 record is possible. The glacial-maximum map developed here for the Atlantic Ocean suggests that glacial-interglacial productivity changes are of a size and direction consistent with those predicted by some carbon cycle models. A further test must come from comparing the global-average time history of productivity with time series of $\Delta\delta^{13}C$ and paleo-CO_2. This can not yet be done, as the Atlantic may not represent the rest of the world's oceans. Eppley and Peterson (1979) suggest that the Atlantic accounts for about one third of the total oceanic production (both new and primary productivity). If the productive regions of the Pacific,

Indian, and Southern Oceans behave differently from the Atlantic, the total oceanic budget could be misconstrued from the present data.

Whatever the implications for the global budget, the present reconstruction emphasizes the importance of understanding both equatorial upwelling variations and oceanographic mechanisms for enhanced productivity away from the upwelling regions. It supports modeling efforts on the role of tropical paleoproductivity on CO_2. It also suggests the need for global studies. Only through analysis of broad spatial patterns of productivity change will we constrain the productivity budgets and mechanisms responsible for change.

The disagreement between the organic carbon and foraminiferal reconstructions raises important questions. The optimistic view is that both types of estimates may be valuable, but examine different ranges of production. Available organic carbon equations are best calibrated at high productivities of >75 gC m^{-2} yr^{-1}. The foraminiferal equation presented here is calibrated mostly at lower productivities of 30–120 gC m^{-2} yr^{-1} (Fig. 2). The downcore records used here for comparing the techniques are in low productivity regimes of 40–60 gC m^{-2} yr^{-1} at present. They also have relatively low sedimentation rates, near 2 cm ky^{-1}. This is in the range where organic carbon preservation may be very sensitive to small changes in sedimentation rate (Müller et al. 1983). More comparisons of techniques must be made at a range of productivities, sedimentation rates, and deep water oxygen values to assess the similarities and differences of the two methods.

Both techniques have advantages and disadvantages. They are similar in being empirical fits to modern data. These regression models are useful, as they provide insight and integrate complex and poorly understood processes into simple indices. Their weakness is that they do not include explicit physical/chemical processes. Any mechanisms not reflected in the modern data, perhaps because they vary little in the modern ocean, are left out of this type of model. Beyond these similarities, the foraminiferal transfer function has the advantages of being relatively insensitive to preservation problems and independent of sedimentation rates. It has the disadvantage of indirect linkage to productivity. The organic indicator has the advantage of being directly related to new productivity, which is the property most relevant to carbon cycle models. It has the disadvantages of highly complex and poorly understood preservation properties, and sensitivity to errors in determining sedimentation rates and bulk densities. Organic carbon indices also have the potential for contamination with terrestrial carbon. Müller and Suess (1979) and Sarnthein et al. (1987) raised this issue as a caveat in their studies. Müller et al. (1983) estimate a worst case contribution of 40% contamination with terrestrial carbon off N.W. Africa, but discount this possibility with stable isotope evidence. Organic biomarker data, however

(Prahl, this volume), suggest larger contributions of terrestrial carbon to marine sediments at many sites.

Thus, although we have learned much from reconstructing paleoproductivity, future research must reconcile the conflicting estimates. To constrain the CO_2 models, we must refine our ability to measure productivity (both primary and "new") from the geologic record and work toward global reconstructions with regional resolution.

Acknowledgements. I thank the Dahlem Foundation for making attendance at this meeting possible, and the National Science Foundation for research funding (ATM83-19371 and OCE87-16856). Discussions with many colleagues, notably S. Emerson, M. Lyle, N. Pisias, F. Prahl, E. Suess, and R. Zahn added to this paper. Reviews of the original submission by W. Berger, S. Emerson and M. Sarnthein improved the paper. I appreciate assistance from C. Peterson on foraminiferal species counts and data analysis, and from C.A. Ungerer on carbonate and organic carbon analyses.

REFERENCES

Barnola, J.M.; Raynaud, D.; Korotkevich, Y.S.; and Lorius, C. 1987. Vostok ice core provides 160,000-year record of atmospheric CO_2. *Nature* **329**: 408–414.

Bé, A.W.H. 1977. An ecological, zoogeographic, and taxonomic review of recent planktonic foraminifera. In: Oceanic Micropaleontology 1, ed. A.T.S. Ramsay, pp. 1–100. London: Academic.

Berger, W.H.; Fischer, K.; Lai, C.; and Wu, G. 1987. Oceanic productivity and organic carbon flux, Part I. Overview and maps of primary production and export production. *Scripps Inst. Ocean. Ref.* **87-30**.

Berger, W.H., and Killingley, J. 1977. Glacial-Holocene transition in deep-sea carbonates: selective dissolution and the stable isotope signal. *Science* **197**: 563–566.

Berger, W.H., and Vincent, E. 1986. Deep-sea carbonates: reading the carbon-isotope signal. *Geol. Rundschau* **75/1**: 249–269.

Bertine, K.K., and Turekian, K.K. 1973. Molybdenum in marine deposits. *Geochim. Cosmochim. Acta* **37**: 1415–1434.

Boyle, E.A. 1986. Deep ocean circulation, preformed nutrients, and atmospheric carbon dioxide: theories and evidence from oceanic sediments. In: Mesozoic and Cenozoic Oceans, ed. K.J. Hsu, pp. 337–351. Washington, D.C.: Am. Geophys. Union.

Boyle, E.A. 1988. Late Quaternary atmospheric carbon dioxide: control by ocean circulation and vertical chemical fractionation. *J. Geophys. Res.*, in press.

Boyle, E.A., and Keigwin, L.D. 1988. North Atlantic thermohaline circulation during the last 20,000 years and a link to high-latitude surface temperature. *Nature* **330**: 35–40.

Broecker, W.S. 1982. Ocean chemistry during glacial time. *Geochim. Cosmochim. Acta* **46**: 1689–1705.

Broecker, W.S., and Peng, T.-H. 1986. Global carbon cycle: 1985. *Radiocarbon* **28**: 309–327.

Broecker, W.S., and Peng, T.-H. 1987. The role of $CaCO_3$ compensation in the glacial to inter-glacial atmospheric CO_2 change. *Glob. Biogeochem. Cyc.* **1**: 15–20.

CLIMAP. 1981. Seasonal reconstruction of the earth's surface at the last glacial maximum. GSA Map and Chart Series MC-36.

Curry, W.B., and Crowley, T.J. 1987. The $\delta^{13}C$ of equatorial Atlantic surface waters: implications for ice age pCO_2 levels. *Paleocean.* **2**: 489–518.

Curry, W.B., and Lohmann, G.P. 1985. Carbon deposition rates and deep water residence time in the equatorial Atlantic Ocean throughout the last 160,000 years. In: The Carbon Cycle and Atmospheric CO_2: Natural Variations Archean to Present, eds. E.T. Sundquist and W.S. Broecker. *Geophys. Monog.* **32**: 285–301. Washington, D.C.: Am. Geophys. Union.

Delmas, R.J.; Ascencio, J.M.; and Legrand, M. 1980. Polar ice evidence that atmospheric CO_2 20,000 years ago was 50% of present. *Nature* **284**: 155–157.

Emerson, S. 1985. Organic carbon preservation in marine sediments. In: The Carbon Cycle and Atmospheric CO_2: Natural Variations Archean to Present, eds. E.T. Sundquist and W.S. Broecker. *Geophys. Monog.* **32**: 78–89. Washington D.C.: Am. Geophys. Union.

Emerson, S.; Fischer, K.; Reimers, C.; and Heggie, D. 1985. Organic carbon dynamics and preservation in deep-sea sediments. *Deep-Sea Res.* **32**: 1–22.

Emerson, S., and Hedges, J.I. 1988. Processes controlling the organic carbon content of open ocean sediments. *Paleocean.*, in press.

Emerson, S.; Stump, C.; Grootes, P.M.; Stuiver, M.; Farwell, G.W.; and Schmidt, F.H. 1987. Organic carbon in surface deep-sea sediments: C-14 concentration. *Nature* **329**: 51–54.

Eppley, R.W., and Peterson, B.J. 1979. Particulate organic matter flux and planktonic new production in the deep ocean. *Nature* **282**: 677–680.

Fairbanks, R.G.; Sverdlove, M.; Free, R.; Wiebe, P.H.; and Bé, A.W.H. 1982. Vertical distribution and isotopic fractionation of living planktonic foraminifera from the Panama Basin. *Nature* **298**: 841–844.

Ganssen, G., and Sarnthein, M. 1983. Stable-isotope composition of foraminifers: the surface and bottom-water record of coastal upwelling. In: Coastal Upwelling, Its Sediment Record, Part A: Responses of the Sedimentary Regime to Present Coastal Upwelling, eds. E. Suess and J. Thiede, NATO Conference Series 10A, pp. 99–121. New York: Plenum.

Grundmanis, V., and Murray, J.W. 1982. Aerobic respiration in pelagic marine sediments. *Geochim. Cosmochim. Acta* **46**: 1101–1120.

Imbrie, J.; Hays, J.D.; Martinson, D.G.; McIntyre, A.; Mix, A.C.; Morley, J.J.; Pisias, N.G.; Prell, W.L.; and Shackleton, N.J. 1984. The orbital theory of Pleistocene climate: support from a revised chronology of the marine $\delta^{18}O$ record. In: Milankovitch and Climate, Part 1, eds. A.L. Berger et al., pp. 269–305. Dordrecht: Reidel.

Imbrie, J., and Kipp, N.G. 1971. A new micropaleontological method for quantitative paleoclimatology: application to a late Pleistocene Caribbean core. In: The Late Cenozoic Glacial Ages, ed. K.K. Turekian, pp. 77–181. New Haven: Yale Univ. Press.

Kipp, N.G. 1976. New transfer-function for estimating past sea-surface conditions from sea-bed distributions of planktonic foraminiferal assemblages in the North Atlantic. In: Investigations of Late Quaternary Paleoceanography and Paleoclimatology, eds. R.M. Cline and J.D. Hays. *Geol. Soc. Amer. Mem.* **145**: 3–42. Boulder, CO: GSA.

Kroopnick, P.M. 1985. The distribution of ^{13}C of ΣCO_2 in the world oceans. *Deep-Sea Res.* **32**: 57–84.

Lyle, M.; Murray, D.W.; Finney, B.P.; Dymond, J.; Robbins, J.M.; and Brooksforce, K. 1988. The record of Late Pleistocene biogenic sedimentation in the eastern tropical Pacific Ocean. *Paleocean.* **3**: 39–60.

Martin, J.H.; Knauer, G.A.; Karl, D.M.; and Broenkow, W.W. 1987. VERTEX: carbon cycling in the northeast Pacific. *Deep-Sea Res.* **34**: 267–285.

McCorkle, D.C.; Emerson, S.R.; and Quay, P.D. 1985. Stable carbon isotopes in marine porewaters. *Earth Planet. Sci. Lett.* **74**: 13–26.

Mix, A.C.; Ruddiman, W.F.; and McIntyre, A. 1986. Late Quaternary paleoceanography of the tropical Atlantic. I. Spatial variability of annual mean sea-surface temperatures, 0–20,000 years B.P. *Paleocean.* **1**: 43–66.

Mix, A.C., and Shackleton, N.J. 1986. $\delta^{13}C$ analyses of foraminifera and atmospheric pCO_2 variations. Second International Conference on Paleoceanography (Abstr.), Woods Hole Oceanographic Institution, Woods Hole, MA.

Molfino, B.E.; Kipp, N.G.; and Morley, J.J. 1982. Comparison of foraminiferal, coccolithophorid, and radiolarian paleotemperature estimates: assemblage coherency and estimate concordancy. *Quatern. Res.* **17**: 279–313.

Müller, P.J.; Erlenkeuser, H.; and von Grafenstein, R. 1983. Glacial-interglacial cycles in oceanic productivity inferred from organic carbon contents in eastern north Atlantic sediment cores. In: Coastal Upwelling, Its Sediment Record, Part B: Sedimentary Records of Ancient Coastal Upwelling, eds. J. Thiede and E. Suess, NATO Conference Series 10B, pp. 365–398. New York: Plenum.

Müller, P.J., and Mangini, A. 1980. Organic carbon decomposition rates in sediments of the Pacific manganese nodule belt dated by ^{230}Th and ^{231}Pa. *Earth Planet. Sci. Lett.* **51**: 94–114.

Müller, P.J., and Suess, E. 1979. Productivity, sedimentation rate, and sedimentary organic matter in the oceans. I. Organic carbon preservation. *Deep-Sea Res.* **26A**: 1347–1362.

Neftel, A.; Oeschger, H.; Schwander, J.; Stauffer, B.; and Zumbrunn, R. 1982. Ice core sample measurements give atmospheric CO_2 content during the past 40,000 years. *Nature* **295**: 220–223.

Pedersen, T.F. 1983. Increased productivity in the eastern equatorial Pacific during the last glacial maximum (19,000 to 14,000 yr bp). *Geology* **11**: 16–19.

Pedersen, T.F.; Pickering, M.; Vogel, J.S.; Southon, J.N.; and Nelson, D.E. 1988. The response of benthic foraminifera to productivity cycles in the eastern equatorial Pacific: faunal and geochemical constraints on glacial bottom-water oxygen levels. *Paleocean.* **3**: 157–168.

Pisias, N.G.; Mix, A.C.; Shackleton, N.J.; and Imbrie, J. 1988. Response of the carbon system over the Milankovitch frequency bands. *EOS* **69**: 299.

Pisias, N.G., and Shackleton, N.J. 1984. Modelling the global climate response to orbital forcing and atmospheric carbon dioxide changes. *Nature* **310**: 757–759.

Prell, W.L. 1984. Monsoonal climate of the Arabian Sea during the late Quaternary: a response to changing solar radiation. In: Milankovitch and Climate, eds. A. Berger et al., pp. 349–366. Dordrecht: Reidel.

Prell, W.L., and Curry, W.B. 1981. Faunal and isotopic indices of monsoonal upwelling: Western Arabian Sea. *Oceanol. Acta* **4**: 91–98.

Prell, W.L., and Kutzbach, J.E. 1987. Monsoon variability over the past 150,000 years. *J. Geophys. Res.* **92**: 8411–8425.

Romankevich, E.A. 1984. Geochemistry of Organic Matter in the Ocean. Berlin: Springer.

Sarnthein, M.; Winn, K.; Duplessy, J.-C.; and Fontugne, M.R. 1988. Global variations of surface ocean productivity in low- and mid-latitudes: influence on CO_2 reservoirs of the deep ocean and atmosphere during the last 21,000 years. *Paleocean.* **3**: 361–399.

Sarnthein, M.; Winn, K.; and Zahn, R. 1987. Paleoproductivity of oceanic upwelling and the effect on atmospheric CO_2 and climatic change during deglaciation times. In: Abrupt Climatic Change, eds. W.H. Berger and L. Labeyrie, pp. 311–337. Dordrecht: Reidel.

Shackleton, N.J. 1977. Carbon-13 in Uvigerina: tropical rainforest history and the equatorial Pacific carbonate dissolution cycles. In: The Fate of Fossil Fuel CO_2 in the Oceans, eds. N.R. Andersen and A. Malahoff, pp. 401–427. New York: Plenum.

Shackleton, N.J., and Pisias, N.G. 1985. Atmospheric carbon dioxide, orbital forcing, and climate. In: The Carbon Cycle and Atmospheric CO_2: Natural Variations Archean to Present, eds. E.T. Sundquist and W.S. Broecker. *Geophys. Monog.* **32**: 303–317. Washington, D.C.: Am. Geophys. Union.

Suess, E., and Müller, P.J. 1980. Productivity, sedimentation rate and sedimentary organic matter in the oceans. II. Elemental fractionation. *Colloques Inter. C.N.R.S.* **293**: 17–26. Paris: Editions du Centre National de la Recherche Scientifique.

Thiede, J. 1983. Skeletal plankton and nekton in upwelling water masses off northwestern South America and Northwest Africa. In: Coastal Upwelling, Its Sediment Record, Part A: Sedimentary Records of Ancient Coastal Upwelling, eds. E. Suess and J. Thiede, NATO Conference Series 10B, pp. 183–207. New York: Plenum.

Toggweiler, J.R., and Sarmiento, J.L. 1985. Glacial to interglacial changes in atmospheric carbon dioxide: the critical role of ocean surface-water in high latitudes. In: The Carbon Cycle and Atmospheric CO_2: Natural Variations Archean to Present, eds. E.T. Sundquist and W.S. Broecker. *Geophys. Monog.* **32**: 163–184. Washington, D.C.: Am. Geophys. Union.

Wefer, G.; Dunbar, R.B.; and Suess, E. 1983. Stable isotopes of foraminifers off Peru recording high fertility and changes in upwelling history. In: Coastal Upwelling, Its Sediment Record, Part B: Sedimentary Records of Ancient Coastal Upwelling, eds. J. Thiede and E. Suess, NATO Conference Series 10B, pp. 295–308. New York: Plenum.

Weliky, K.; Suess, E.; Ungerer, A.; Müller, P.J.; and Fischer, K. 1983. Problems with accurate carbon measurements in marine sediments and water column particulates: a new approach. *Limnol. Ocean.* **28**: 1252–1259.

Zahn, R.; Winn, K.; and Sarnthein, M. 1986. Benthic foraminiferal $\delta^{13}C$ and accumulation rates of organic carbon: Uvigerina peregrina group and Cibicidoides wuellerstorfi. *Paleocean.* **1**: 27–64.

Appendix: Calcium carbonate and organic carbon data.

CORE V32-08

Depth (cm)	Age (ky)	%$CaCO_3$	%C-org	Depth (cm)	Age (ky)	%$CaCO_3$	%C-org
4.	0.882	86.2	0.195	188.	70.526	79.5	0.190
8.	1.765	88.5	0.187	192.	72.105	80.6	0.180
12.	2.647	88.6	0.184	200.	75.263	84.7	0.190
16.	3.529	88.9	0.177	204.	76.842	85.5	0.200
20.	4.412	88.1	0.161	208.	78.421	86.0	0.170
24.	5.294	83.6	0.159	212.	80.000	90.5	0.190
28.	6.176	83.9	0.165	216.	81.750	89.2	0.210
32.	7.059	84.0	0.164	220.	83.500	88.6	0.200
36.	7.941	83.1	0.149	224.	85.250	83.0	0.180
40.	8.824	81.8	0.143	228.	87.000	86.3	0.200
44.	9.706	80.4	0.156	232.	88.714	90.3	0.180
48.	10.588	79.6	0.152	236.	90.429	89.8	0.190
52.	11.471	79.3	0.194	240.	92.143	90.8	0.180
56.	12.353	69.6	0.179	244.	93.857	89.0	0.190
60.	13.235	73.2	0.179	248.	95.571	91.0	0.190
64.	14.118	74.4	0.160	252.	97.286	91.4	0.190
68.	15.000	74.6	0.162	256.	99.000	91.3	0.170
72.	16.286	73.1	0.149	260.	100.000	88.0	0.190
76.	17.571	68.6	0.149	264.	101.000	91.5	0.160
80.	18.857	73.1	0.162	268.	102.000	88.0	0.170
84.	20.143	72.9	0.146	272.	103.000	89.3	0.180
88.	21.429	75.5	0.155	276.	104.000	90.3	0.190
92.	22.714	74.7	0.187	280.	105.000	89.9	0.190
96.	24.000	77.9	0.147	284.	106.000	88.3	0.180
100.	25.933	81.2	0.148	288.	107.000	90.1	0.170
104.	27.867	84.2	0.150	292.	109.500	89.0	0.170
108.	29.800	85.4	0.143	296.	112.000	89.4	0.180
112.	31.733	84.5	0.132	300.	114.500	89.4	0.160
116.	33.667	83.1	0.180	304.	117.000	88.8	0.160
120.	35.600	79.3	0.210	308.	119.500	89.6	0.150
124.	37.533	82.4	0.190	312.	122.000	87.8	0.170
128.	39.467	84.0	0.220	316.	123.857	85.0	0.179
132.	41.400	84.5	0.170	320.	125.714	83.0	0.143
140.	45.267	85.1	0.210	324.	127.571	79.4	0.150
144.	47.200	86.8	0.193	328.	129.429	78.4	0.154
148.	49.133	83.3	0.190	332.	131.286	81.5	0.165
152.	51.067	83.7	0.179	336.	133.143	83.0	0.184
156.	53.000	78.7	0.220	340.	135.000	81.0	0.167
164.	58.333	75.1	0.190	344.	136.600	79.9	0.154
168.	61.000	76.8	0.220	352.	139.800	78.7	0.151
172.	63.667	70.0	0.170	356.	141.400	79.1	0.169
176.	65.789	70.1	0.190	360.	143.000	77.8	0.182
180.	67.368	68.9	0.200	364.	144.600	77.3	0.168
184.	68.947	74.4	0.220	368.	146.200	63.1	0.180

CORE V32-08 (continued)

Depth (cm)	Age (ky)	%$CaCO_3$	%C-org	Depth (cm)	Age (ky)	%$CaCO_3$	%C-org
372.	147.800	69.0	0.182	528.	220.000	82.1	0.156
380.	151.000	66.9	0.156	532.	224.000	70.7	0.147
380.	151.000	66.9	0.157	536.	228.000	67.2	0.150
384.	152.818	68.5	0.187	540.	229.250	68.9	0.158
388.	154.636	72.5	0.186	544.	230.500	72.0	0.171
392.	156.455	72.5	0.174	548.	231.750	76.3	0.157
400.	160.091	72.2	0.180	552.	233.000	79.2	0.178
404.	161.909	76.0	0.191	556.	234.250	78.5	0.136
408.	163.727	76.8	0.219	560.	235.500	87.6	0.146
412.	165.545	80.1	0.194	564.	236.750	89.0	0.161
416.	167.364	85.0	0.180	568.	238.000	85.3	0.148
420.	169.182	84.3	0.187	572.	239.833	90.3	0.153
424.	171.000	82.7	0.193	576.	241.667	89.1	0.146
428.	174.000	80.3	0.208	580.	243.500	85.7	0.216
432.	177.000	81.0	0.167	584.	245.333	85.3	0.184
436.	180.000	81.5	0.179	588.	247.167	83.0	0.141
440.	183.000	81.5	0.160	592.	249.000	83.7	0.153
448.	184.956	82.7	0.156	596.	251.667	84.5	0.150
452.	185.933	83.6	0.155	600.	254.333	84.7	0.181
456.	186.911	83.1	0.159	604.	257.000	90.6	0.172
460.	187.889	83.7	0.187	608.	258.500	90.3	0.149
464.	188.867	85.2	0.158	612.	260.000	89.4	0.148
468.	189.844	84.7	0.159	616.	261.500	86.4	0.167
472.	190.822	88.4	0.172	620.	263.000	84.1	0.160
476.	191.800	89.2	0.160	624.	264.500	76.4	0.154
480.	192.778	88.7	0.180	628.	266.000	68.0	0.170
484.	193.756	89.0	0.169	632.	267.500	65.2	0.205
488.	205.000	87.9	0.153	636.	269.000	66.2	0.161
492.	206.222	87.3	0.165	640.	271.250	63.5	0.194
496.	207.444	89.0	0.158	644.	273.500	59.2	0.220
500.	208.667	89.5	0.176	652.	278.000	74.1	0.142
504.	209.889	89.6	0.148	656.	280.250	74.5	0.177
508.	211.111	87.9	0.155	660.	282.500	77.9	0.179
512.	212.333	86.4	0.152	664.	284.750	86.4	0.143
520.	214.778	82.4	0.186	668.	287.000	86.5	0.172
524.	216.000	86.5	0.156	676.	291.500	94.8	0.149

CORE V30-36

Depth (cm)	Age (ky)	%$CaCO_3$	%C-org	Depth (cm)	Age (ky)	%$CaCO_3$	%C-org
6.	3.394	76.0	0.340	27.	12.723	55.3	0.218
9.	5.090	77.3	0.286	30.	13.753	51.7	0.212
15.	8.153	76.9	0.250	33.	14.783	44.7	0.265
24.	11.693	62.2	0.222	36.	15.813	48.1	0.239

CORE V30-36 (continued)

Depth (cm)	Age (ky)	%$CaCO_3$	%C-org	Depth (cm)	Age (ky)	%$CaCO_3$	%C-org
45.	18.886	42.5	0.275	193.	109.250	48.3	0.130
48.	19.909	40.2	0.248	197	112.250	48.8	0.137
54.	21.955	44.7	0.264	203.	116.750	47.8	0.153
57.	22.977	50.9	0.255	206.	119.000	54.1	0.151
60.	24.000	52.4	0.246	212.	123.318	57.2	0.172
63.	25.933	48.5	0.258	215.	125.295	75.5	0.120
66.	27.867	57.0	0.244	218.	127.273	68.9	0.159
69.	29.800	55.8	0.264	222.	129.909	55.4	0.141
72.	31.733	54.9	0.282	226.	132.545	54.3	0.153
75.	33.667	50.8	0.222	229.	134.523	50.9	0.160
78.	35.600	49.1	0.223	232.	136.500	43.8	0.163
81.	37.533	52.3	0.227	236.	139.136	41.8	0.177
84.	39.467	54.4	0.230	242.	143.091	45.7	0.206
87.	41.400	53.3	0.207	245.	145.068	39.1	0.231
90.	43.333	49.7	0.217	248.	147.045	41.1	0.295
93.	45.267	50.0	0.296	250.	148.364	41.8	0.228
96.	47.200	58.7	0.277	254.	151.000	42.2	0.255
99.	49.133	66.0	0.197	257.	152.333	39.4	0.281
102.	51.067	57.1	0.226	260.	153.667	39.0	0.242
104.	52.356	57.7	0.244	265.	155.889	34.4	0.315
109.	55.526	55.8	0.226	269.	157.667	35.7	0.260
111.	56.789	35.1	0.252	271.	158.556	35.6	0.253
114.	58.684	32.3	0.236	274.	159.889	33.7	0.208
117.	60.579	29.1	0.251	276.	160.778	31.2	0.211
120.	62.474	18.1	0.224	279.	162.111	33.9	0.176
123.	64.368	16.0	0.190	282.	163.444	28.6	0.203
126.	66.875	17.9	0.169	284.	164.333	30.8	0.186
129.	69.687	23.5	0.199	288.	166.111	24.4	0.209
133.	73.437	38.7	0.245	291.	167.444	34.2	0.271
135.	75.312	43.9	0.227	294.	168.778	46.1	0.258
138.	78.125	49.9	0.190	295.	169.222	52.1	0.249
141.	80.700	64.2	0.189	301.	172.714	69.3	0.242
145.	83.500	65.9	0.172	303.	174.429	68.4	0.314
147.	84.900	60.2	0.193	306.	177.000	66.8	0.250
150.	87.000	62.4	0.207	309.	179.571	60.4	0.241
153.	88.440	55.4	0.228	312.	182.143	58.5	0.214
156.	89.880	58.4	0.167	315.	183.710	49.5	0.205
160.	91.800	60.5	0.204	318.	184.774	43.8	0.239
162.	92.760	60.5	0.164	321.	185.839	40.2	0.177
165.	94.200	60.0	0.152	324.	186.903	36.7	0.189
168.	95.640	58.2	0.206	327.	187.968	33.9	0.183
174.	98.520	62.0	0.148	330.	189.032	35.5	0.266
177.	100.067	66.4	0.168	333.	190.097	36.8	0.202
180.	101.667	66.5	0.172	336.	191.161	39.2	0.318
183.	103.267	53.8	0.147	342.	193.290	32.4	0.322
185.	104.333	55.8	0.152	345.	195.571	35.6	0.312
188.	105.933	55.6	0.152	347.	198.714	45.2	0.333

CORE V30-36 (continued)

Depth (cm)	Age (ky)	%$CaCO_3$	%C-org	Depth (cm)	Age (ky)	%$CaCO_3$	%C-org
353.	208.143	65.7	0.192	437.	250.667	62.1	0.153
356.	212.857	60.0	0.183	440.	253.381	56.8	0.112
359.	216.343	55.9	0.159	443.	256.095	51.5	0.132
362.	217.371	52.5	0.182	446.	259.182	56.5	0.159
365.	218.400	54.5	0.160	449.	262.455	56.7	0.128
368.	219.429	62.1	0.143	452.	265.727	57.6	0.162
371.	220.457	60.3	0.139	455.	269.000	69.0	0.188
374.	221.486	73.0	0.213	458.	269.931	68.6	0.175
377.	222.514	69.3	0.181	461.	270.862	65.8	0.182
383.	224.571	69.3	0.183	464.	271.793	70.3	0.157
386.	225.600	55.7	0.194	467.	272.724	72.5	0.158
389.	226.629	52.6	0.169	470.	273.655	67.9	0.183
392.	227.657	38.1	0.164	473.	274.586	64.3	0.192
395.	228.667	34.2	0.171	476.	275.517	60.6	0.177
398.	229.667	37.6	0.179	479.	276.448	56.5	0.179
401.	230.667	24.1	0.140	482.	277.379	54.1	0.240
404.	231.667	24.2	0.146	485.	278.310	41.0	0.196
407.	232.667	25.7	0.180	488.	279.241	21.3	0.225
410.	233.667	45.1	0.221	506.	284.828	19.3	0.272
413.	234.667	63.0	0.190	509.	285.759	33.5	0.200
416.	235.667	58.3	0.217	512.	286.690	55.4	0.184
419.	236.667	71.2	0.177	515.	288.500	68.3	0.166
422.	237.667	80.1	0.141	518.	290.750	69.1	0.160
425.	239.810	75.7	0.132	521.	293.000	66.0	0.151
428.	242.524	73.5	0.119	524.	295.250	55.9	0.169
431.	245.238	70.9	0.125	527.	297.500	45.2	0.153
434.	247.952	61.4	0.123	530.	299.647	52.8	0.143

Productivity of the Ocean: Present and Past
eds. W.H. Berger, V.S. Smetacek and G. Wefer, pp. 341–354
John Wiley & Sons Limited

Tertiary Cooling Steps and Paleoproductivity as Reflected by Diatoms and Biosiliceous Sediments

J.A. Barron[1] and J.G. Baldauf[2]

[1] *U.S. Geological Survey*
Menlo Park, CA 94025, U.S.A.

[2] *Ocean Drilling Program*
College Station, TX 77840, U.S.A.

Abstract. Depositional patterns of diatomaceous sediments are good monitors of high surface water productivity. Progressive, stepwise polar cooling during the Tertiary (66–1.6 Ma) increased the thermal contrast between high and low latitudes which, in turn, affected global oceanic circulation, upwelling, and biogenic opal production.

The early Tertiary (66–53 Ma) was characterized by an unobstructed flow of equatorial surface waters through the ocean basins and a broad band of tropical biosiliceous sediments. Major cooling steps during the early-middle Eocene (52–48 Ma), earliest Oligocene (36–35 Ma), the middle Miocene (15–13 Ma), and the late Pliocene (2.5–2.4 Ma) caused the partitioning of both surface and deep water masses and resulted in regional differentiation of areas of strong upwelling and high rates of biogenic opal production.

INTRODUCTION

Ultimately, high surface productivity in the ocean is dependent on the rate of resupply of nutrients (nitrogen, phosphorus) to surface waters where they can be utilized by phytoplankton. Deeper waters (>500 m) are naturally enriched in such nutrients through bacterial degradation of organic matter settling through the water column; however, temperature and salinity contrasts between surface and deep waters, as well as basin-basin fractionation, influence nutrient supply and surface productivity.

Oceanic regions such as those along the western edges of continents and within 5° of the equator are characterized by surface water divergence and upwelling of nutrient-enriched intermediate waters (ca. 500–1500 m). Active resupply of nutrients to surface waters is also characteristic of the high latitudes, where little temperature difference exists between surface and deep waters. Consequently, in the modern ocean there is a strong correlation between areas of high concentrations of nutrients in surface waters and increased productivity.

As pointed out by Lisitzin (1971) and Calvert (1974), these areas of high surface water productivity are also characterized by rich accumulations of biosiliceous oozes on the seafloor; specifically, the skeletal remains of organisms which secrete opaline tests (diatoms, radiolarians, and silicoflagellates). It should be remembered, however, that generally only 1–5% of the skeletal remains of those organisms produced in surface waters actually are preserved in the sedimentary record (Calvert 1974). Dissolution of diatom frustules as they settle through the water column and in the upper few centimeters of sediment effectively removes most of the record of diatom surface water productivity from incorporation into the geologic record. Nevertheless, there is a strong relationship between surface productivity and frustule abundance with higher rates of opal accumulation corresponding closely to areas with higher rates of production of opal-secreting organisms (mainly diatoms) (Lisitzin 1971; Calvert 1974).

Just as surface and near-surface circulation patterns affect biosiliceous production, the characteristics of deep waters and their circulation patterns between ocean basins also affect biosiliceous sedimentation. Bacterial degradation of organic matter settling through the water column releases nutrients such as nitrogen and phosphorus, so that deep waters are enriched in such nutrients. Similarly, deep waters are enriched in silica due to dissolution of diatom tests (and radiolarian tests to a lesser degree), both as they settle through the water column and as they accumulate at the sediment/water interface on the seafloor.

Waters of the modern Pacific Ocean are enriched in silica by a factor of 4 compared to those of the North Atlantic; the factors for nitrate and phosphate are 2 to 3 in favor of the Pacific (Berger 1970). These differences are due to the fact that the North Atlantic receives both young, northern-sourced deep waters (North Atlantic Deep Water) and older, southern-sourced deep waters (Antarctic Bottom Water and Circumpolar Deep Water). The southern-sourced deep waters of the Pacific are older than those of the Atlantic, and they are enriched in nitrogen, phosphorus, and silica compared to those of the Atlantic (Berger 1970). Upward diffusion and upwelling of this nutrient-rich water causes a greater production of opal-secreting organisms in the North Pacific compared to the Atlantic, and the increased concentration of silica in Pacific deep and intermediate waters enhances preservation of biosiliceous oozes on the seafloor of the Pacific.

In the modern ocean, biogenic opal is concentrated as sediments on the seafloor in three main geographic provinces (Lisitzin 1971; Calvert 1974).

Approximately 80% of the total biogenous silica that is incorporated into the geologic record accumulates in a latitudinal belt between 45° and 65°S around Antarctica. A second area of high biogenous silica accumulation occurs in the northernmost Pacific including the Sea of Okhotsk, the Sea of Japan, and the Bering Sea. An equatorial belt, which is well defined in the Pacific and Indian oceans but less defined in the Atlantic, is the third main region of high biogenous silica accumulation in the ocean basins (Calvert 1974). Whereas the southern and northern provinces are dominated by diatom frustules, radiolarian skeletons constitute the dominant component (by weight) of equatorial siliceous oozes. In addition, much of the biogenous silica deposited in equatorial regions is masked by carbonate, and it is only where the seafloor lies below the calcite compensation depth (CCD) that relatively pure siliceous oozes accumulate.

Restricted belts of high biogenic silica surface water productivity also occur in areas of coastal upwelling, generally along the western continental margins which are influenced by equatorward-flowing, eastern boundary currents. The coasts of California, Peru, and southwest Africa are examples, but biogenic opal in the sediments of such areas may be masked by increased terrigenous components.

CENOZOIC CLIMATES

The response of the oceans to Cenozoic changes in basin configuration and communication with one another is reflected by the Cenozoic benthic foraminiferal oxygen isotope records of the Atlantic and the Pacific (Fig. 1). Inasmuch as deep waters are formed at high latitudes, isotopic records provide both an estimate of the paleotemperatures of high-latitude surface waters, as well as an estimate of the amount of water stored in polar ice caps. It is generally agreed that the paleotemperatures of tropical surface waters underwent little change during the Cenozoic (Miller et al. 1987).

At the beginning of the Cenozoic, warm, equitable conditions prevailed in the Paleocene (66–58 Ma), with bottom water (and high-latitude surface waters) temperatures of about 10°C (Fig. 1). Following a brief period in the earliest Eocene (ca. 56–52 Ma), marked by the warmest bottom water temperatures of the Cenozoic (about 12–14°C), a cooling trend began, which has characterized the last 50 m.y. of Earth history. This trend has been marked by a series of rapid cooling steps, which are separated by intervals of more stable conditions and occasional brief reversals (warmings) in the trend (Berger 1982).

Miller et al. (1987) identify nine Tertiary cooling events between the early Eocene and late Pliocene (Table 1). These cooling events vary in their duration and intensity, with four events standing out as major cooling steps: the early-middle Eocene event (52–48 Ma), the earliest Oligocene event

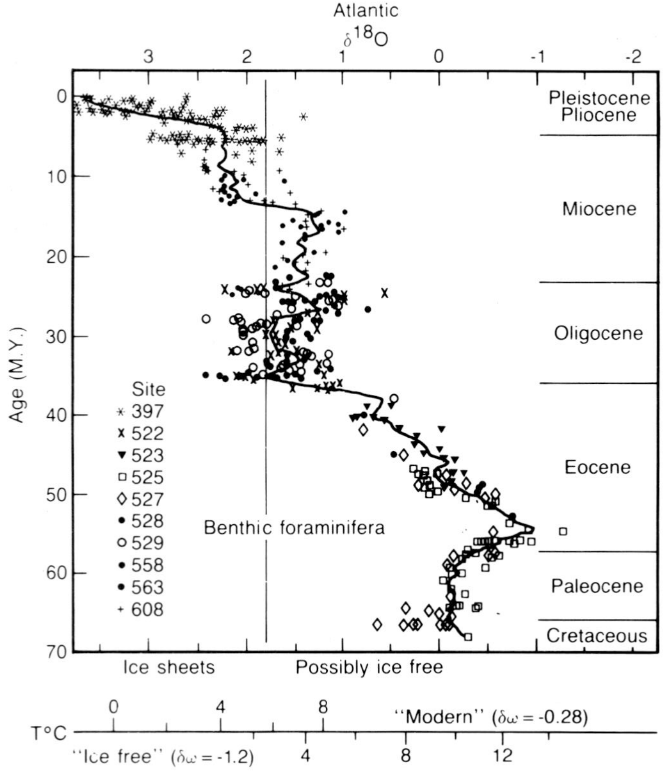

Atlantic			Indian		Pacific		Age
High N	EQ	High S	EQ	High S	EQ	High N	
−	+	+	+	+	+	+	Pleistocene Pliocene
−	−	+	+	+	+	+	Miocene
+	−	+	+	+	+	+	
+	+	+		+	+	−	
+	+	+		+	−	+	Oligocene
−	−	+		+	+	−	

Fig. 1—Cenozoic paleotemperature record of Atlantic deep waters (approximates polar surface water temperature record) generalized from benthic foraminiferal oxygen isotopic data (after Miller et al. 1987: Copyright, AGU), compared with the distribution of widespread (+) biosiliceous sediments in the oceans for the Oligocene through Quaternary. (−) = dissolution or reduced presence.

(36–35 Ma), the middle Miocene event (15–13 Ma), and the late Pliocene event (2.5–2.4 Ma) (Table 1; Fig. 1).

The response of the polar regions to increased cooling has included the initiation of glaciation in East Antarctica by earliest Oligocene (36 Ma), followed by the development of a stable ice cap there by middle Miocene (14 Ma; Miller et al. 1987; Barker et al. 1987). Further polar cooling resulted in a stable West Antarctic ice cap by earliest Pliocene (4.8 Ma; Barker et al. 1987) and the initiation of Northern Hemisphere continental glaciation during the late Pliocene (2.4 Ma; Miller et al. 1987).

One might expect that the steepened latitudinal thermal gradients that would result from high-latitude cooling would cause intensified gyral circulation of surface currents, leading to increased coastal and trade winds which would promote coastal and equatorial upwelling. Brewster (1980) argues that periods of higher opal production in the Southern Ocean correspond to periods of climatic cooling. In addition, Pedersen (1983) demonstrates that surface water productivity was enhanced in the eastern equatorial Pacific (an area of high diatom productivity) during the last glacial interval of the Pleistocene.

The later part of the Cenozoic was a time when significant ice was present in the Antarctic ice cap, and global sea level falls during glacial intervals exposed large areas of the continental shelves. Sediments deposited on the shelves during sea level highstands were subject to erosion, and nutrients stored in those sediments could be returned to the deep sea (through riverine and downslope transport) (Broecker 1982). With increased supplies of available nutrients and increased mixing rates (upwelling), one would expect increased fertility and higher production of diatoms in upwelling regions during times of polar cooling and reduced sea levels. Such may not be the case in polar regions during the late Cenozoic, where polar cooling led to more sea ice which would have reduced the season for diatom growth in surface waters.

CENOZOIC BIOSILICEOUS SEDIMENTATION PATTERNS

The modern patterns of surface water productivity and biosiliceous sedimentation have evolved during the Cenozoic (the last 66 m.y.) in response to increasing temperature contrasts between relatively stable, warm tropical regions and successively cooler polar regions. Early Cenozoic oceans were characterized by an uninterrupted flow of warm equatorial surface currents, an isolated or semi-isolated Arctic Ocean, and the presence of barriers to surface and deep water circulation around Antarctica due to both a more southerly location of Australia and India and to the absence of the Drake Passage between South America and the Antarctic Peninsula. Consequently, the equatorial ocean exerted a major influence on global climate patterns during the early Cenozoic. As the Cenozoic progressed, however, movement of the Earth's crustal plates imposed barriers to that free equatorial circulation and caused changes in the configuration of ocean basins and gateways, thermal gradients, and atmospheric circulation, thereby increasing the importance of high-latitude oceans on global oceanic circulation patterns and climate.

Circulation between the Arctic and the North Atlantic was enhanced through the opening of the Norwegian Sea in the early Eocene (55 Ma), followed by the opening of a deep water passage to the Arctic between Greenland and Spitzbergen in the earliest Oligocene (35 Ma) (Talwani and Eldholm 1977). Similarly, northward movement of Australia and the Indian plate allowed the initiation of a circum-Antarctic current by the earliest Oligocene. The opening of the Drake Passage in the latest Oligocene (ca. 24 Ma) (Barker and Burrell 1977) further promoted the development of that current (Kennett 1978). Similarly, elevation of a land barrier between Africa and Asia sealed off the eastern end of the Mediterranean Sea in the latest middle Miocene, and collision of Australia with Indonesia caused destruction of the Indonesian seaway between the equatorial Indian and Pacific oceans near the end of the middle Miocene. Surface water communication between the eastern equatorial Pacific Ocean and the Caribbean Sea was finally discontinued by the emergence of the Isthmus of Panama in the late Pliocene, but deep water communications across that barrier must have been disrupted earlier in the Miocene.

Quantitative studies of opal accumulation rates through the Cenozoic have been completed in the equatorial Pacific (Leinen 1979), the Southern Ocean south of Australia (Brewster 1980), the North Atlantic (Ehrmann and Thiede 1985), and the eastern Indian Ocean (Schmitz 1987). General regional trends in opal accumulation rates from these studies are compared in Table 1 with Tertiary cooling steps, paleoceanographic events, and observations by the authors of biosiliceous sediment distribution patterns.

These studies indicate that the early Paleogene was a time of reduced latitudinal and vertical thermal gradients, and patterns of upwelling and biosiliceous sedimentation were not regionally complex. Uniformly high equatorial opal accumulation rates during the middle Eocene signal a broad tropical and subtropical band of upwelling. As pointed out by Schmitz (1987), however, Paleogene radiolarians and diatoms possessed considerably thicker-walled siliceous frustules and tests than their Neogene counterparts, so that differences in Paleogene and Neogene opal accumulation rates may not completely reflect differences in biologic productivity of opal-secreting organisms, but rather preservation patterns. Whereas only 1–5% of diatom frustules produced in surface waters of the modern ocean is incorporated into the sediment record on the seafloor (Calvert 1974), this ratio may have been higher in the Paleogene.

Partitioning of biosiliceous sedimentation between different areas of the oceans was enhanced during the late middle Eocene, when polar cooling caused a decoupling of near-surface and bottom temperatures and strengthened latitudinal thermal gradients (Boersma et al. 1987). Provincialism increased in diatom assemblages (Fenner 1985), and diatomaceous sediments appeared in coastal upwelling settings in California, Kamchatka, Peru, and New Zealand. As temperature contrasts between surface and deep waters increased, the depth of the thermocline marking the transition between surface and deep waters in any given area became crucial in determining whether significant biosiliceous productivity could occur in surface waters—a deep thermocline inhibits upwelling of nutrient-rich (and silica-rich) intermediate waters. As polar cooling progressed in the later Cenozoic, further decoupling of water masses occurred. Decoupling of surface water masses enhanced latitudinal thermal gradients and resulted in increased provincialism in diatom assemblages. Decoupling of vertical water masses caused increased regional differences in the profile and depth of the thermocline.

Coinciding with the earliest Oligocene cooling step (36–35 Ma), there was a decreased abundance of diatoms in sediments from the equatorial Atlantic and the Bay of Biscay (Fenner 1985; Boersma et al. 1987). At the same time, diatoms increased in abundance in sediments of the equatorial Pacific (Fenner 1985; Barron 1985b), and the biosiliceous component increased in Southern Ocean sediments (Barker et al. 1987). One should note, however, that diatomaceous sediments are recorded from lower Oligocene sediments from the Labrador Sea (Baldauf and Monjanel 1988), suggesting that the decreased abundance of diatoms in lower Oligocene sediments from the North Atlantic did not include the higher latitudes of the northwest North Atlantic.

The change in depositional patterns of biosiliceous ooze from the Atlantic to the Pacific and Southern Ocean (i.e., "silica switch") may signal enhanced basin-basin fractionation. Miller and Fairbanks (1985) use benthic foraminiferal carbon isotope data to argue for increased formation of North Atlantic Deep Water in the early Oligocene, causing increased contrasts in the ages of North Atlantic and Pacific deep waters (with the North Atlantic being younger). Introduction of "young" deep water into the North Atlantic during the early Oligocene would inhibit the upward diffusion of silica from silica-rich, southern-sourced deep water in the North Atlantic and lead to reduced silica concentrations in North Atlantic intermediate waters. This could lead to reduced surface water concentrations and reduced diatom production. At the same time, the presence of "young" deep water in the North Atlantic would increase dissolution on the seafloor at the sediment/seawater interface. The accumulation of diatomaceous sediments in the lower Oligocene of the Labrador Sea (Baldauf and Monjanel 1988) may reflect more intense upwelling there or may be evidence that "young" deep

TABLE 1 Comparison of Cenozoic cooling events with paleoceanographic events, patterns of biogenic opal distribution (+ = increasing; − = decreasing), provincialism in diatom assemblages, latitudinal thermal gradients, and evolutionary turnover in diatom assemblages.

Age (Ma)	Cool Event	Paleoceanographic Event	Opal Distribution	Provincialism/Lat. Therm. Grad.	Turnover
2.5–2.4	Major	Onset N. Hemisphere Glaciation	+ mid-lat N.E. Pacific + mid-lat N. Atlantic	Very High/Sharp	Moderate – Extinctions
5.5–5.0	Brief	Messinian desiccation Stable W. Antarctic Ice cap	++ Southern Ocean + N.W. Africa − Eq. Pacific − mid-lat N.E. Pacific	Increasing	Moderate – Endemism
10–8	Medium	Close Indonesian Passage? Pacific – Equatorial Undercurrent est. Extensive ice W. Antarctic	+ California + S.W. Africa + Eq. Pacific − mid-lat Atlantic	Increasing/ Decoupling eq. and temperate water masses	Moderate – Extinctions
15–13	Major	Permanent E. Antarctic Ice cap est. W. Mediterranean close	+ N. Pacific − N.W. Africa − Eq. Atlantic (17 Ma)	Low – expansion temp. assem/ Moderate	Major
25–24	Brief	Drake Passage open	+ Southern Ocean − N. Pacific?	Low	Low

31–28	Brief	Glaciation E. Antarctica	+ Bering Sea + N.W. N. Atlantic + N.W. Africa – Eq. Pacific*	High Evol. finely silicified upwelling assem.	Major
36–35	Major	Glaciation E. Antarctica Arctic/N. Atlantic deep water connection	+ Eq. Pacific* + Southern Ocean + Labrador Sea – Eq. Atlantic – N. Pacific?	Moderate	Moderate
45–40	Trend	Isolation Antarctic from warm currents Decoupling surface-bottom paleotemp. records	+ California + Eq. Pacific – N.W. Africa	Increasing	Low to Moderate
52–48	Major	Slightly after Norwegian Sea opening	+ N. Atlantic + S. Atlantic + Southern Ocean	Low/Low	Major

Source: Leinen (1979)
Brewster (1980)
Ehrmann & Thiede (1985)
* = conflict with Leinen

waters were mainly formed in the eastern rather than the western basin of the North Atlantic. Reduced deposition of silica and/or increased dissolution of silica in seafloor sediments in the North Atlantic, combined with increased outward flow of deep waters from the Atlantic to the Indian and finally to the Pacific basins, would lead to increased silica concentrations in North Pacific deep and intermediate waters and favor biosiliceous sedimentation there.

Such an Atlantic to Pacific "silica switch," however, apparently was relatively short-lived, because improved diatom preservation rates increased off northwest Africa by the late Oligocene, at about the same time that equatorial Pacific sediments became diatom-poor (Ehrmann and Thiede 1985; Barron 1985b). It should be pointed out, however, that diatomaceous sediments are present in the upper Oligocene of high-latitude areas of the North Pacific such as the Bering Sea and in coastal upwelling settings such as southern Baja California, Mexico. Again, it is interesting to note that Miller and Fairbanks (1985) report reduced age contrasts in the ages of North Atlantic and Pacific deep waters during the late Oligocene, so one might expect reduced basin-basin fractionation.

A substantial evolutionary turnover occurred in diatoms during the early Oligocene, with the demise of many thickly silicified Paleogene forms and the evolution of more thinly silicified taxas which have characterized Neogene open ocean assemblages (Fenner 1985; Baldauf and Monjanel 1988). Presumably, this evolutionary trend, which is also present in radiolarians, reflects increasing competition for silica during the late Cenozoic, although it might also reflect more rapid recycling of silica by organisms in the well-mixed oceans, which have characterized the later Cenozoic.

The opening of the Drake Passage in the latest Oligocene resulted in an enhanced circum-Antarctic current and increased biosiliceous sedimentation in the Southern Ocean (Barker et al. 1987). During the early Miocene, diatomaceous sediments were rare in the North Pacific and restricted in the equatorial Pacific but widespread in the Caribbean Sea and low-latitude North Atlantic (Keller and Barron 1983; Ehrmann and Thiede 1985). Consequently, one would expect weakened basin-basin fractionation and reduced age contrasts between North Atlantic and Pacific deep waters. This conflicts with the interpretation of Miller and Fairbanks (1985) but is consistent with the study of Savin and Woodruff (1987), which shows reduced age contrasts between the North Atlantic and Pacific during the early Miocene. It should be noted that Savin and Woodruff's (1987) results incorporate more isotopic measurements and more deep-sea sections than does Miller and Fairbanks' (1985) study.

The end of the early Miocene brought about the start of another "silica switch" from the North Atlantic to the Pacific, which began at about 17–15 Ma with the disappearance of diatomaceous sediments from the Caribbean

Sea and low-latitude North Atlantic and accelerated at the end of the middle Miocene (about 10 Ma) with the decline in biosiliceous sedimentation in the middle-latitude North Atlantic (Keller and Barron 1983; Ehrmann and Thiede 1985; Baldauf 1986). Coinciding with these North Atlantic declines in biosiliceous sediments, the North Pacific underwent successive enrichments in the abundance and distribution of diatomaceous sediment (Keller and Barron 1983; Barron 1986).

The equatorial Pacific was characterized by very high rates of opal accumulation during the late Miocene, especially in the eastern equatorial Pacific (Leinen 1979). A strong east-west gradient in opal productivity (higher production in the east) developed in the equatorial Pacific at about 10 Ma, presumably in response to the establishment of the Equatorial Undercurrent (Leinen 1979). The Equatorial Undercurrent is a subsurface current, which flows eastward beneath the equator and is 300 km wide and 200 m thick. Near the Galapagos Islands (90°W), this undercurrent surfaces and enhances upwelling, promoting increased biologic productivity. Collision of Australia with Indonesia in the late middle Miocene placed a barrier in the path of westward-flowing equatorial Pacific surface currents; the return, subsurface flow of waters, which were piling up in the western equatorial Pacific, may have initiated the Equatorial Undercurrent (Leinen 1979). Interestingly, provincialism was enhanced by 8.5 Ma between equatorial and temperate Pacific diatom assemblages, presumably marking the steepening of latitudinal thermal gradients (Barron 1985a).

A major increase in opal accumulation rates in the Southern Oceans at about 5 Ma (near the Miocene/Pliocene boundary) coincided with a substantial decline in opal accumulation rates in the equatorial Pacific (Leinen 1979; Brewster 1980) and a near disappearance of diatoms in sediments of the middle-latitude northeastern Pacific (Barron 1986). This "silica switch" from the Pacific to the Southern Ocean marked the beginning of the modern status of the Antarctic as a major silica sink. The West Antarctic ice cap became a stable feature at this time and there was a northward expansion of the Antarctic Convergence (Kennett 1978; Barker et al. 1987).

Biogenic opal surface productivity and sediment distribution patterns became similar to those of the present day after the establishment of the present oceanographic circulation patterns in the late Pliocene. Since that time, diatomaceous accumulation patterns in part reflect the oceanic response to the onset of Northern Hemisphere glaciation at 2.4–2.5 Ma and subsequent climatic fluctuations. In the North Atlantic, changes in abundance of diatom frustules in the sediments suggest fluctuating surface productivity in response to the latitudinal migration of the polar front (Baldauf 1986). North of about 40°N, diatoms are generally present in sediment deposited during interglacial conditions and absent in sediment deposited during glacial conditions. South of about 40°N, diatom frustules are present in sediments deposited during both glacial and interglacial conditions. Baldauf (1986), however, also provides evidence to indicate that in the Rockall Plateau region of the North Atlantic, the deposition of biogenic silica is strongly influenced by silica dissolution between 1.8 to 0.44 Ma.

Following the onset of Northern Hemisphere glaciation, the Southern Ocean continued to be the silica sink for the world's oceans, as did regions such as the Sea of Okhotsk, the Bering Sea, and the Norwegian-Greenland Sea. Rainer Gersonde (personal communication, 1988) reports an interval of silica dissolution and/or decreased diatom production in the Weddell Sea between about 2.0 and 0.6 Ma, indicating that changes did occur in Southern Ocean biosiliceous sedimentation during the last 5 m.y. With the exception of the eastern equatorial Pacific, and other coastal areas of high upwelling enhanced surface productivity, diatom surface productivity is presently concentrated in the high latitudes of the modern ocean.

SUMMARY

Global distribution patterns of biosiliceous sediments are determined by oceanographic processes such as upwelling, which actively resupply nutrients to surface waters where they can be utilized by phytoplankton and by circulation patterns of deep waters, which influence the preservation of biogenic opal on the seafloor. The Tertiary has been characterized by a series of polar cooling steps, which have resulted in progressively greater thermal contrasts between high- and low-latitude regions of the oceans. These cooling steps, in part, reflect movement of the Earth's crustal plates during the Tertiary, which have both placed obstructions in the path of equatorial surface currents and opened paths for the free flow of surface and deep waters around Antarctica. Circulation patterns of deep waters within the major ocean basins have also been affected by tectonic movements, resulting in changes in the age of deep waters within the various basins and their corrosiveness to biogenic opal settling through the water column.

These Tertiary paleoceanographic and paleoclimatic changes have resulted in increased regional differentiation of areas of high upwelling and associated high productivity of biogenic opal, reflecting both regional variability in the depth and profile of the thermocline and a more regional differentiation of wind-induced coastal upwelling. Overprinting this Tertiary trend are basin-to-basin changes in the accumulation of rich biosiliceous oozes on the seafloor, shifting back and forth between the North Atlantic and Pacific and Southern Oceans and reflecting changes in basin-basin fractionation. The relatively recent (last 5 m.y.) development of the Southern Ocean as a major silica sink has been the latest outcome of these Tertiary changes in deep and surface water circulation patterns.

Acknowledgements. We thank Rainer Gersonde, Alan Mix, and Hans-Joachim Schrader for their review of this paper. The manuscript also benefitted from the comments of Wolfgang Berger. Special thanks are extended to Amy Russell and Judy Duke for typing, and to Doris Cooley for typing and coordinating the manuscript while the authors were at sea.

REFERENCES

Baldauf, J.G. 1986. Diatom biostratigraphic and paleoceanographic interpretations for the middle to high latitude North Atlantic Ocean. In: North Atlantic Paleoceanography, eds. C.P. Summerhayes and N.J. Shackleton. *Geol. Soc. Spec. Publ.* **21**: 243–252. Oxford: Blackwell Scientific Publications.

Baldauf, J.G., and Monjanel, A.L. 1988. An Oligocene diatom biostratigraphy for the Labrador Sea; DSDP Site 112 and ODP Site 647A. In: Init. Repts. ODP, eds. S. Srivastiva, M. Arthur, B. Clement, et al., 105(B). Washington, D.C.: US Govt. Printing Office, in press.

Barker, P.F., and Burrell, J. 1977. The opening of the Drake Passage. *Marine Geol.* **25**: 15–34.

Barker, P.F.; Kennett, J.P., and the ODP Leg 113 shipboard scientific party. 1987. Glacial history of Antarctica. *Nature* **328**. 115–116.

Barron, J.A. 1985a. Diatom paleoceanography and paleoclimatology of the central and eastern equatorial Pacific between 18 and 6.2 Ma. In: Initial Reports of the Deep Sea Drilling Project, eds. L. Mayer and F. Theyer, vol. 85, pp. 935–945. Washington, D.C.: US Govt. Printing Office.

Barron, J.A. 1985b. Late Eocene to Holocene diatom biostratigraphy of the equatorial Pacific Ocean, DSDP Leg 85. In: Initial Reports of the Deep Sea Drilling Project, eds. L. Mayer and F. Theyer, vol. . 85, pp. 413–456. Washington, D.C.: US Govt. Printing Office.

Barron, J.A. 1986. Paleoceanographic and tectonic controls on deposition of the Monterey Formation and related siliceous rocks in California. *Palaeogeog. Pal. Pal.* **53**: 27–45.

Berger, W.H. 1970. Biogenous deep-sea sediments: fractionation by deep-sea circulation. *Geol. Soc. Amer. Bull.* **81**: 1385–1402.

Berger, W.H. 1982. Deep-sea stratigraphy: Cenozoic climate steps and the search for chemo-climatic feedback. In: Cyclic and Event Stratification, eds. G. Einsele and A. Seilacher, pp. 212–257. Berlin: Springer.

Boersma, A.; Premoli-Silva, I.; and Shackleton, N.J. 1987. Atlantic Eocene planktonic foraminiferal paleohydrographic indicators and stable isotope paleogeography. *Paleocean.* **2(3)**: 287–331.

Brewster, N.A. 1980. Cenozoic biogenic silica sedimentation in the Antarctic Ocean. *Geol. Soc. Amer. Bull.* **91**: 337–347.

Broecker, W.S. 1982. Glacial to interglacial changes in ocean chemistry. *Prog. Ocean.* **11**: 151–197.

Calvert, S.E. 1974. Deposition and diagenesis of silica in marine sediments. *Spec. Publ. Int. Assn. Sediment.* **1**: 273–299.

Ehrmann, W.U., and Thiede, J. 1985. History of Mesozoic and Cenozoic sediment fluxes to the North Atlantic Ocean. In: Contributions to Sedimentology, eds. H. Füchtbauer, A.P. Lisitzin, J.D. Milliman, and E. Seibold, vol. 15, pp. 1–109. Stuttgart: E. Schweizerbart'sche Verlagsbuchhandlung.

Fenner, J. 1985. Late Cretaceous to Oligocene planktonic diatoms. In: Plankton

Stratigraphy, eds. H.M. Bolli, J.B. Saunders, and K. Perch-Nielsen, pp. 713–762. Cambridge: Cambridge Univ. Press.

Keller, G., and Barron, J.A. 1983. Paleoceanographic implications of Miocene deep-sea hiatuses. *Geol. Soc. Amer. Bull.* **94**: 590–613.

Kennett, J.P. 1978. The development of planktonic biogeography in the Southern Ocean during the Cenozoic. *Mar. Micropal.* **3**: 301–345.

Leinen, M. 1979. Biogenic silica accumulation in the central equatorial Pacific and its implications for Cenozoic paleoceanography. *Geol. Soc. Amer. Bull. Part II* **90**: 1310–1376.

Lisitzin, A.P. 1971. Distribution of siliceous microfossils in suspension and in bottom sediments. In: The Micropaleontology of the Oceans, eds. B.M. Funnell and W.R. Reidel, pp. 173–195. Cambridge: Cambridge Univ. Press.

Miller, K., and Fairbanks, R.G. 1985. Oligocene to Miocene global isotope cycles and abyssal circulation changes. In: The Carbon Cycle and Atmospheric CO_2: Natural Variations Archean to Present, eds. E.T. Sundquist and W.S. Broecker. *Geophys. Monog.* **32**: 469–486. Washington, D.C.: Amer. Geophys. Union.

Miller, K.G.; Fairbanks, R.G.; and Mountain, G.S. 1987. Tertiary isotope synthesis, sea level history, and continental margin erosion. *Paleocean.* **2(1)**: 1–19.

Pedersen, T.F. 1983. Increased productivity in the eastern equatorial Pacific during the last glacial maximum (19,000 to 14,000 yr. BP). *Geology* **11**: 16–19.

Savin, S.M., and Woodruff, F. 1987. Miocene thermohaline circulation and speculation on the origin of the Antarctic ice cap. Abstracts with Programs, Geol. Soc. Amer. 1987 Annual Meeting, **19(7)**: 830.

Schmitz, B. 1987. Barium, equatorial high productivity, and the northward wandering of the Indian continent. *Paleocean.* **1(2)**: 63–77.

Talwani, N., and Eldholm, O. 1977. Evolution of the Norwegian Greenland Sea. *Geol. Soc. Amer. Bull.* **88**: 969–999.

Productivity of the Ocean: Present and Past
eds. W.H. Berger, V.S. Smetacek and G. Wefer, pp. 355–375
John Wiley & Sons Limited

Inventory of Paleoproductivity Records: The Mid-Cretaceous Enigma

H.R. Thierstein

Geological Institute, ETH-Zentrum
8092 Zürich, Switzerland

Abstract. Mid-Cretaceous pelagic deposits are characterized globally by enrichments in organic carbon and depletion in biogenous carbonate when compared to sediments of any other late Phanerozoic period. This points to a significant anomaly in the planetary cycling of carbon and thus must have been linked to changes in the biosphere. Overprinted onto this longer-term anomaly are frequent short-term environmental oscillations that resulted in bedded sequences suggesting periodic, possibly orbital forcing. The mid-Cretaceous time interval therefore represents an alternative planetary state to that of the late Neogene and as such is a major challenge to our understanding of global environmental modeling.

INTRODUCTION

The mid-Cretaceous interval, about 120–90 Ma (million years) ago, is considered by many to represent a time period when global environments were the most different from those of the current state of polar refrigeration, which has prevailed through the late Neogene and for which a reasonably complete sedimentological record is still preserved on land and in the oceans. A number of findings and features can be cited in support of this view.

The distribution of land and sea was characterized by a closer assembly of the continents on one hemisphere and a considerably larger Pacific Ocean which reached closer to both poles than at any time since (Fig. 1). The widespread occurrence of marine sediments on continental platforms, which covered an area close to 31×10^6 km^2 (Brass et al. 1982) 100 Ma ago, would have increased the proportion of low albedo ocean area to 77% of the global surface, compared to 71% today. A global sea level rise of 130 to 350 m has also been inferred from changes in ocean volume caused by changing spreading rates and mid-ocean ridge lengths (Kominz 1984).

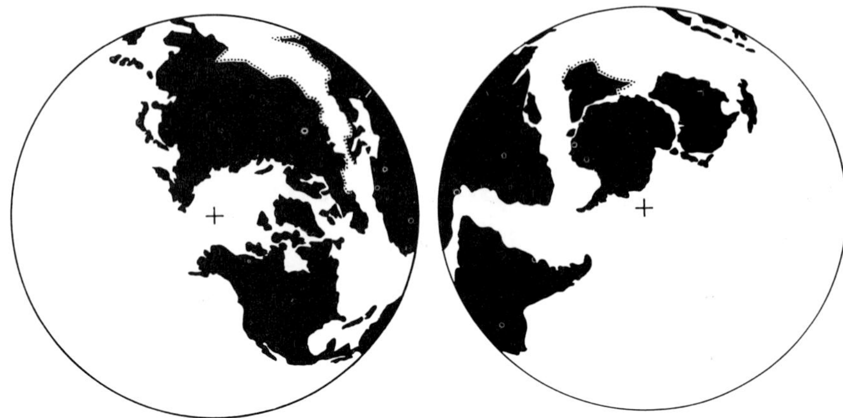

Fig. 1—Mid-Cretaceous paleogeography, 100 Ma (after Firstbrook et al. 1979).

There is sedimentological, micropaleontological, paleobotanical, and stable isotopic evidence to suggest that the latitudinal temperature gradients were considerably lower, mostly caused by only moderately cool temperatures at the poles, and that oceanic deep water temperatures may have been considerably higher than today (Barron 1983; Hallam 1985).

The mid-Cretaceous was a time of widespread organic carbon deposition in marine environments (Schlanger and Jenkyns 1976; Ryan and Cita 1977; Thierstein, in Talwani et al. 1979; Jenkyns 1980), during which the source rocks of over 60% of the currently known large hydrocarbon reservoirs were deposited (Irwing et al. 1974). As potential causes for this global geochemical anomaly a number of explanations have been offered, based on studies in various areas and stratigraphic intervals. The proposed models include changes in the supply of terrestrial organic matter to the oceans (e.g., Stein et al. 1986), exceptionally high oceanic fertility (e.g., Weissert et al., in Sundquist and Broecker 1985), and slow deep water renewal rates with concurrent low global oceanic surface productivity (e.g., Bralower and Thierstein 1984).

Most marine mid-Cretaceous sequences exhibit unusually consistent and lithologically obvious bedding features, which have been linked to periodic changes in the orbital characteristics of the Earth (Fischer 1980). Studies of the micropaleontological, sedimentological, and geochemical variabilities are few, and chronological constraints on suspected rhythmicities are difficult to come by because of the absence of magnetic field reversals in the Cretaceous magnetic quiet zone between 119–84 Ma ago (Kent and Gradstein 1985). It remains unclear, however, how orbitally induced changes in insolation distribution could be climatically and ecologically amplified to be recorded in deep-sca deposits in the absence of significant continental ice caps.

EVALUATION OF POTENTIAL GEOLOGICAL INVENTORIES OF CRETACEOUS PALEOPRODUCTIVITY

A number of proxy data have been used to evaluate patterns and changes of marine and terrestrial productivity in the geological past. Some of these have been applied on a rather ad hoc basis, often with insufficient evidence or data to evaluate their usefulness in recent environments. If the goal is to understand and model the sensitivity of various environmental states of the Earth's system, knowledge of the significance and reliability of many of the generally used geological proxy data in recent environments will have to be maximized. The more commonly used of these methods are therefore briefly reviewed.

Lithology and Sedimentology

The widespread occurrence of "black shales" in the mid-Cretaceous has become widely recognized after the initiation of sediment sampling in the deep ocean basins (Weissert, in Warme et al. 1981). The sedimentology of commonly occurring bedding alternations of shales, marls, and limestones has been investigated in only a few sections (e.g., De Boer, in Einsele and Seilacher 1982; Arthur and Premoli Silva, in Schlanger and Cita 1982). Many of these studies have focused on the development of particular techniques. Among these mostly monodisciplinary approaches are the following: counts of bed thicknesses suggesting possible orbital frequencies (Schwarzacher and Fischer, in Einsele and Seilacher 1982); measurements of grain size distributions in the silt fraction which may be caused by either turbidite deposition or variable supply from the photic zone or changing preservation of nannofossils, pyrite and calcite casts of radiolarians, and foraminifera (McCave, in Tucholke et al. 1979); trace fossil tiering documenting changing paleo-oxygenation of bottom waters (Savrda and Bottjer 1986); stable isotopic analyses of bulk sediments with multiple interpretation possibilities (Scholle and Arthur 1980; Weissert et al., in Sundquist and Broecker 1985); and identification of laminated or nonbioturbated sediment intervals, suggesting deep-water anoxia with concomitant high organic carbon preservation conditions (Bralower and Thierstein 1984). Even in multidisciplinary geochemical studies the meaning of the obtained measurements has often remained obscure, sometimes because direct light or electron microscopic examination and documentation of the kinds of organic and inorganic particles analyzed have been lacking, or because the sediments at hand were not analyzed by more than one of the available techniques and alternative explanations for the phenomena remained highly likely.

Carbonate Contents and Accumulation Rates

Deep-sea sediments of the mid-Cretaceous time interval are characterized by their relatively low carbonate contents ("rise of the CCD" in Tucholke, Vogt et al. 1979), as well as by their elevated organic carbon contents ("oceanic anoxic events" in Schlanger and Jenkyns 1976). Plate-stratigraphic concepts first developed for reconstructions of the Cenozoic history of pelagic sedimentation soon were expanded to the Mesozoic, and it became apparent that dramatic changes in either the supply or the preservation of biogenous carbonate particles, or both, had to be invoked in order to explain the observed facies patterns (Thierstein, in Talwani et al. 1979). However, the hypothesized processes responsible for these changes remain controversial, largely because the patterns in the mid-Cretaceous are unlike any patterns observed in the latest Cretaceous or Cenozoic, where biogenous oozes of more consistently high (i.e., >80%) carbonate contents, consisting of well-preserved nannofossil and planktonic foraminifera shells, are encountered at intermediate paleodepths (Thierstein, in Talwani et al. 1979; Pisias and Prell, in Sundquist and Broecker 1985). That this is not so in the mid-Cretaceous is documented by the globally known occurrence of black shales at all paleodepths studied, as well as by the fact that mid-Cretaceous deep-sea sediments with high carbonate contents (i.e., >50–60%) are dominated by diagenetic micarb or micrite and do not consist of dominantly biogenous particles, as they do from the late Cretaceous onward (Thierstein 1983; Thierstein and Roth, in preparation). Samples with carbonate contents of less than about 40% constantly show strong signs of etching on the ultrastructure of the preserved nannofossils. Often only imprints of calcareous nannofossils or pyrite or calcite casts of radiolarians are left (Noël and Melguen 1978). A great need for careful studies of the influence of diagenetic alteration versus original water column processes seems apparent. In the absence of any known mid-Cretaceous high-carbonate pelagic sediment primarily composed of identifiable biogenous plankton particles, the possibility must be seriously considered that many mid-Cretaceous oceanic chalks and limestones are of diagenetic origin (Eder, in Einsele and Seilacher 1982; Ricken 1986). A major difficulty arises because of the uncertainty of possible sources, lengths of migration paths, and multiple recrystallization episodes of the diagenetic micarb (e.g., Perry et al. 1976; Baker et al. 1982; Talbot and Kelts 1986).

As a consequence, the question arises whether during that time interval the oceanic shell carbonate production was lower or the noncarbonate supply was higher. In order to differentiate between these two alternatives the bathymetric patterns of accumulation rate changes have to be known and this, in turn, requires a reasonably accurate chronology. Comparisons of recently proposed Mesozoic chronologies attest to persistent uncertainties due to the lack of reliable radiometric age determinations on high temperature minerals (Kent and Gradstein 1985). Despite the considerable uncertainties in mid-Cretaceous biochronology, a number of average carbonate accumulation rate estimates for black shale intervals have been made. For nonturbiditic and biostratigraphically datable intervals these average around 0.5–1.4 g $CaCO_3$ cm^{-2} ka^{-1} (Arthur et al., in Sundquist and Broecker 1985; Bralower and Thierstein, in Brooks and Fleet 1987), which is similar to an estimate of 0.5–0.8 g $CaCO_3$ cm^{-2} ka^{-1}, based on an inferred orbital chronology in a core from central Italy (Herbert et al. 1986). For bathyal and abyssal depositional areas, these values seem to be within the range of late Quaternary calcite accumulation rates in tropical and subtropical oceanic environments (e.g., Curry and Lohmann, in Sundquist and Broecker 1985; and Fig. 2).

Organic Carbon Contents and Accumulation Rates

The observed increased organic carbon contents in many Cretaceous marine sediments show distinct stratigraphic, paleobathymetric, and paleogeographic distribution patterns. After a prolonged period of intermittent, minor organic

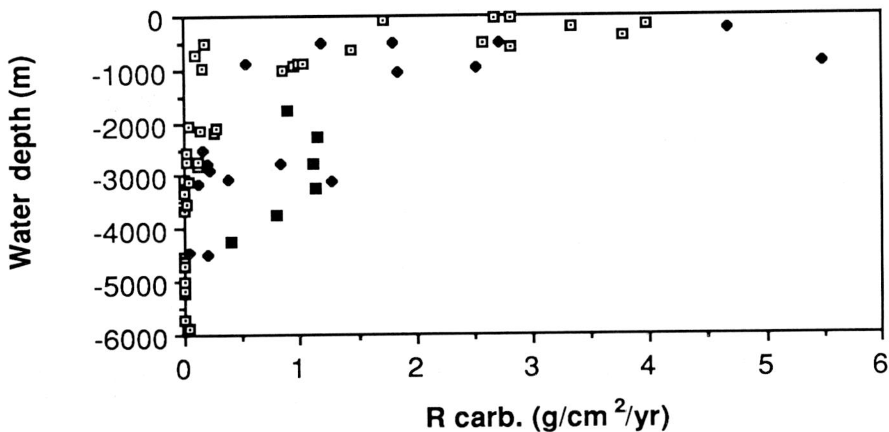

Fig. 2—Bathymetric distribution of calcium carbonate accumulation rates in mid-Cretaceous and Holocene marine sequences. Mid-Cretaceous (diamonds) and global Holocene data (open squares) are from Bralower and Thierstein in Brooks and Fleet (1987); ERDC Holocene data (solid squares), from the equatorial Pacific, are averages for 500 m depth intervals (Thierstein et al., unpublished).

carbon enrichments, starting in the lowermost Cretaceous of the western North Atlantic, a prolonged interval of black shale deposition is observed in the late Aptian to early Albian in the deep North and South Atlantic, the deep Indian Ocean, and the shallow Pacific (Schlanger and Jenkyns 1976; Thierstein, in Talwani et al. 1979). The worldwide occurrence of a distinct, usually black, anoxic sediment layer of a few decimeters to a few meters thickness at the Cenomanian-Turonian boundary has led to the proposition of so-called oceanic anoxic events (Jenkyns 1980).

Accumulation rates of organic carbon are more instructive than concentration measurements for several reasons. First, their geographic and bathymetric distributions provide insights into the effects of supply versus those of preservation; second, they allow comparisons with carbon fluxes during other geologic time periods, particularly the present; finally, they form the basis for any global carbon cycle model. Determination of accumulation rates, however, is only possible if complete stratigraphic sections are available, if the chronology is reasonably accurate, and if the physical properties of the sediments are known. It is particularly the first and second of these requirements which have been the most limiting in the computation and interpretation of organic carbon accumulation rates and their incorporation into models of the global carbon cycle. The published estimates of organic carbon accumulation rates in various mid-Cretaceous intervals are averages over a few million years, and they range from a low of 0.003 g C cm^{-2} ka^{-1} to a maximum of 0.12 g C cm^{-2} ka^{-1} (Bralower and Thierstein 1984, and in Brooks and Fleet 1987; Arthur et al., in Brooks and Fleet 1987; Herbert et al. 1986; Summerhayes, in Brooks and Fleet 1987). In Holocene oceanic sediments, organic carbon accumulation rates are generally lower by at least an order of magnitude; however, they may exceed 0.1 g C cm^{-2} ka^{-1} in areas of high bulk sedimentation rates such as in marginal basins and along continental margins (Heath et al. 1977; Müller and Suess 1979; Bralower and Thierstein, in Brooks and Fleet 1987). A special mechanism to explain the much higher average organic carbon accumulation in the mid-Cretaceous, compared with the Holocene, is therefore clearly required.

An additional interpretative difficulty, however, arises from the fact that the organic carbon contents are highly variable within the stratigraphic intervals that are used for sedimentation rate calculations (usually biostratigraphic zones of a few meters to a few tens of meters thickness). Only if representative distributions of organic carbon measurements within these intervals have been used can such averaged organic carbon accumulation rates have any real meaning. The application of high resolution stratigraphy using orbital frequencies possibly represented by bedding alternations would, if confirmed as such, provide a great improvement in our ability to analyze the rapid variability in organic carbon accumulation rates in mid-Cretaceous sequences (Herbert and Fischer 1986). For the time being, comparisons of Holocene with mid-Cretaceous global carbon budgets must remain highly uncertain because of the great temporal variability and the poor geographic coverage of mid-Cretaceous data.

Kerogen Analyses and Organic Geochemistry

A growing number of techniques have been developed and successfully employed to analyze sources and diagenetic history of the organic carbon preserved in ancient deep-sea sediments. The elementary composition (H/C and O/C ratios) of the insoluble fraction (kerogen) can be used to characterize hydrocarbon maturity and, in reasonably unaltered material, the continental or marine origin of the kerogen (Tissot and Welte 1984). Similarly, microscopic analyses of the insoluble kerogen fractions have proven very successful in distinguishing kerogen of marine planktonic origin (amorphous and structured phytoplankton remains) from terrestrially derived kerogen (spores, pollen, wood, charcoal, etc.). More recently, so-called biomarker compounds, which are characteristic of certain algal or bacterial groups, have been identified; they can provide information on the sources and diagenetic histories of organic compounds in sediments up to a few hundred million years old (e.g., Brassel et al., in Brooks and Fleet 1987). The success and applicability of all of these methods decreases, however, with increasing diagenesis and microbial alteration.

Results from such analyses in mid-Cretaceous sediments imply that there are distinct spacial and temporal patterns observable in the composition of the preserved organic material. In samples with relatively low organic carbon contents (i.e., less than 1%), the carbon enrichment consists dominantly of terrestrially derived or diagenetically altered kerogen. Warm, humid climates promoting luxuriant growth of coastal forests and high river or aeolian supply to oceanic environments have been proposed as an explanation for the widespread occurrence of slightly elevated kerogen contents of often identifiably terrestrial origin throughout lower and mid-Cretaceous black shales (Ryan and Cita 1977; Stein et al. 1986; Summerhayes, in Brooks and Fleet 1987). The very highest accumulation rates of terrestrial organic carbon in Cretaceous sediments are around 0.2 g C cm^{-2} ka^{-1} (Bralower and Thierstein, in Brooks and Fleet 1987; Summerhayes, in Brooks and Fleet 1987). This is lower (by a factor of 10 or more) than any terrestrial organic carbon accumulation rate observed in Holocene or Pleistocene deep-sea sediments, including the turbiditic, high-sedimentation rate deposits of the Amazon cone, where the kerogen is known to be derived from one of the largest highly productive tropical rain forest areas (Pelet 1983). Since it is difficult to visualize a more highly productive terrestrial environment than present-day tropical rain forests, the abundance of terrestrially derived kerogen in many Cretaceous deep-sea sediments need therefore not be explained by higher continental supply, but rather by processes which lead to a significantly higher proportion of the supplied organic material being preserved, such as widespread oxygen deficiency.

At higher organic carbon contents the proportion of planktonic kerogen generally increases much more rapidly than the terrigenous fraction (Summerhayes, in Brooks and Fleet 1987). In black shales deposited below paleogeographic upwelling settings, and particularly in those found at the Cenomanian-Turonian boundary, plankton-derived organic matter strongly dominates (Dean et al. 1981; Stein et al. 1986). In these cases, higher marine surface productivity, leading to intensification and expansion of an oxygen minimum layer, has usually been advocated as an explanation for the observed organic carbon enrichments (Tissot et al., in Talwani et al. 1979; Summerhayes 1981; Arthur et al., in Sundquist and Broecker 1985).

Carbon Isotopes

The potential pointed out by Tappan (1968), of using carbon isotopes from calcitic as well as organic biogenous remains for reconstructing past productivity and carbon reservoir changes, has only recently been applied to the mid-Cretaceous black shale problem. The available analyses of carbonates are almost exclusively from bulk rocks, in which considerable carbon isotopic fluctuations have been documented (Scholle and Arthur 1980; Weissert et al., in Sundquist and Broecker 1985; Renard 1986). The significance and causes of these fluctuations, however, remain largely obscure, because neither the carriers of the signals nor the degree of postburial alteration are known. In particular, the mid-Cretaceous carbonate oozes, all of which may originally have been deposited with at least minor proportions of organic matter, may likely have been subject to postdepositional organic matter decomposition and consequent diagenetic carbonate alteration. The succession of known microbial degradation processes after burial, leading from oxidation to sulfate reduction and eventually to methanogenesis, may have led to both enrichment and depletion of ^{12}C in the dissolved bicarbonate. This, in turn, may have reprecipitated during cementation of the rocks (Irwin et al. 1977). Detailed microscopic studies have shown that the proportion of preserved biogenous particles decreases relative to the proportion of diagenetic micarb in sediments with carbonate contents of less than 40%, or more than 60%. In addition, a frequent correlation can be documented between carbonate contents, carbon isotopic ratios of the carbonate particles, and preservational features of calcareous nannofossils in lower and mid-Cretaceous deep-sea carbonate oozes from numerous bedded intervals (Thierstein 1983; Baltuck 1987; Thierstein and Roth, in preparation). Thus, the crude similarity between the averaged organic carbon accumulation rates and the averaged carbon isotopic record (Arthur et al., in Sundquist and Broecker 1985) may be influenced by the variable occurrence and importance of microbial methanogenic organic carbon degradation. There is strong evidence that a variety of diagenetic microbial organic carbon decomposition processes have led to enrichments and depletions of ^{12}C in reprecipitated micritic cements in limestones (Thierstein and Roth, in preparation).

Interpretations based on the carbon isotopic composition of organic matter have been plagued by a serious, apparently nonanalog situation in the Cretaceous. In modern plankton, as well as in Neogene kerogen compounds, the $\partial^{13}C$ (relative to PDB) is on the average about −22‰; that is 5‰ heavier than that of average terrestrial plant material (excluding C-4 plants). In Cretaceous marine sediments, plankton-derived kerogen has an average $\partial^{13}C$ value of −27‰, very similar to modern terrestrial plant material, whereas heavier values of about −24‰ have only been reported from organic carbon that is mostly terrestrially derived (Dean et al. 1986; Pratt and King 1986). Exceptions are two samples measured from a lower Albian black shale layer in a core from central Italy which showed $\partial^{13}C$ values of −22.2‰ and −24.1‰ (Pratt and King 1986). The most likely explanation offered for the discrepancy in the carbon isotopic ratios between the Cretaceous and Neogene marine organic matter is elevated ΣCO_2 contents in surface waters, caused by a combination of higher pCO_2 pressures in the atmosphere and lower surface water productivity (Dean et al. 1986). The minor effects of temperature on the carbon isotopic composition observable in recent organic carbon compounds, and the occasionally major effects of diagenesis observable in Neogene sections, seem to go in the opposite direction from the anomalies observed in the Cretaceous and can therefore be discounted as explanations. Too little is known about carbon isotopic fractionation among various plankton groups to evaluate the potential influence of vital effects caused by significant evolutionary changes in the composition of the marine plankton since the Cretaceous (Deines 1980).

Metal and Sulfide Enrichments

Inorganic metal sulfide precipitation along a migrating redoxcline overlying a euxinic (hydrogen sulfide-containing) deep water mass has been invoked as an explanation for the generally observed enrichment of transition metals, such as nickel, copper, zinc, and others, in sediments rich in organic carbon (Brumsack 1980). Euxinic deep waters would lead to the preservation in the sedimentary deposits of a relatively high proportion of the organic matter produced in surface waters, i.e., to a relatively high preservation factor of organic carbon (Müller and Suess 1979; Bralower and Thierstein 1984). However, it has not been possible to demonstrate that such euxinic hydrogenous transition metal fluxes, in excess of normal biogenous and terrigenous background fluxes, do indeed occur in recent euxinic environments (Bralower and Thierstein, in Brooks and Fleet 1987). Although euxinic deep water conditions and hydrogenous metal sulfide precipitation may have occurred during certain intervals of the Cretaceous, there is abundant evidence (such as the presence of benthic fauna and burrow structures) of at least intermittent re-oxygenation in many oceanic sections. Instead of these excess metal accumulation rates, the absence of bioturbation features in anoxic Holocene environments has been used to infer depositional anoxia in certain nonbioturbated Cretaceous deep-sea sections where, as a direct consequence, relatively high organic carbon preservation factors can be inferred (Bralower and Thierstein 1984).

Opal and Phosphate

The accumulation of opaline silica in Holocene deep-sea sediments closely mirrors present-day surface water productivity patterns (Berger 1970). The deposition of phosphate in today's oceans seems to be tied to areas of high primary productivity where they lead to the development of an oxygen minimum zone which intersects the ocean floor (Glenn and Arthur 1985). In general, siliceous microfossils, as well as chert and phosphate deposits, are very rare in mid-Cretaceous oceanic deposits (Arthur and Jenkyns 1981). Nevertheless, their occasional occurrence and changing abundance have been interpreted as indicating higher paleoproductivity (Leckie 1984).

A few estimates of silica accumulation rates in mid-Cretaceous sequences indicate moderate values by Holocene standards (Pisciotto, in Warme et al. 1981; Miskell et al. 1985; Herbert et al. 1986), but their validity as a paleoproductivity indicator may be plagued by the fact that clay mineral alteration often leads to diagenetic silica migration and reprecipitation (Kastner et al. 1977; Mélières, personal communication, 1986).

Plankton Ecology

Quantitative biogeographic studies of calcareous nannofossil distributions have revealed comparatively weak paleolatitudinal biogeographic gradients but have identified elevated abundances of a few nannofossil taxa in what must be considered paleo-upwelling areas based on paleogeographic setting (Roth and Bowdler, in Warme et al. 1981). However, detailed taxonomic counts in numerous cyclic Cretaceous deep-sea sequences show very minor relative abundance changes in beds within a given cyclic interval at any one site, suggesting stability of surface water environments through the cyclic alternations. This is particularly true when these minor fluctuations are compared to the known mid-Cretaceous paleobiogeographic variability of nannofossil assemblages (Roth 1986; Thierstein and Roth, in preparation). First, this result demonstrates that the often obvious, rapidly changing dissolution and overgrowth features in bedded intervals did not lead to significant taxonomic changes in these mid-Cretaceous nannofossil assemblages, with the exception of one resistant taxon (*Watznaueria*), which increases relative to all other taxa (Roth and Krumbach 1986). This clearly contrasts with the differential dissolution patterns observed in uppermost Cretaceous and lowermost Tertiary assemblages (Thierstein 1980). Second, the result also implies that there were no significant ecologic changes occurring in the surface waters overlying these deep-sea sections through the bedding alternations between laminated, dark, organic carbon-rich shales and the bioturbated, carbonate-rich, light grey-to-white marls and limestones (Thierstein and Roth, in preparation). These bedding alternations therefore seem to be caused mostly by periodic changes in the dissolved oxygen contents of the local bottom waters which determined bioturbation intensity and thus differential diagenetic history.

Paleobiogeographic studies of mid-Cretaceous planktonic and benthic foraminifera are few and regionally restricted, mostly due to the scarcity of foraminifera and of suitably preserved samples (Moullade and Guerin 1981). Other studies have concentrated on evolutionary, paleobathymetric, and ecological interpretations in restricted areas or in sections with as yet little direct impact on paleoproductivity estimates (Butt 1982; Darmedru et al. 1982; Leckie 1984).

MODELS AND BUDGETS

Atmospheric and Oceanic Circulation

Cretaceous climates are thought to have been significantly different from today's climate for a number of geological reasons (Barron 1983). Tropical, subtropical, and temperate fossil plants and animals existed at considerably higher latitudes than at any time since, suggesting warmer polar temperatures and decreased latitudinal temperature gradients. Evidence for permanent ice cover in polar areas is either lacking or controversial and is limited to very short time intervals (Kemper 1987). The area covered by ocean water may have been considerably larger during the global Cretaceous transgression. Continental areas showed a more even distribution with respect to latitudinal belts than today, but they were concentrated in the eastern hemisphere, leaving a hemispherical Pacific super-ocean which probably extended from pole to pole. Sensitivity studies using a variety of heat-budget and atmospheric circulation models suggest that an elevated sea level and changed latitudinal distributions of continents, even in combination, may have been insufficient to account for the elevated polar temperatures (Schneider et al., in Sundquist and Broecker 1985). Various authors have proposed that higher pCO_2 contents in the Cretaceous contributed to higher atmospheric heat retention. The increased paleolatitudinal width of biogeographic realms and the lack of strong E-W bioprovinciality of calcareous phytoplankton (Roth 1981) suggest a diminished subtropical Hadley circulation. The development of pronounced monsoonal wind patterns, however, could be expected from the assembly of continental masses (Barron et al. 1985).

A significantly different global oceanic deep water circulation from today's has been invoked by numerous authors to account for the organic carbon enrichment of most mid-Cretaceous deep-sea sediments (Thierstein and Berger 1978; Brass et al. 1982). With increased polar temperatures during most of mid-Cretaceous, the production of cold, polar deep water would have been considerably decreased, whereas the production of warm, saline deep waters in low-latitude marginal seas might have been more important than today. This, together with the decreased solubility of oxygen in warm, saline waters, may have strongly promoted the development of deep water oxygen deficiency (Wilde and Berry, in Schlanger and Cita 1982; Brass et al. 1982).

A quite dramatic slowdown of mid-Cretaceous deep water renewal rates is implied from the consideration of the deep water oxygen budget (Bralower and Thierstein 1984). The sedimentological evidence, such as the widespread absence of bioturbation features and the concurrent organic carbon enrichment observed in virtually all deep ocean basins sampled so far, strongly suggests that oxygen consumption frequently exceeded the dissolved oxygen supply in the global deep water reservoir, at least intermittently between the Barremian to the latest Cenomanian.

In today's oceans, the supply of oxygen to the global deep water reservoir is estimated at $0.635 \cdot 10^{15}$ mol O_2 a^{-1}, whereas the consumption is estimated at $0.106 \cdot 10^{15}$ mol O_2 a^{-1} (Fig. 3). To balance the oxygen budget in the present-day global deep water reservoir, a decrease in the amount of sinking surface waters by a factor of six would be necessary. Applying organic carbon preservation factors of 2% in a few continuously laminated or nonbioturbated mid-Cretaceous sediment sections, a former average primary productivity estimate of about a tenth of the present primary productivity seems to be implied (Bralower and Thierstein 1984). Variability of lithology and of the kerogen and nannofossil composition indicates that there still existed considerable spacial and temporal fertility gradients within these generally slowly convecting mid-Cretaceous oceans. As a consequence of the lowered organic matter supply, the oxygen demand in the deep water reservoir would also have been lower by a factor of about ten. To meet this lowered oxygen demand, just about one Mediterranean outflow would be sufficient (Fig. 4). Such a slowly convecting global ocean would imply considerably longer oceanic deep water residence times for nutrients and

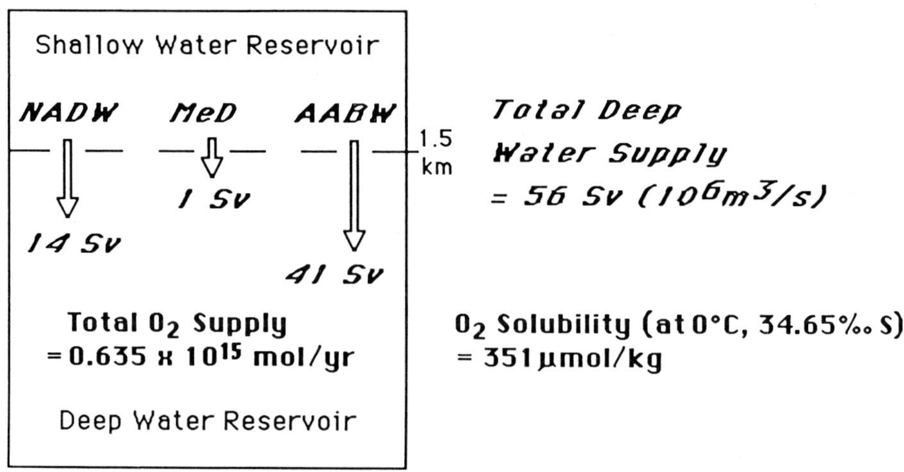

Fig. 3—Box model of present-day global oceanic deep water circulation showing the deep water supply from North Atlantic (NADW), Antarctic (AABW), and Mediterranean and Red Seas (MeD) in Sverdrups (1 Sv = 10^6 m^3/s). Given saturation with atmospheric oxygen at the indicated solubility, the total supply of dissolved oxygen to the deep water reservoir is 0.635×10^{15} mol/yr (data from Bralower and Thierstein 1984).

other elements. Because of the increased age of the deep waters, their total dissolved carbon dioxide content would also have been higher than today.

Global Carbon Budgets

How would such a slowly convecting ocean influence the global cycling of carbon? Given the uncertainties of the present-day controls dominating the exchange rates of carbon dioxide between the oceans and atmosphere, even qualitative reconstructions of mid-Cretaceous global carbon cycles must necessarily remain speculative. In an evaluation of the dependence of atmospheric carbon dioxide contents on oceanic circulation and productivity, Broecker and Takahashi (1984) inferred that a slowdown of deep water ventilation rates would tend to push the current global ocean circulation towards their "Thermodynamic Ocean" model, where the rate of CO_2 gas exchange between surface water and atmosphere dominates the rate of

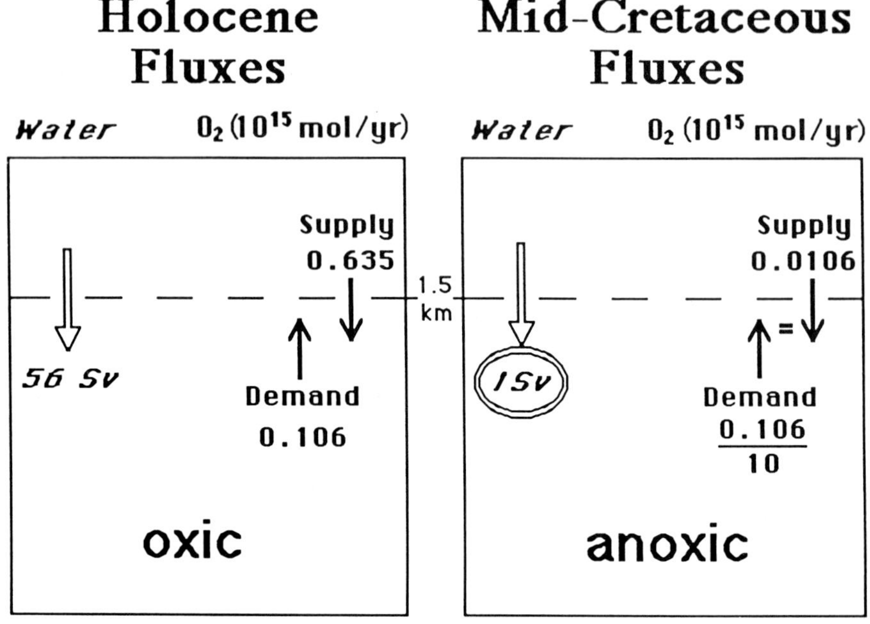

Fig. 4—Box models of Holocene and mid-Cretaceous oceanic deep water circulation and dissolved oxygen budgets. Widespread oxygen deficiency in mid-Cretaceous deep waters must be interpreted as a balanced oxygen budget (demand = supply). The required oxygen supply to the mid-Cretaceous deep water reservoir could be provided by about one present-day Mediterranean outflow (data from Bralower and Thierstein 1984).

exchange of carbon between the surface water and deep water reservoirs. For the "Thermodynamic Ocean" model, an increase of the atmospheric pCO_2 content was inferred to a level corresponding to the pCO_2 pressure of the cold deep water reservoir, which would leak CO_2 to the atmosphere in high-latitude outcrops. In the opposite mode, the so-called "Redfield Ocean" model, the atmospheric pCO_2 would be controlled by relatively slow carbon dioxide exchange rates between surface waters and the atmosphere and relatively rapid transport of carbon between the warm and cold water reservoirs (assuming a C/P ratio of 130 for the falling organic debris). The "Redfield Ocean" model would lead to lowered atmospheric pCO_2 contents because of the effectiveness of the biological pump in the highly productive present-day oceans.

The measured differences betwen the CO_2 partial pressure in the surface ocean water and that in the atmosphere suggest that surface waters today act as a source of CO_2 in the warm tropical and subtropical areas and as a sink in high latitudes, mostly because of high primary production at high latitudes in summertime (Broecker and Takahashi 1984).

For the mid-Cretaceous, both the inferred lowered primary productivity and the inferred lowered deep water renewal rates would suggest a paleocirculation mode closer to the "Thermodynamic Ocean" model than is observed today. Atmospheric pCO_2 contents would thus have been controlled mostly by the higher average surface water temperatures leading to higher atmospheric pCO_2 contents (Broecker and Peng 1982). Uncertainties concerning the former ratio of organic carbon to carbonate in the particle rain and thus alkalinity further complicate such inferences.

Recent estimates of the variability of the global carbon budgets have focused mainly on the changes in the organic carbon fluxes and their effects on the carbon isotopic composition of the individual reservoirs (Berner et al. 1983; Holland 1984; Arthur et al., in Sundquist and Broecker 1985). In the BLAG model, a gradual decrease in global pCO_2 and temperature over the past 100 Ma has resulted predominantly from a decrease in average sea-floor spreading rates (Berner et al. 1983). Recalibrations of the Cretaceous magnetochronology and reevaluations of the inferred spreading rate changes, however, suggest that the variability of CO_2 production by changes in subduction rates may have been considerably less than previously assumed (Kominz 1984). The BLAG model was based on present-day global material flux estimates and quantification of rate law expressions for weathering as a function of continental area, global temperature, runoff, and pCO_2, as well as quantification of rate law expressions for volcanic rock-seawater reactions and metamorphism-magmatism as a function of spreading rate changes. Steady-state chemical equilibrium in the oceans and invariability of the global biosphere response were assumed. Variability in the global organic carbon and sulfur burial rates was included in a subsequent refinement of the BLAG model, which resulted in mid-Cretaceous pCO_2 values 16 times higher than present-day values, rather than 26 times as in the BLAG model (Lasaga et al., in Sundquist and Broecker 1985). With this latter model, however, an unreasonably large increase of 300% of the atmospheric CO_2 contents and a 30% decrease of the carbonate flux during the past 100 Ma were obtained, suggesting that considerable uncertainties still exist concerning the appropriateness of the average global carbon isotopic values used in the model. These uncertainties are also reflected in the great variance of the averaged carbon and sulfur isotopic input data used in the various modeling attempts (Garrels and Lerman 1981; Lasaga et al., in Sundquist and Broecker 1985; Arthur et al., in Sundquist and Broecker 1985), as well as by the conflict between the high mid-Cretaceous carbon fluxes predicted by the models and the much lower ones determined stratigraphically in mid-Cretaceous sequences (Arthur et al., in Sundquist and Broecker 1985; Bralower and Thierstein, in Brooks and Fleet 1987).

Orbital Forcing

Orbital control has been suggested as a potential cause for the widely observed bedding cycles in mid-Cretaceous sequences (Einsele and Seilacher 1982). However, two major pieces of evidence, which would be necessary for confirmation of such a hypothesis, have so far been elusive. The first would be a reasonably convincing calibration of the observed frequency spectra with a reliable bio- or magnetochronology. This has been impossible because most bedded intervals analyzed so far were deposited during the Cretaceous magnetic quiet zone and because of the uncertainties of radiometric calibration of the Cretaceous time scale (De Boer and Wonders 1984). The second piece of evidence relates to the difficulty in interpreting the available proxy data to understand the possible climatic and oceanographic mechanisms which imparted the primary imprint for the bedding cycles to develop (Arthur et al. 1984).

In the late Neogene, the rather minor changes in the distribution of solar insolation seem to be amplified by feedback of rapidly changing snow and ice cover. In the absence of polar ice caps in the Cretaceous, an alternate amplification mechanism may have operated. Fluctuating productivity in the oceans as well as on land have been advocated by some (De Boer, in Einsele and Seilacher 1982; Herbert and Fischer 1986), while changing monsoonal wind and precipitation patterns have been advocated by others (Barron et al. 1985). The proxy data necessary to confirm such mechanisms have not been available in the literature, or have been altered significantly by diagenesis. The nannofloral evidence specifically shows invariable relative abundances of calcareous nannofossil taxa in cyclic sequences, so that significant ecological changes in surface waters in the oceanic areas studied seem unlikely (Roth 1986).

An alternative forcing mechanism for the observed bedding alternations, which deserves closer scrutiny, is the oxic/anoxic threshold of a delicately balanced oxygen budget in the deep sea. The redoxcline leaves a very dramatic imprint in the sediments when crossed, because the presence or absence of bioturbation strongly influences sediment fabric and the diagenetic history of organic carbon and of the mineralized sediment particles. Detailed paleontological, sedimentological, and geochemical studies in numerous cyclic mid-Cretaceous intervals seem to suggest such a mechanism of rather stable oceanic surface water environments and repeated crossings of the redoxcline at depth (Thierstein and Roth, in preparation). The source areas for the inferred, relatively small amounts of warm saline deep waters could have been in low-latitude marginal seas or coastal lagoons, where relatively minor, orbitally induced regional insolation changes could have led to periodic shifts in the intensity or location of deep water formation.

A necessary consequence of such a scenario would of course be significantly slower average global upwelling rates, lowered global primary productivity, and thus changes in carbon cycling (Bralower and Thierstein 1984; De Boer 1986).

Prospects for Progress?

The question of paleofertility fluctuations and anomalies in the Cretaceous seems to be on the threshold of a better understanding. Several of the currently pursued methods and models, if integrated in the future, promise to further our understanding of these mid-Cretaceous events significantly. Among the most promising and important seem to be:

1. A significant improvement in the radiometric calibration of the magneto- and biochronologies, combined with frequency analyses of fluctuating sedimentary parameters, will allow the hypothesis of orbital forcing to be confirmed or rejected.
2. From precise, high resolution correlation of bedded sequences we may learn whether the observed variable oxygen supply to the deep water reservoirs was local, basin-wide, or globally synchronous.
3. The increasing number of multidisciplinary sedimentological studies, including close scrutiny of processes of diagenetic alteration of sedimentary particles, will lead to more reliable interpretations of paleoenvironmental proxy data.
4. Comparisons of oceanic and terrestrial sequences from widely spread areas will be necessary to evaluate the importance of changes in global volcanism, the rapid evolutionary changes of the marine and terrestrial biosphere, the impact of the rise of angiosperms, and the magnitude and synchroneity of sea-level changes.
5. A better understanding of the importance of biological versus thermodynamic processes controlling present-day carbon dioxide exchange rates between oceans and atmosphere is necessary to constrain models of global carbon budgets for the mid-Cretaceous.

REFERENCES

Arthur, M.A.; Dean, W.E.; and Stow, D.A.V. 1984. Models for the deposition of Mesozoic-Cenozoic fine-grained organic-carbon-rich sediment in the deep sea. In: Fine-Grained Sediments: Deep-Water Processes and Facies, eds. D.A.V. Stow and D.J.W. Piper, pp. 527–560, Geol. Soc. of London, Spec. Publ. 15.

Arthur, M.A., and Jenkyns, H.C. 1981. Phosphorites and paleoceanography. *Oceanol. Acta, spec. vol.* **4**: 83–96.

Baker, P.A.; Gieskes, J.M.; and Elderfield, H.J. 1982. Diagenesis of carbonates in deep-sea sediments – evidence from Sr/Ca ratios and interstitial dissolved Sr^{2+} data. *J. Sed. Petrol.* **52**: 71–82.

Baltuck, M. 1987. Geochemistry, carbon and oxygen stable isotope composition, and diagenetic textural features of Lower Cretaceous pelagic cyclic sediments from the western North Atlantic, Deep Sea Drilling Project Hole 603B. In: Initial Reports of the Deep Sea Drilling Project, vol. 93/2, pp. 989–995. Washington, D.C.: US Govt. Printing Office.

Barron, E.J. 1983. A warm equable Cretaceous: the nature of the problem. *Earth Sci. Rev.* **19**: 305–338.

Barron, E.J.; Arthur, M.A.; and Kauffman, E.G. 1985. Cretaceous rhythmic bedding sequences: a plausible link between orbital variations and climate. *Earth Planet. Sci. Lett.* **72**: 327–340.

Berger, W.H. 1970. Biogenous deep-sea sediments: fractionation by deep-sea circulation. *Geol. Soc. Amer. Bull.* **81**: 1385–1402.

Berner, R.A.; Lasaga, A.C.; and Garrels, R.M. 1983. The carbonate-silicate geochemical cycle and its effect on atmospheric carbon dioxide over the past 100 million years. *Am. J. Sci.* **283**: 641–683.

Bralower, T., and Thierstein, H.R. 1984. Low productivity and slow deep-water circulation in mid-Cretaceous oceans. *Geology* **12**: 614–618.

Brass, G.W.; Southam, J.R.; and Peterson, W.H. 1982. Warm saline bottom water in the ancient ocean. *Nature* **296**: 620–623.

Broecker, W.S., and Peng, T.-H. 1982. Tracers in the Sea. New York: Eldigio.

Broecker, W.S., and Takahashi, T. 1984. Is there a tie between atmospheric CO_2 content and ocean circulation. In: Climate Processes and Climate Sensitivity, eds. J.E. Hansen and T. Takahashi. *Geophys. Monog.* **29**: 314–326. Washington, D.C.: Amer. Geophys. Union.

Brooks, J., and Fleet, A.J. 1987. Marine Petroleum Source Rocks. *Geol. Soc. Spec. Publ.* **26**. Oxford: Blackwell Scientific Publications.

Brumsack, H.-J. 1980. Geochemistry of Cretaceous black shales from the Atlantic Ocean (DSDP Legs 11, 14, 36, and 41). *Chem. Geol.* **31**: 1–25.

Butt, A. 1982. Micropaleontological bathymetry of the Cretaceous of western Morocco. *Palaeogeog. Pal. Pal.* **37**: 235–275.

Darmedru, C.; Cotillon, P.; and Rio, M. 1982. Rythmes climatiques et biologiques en milieu marin pélagique. Leurs relations dans les dépots crétacés alternants du bassin vocontien (Sud-Est de la France). *Bull. Soc. Géol. France* **24**: 627–640.

Dean, W.E.; Arthur, M.A.; and Claypool, G.E. 1986. Depletion of ^{13}C in Cretaceous marine organic matter: source, diagenetic, or environmental signal? *Marine Geol.* **70**: 119–157.

Dean, W.E.; Claypool, G.E.; and Thiede, J. 1981. Origin of organic-carbon-rich mid-Cretaceous limestones, mid-Pacific Mountains and southern Hess Rise. In: Initial Reports of the Deep Sea Drilling Project, vol. 62, pp. 877–890. Washington, D.C.: US Govt. Printing Office.

DeBoer, P.L. 1986. Changes in the organics carbon burial during the Early Cretaceous. In: North Atlantic Palaeoceanography, eds. C.P. Summerhayes and N.J. Shackleton. *Geol. Soc. Spec. Publ.* **21**: 321–331.

DeBoer, P.L., and Wonders, A.A.H. 1984. Astronomically induced rhythmic bedding in Cretaceous pelagic sediments near Moria (Italy). In: Milankovitch and Climate, Part 1, eds. A.L. Berger et al., pp. 177–190. Dordrecht: Reidel.

Deines, P. 1980. The isotopic composition of reduced organic carbon. In: Handbook of Environmental Isotope Geochemistry 1, eds. P. Fritz and J.C. Fontes, pp. 329–406. Amsterdam: Elsevier.

Einsele, G., and Seilacher, A. 1982. Cyclic and Event Stratification. Berlin: Springer.

Firstbrook, P.L.;. Funnell, B.M.; Hurley, A.M., and Smith, A.G. 1979. Paleoceanic reconstructions, 160–0 Ma. La Jolla, CA: National Science Foundation (Univ. California, San Diego).

Fischer, A.G. 1980. Gilbert-Bedding rhythms and geochronology. *Geol. Soc. Amer. Spec. Paper* **183**: 93–104.
Garrels, R.M., and Lerman, A. 1981. Phanerozoic cycles of sedimentary carbon and sulfur. *Proc. Natl. Acad. Sci. USA* **78**: 4652–4656.
Glenn, C.R., and Arthur, M.A. 1985. Sedimentary and geochemical indicators of productivity and oxygen contents in modern and ancient basins: the Holocene Black Sea as the "type" anoxic basin. *Chem. Geol.* **48**: 325–354.
Hallam, A. 1985. A review of Mesozoic climates. *J. Geol. Soc. Lon.* **142**: 433–445.
Heath, G.R.; Moore, T.C. Jr.; and Dauphin, J.P. 1977. Organic carbon in deep-sea sediments. In: The Fate of Fossil Fuel CO_2 in the Oceans, eds. N.R. Andersen and A. Malahoff, pp. 605–625. New York: Plenum.
Herbert, T.D., and Fischer, A.G. 1986. Milankovitch climatic origin of Mid-Cretaceous black shale rhythms in central Italy. *Nature* **321**: 739–743.
Herbert, T.D.; Stallard, R.F.; and Fischer, A.G. 1986. Anoxic events, productivity rhythms, and the orbital signature in a Mid-Cretaceous deep-sea sequence from central Italy. *Paleocean.* **1**: 495–506.
Holland, H.D. 1984. The Chemical Evolution of the Atmosphere and Oceans. Princeton, NJ: Princeton Univ. Press.
Irwin, H.; Curtis, C.; and Coleman, M. 1977. Isotopic evidence for source of diagenetic carbonates formed during burial of organic-rich sediments. *Nature* **269**: 209–213.
Irwing, E.; North, F.K.; Couillard, R. 1974. Oil, climate and tectonics. *Can. J. Earth Sci.* **11**: 1–17.
Jenkyns, H.C. 1980. Cretaceous anoxic events: from continents to oceans. *J. Geol. Soc. Lon.* **137**: 171–188.
Kastner, M.; Keene, J.B.; and Gieskes, J.M. 1977. Diagenesis of siliceous oozes. 1. Chemical controls on the rate of opal-A to opal-CT transformation – an experimental study. *Geochim. Cosmochim. Acta* **41**: 1041–1059.
Kemper, E. 1987. Das Klima der Kreide-Zeit. *Geol. Jahrbuch.* **96A**: 1–185.
Kent, D.V., and Gradstein, F.M. 1985. A Cretaceous and Jurassic geochronology, *Geol. Soc. Amer. Bull.* **96**: 1419–1427.
Kominz, M.A. 1984. Oceanic ridge volumes and sea-level change: an error analysis. In: Interregional Unconformities and Hydrocarbon Accumulation. *Am. Assn. Petrol. Geol. Mem.* **36**: 109–127.
Leckie, R. 1984. Mid-Cretaceous planktonic foraminiferal biostratigraphy off central Morocco, DSDP Leg 79, Sites 545 and 547. In: Initial Reports of the Deep Sea Drilling Project vol. 79, pp. 579–620. Washington, D.C.: US Govt. Printing Office.
Miskell, K.J.; Brass, G.W.; and Harrison, C.G.A. 1985. Global patterns in opal deposition from late Cretaceous to late Miocene. *Am. Assn. Petrol. Geol. Bull.* **69**: 996–1012.
Moullade, M., and Guerin, S. 1981. Correlations biostratigraphiques dans le Crétacé moyen de l'Atlantique sud et de la marge africaine de l'Atlantique nord (Sites 137, 327A, 356, 363, 364, 369A, 370, 415A, legs D.S.D.P. 14, 36, 39, 40, 41, 50). In: Corrélations Microbiostratigraphiques du Méso-Cénozoique des Provinces Mésogéennes et Africaines. *Trav. Centre Recherches Micropal. "Jean Cuvillier"* **1**: 30–35.
Müller, P.J., and Suess, E. 1979. Productivity, sedimentation rate, and sedimentary organic matter in the oceans. 1. Organic carbon preservation. *Deep-Sea Res.* **26A**: 1347–1362.
Noël, D., and Melguen, M. 1978. Nannofacies of Cape Basin and Walvis Ridge

sediments, Lower Cretaceous to Pliocene (Leg 40). In: Initial Reports of the Deep Sea Drilling Project, vol. 40, pp. 487–524. Washington, D.C.: US Govt. Printing Office.

Pelet, R. 1983. Connaissance de la sédimentation organique actuelle et récente: vue d'ensemble sur les missions "ORGON." Géochimie organique des sédiments marins. D'ORGON à MISEDOR. Ed. C.N.R.S., Paris: 453–480.

Perry, E.A.; Gieskes, J.M.; and Lawrence, J.R. 1976. Mg, Ca and $0^{18}/0^{16}$ exchange in the sediment-pore water system, Hole 149, DSDP. *Geochim. Cosmochim. Acta* **40**: 413–423.

Pratt, L.M., and King, J.D. 1986. Variable marine productivity and high eolian input recorded by rhythmic black shales in Mid-Cretaceous pelagic deposits from central Italy. *Paleocean.* **1**: 507–522.

Renard, M. 1986. Pelagic carbonate chemostratigraphy (Sr, Mg, ^{18}O, ^{13}C). *Mar. Micropal.* **10**: 117–164.

Ricken, W. 1986. Diagenetic Bedding. Berlin: Springer.

Roth, P.H. 1981. Mid-Cretaceous calcareous nannoplankton from the central Pacific: implications for paleoceanography. Initial Reports of the Deep Sea Drilling Project, vol. 62, pp. 471–489. Washington, D.C.: US Govt. Printing Office.

Roth, P.H. 1986. Mesozoic paleoceanography of the North Atlantic and Tethys Oceans. In: North Atlantic Paleoceanography, eds. C.P. Summerhayes and N.J. Shackleton. *Geol. Soc. Spec. Publ.* **21**: 299–320.

Roth, P.H., and Krumbach, K.R. 1986. Middle Cretaceous calcareous nannofossil biogeography and preservation in the Atlantic and Indian oceans: implications for paleoceanography. *Mar. Micropal.* **10**: 235–266.

Ryan, W.B.F., and Cita, M.B. 1977. Ignorance concerning episodes of ocean-wide stagnation. *Marine Geol.* **23**: 197–215.

Savrda, C.E., and Bottjer, D.J. 1986. Trace-fossil model for reconstruction of paleo-oxygenation in bottom waters. *Geology.* **14**: 3–6.

Schlanger, S.O., and Cita, M.B. 1982. Nature and Origin of Cretaceous Carbon-rich Facies. London: Academic Press.

Schlanger, S.O., and Jenkyns, H.C. 1976. Cretaceous oceanic anoxic events: causes and consequences. *Geologie en Mijnbouw* **55**: 179–184.

Scholle, P.A., and Arthur, M.A. 1980. Carbon isotope fluctuations in Cretaceous pelagic limestones: potential stratigraphic and petroleum exploration tool. *Am. Assn. Petrol. Geol. Bull.* **64**: 67–97.

Stein, R.; Rullkötter, J.; and Welte, D.H. 1986. Accumulation of organic-carbon-rich sediments in the Late Jurassic and Cretaceous Atlantic Ocean – a synthesis. *Chem. Geol.* **56**: 1–32.

Summerhayes, C.P. 1981. Organic facies of Middle Cretaceous black shales in deep North Atlantic. *Am. Assn. Petrol. Geol. Bull.* **65**: 2364–2380.

Sundquist, E.T., and Broecker, W.S. 1985. The Carbon Cycle and Atmospheric CO_2: Natural Variations Archean to Present. *Geophys. Monog.* 32. Washington, D.C.: Amer. Geophys. Union.

Talbot, M.R., and Kelts, K. 1986. Primary and diagenetic carbonates in the anoxic sediments of Lake Bosumtwi, Ghana. *Geology* **14**: 912–916.

Talwani, M.; Hay, W.; and Ryan, W.B.F. 1979. Deep Drilling Results in the Atlantic Ocean: Continental Margins and Paleoenvironment. M. Ewing Series 3. Washington, D.C.: Amer. Geophys. Union.

Tappan, H. 1968. Primary production, isotopes, extinctions and the atmosphere. *Paleogeog. Pal. Pal.* **4**: 187–210.

Thierstein, H.R. 1980. Selective dissolution of Late Cretaceous and earliest Tertiary calcareous nannofossils: experimental evidence. *Cretac. Res.* **2**: 165–176.

Thierstein, H.R. 1983. Oxygen and carbon isotopic fluctuations in cyclic mid-Cretaceous deep-sea sediments: dominance of diagenetic effects. *EOS* **64/45**: 733.
Thierstein, H.R., and Berger, W.H. 1978. Injection events in ocean history. *Nature* **276**: 461–466.
Tissot, B.P., and Welte, D.H. 1984. Petroleum formation and occurrence. Berlin: Springer.
Tucholke, B.E.; Vogt, P.R.; et al. 1979. Initial Reports of the Deep Sea Drilling Project. Washington, D.C.: US Govt. Printing Office.
Warme, J.E.; Douglas, R.G.; and Winterer, E.L. 1981. The Deep Sea Drilling Project: A Decade of Progress. Spec. Publ. No. 32. Tulsa, OK: Soc. Econ. Paleon. Miner.

Productivity of the Ocean: Present and Past
eds. W.H. Berger, V.S. Smetacek and G. Wefer, pp. 377–394
John Wiley & Sons Limited

Phosphorus vs. Nitrogen Limitation of New and Export Production

L.A. Codispoti

Monterey Bay Aquarium Research Institute
Pacific Grove, California 93950, U.S.A.

Abstract. Nitrogen and phosphorus distributions, the relative rates of oceanic nitrogen fixation and denitrification, and the physiology and biochemistry of phytoplankton growth suggest that nitrate is the major nutrient limiting export and new production in today's ocean. This condition exists largely because oceanic denitrification can respond rapidly to low oxygen levels and high fluxes of organic matter while a shortage of combined nitrogen is not sufficient for stimulating nitrogen fixation. Over several million year periods, nitrogen fixation has probably kept up with the demand, but imbalances in which denitrification exceeds nitrogen fixation for periods of several thousand years could cause changes of 20 to 30% in oceanic export (new) production. Such differences might contribute to the variation in the atmospheric carbon dioxide record. Phosphate may have been the major limiting nutrient in past oceans, during epochs of giant phosphorite deposit formation and also during warmer epochs, since temperatures >20°C favor nitrogen fixation.

An examination of the literature suggests that the Redfield ratio of change for nitrogen and phosphorus uptake by marine phytoplankton (16N:1P, by atoms) may vary appreciably from this value only in atypical marine ecosystems.

INTRODUCTION

Classic studies demonstrate that marine phytoplankton growth causes both phosphate and nitrate concentrations to fall to low levels whenever there is favorable light. These observations have been repeated many times but the wealth of observations fails to settle a difference in perceptions between biologists and geochemists. If you ask the former whether nitrate or phosphate is the master limiting variable, the biologist is likely to say, "Your question makes a complex situation too simple, but if you want an either/or type of response, then the answer is nitrate." When asked the same question, the geochemist is most likely to say "phosphate" and if pressed will back up this answer by invoking aspects of marine biology about which he/she should know less than the biologist! This paper is going to examine this difference in perceptions.

First let us dispose of the following caveats: (*a*) In setting up the discussion as phosphate versus nitrate, we are oversimplifying. For example, we are neglecting light limitation of phytoplankton growth. (*b*) We also neglect cases where nitrate, phosphate, and light are abundant but which do not display elevated primary production rates. One such region occurs in the South Pacific, where high surface nutrient concentrations arising from Equatorial Upwelling spread over a wide area (Fig. 1). A similar situation can be found in the N.E. Pacific and in the Southern Ocean (J. Martin, personal communication). Explanations for the occurrence of low primary production rates despite adequate nitrate, phosphate, and light abound, and the resurgence in interest of iron as a limiting factor is discussed elsewhere in this volume. Whether such situations have a significant effect on the ocean's export production is not clear at this time, but this problem should not be ignored by paleoceanographers.

BACKGROUND

Before getting into the meat of the debate, it will be useful to review aspects of the nitrogen and phosphorus cycles. Doing this will uncover facts that most can agree on and allow us to focus on the uncertainties.

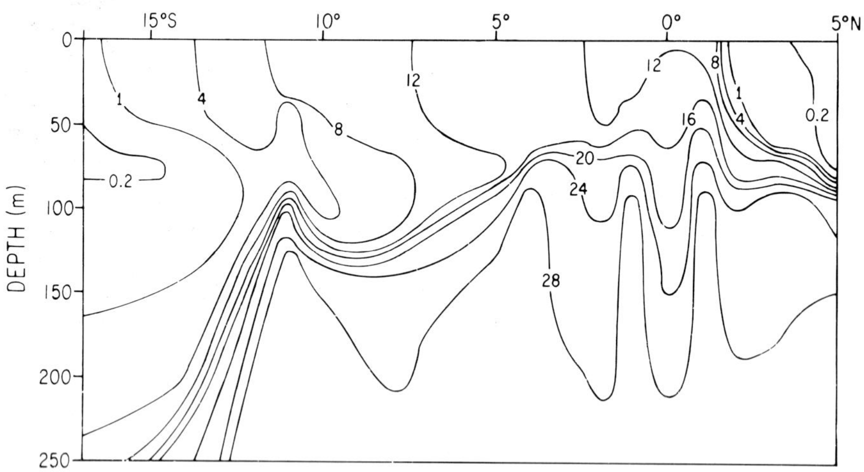

Fig. 1—A cross-equatorial section of NO_3^- (in μM) at 110° W made in November 1985 illustrates the meridional extent of high nutrient levels in the equatorial Pacific and the spillover of significant levels of nutrients into the South Pacific gyre (from Chavez and Barber 1987).

The Nitrogen Cycle

For this discussion, the two most important portions of the nitrogen cycle are denitrification and nitrogen fixation. The former reduces oxidized forms of combined nitrogen (nitrate, nitrite, and nitrous oxide) to free nitrogen gas and is the major sink for marine combined nitrogen while the latter process produces combined nitrogen from free nitrogen gas. Although oceanic nitrogen fixation is only one of three major sources (e.g., Codispoti and Christensen 1985) for marine combined nitrogen (the others being runoff and input from the atmosphere), it is the input term most likely to be subject to major revision and the term most often invoked as being able to compensate for increased denitrification.

The thermodynamically stable form of nitrogen in the present-day atmosphere is nitrate, but the stability of the N = N triple bond and the conversion of combined nitrogen to free nitrogen during denitrification prevents attainment of the equilibrium state. The atmospheric nitrogen reservoir is large (3.95×10^{21}g), and when scaled against the global nitrogen fixation and denitrification rate (100–400 Tg N yr^{-1}; Tg=10^{12}g) the residence time of nitrogen in the atmosphere is ~10^7 yr (Carpenter and Capone 1983; see below). Denitrification and nitrogen fixation may be approximately balanced on this time scale (Holland 1978).

Denitrification is a respiratory process that provides less free energy than oxygen respiration but considerably more than sulfate respiration. Consequently, it becomes a prominent respiratory process when oxygen concentrations fall below ~0.05 ml/l (Riley and Skirrow 1965) and before the onset of sulfate respiration. The five electrons transferred during the reduction of nitrate to free nitrogen and the mean oceanic nitrate concentration of 30 to 35 μM give the oceanic nitrate inventory the ability to oxidize about 25% as much organic material as can be consumed by the oxygen inventory. As a consequence, high denitrification rates in the ocean require a delicate balance between the supply of oxygen, nitrate, and organic material. If the oxygen supply is too high, denitrification will not be important. If the organic supply is too high or the nitrate supply too low, sulfate respiration will predominate since sulfate is abundant (~28,000 μM). Sites that are favorable for denitrification appear to be oxygen-deficient portions of the water column in the open ocean (where advection and mixing supply water low in oxygen and high in nitrate), and shallow or hemipelagic sediments (e.g., Carpenter and Capone 1983; Christensen et al. 1987). Codispoti and Christensen (1985) suggest that the total oceanic denitrification rate is ~120 Tg N yr^{-1} (Tg=10^{12}g).

The most important water column rates occur in oxygen-deficient waters found in the eastern tropical Pacific (~50 Tg N yr^{-1}; Carpenter and Capone 1983; Codispoti and Christensen 1985) and in the Arabian Sea, regions of relatively high export production. Naqvi et al. (1986) recently revised the denitrification rate estimate for the Arabian Sea upwards to ~30 Tg N yr^{-1} and suggested that Codispoti and Christensen's (1985) estimates for oceanic denitrification (120 Tg N yr^{-1}) should be revised upward by ~25 Tg N yr^{-1}. Their rate also did not include denitrification in estuaries, the possibility of denitrification in microenvironments within the oxygenated portion of the water column, and some smaller denitrification sites.

Only small changes in the oxygen supply and/or amount and distribution of export productivity are required to drive globally significant changes in denitrification (Table 1), and evidence from Peru suggests globally significant temporal variability in the regional denitrification rate over several-month

TABLE 1 Mechanisms for doubling the denitrification rate off Peru*.

A. *O_2 Decrease in Source Waters*

1. Volume of MSNM ~1.4×10^{14} m^3
2. Denitrification rate in MSNM ~2×10^{13} g N yr^{-1}
3. Nitrate → N_2 rate ~10 μg-atoms l^{-1} yr^{-1}
4. 10 Nitrate × 5/2 – 25 μg-atoms O_2 l^{-1} yr^{-1}
5. Therefore, 25 μg-atoms O_2 l^{-1} (0.3 ml/l) decrease in source waters could double the rate

B. *Increase in Carbon Flux*

1. Area of MNSM ~1.0×10^{12} m^2 (see Fig. 2)
2. "New" productivity of ~400 g C m^{-2} yr^{-1}
3. Denitrification rate of 20 g N m^{-2} yr^{-1} (Codispoti and Christensen 1985)
4. This requires ~20 g C m^{-2} yr^{-1} (See Fig. 1)
5. Therefore, an additional ~5% of "new" primary productivity is all that is required to double the denitrification rate

C. *Decrease in Downward O_2 Flux*

1. Range of possible downward O_2 fluxes through the bottom of the photic zone using k_z – 0.3 cm^2 s^{-1} for weak winds and 1.0 cm^2 s^{-1} for strong winds and an average O_2 gradient
   min: 15 g-atoms O_2 m^{-2} yr^{-1}
   max: 50 g-atoms O_2 m^{-2} yr^{-1}
2. Difference max-min = 35 g-atoms O_2 m^{-2} yr^{-1} = 13 g-atoms of C m^{-2} yr^{-1} with a Redfield ratio of O_2 to C of 276:106 by atoms
3. A C:N ratio of 1:1 by atoms during denitrification (Codispoti and Christensen 1985)
4. Hence 13 g-atoms C m^{-2} yr^{-1} ≈ 13 g-atoms N m^{-2} or 182 g N m^{-2} yr^{-1}
5. Therefore, changes in downward flux of O_2 only 10% as large as the one estimated above could double the denitrification rate.

* These calculations assume a doubling of the rate that Codispoti and Packard estimated for the main secondary nitrite maximum (MSNM) off Peru (see Codispoti and Christensen 1985).

to several-year periods (Codispoti et al. 1986). Christensen et al. (1987) suggest that exposure and re-flooding of continental shelves during glacial cycles may cause massive changes in the sedimentary denitrification rate. With a denitrification rate of ~100 Tg N/yr^{-1}, the residence time for combined nitrogen in the ocean is ~10^4 yr. This time scale is similar to some of the oscillations in the atmospheric carbon dioxide inferred from ice cores (McElroy 1983). The residence time for phosphorus is considerably longer, ~10^5 yr (Froelich et al. 1982).

Despite the deficiencies in our knowledge of marine denitrification, we can make the following assertions that are important for this discussion. First, denitrification is a major sink for oceanic combined nitrogen, and the rate for this process in the present-day ocean probably exceeds 100 Tg N yr^{-1} and might exceed 200 Tg N yr^{-1}. Second, the oceanic denitrification rate can vary significantly over several-month to several-year periods, and the possibility for massive changes exists during glacial cycles. Thus, *denitrification rates appear to respond quickly to changes in the supplies of oxygen, nitrate, and organic material.*

It might seem reasonable to expect that the rate of nitrogen fixation would increase as soon as combined nitrogen limits autotrophic growth. Certainly, the potential for nitrogen fixation is often present (e.g., Carpenter and Capone 1983). The available information, however, suggests that significant biochemical barriers prohibit such a simple response. Among these barriers is a requirement for low oxygen tensions (Carpenter and Capone 1983; Paerl and Prufert 1987). This requirement exists despite the fact that nitrogen limitation usually occurs in the well-oxygenated photic zone and that most of the nitrogen fixation that occurs in the sea is associated with photosynthetic bacteria that produce oxygen. It might be reasonable to suggest that high nitrogen fixation rates could occur in anoxic waters such as those that occur in the Cariaco Trench and the Black Sea. These environments, however, are rich in available combined nitrogen in the form of ammonium, and the data do not suggest significant nitrogen fixation (Riley and Skirrow 1965). A frequent problem for oceanic nitrogen fixation is, therefore, a necessity for reducing microzones within the well-oxygenated photic zone, and some studies suggest that this is difficult in the oceanic surface layer (e.g., Paerl and Prufert 1987). In addition to the requirement for anoxic microzones, oceanic nitrogen fixation may be limited by the following factors: (*a*) turbulence which can destroy the structure needed to provide anoxic microzones, (*b*) the availability of molybdenum (Howarth and Cole 1985) or iron (this volume), (*c*) organic matter availability (Paerl and Prufert 1987), and (*d*) a need for high light intensities and warm temperatures (≥20°C) at least in the most commonly studied oceanic nitrogen-fixing bacteria (*Oscillatoria*; Carpenter and Capone 1983).

Carpenter and Capone (1983) estimate a total marine nitrogen fixation rate of only ~20 Tg N yr^{-1}. This rate includes estuarine environments such as mangrove swamps, but they suggest that the rate could be an underestimate. There has been considerable discussion in the literature concerning the relatively low rates of nitrogen fixation that are generally encountered and some controversy surrounds the issue of how to measure oceanic nitrogen fixation. One important bit of information, however, is already apparent: many studies suggest that *a situation in which phosphate is available and nitrate is limiting is not a sufficient condition for the onset of nitrogen fixation.*

The Phosphorus Cycle

The following aspects of the oceanic phosphorus cycle that are pertinent to this discussion have been taken from Froelich et al. (1982):

1. Unlike nitrogen, which has important transfers between the ocean and atmosphere, phosphorus is added to the ocean primarily by runoff and is removed by sedimentation.
2. Biologically available phosphorus (dissolved + species that can dissolve or desorb and act as a buffer or reservoir) or BAP is removed from the ocean by biological activity. Approximately 40% is buried in the form of organic matter and approximately another 40% is buried in the form of biogenic calcium carbonate, primarily coccoliths.
3. BAP may also be deposited as phosphorite. Such deposits often occur in or near zones of enhanced denitrification but this probably does not permit a significant co-variance between BAP and combined nitrogen in today's ocean. This is because phosphorite deposition accounts for only about 10% of BAP removal which has a much longer residence time ($\sim 10^5$yr) than combined nitrogen.
4. During some past epochs giant phosphorite deposits were formed. For example, the Permian Phosphoria Formation contains several times as much phosphorus as is present in the present-day ocean (Froelich et al. 1982). It is possible, therefore, that the average BAP concentration of the ocean was significantly lower during such periods.

DISCUSSION

The Biologist's Case

Why does my hypothetical biologist say that if one has to pick a master chemical variable that limits oceanic export productivity, then the answer is nitrate? He or she might start by looking at photic zone nutrient concentrations (Table 2). In examining these values, the biologist would point out that organisms require approximately an order of magnitude more

TABLE 2 Selected nutrient values observed at the sea surface.

Region		$NO_3^- + NO_2^-$	NH_4^+ (μM)	PO_4^{3-}	$PO_4^{3-} \times 16$	Reference
Aluetian Island Upwelling, −53°N & 170°W		30	–	2.5	40	Hood & Kelley (1976)
NW African Upwelling ~22°N:[1]						Codispoti
	Sta. 14:	13.0	0.88	0.88	14.1	
	Sta. 161:	11.6	1.96	0.86	13.8	
Peruvian Upwelling:						Codispoti
	Sta. 378	0.5	0.11	0.52	8.3	
	Sta. 1	16.9	0.40	2.18	34.9	
	Sta. 107[2]	9.7	0.10	2.59	41.4	
Arctic Ocean ~83°N & 7°E:		3.0	0.04	0.32	5.1	Codispoti
Atlantic Ocean off N. Carolina:						
33°54′W, 75°26′W	Sta. 4897	0.3	–	0.06	0.96	Stefánsson & Atkinson (1967)
35°12′N, 75°36′W	Sta. 6379	0.6	–	0.19	3.0	Stefánsson & Atkinson (1967)
34°34′N, 75°01′W	Sta. 6547	0.5	–	0.06	0.96	Stefánsson & Atkinson (1967)

[1] It is possible that the N.W. African upwelling is influenced by Mediterranean waters that should be relatively rich in combined nitrogen (see text).

[2] This was a station taken close to the coast during strong coastal upwelling. The bottom water over the shelf at this location had experienced complete denitrification and contained hydrogen sulfide.

nitrogen than phosphorus for the production of biogenic material. Because of this, a "phosphate times 16" column is included in Table 2 to account for the fact that the ratio of change for $\triangle N:\triangle P$ (by atoms) arising from biological processes in the ocean is ~16:1 (e.g., Redfield et al. 1963). This ratio may be maintained even when autotrophic production in the photic zones has stripped nitrate concentrations down to extremely low levels. For example, Goldman et al. (1979), and Goldman (1986), suggest that in a "healthy" marine ecosystem, phytoplankton grow at rates that are close to their maximum-possible growth rates even when ambient nutrient concentrations are low, and that under these conditions C:N:P ratios approach the "Redfield" ratios. This does not mean that the absolute community growth rates have to be high. Different species can have differing maximum growth rates, and the biomass of phytoplankton per unit volume of seawater can vary. Consequently, large differences in total phytoplankton production and in export (new) production can occur even though individual members of the phytoplankton community may be growing at or near their maximum rate. The biologist would also state that nitrate concentrations in some cases have been shown to fall below the nanomolar level (Garside 1985) and that the data in Table 2 may understate the case for nitrogen limitation because phosphate may be recycled more rapidly than combined nitrogen (e.g., Ammerman and Azam 1985). The biologist would also cite a number of coastal and open ocean studies that present physiological and biochemical data suggesting that marine phytoplankton are typically nitrogen-limited (e.g., Ryther and Dunstan 1971).

A biologist with a philosophical bent might ask you if a healthy jack rabbit living in the desert contained markedly different elemental ratios than a cottontail rabbit living in a well-watered open forest, or whether or not the difference in water availability was not expressed primarily as the number of rabbits per hectare? This line of thought suggests that it is what an organism must do to compete that controls its N:P ratio more than what is most available in the immediate environment.

The Geochemist's Response

In reply, the geochemist might say that the biologist was correct when describing the ocean as sensed on the time scale of a plant, but that geochemists take a longer view. Since some bacteria can fix nitrogen and since the photic zone is in contact with an atmosphere that is mostly nitrogen, there is no way that combined nitrogen can really be a master chemical variable that controls export productivity. He or she would say that the "excess phosphate" is small and that a little excess must be present to stimulate nigrogen fixation. Obviously, a plant isn't going to fix much nitrogen if it is also deficient in other essential nutrients such as phosphate.

Given enough time, the geochemist's response has to be correct. The real problem, however, is to identify the time scale over which the geochemist is correct. If nitrogen fixation can increase rapidly whenever phosphorus concentrations of $>\sim 0.1 \mu M$ occur in the absence of combined nitrogen, then combined nitrogen can be cast aside as a candidate for the master chemical variable that controls export and new production. If, on the other hand, the characteristic time scale for nitrogen fixation to respond is long, then changes in the marine combined nitrogen inventory could become large enough to have globally significant effects.

Time Scales and Feed-back Mechanisms

Two negative feed-back mechanisms exist that can compensate for increased denitrification. One is, of course, increased nitrogen fixation. The other is decreased subsurface metabolism arising from a decrease in the NO_3^- transport to the surface layer. The latter feed-back mechanism will be referred to as the "slow feed-back" and can be summarized as follows: 1) A decreased NO_3^- flux to the photic zone causes a decrease in export and new productivity. 2) A decreased flux of export productivity to depth reduces subsurface respiration. 3) This reduces oxygen demand at depth and reduces the extent of oxygen deficient environments where denitrification flourishes. 4) As a consequence, oceanic denitrification diminishes and the oceanic combined nitrogen inventory begins to increase if N-fixation rates are constant. This feed-back mechanism is not closely coupled to increased denitrification (Piper and Codispoti 1975). This is because the nitrate removal by denitrification occurs beneath the photic zone and often beneath the depths that provide source waters for local coastal upwelling. Thus, the reduced NO_3^- flux to the photic zone arising from denitrification may be fully expressed only after one or more complete oceanic circulation cycles (~1000 yr). In addition, this feed-back mechanism could be overwhelmed by other processes for long periods. Consider, for example, the sensitivity calculations presented in Table 1. These data suggest that any process that can cause even modest rearrangements in the oceanic distributions of dissolved oxygen and organic material can temporarily overwhelm this slow feed-back mechanism.

Among the processes that could help to overwhelm the slow feed-back mechanisms are changes in the flux of O_2 across the sea surface due to a reduction in wind velocities and decreases in the 100% saturation value of oxygen due to rises in seawater temperature. Such changes may have occurred in going from the last glacial maximum to the present time. Changes in ecosystem structure (Codispoti et al. 1986 and references therein), and increases in the area of the continental shelf (Christensen et al. 1987), could also oppose the slow feed-back. Finally, intra- and inter-ocean fractionation of nutrients and dissolved oxygen by the process described by Berger (1970) could lead to enlargements of the water column sites of denitrification by "concentrating' the upwards nitrate flux and subsurface oxygen consumption in these regions. These oxygen deficient sites occupy only ~0.1% of the oceanic volume and they would not exist if subsurface oxygen concentrations and metabolism were distributed uniformly since, *on the average*, enough oxygen is present to oxidize all of the ocean's new and export production.

Since the total combined N content of the ocean is $\sim 1100 \times 10^{15}$g (Riley and Skirrow 1965), removal of ~20% of the total with excess denitrification of ~100 Tg N yr^{-1} would take ~2000 years. Some of the combined nitrogen is present as an unreactive DON fraction, so removal of 20% of the biologically active inventory would take < 2000 yr. *My conclusion, therefore, is that significant variations in the marine export productivity can be driven by changes in the marine combined nitrogen unless nitrogen deficiencies can rapidly be compensated by increased N-fixation.* It is also worth noting that imbalances in the oceanic nitrate inventory can influence atmospheric carbon dioxide levels (McElroy 1983).

A central and presumably testable question that relates to time scales is the response rate for oceanic N-fixation. Above, we have established that oceanic denitrification rates can change rapidly but that the marine nitrogen fixation rate appears to be low and may be inhibited by a number of factors that have little or nothing to do with the N:P ratio. If nitrogen fixation could be shown to respond rapidly, then many of the questions raised in this paper would disappear because of the existence of a "fast" feed-back mechanism.

There are, in fact, some reports that suggest that oceanic nitrogen fixation can be fast enough to meet the nitrogen requirements of the *local* plant communities. For example, studies by Smith (1984) and co-workers suggest that reactive phosphorus is the master limiting nutrient in some tropical embayments. Naqvi et al. (1986) suggest that high nitrogen fixation rates occur in the Red Sea and also suggest that if the Red Sea nitrogen fixation rate were extended to the rest of the ocean, enough N-fixation would occur to balance the marine combined N-budget. Finally, Bethoux and Copin-Montegut (1986) claim that phosphorus is the master limiting nutrient in the Mediterranean Sea due to enhanced nitrogen fixation associated with seagrasses in the bordering bays and estuaries. In interpreting these views *vis a vis* the problem of phosphorus vs. nitrogen limitation, however, these regional studies have to be put into a global context. When this is done, it becomes obvious that the three environments mentioned above share a number of characteristics that make them favorable for the "fast feed-back mechanism." These characteristics are as follows: (*a*) they are warm; (*b*) reduced turbulence exists due to their morphology which reduces "fetch" and the passage of swells from the open ocean; (*c*) large amounts of organic material (including abundant tar balls in the Red Sea) may be present; (*d*) there is the possibility for trace metal enrichment from nearby land sources and from shallow sediments; (*e*) relatively high rates of phosphorus removal may occur because these warm oligotrophic environments favor coccolithophores and other $CaCO_3$ secreting organisms (Berger and Keir 1984); and (*f*) they are zones of low export and new productivity. This makes it easier for nitrogen fixation to keep up with the demand. The Red Sea, after all, is so-named because this coloration arises from blooms of *Oscillatoria* (Naqvi et al. 1986), which is a nitrogen fixer.

The above authors suggest total rates of nitrogen fixation of ~2 Tg N yr^{-1} for the Red sea and about ~1 Tg N yr^{-1} for the Mediterranean Sea. so the favorable conditions for nitrogen fixation would have to be widespread in order for the oceanic fixation rate to be comparable to the denitrification rate. The potential for cold temperatures to inhibit nitrogen fixation more than denitrification are suggested by data from the Bering Sea and the Arctic Ocean (Table 2; Fig. 2), where the Pacific waters entering the Arctic via Bering Strait have much lower nitrate:phosphate ratios than the oceanic average of ~15:1 (by atoms). The Bering Sea and marginal seas of the Arctic are zones of high export production (Hood 1986) unlike the Mediterranean and Red Seas. Our tentative conclusion, therefore, is that the weight of the evidence suggests that the fast feed-back mechanism is only important in warm, nearshore zones of low export productivity.

Uptake Ratios and the Distribution of N:P Ratios

We have already summarized information (e.g., Goldman 1986; Goldman et al. 1979) that suggests that phytoplankton uptake ratios of N:P should be close to the 16:1 Redfield ratio (by atoms) in a healthy marine ecosystem. Although the plasticity of this ratio deserves closer scrutiny, we shall accept it (16:1 by atoms) as a working hypothesis for the following reasons:

1. The works of Goldman (cited above) provides support; he also cites the large scale study of the ratio in particulate matter by Copin-Montegut and Copin–Montegut (1983) as supporting his results.
2. Takahashi et al.'s (1976) revision of the Redfield ratios is controversial (see Shaffer 1987), but even so, their ratio for changes in nitrogen to phosphorus was still about 16:1, based on the GEOSECS data from the Atlantic and Indian oceans. Indeed, the Atlantic average was a bit higher than 16:1, and the overall average of 16:1 could have been influenced by the denitrification that occurs in the Arabian Sea.
3. Studies of changing water mass properties in productive coastal upwelling regions suggest a 16:1 ratio of uptake during phytoplankton growth (e.g., Minas et al. 1986), and deviations from this ratio during

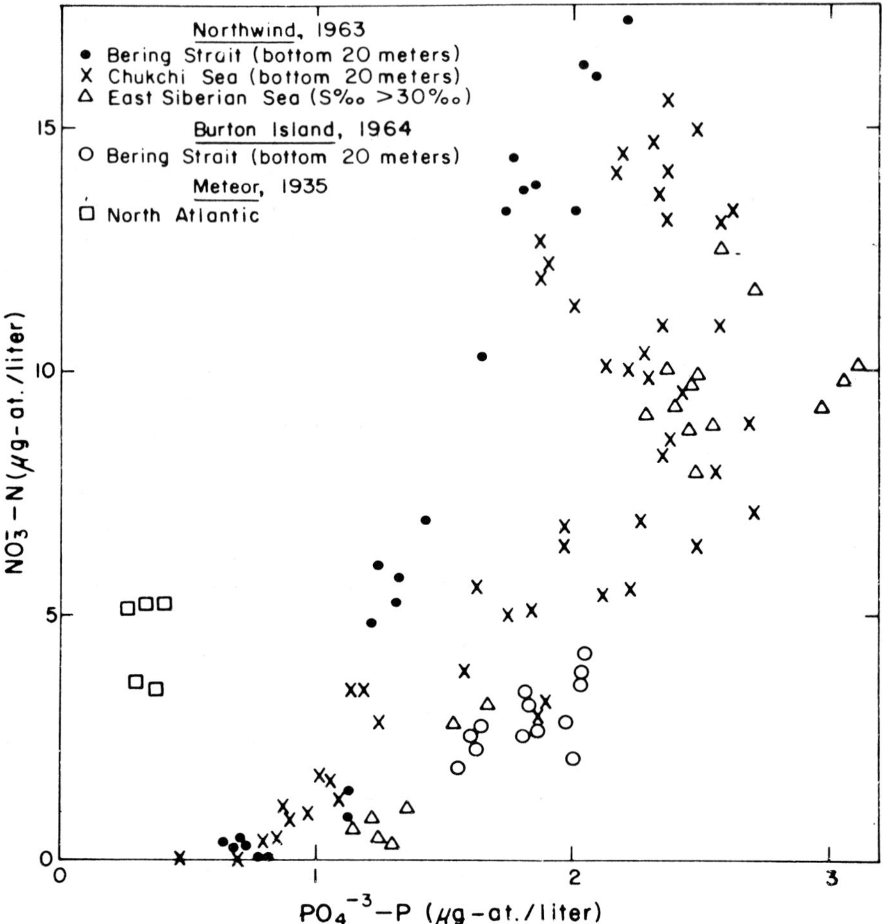

Fig. 2—Data from the vicinity of Bering Strait showing some extremely low nitrate/phosphate ratios.

regeneration may be relatively confined in time and space and arise from temporary imbalances.

4. Some variations could also arise from "luxury" consumption of phosphorus after nitrogen becomes limiting (Redfield et al. 1963) and from the incorporation of phosphorus into biogenic calcium carbonate, processes which are essentially de-coupled from nutrient limitation.
5. Studies of the ratios in British Columbian fjords and the Baltic, regions that are heavily influenced by denitrification (Shaffer 1987; Smethie et al. 1987), suggest a lower ratio of uptake, but these regions are not typical of most of the ocean. For example, salinities are relatively low

and it is possible, for example, that there could be significant terrestrial influences.

Inspection of the data presented in Fig. 3, Table 2, and in GEOSECS data from the Atlantic suggests that the overall nitrate to phosphate *concentration* ratio in the ocean is about 15:1 or perhaps a bit less, but there are significant geographical variations. High ratios (~22:1) are found in the Mediterranean and Red Seas, where relatively high nitrogen fixation rates have been suggested by Bethoux and Copin-Montegut (1986) and by Naqvi et al. (1986). The former authors suggest that the high ratios arise from the decomposition of seagrasses, which they suggest are primarily responsible for the relatively high nitrogen fixation rates which they believe occur in the Mediterranean Sea. They further suggest that the pelagic biota have the "normal" ratio of 16:1. Lowest ratios occur in portions of the Pacific and Indian Oceans that are influenced by denitrification within the water column. Off Peru, extremely low ratios can be encountered during episodes of complete denitrification (Table 2).

Within the photic zone, extreme "nitrate"/"phosphate" ratios (N + N in Fig. 3 = $NO_3^- + NO_2^-$ and P = phosphate) can occur when either available P or N is exhausted or is close to exhaustion, and the tendency for generally low photic zone N:P ratios in most of the oceanic surface layer is suggested by the data in Table 2 and Fig. 3. If we neglect the low ratios near the surface, the Atlantic ocean proper tends to have a ratio of about 15:1 and the Indo-Pacific ratio is a bit lower (GEOSECS data and Fig. 3). It is important to note that the high ratios tend to occur in zones where overall nutrient concentrations are low and that the lower ratios occur in zones where nutrient concentrations are high; just examining the "raw" ratios tends to overstate the importance of the high ratios in the Mediterranean and Red Seas.

For the sake of argument, let us assume that the overall 16:1 ratio of uptake cannot vary and that the nitrate:phosphate ratio for the ocean is ~14.5:1 with a ratio of ~15:1 in the Atlantic. A ratio of ~14.5:1 suggests a potential inhibition of export and new production in the present-day ocean of ~10%, primarily due to an excess of denitrification over nitrogen fixation. Berger and Keir (1984) suggest that only a 30% change is necessary to explain the pCO_2 variations in ice-cores.

A plausible explanation for the higher N:P ratio in the Atlantic relative to the rest of the ocean is that the Atlantic has a higher nitrogen fixation/denitrification ratio. This is a reasonable assumption given the generally high oxygen concentrations in the Atlantic and the evidence suggesting high nitrogen fixation in the Mediterranean. If we assume that waters with relatively low N:P ratios enter the Atlantic and have their ratios increased from about 14.4:1 to 15:1 by an excess of nitrogen fixation over denitrification,

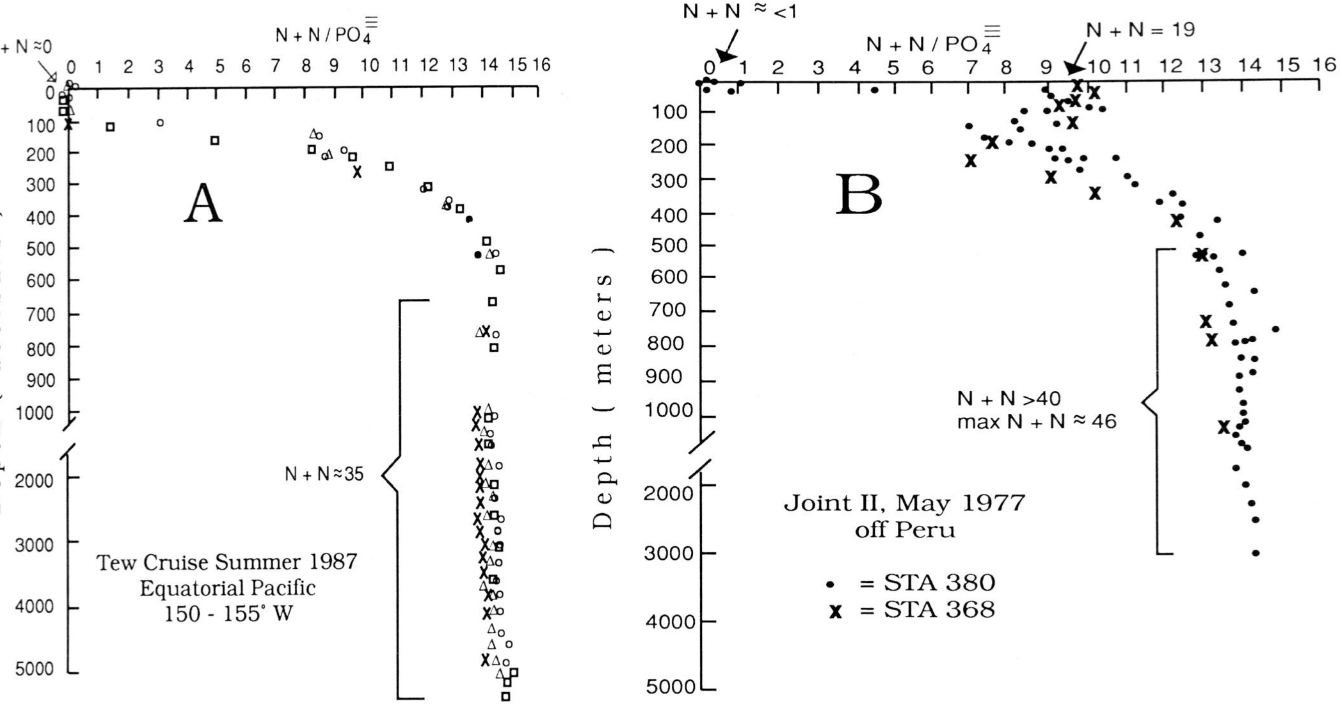

Fig. 3—Nitrate + Nitrite (N + N)/Phosphate ratios (by atoms of N and P) from the equatorial Pacific Ocean taken during the TEW experiment in summer 1987 (A) and from the ocean off Peru during the JOINT II experiment (B). The latter data are from May 1977 and are reported by Hafferty et al. (1978). Low ratios occur in the photic zone/mixed surface layer which is deep (~100 m) in the equatorial Pacific but very shallow (~20 m) off Peru. Our assumption is that the relatively low ratios arise ultimately from denitrification. The stations most heavily influenced by local denitrification are station 368, which occurs over the continental slope near 15°S off Peru, and station 380, about 350 km from shore (B).

we can estimate the rate of this excess nitrogen fixation in the following way:

1. Broecker and Li (1970) estimate the rate at which water is traded between the Atlantic and the rest of the ocean as ~15 Sv (Sv=10^6 m s^{-1}).
2. The average phosphate concentration in the North Atlantic is ~ 1.5 μM.
3. Using the change in ratios from 14.4 to 15:1 gives a net nitrate increase in the North Atlantic of ~1 μM, and multiplying by the volume of water traded (15 Sv) gives an annual excess of nitrogen fixation in the North Atlantic of about 7 Tg N yr^{-1}. Interestingly, this is similar to the rate of nitrogen fixation one might expect for the Atlantic based on Carpenter and Capone (1983).

A problem in interpreting the N:P ratio information arises from the recent work of Sugimura, Suzuki, and Itoh. As reported elsewhere in this volume (see Toggweiler), they found higher values for dissolved organic nitrogen (DON) and dissolved organic carbon (DOC) than are typical of much of the existing literature and suggest that the fraction that has been "missed" may be labile. An important point is that the new data suggest that DON is maximal in the photic zone. Therefore, if the new data "stand up," one effect on the previously "accepted" paradigm will be to downplay the loss of organic nitrogen from the photic zone by a particulate flux since a significant loss term would also arise from the vertical DON gradient. *The total loss should still, however, be supported predominantly by the upward flux of nitrate.*

CONCLUSIONS

The existing data do not permit a conclusive resolution to the question posed in the title of the paper, but I was invited to be controversial and have tried to make a case for nitrate as the master chemical variable limiting export productivity throughout much of the ocean. To be sure, there are environments where nitrogen fixation keeps up with the demand, such as the Red and Mediterranean Seas and the tropical lagoons studied by S.V. Smith and colleagues. As noted above, however, these are environments of low export production and probably of enhanced phosphate removal. The underlying cause for nitrogen limitation, in my opinion, is the demonstrated ability of denitrification rates to change rapidly, versus the factors that appear to inhibit nitrogen fixation in the open ocean.

To prove me wrong, it must be demonstrated that nitrogen fixation rates in the ocean are much higher than reported in the existing literature.

Martinez et al. (1983) suggest that open ocean nitrogen fixation rates have been underestimated due to the destruction of needed structure with the "normal" bottle techniques, but their results have been challenged (Scranton et al. 1987) and not yet reproduced. Clearly, further attempts to measure oceanic nitrogen fixation with the careful collection techniques used by Martinez et al. are a high priority item if we are to resolve this issue. Similarly, a closer look at nitrogen fixation associated with seagrasses in the Mediterranean may be in order. Refining my crude calculation for the "excess" nitrogen fixation rate in the Atlantic is also feasible and might prove useful.

Our review of the factors that favor nitrogen fixation suggests that periods of oceanic evolution that favor the existence of warmer temperatures, weaker winds, and reasonably warm shallow seas with seagrasses would favor phosphate limitation of export production. Since some of the great phosphorite deposits are thought to have formed during such periods (Froelich et al. 1982), I believe that there were stages of oceanic evolution when phosphate may have been the master chemical controlling export production. Warm temperatures and low winds, however, should also reduce the oxygen flux from the atmosphere into the ocean which could favor enhanced denitrification, so it may be difficult to "sort out" the effects of the competing processes in past oceans. There is no question that denitrification has at least a local effect on new and export production in the present-day ocean in regions such as Peru (Table 2).

I would have liked to conclude this paper with a comment on whether or not the evidence suggests that the marine combined nitrogen inventory was higher or lower than at present during the last glaciation, but I cannot. My instinct is that the inventory was higher since this would be in line with the atmospheric carbon dioxide record, and with a reduction in marine denitrification that could be attributed to the reduced area of shelves (Christensen et al. 1987). However, this notion seems contrary to those models that suggest more oxygen depletion in the glacial ocean, and I do not know how the input terms may have varied. Since high denitrification rates appear to require almost zero oxygen concentration and a good supply of nitrate and organic material, an *average* water column oxygen concentration that is lower than the present value does not automatically signal an increase in marine denitrification. With extreme stagnation, for example, such as we see in the present-day Black Sea, sulfate respiration dominates and produces ammonium, which is readily available to plants but an unsuitable substrate for denitrification.

Acknowledgements Financial support for my studies of the marine nitrogen cycle has been provided by the National Science Foundation, The Office of Naval Research, and the Monterey Bay Aquarium Research Institute.

Technical and typing support for this paper has been generously given by P. Colby, A. Sanico, and J. Rollins who have put up with my impossible demands with both humor and grace. I am particularly indebted to Drs. J. Ammerman, W. Berger, P. Bienfang, G. Jackson, B. Mycke, V. Smetacek, and J. Walsh for helpful criticisms and advice.

REFERENCES

Ammerman, J.W., and Azam, F. 1985. Bacterial 5' – nucleotidase in aquatic ecosystems: a novel mechanism of phosphorus regeneration. *Science* **227**: 1338–1340.

Berger, W.H. 1970. Biogenous deep-sea sediments: fractionation by deep-sea circulation. *Geol. Soc. Amer. Bull.* **81**: 1385–1402.

Berger, W.H., and Keir, R.S. 1987. Glacial-holocene changes in atmospheric CO_2 and the deep-sea record. In: Climate Processes and Climate Sensitivity, eds. J.E. Hansen and T. Takahashi. Amer. Geophys. Union. *Geophys. Monogr.* **29**: 337–351.

Bethoux, J.P., and Copin-Montegut, G. 1986. Biological fixation of atmospheric nitrogen in the Mediterranean Sea. *Limnol. Ocean* **31**: 1353–1358.

Broecker, W.S., and Li, Y.H. 1970. Interchange of water between the major oceans. *J. Geophys. Res.* **75**: 3545–3552.

Carpenter, E.J., and Capone, D.G. 1983. Nitrogen in the Marine Environment. New York: Academic Press.

Chavez, F.P., and Barber, R.T. 1987. An estimate of new production in the equatorial Pacific. *Deep-Sea Res.* **34**: 1229–1244.

Christensen, J.P.; Murray, J.W.; Devol, A.H.; and Codispoti, L.A. 1987. Denitrification in continental shelf sediments has a major impact on the oceanic nitrogen budget. *Glob. Biogeochem. Cyc.* **1**: 97–116.

Codispoti, L.A., and Christensen, J.P. 1985. Nitrification, denitrification, and nitrous oxide cycling in the eastern tropical South Pacific Ocean. *Marine Chem.* **16**: 277–300.

Codispoti, L.A.; Friederich, G.E.; Packard, T.T.; Glover, H.T.; Kelly, P.J.; Spinrad, R.W.; Barber, R.T.; Elkins, J.W.; Ward, B.B.; Lipschultz, F.; and Lostaunau, N. 1986. High nitrite levels off Northern Peru: a signal of instability in the marine denitrification rate. *Science* **233**: 1200–1202.

Copin-Montegut, C., and Copin-Montegut, G. 1983. Stoichiometry of carbon, nitrogen, and phosphorus in marine particulate matter. *Deep-Sea Res.* **30**: 31–46.

Froelich, P.N.; Bender, M.L.; Luedtke, N.A.; Heath, G.R.; and De Vries, T. 1982. The marine phosphorus cycle. *Am. J. Sci.* **282**: 474–511.

Garside, C. 1985. The vertical distribution of nitrate in open ocean surface water. *Deep-Sea Res.* **32**: 723–732.

Goldman, J.C. 1986. On phytoplankton growth rates and particulate C:N:P ratios at low light. *Limnol. Ocean* **31**: 1358–1363.

Goldman, J.C.; McCarthy, J.J.; and Peavey, D.G. 1979.Growth rate influence on the chemical composition of phytoplankton in oceanic waters. *Nature* (Lon.) **279**: 210–215.

Hafferty, A.J.; Codispoti, L.A.; and Huyer, A. 1978. JOINT-II, R/V Melville Legs I, II, and IV; R/V Iselin Leg II Bottle data, March 1977–May 1977. Dept. of Oceanogr. Data Rep. 45. Seattle: Univ. of Washington.

Holland, H.D. 1978. The Chemistry of the Atmosphere and Oceans. New York: Wiley.

Hood, D.W. 1986. Processes and resources of the Bering Sea shelf. (PROBES). *Cont. Shelf Res.* **5(1&2)**.

Hood, D.W., and Kelley, J.J. 1976. Evaluation of mean vertical transports in an upwelling system by CO_2 measurements. *Mar. Sci. Comm.* **2(6)**: 387–411.

Howarth, R.W., and Cole, J.J. 1985. Molybdenum availability, nitrogen limitation, and phytoplankton growth in natural waters. *Science* **229**: 653–655.

Martinez, L.A.; Silver, M.W.; King, J.M.; and Alldredge, A.L. 1983. Nitrogen fixation by floating diatom mats: a source of new nitrogen to oligotrophic ocean waters. *Science* **221**: 152–154.

McElroy, M.F. 1983. Marine biologic controls on atmospheric CO_2 and climate. *Nature* (Lon.) **302**: 328–329.

Minas, H.J.; Minas, M.; and Packard, T.T. 1986. Productivity in upwelling areas deduced from hydrographic and chemical fields. *Limnol. Ocean.* **31**: 1182–1206.

Naqvi, S.W.A.; Hansen, H.P.; and Kureishy, T.W. 1986. Nutrient uptake and regeneration ratios in the Red Sea with reference to the nutrient budget. *Oceanol. Acta* **9(3)**: 261–275.

Paerl, H.W., and Prufert, L.E. 1987. Oxygen-poor microzones as potential sites of microbial N_2 fixation in nitrogen-depleted aerobic marine waters. *Appl. Env. Microbiol.* **53(5)**: 1078–1087.

Piper, D.Z., and Codispoti, L.A. 1975. Marine phosphorite deposits and the nitrogen cycle. *Science* **188**: 15–18.

Redfield, A.C.; Ketchum, B.H.; and Richards, F.A. 1963. The influence of organisms on the composition of sea water. In: The Sea, ed. M.N. Hill, vol. 2. New York: Interscience.

Riley, J.P., and Skirrow, G. 1965. Chemical Oceanography, Vol. I. London: Academic Press.

Ryther, J.G., and Dunstan, W.M. 1971. Nitrogen, phosphorus, and eutrophication in the coastal marine environment. *Science* **171**: 1008–1013.

Scranton, M.L.; Novelli, P.C.; Michaels, A.; Horrigan, S.G.; and Carpenter, E.J. 1987. Hydrogen production and nitrogen fixation by *Oscillatoria thiebautii* during *in situ* incubations. *Limnol. Ocean.* **32**: 998–1106.

Shaffer, G. 1987. Redfield ratios, primary production, and organic carbon burial in the Baltic Sea. *Deep-Sea Res.* **34(5/6)**: 769–784.

Smethie, W.M. 1987. Nutrient regeneration and denitrification in low oxygen fjords. *Deep-Sea Res.* **34 (5/6)** 983–1006.

Smith, S.V. 1984. Phosphorus versus nitrogen limitation in the marine environment. *Limnol. Ocean.* **29**: 1149–1160.

Stefánsson, U., and Atkinson, L.P. 1967. Physical and chemical properties of the shelf and slope waters off North Carolina. Beaufort, NC: Duke Univ. Marine Laboratory.

Takahashi, T.; Broecker, W.S.; and Langer, S. 1985. Redfield ratio based on chemical data from isopycnal surfaces. *J. Geophys. Res.* **90**: 6907–6924.

Productivity of the Ocean: Present and Past
eds. W.H. Berger, V.S. Smetacek and G. Wefer, pp. 395–406
John Wiley & Sons Limited

Paleoproduction and Atmospheric CO_2 Based on Ocean Modeling

R.S. Keir

Geological Research Division, A–015, Scripps Institution of Oceanography
University of California, San Diego
La Jolla, CA 92093, U.S.A.

Abstract. Since the discovery that the atmospheric CO_2 was about 80 ppm lower during glacial climates, several hypotheses as to the cause have been proposed. Box models of the atmosphere and ocean have been employed in these investigations, with varying spacial resolution of the ocean reservoir. The models "predict" nutrient-limited, net biological production from the efficient utilization of the upwelling nutrient flux, as well as predicting the atmospheric CO_2 content. These hypotheses produce different pictures of the Pleistocene ocean pattern of productivity.

Increases in ocean nutrient inventory or biomass C/P ratio simply result in global productivity increases. The result of a high-latitude increase in net production is dependent on the model circulation. In 3-box ocean models, with a Mediterranean type of high-latitude circulation, little change in low-latitude productivity results. In 10- to 13-box ocean models, where the Antarctic has an estuarine type of circulation, an increse of net production in this region causes a decrease in low-latitude productivity, and the decrease may be regional in nature. If the high-latitude biological production remains constant, decrease in convection in that region produces little change in low-latitude productivity.

Changes in the meridional circulation and in the vertical exchange of intermediate and surface water do not cause pronounced alteration of the atmospheric CO_2, but do change the low-latitude ocean productivity significantly. Thus, nutrient-limited production in the low latitudes may be the result of a combination of changes in the biogeochemical system, some of which affect the atmospheric CO_2, and some of which do not.

INTRODUCTION

In this paper, the terms *biological production* or *productivity* refer to the net particle flux downwards across the approximately 100 to 200 m depth horizon. Following the discovery that the atmospheric CO_2 content apparently was 80 to 90 ppm less in the last glacial climate than in the preindustrial present (Neftel et al. 1982; Barnola et al. 1987), various box models of the atmosphere and ocean have been developed to investigate possible mechanisms which might be responsible for this decrease. These models have not been constructed for the purpose of predicting the paleocean biological production explicitly, but rather to explore hypotheses which often result in an increase in the efficiency of the overall biological pump relative to the overall vertical ocean mixing. In general, the strategy has been to postulate the ocean circulation and the nonnutrient-limited biological production in high latitudes *a priori*, and then predict the atmospheric CO_2 as well as the global carbon and nutrient geochemical distributions. In addition, the organic carbon production in surface regions with nutrient limitation may be predicted from the model influx of dissolved nitrate or phosphate to that region.

In all but the simple 2-box ocean model (Broecker 1982), even a single parametric change can result in a combination of biological production- and CO_2 solubility-related, nonbiological effects on the atmospheric CO_2 (Volk and Hoffert 1985). The scope of this paper will be to summarize the changes in ocean biological production predicted by model scenarios that roughly reduce the atmospheric CO_2 content to the extent indicated by the ice cores.

THE SEDIMENTARY RECORD

Before discussing the modeling results, it will be useful to consider what one can find in the marine sedimentary record concerning Pleistocene ocean productivity. Cores taken from eastern margin upwelling regions show that higher organic carbon content and accumulation rates of silica and carbonate occur throughout the entire oxygen isotope stage 2–3 period, approximately 11 to 60 kyr B.P. (Arrhenius 1952; Müller and Suess 1979; Pedersen 1983). It thus appears that production of organic carbon was greater along upwelling margins during the glacial periods, although the greater organic carbon content may have, in part, been due to lower deep water oxygen concentrations (Emerson 1985). It is not clear what the organic carbon production was in other regions of the ocean, nor is it obvious what the global picture for calcium carbonate production was. Broecker and Peng (1986, 1987) claim that the glacial $CaCO_3$ production was greater, but this appears to be based on only two high resolution radiocarbon stratigraphies: core V19–188 in the Indian Ocean (Peng et al. 1977) and ERDC 92 in the western equatorial Pacific. Of these two cores, the latter has a large sand fraction, and the sedimentation rates may be altered by winnowing (W.H. Berger, pers. comm.). In contrast, Curry and Lohmann (1985) show that in the east equatorial Atlantic, calcium carbonate sedimentation is significantly reduced throughout the Stage 2–4 glacial period. This decrease shows up uniformly throughout the depth range of their cores, suggesting that a carbonate productivity decrease rather than an increase in dissolution is responsible. Although there is no major change in the average organic accumulation at this site over the last 150 kyr, the organic carbon accumulation rate increases with depth below 3.5 km during the Stage 2–4 period, and this is mirrored by a decrease in the $\delta^{13}C$ of *Planulina wuellerstorfi*. The depth gradient in organic carbon accumulation and benthic foraminiferal $\delta^{13}C$ during the glacial episode is not found during the Holocene or during interglacial Stage 5 (115 to 125 kyr B.P.).

In the Antarctic, the zone of diatomaceous ooze lying between 50°S to 60°S in the present ocean is displaced northward during the glacial climate periods (Cooke and Hays 1982). In the Atlantic and southwest Indian Ocean sectors, these oozes have much greater accumulation rates than other lithologic units over- or underlying them. Thus one finds, south of the present polar front, that a rapid accumulation of diatom ooze in the Holocene overlies glacial marine sediment with lower accumulation rates. North of the polar front the reverse is true, with more slowly accumulating calcareous ooze (above the compensation depth) overlying faster accumulating diatomaceous oozes of glacial age. At present, it does not seem possible to assess directly which climate period had the greater areally integrated diatom production rates in this important high latitude region.

Therefore, the picture that emerges so far is that biological productivity of organic carbon, silica, and possibly calcium carbonate increased along upwelling margins during glacial climate periods. Away from the continental margin, calcium carbonate production decreased in the eastern equatorial Atlantic, whereas the organic carbon productivity may not have changed appreciably. In other regions of the ocean, it simply is not clear how the productivity may have changed during the glacial climate. Broecker's (1971) oxygen isotope stratigraphy on IC-5 and RC11-209 in the Indian and Pacific Oceans shows little change in the carbonate accumulation rates. Since these cores may have been affected by dissolution, which was greater during the interglacial period in the Indian and Pacific Oceans, the implication is that carbonate production may have been slightly greater during the interglacial in the open Indian and Pacific.

PRESENT OCEAN MODEL CIRCULATIONS

In considering the dependence of the various calculations on the nature of the models, it will be useful to compare the idealized ocean circulations employed. The three main types of ocean configurations in use so far are the 2-box, the 3-box, and the approximately 10-box. The 2-box ocean simply consists of a surface and deep ocean with an exchange of water between them. Typical 3-box circulations used for the present ocean are shown in

Fig. 1. These circulations are similar, particularly in the nature of the meridional circulation, which has 15 to 25 Sv (1 Sv = 10^6 m^3 s^{-1}) moving from the warm to the cold surface ocean, sinking to the deep water and then recycling by upwelling into the warm surface. Thus, the 3-box models have assumed a Mediterranean type circulation for the high-latitude compartment.

Standard 10-box ocean circulations, used in conjunction with paleomodeling work (Broecker and Peng 1986; Keir 1988), are shown in Fig. 2. While the 10-box circulations are similar to each other, only the high-latitude North Atlantic has the Mediterranean nature of the 3-box patterns in Figure 1. The meridional circulation through the high-latitude Antarctic has an estuarine character, i.e., the net throughflow is by upwelling and outflow laterally. This is important because in a Mediterranean circulation, an increase in biological production does not significantly alter the subsurface nutrient distribution or the consequent productivity in other nutrient-controlled surface regions. An increase of production in an estuarine regime can cause major redistribution of nutrients, as preformed source concentrations for intermediate waters are depleted.

HYPOTHESES FOR THE GLACIAL ATMOSPHERIC CO_2

Broecker and Peng (1986) have summarized the different hypotheses (through 1985), as to how the atmospheric CO_2 may have been reduced by approximately 70 ppm between interglacial and glacial climates, and they discuss constraints and limitations of these mechanisms. Prior to the advent of their PANDORA model, the investigations had been based on the 2-

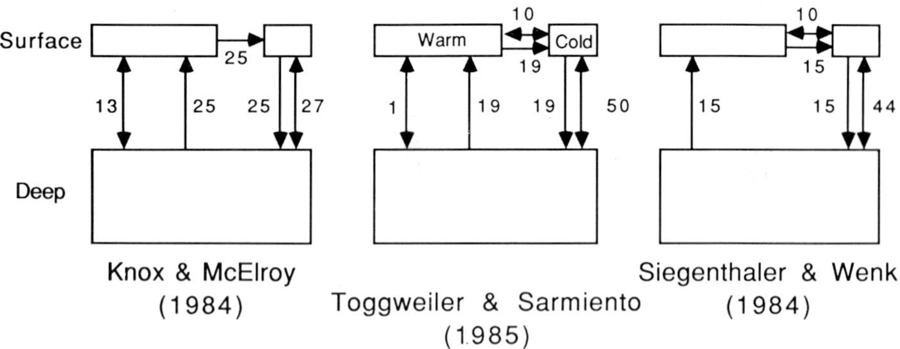

Fig. 1—Typical 3-box models of present ocean circulation. Numbers show transport in 10^6 m^3 s^{-1}.

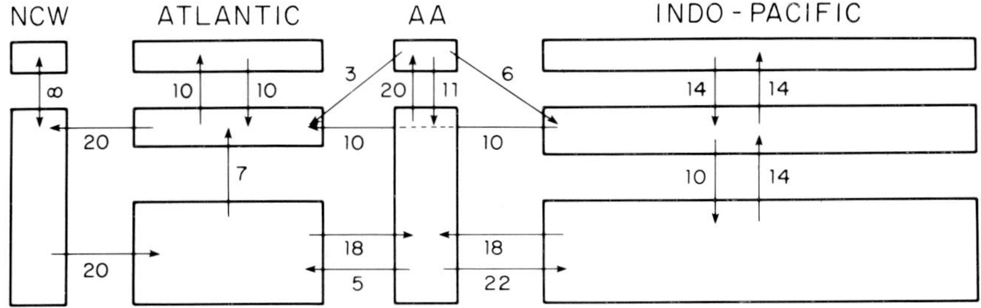

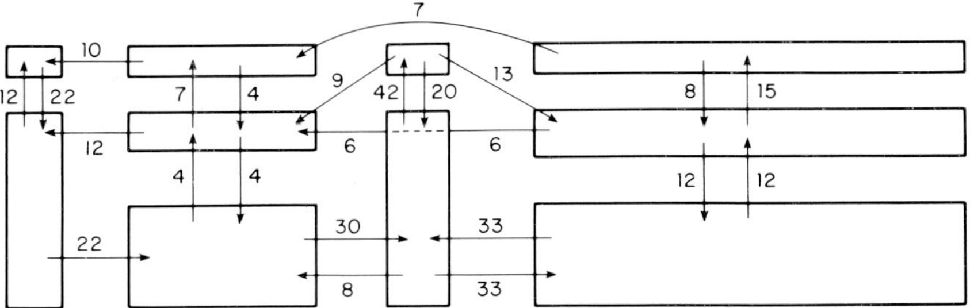

Fig. 2—10-box models of present ocean circulation. PANDORA from Broecker and Peng (1986); CYCLOPS from Keir (1988). The original configurations of both models have been simplified by combining reservoirs together where necessary to produce the common configuration shown. Transports in 10^6 m^3 s^{-1}.

and 3-box divisions of the oceans shown in Fig. 1. The hypotheses include the following:

1. Globally increased organic carbon production, due to an increase in either the ocean nutrient inventory or the C/P ratio of the particulate organic matter in the photosynthetic remineralization cycle (Broecker 1982).
2. A global decrease in calcium carbonate production.
3. An increase in organic carbon production at "high latitudes," where the potential for increased production exists because of light, rather than nutrient limitation (Knox and McElroy 1984; Sarmiento and Toggweiler 1984; Broecker and Peng 1987).
4. Decrease in high-latitude convection or vertical exchange with the underlying deep water (Siegenthaler and Wenk 1984; Ennever and

McElroy 1985; Toggweiler and Sarmiento 1985; Wenk and Siegenthaler 1985).

5. Combination of decrease in high latitude convection and meridional circulation (see references in pts. 3 and 4 above).
6. Combination of an increase in Southern (Antarctic) Ocean organic carbon production with either an increase in Antarctic Bottom Water influx to the Atlantic and/or decrease in North Atlantic Deep Water formation. The last hypothesis is a variation of pt. 3 and is based on a 13-box ocean model (Keir 1988).

MODEL-PREDICTED OCEAN PRODUCTIVITY

Table 1 lists typical model-predicted organic productivity changes together with the ranges of atmospheric CO_2 decreases for the hypotheses listed in the previous section. The model predictions for hypotheses 1 and 2 are elementary. The increase in nutrient inventory or organic C/P necessary to effect a 60 ppm decrease in atmospheric CO_2 is about 40% to 60%. This would cause a global increase in productivity of this magnitude, except possibly in the high latitudes if the production there is not a function of nutrient concentration at all. Support for this has been cited from upwelling margin regions and from the one or two radiocarbon stratigraphies mentioned earlier, but the evidence for a global production increase is lacking and is questionable in view of Curry and Lohmann's (1985) observations.

Broecker and Peng (1986) estimate that in order to effect a full 70 or 80 ppm drop in atmospheric CO_2 via calcium carbonate, the net global production would have to decrease to almost zero. As they observe, the sedimentary record clearly rules out such a possibility.

The effect that an increase in high-latitude production has on low-latitude areas depends significantly on the nature of the model circulation. For the 3-box Mediterranean circulations (Fig. 1), the warm surface ocean productivity changes very little (see Table 1; Knox and McElroy 1984; Toggweiler and Sarmiento 1985; Wenk and Siegenthaler 1985). For the CYCLOPS model in Fig. 2, the northern North Atlantic has a Mediterranean circulation, and increasing production in this region also has negligible effect on productivity in the other ocean regions. This is because surface regions with Mediterranean circulations tend to be upstream in the flow from the subsurface reservoir with the high nutrients. Increasing the productivity simply adds what very small fraction of the total nutrient inventory that the surface region has to the subsurface ocean.

In contrast to the above situation, when a large (factor of 5) increase in organic carbon production was applied to the Antarctic in the CYCLOPS model, nutrient-limited production in the other ocean regions decreased

TABLE 1 Model predicted change in biological production.

	Hypothesis	Ocean Boxes	% Chg in Production: Low Latitude Atl.	Low Latitude Pac.	Low Latitude Ave.	High Lat.	Chg in Atm CO_2 ppm	Refs
1.	Nutrient Inventory	2			+40%		−55 to −90	a,b,c
		4			+58%		−64	g
		10	+50%	+20%	+30%	+30%	−33 to −52	j
		13	+30%	+30%	−30%	+30%	−34	k
	C:P Ratio	2			−35%		−90	c
		10	+30%	+30%	−30%	+30%	−29 to −44	j
2.	Increase HL Production	3,4			~ 0%	+400%	−66	d,e,h,i
		10	0%	−30%	−25%	+80%	−14 to −28	j
		13	−46%	−21%	−27%	+380%	−124	k
3.	Decrease HL Convection							
	by 50 to 90%	3,4			~0%	0%	−45 to −80	f,g,h,i
	by 90%	13	−9%	−4%	−5%	0%	−34	l
4.	Decrease HL Convection and Meridional Circ.							
	37%:80%	3			−39%	0%	−59	h
	50%:50%	10	−75%	−70%	−70%	+5%	−36 to −62	j
5.	Increase AA Production + AABW to Atlantic							
	Const. Downwell	13	−42%	−32%	−36%	+220%	−106	k
	+12% Upwelling	13	−9%	−4%	−5%	+220%	−88	k

[a] Broecker (1982)
[b] Keir and Berger (1983)
[c] Broecker and Peng (1984)
[d] Knox and McElroy (1984)
[e] Sarmiento and Toggweiler (1984)
[f] Siegenthaler and Wenk (1984)
[g] Ennever and McElroy (1985)
[h] Toggweiler and Sarmiento (1985)
[i] Wenk and Siegenthaler (1985)
[j] Broecker and Peng (1987)
[k] Keir (1988)
[l] unpublished results

markedly but unevenly, with the Atlantic production declining the most (−45%) and the North Pacific the least (−8%). This is because of the estuarine nature of the throughput, where the surface reservoir is downstream from the reservoir with the high nutrients. Higher production increases the return of the upwelled nutrients to the circumpolar deep water, reduces the concentration in the surface source for the Antarctic Intermediate Water, and thereby reduces the upwelling flux of nutrients to the warm surface oceans. Boyle and Keigwin (1987) found that the Cd/Ca ratio of benthic foraminifera in the Caribbean decreased significantly during the last glacial period. They interpret this result as indicating a substantial decrease (about $-0.6\ \mu mol\ kg^{-1}$) in the Atlantic intermediate water phosphate concentration.

Broecker and Peng (1987) do not explicitly give the changes in production that result from sensitivity experiments performed on the PANDORA model. Results can be inferred for the Atlantic, Indo-Pacific, and Antarctic Oceans from the tabulated changes in surface phosphate concentration given in their paper. However, for the warm surface oceans, these are only given to one significant figure, and therefore the productivity change can only be discerned to within a range of ±17%. In an experiment where the Antarctic production was increased by 80%, it appeared that the model Pacific production dropped markedly (~ −50%), but the Atlantic productivity appeared not to change too drastically. Thus, a decrease in warm surface ocean production as a consequence of an increase in Antarctic production appears to be common to both models in Fig. 2, but in which ocean the decrease is more pronounced depends on the intermediate level circulation.

In contrast to an increase in high-latitude biological production, the effect on productivity of decreasing the high-latitude vertical exchange of water is only very weakly model dependent. Assuming that the high-latitude production remains constant, the average result for all models is that very little change in low-latitude nutrient limited production occurs (Table 1). In 3-box models with no warm-cold surface horizontal exchange, no change in low-latitude production occurs (Ennever and McElroy 1985; Toggweiler and Sarmiento 1985). In the 3-box model with warm-cold exchange, a small decrease results (Siegenthaler and Wenk 1984). Decreasing the Antarctic convention in the CYCLOPS model by 90% causes a small, unevenly distributed productivity decrease, ranging from −9% in the Atlantic to −1% in the North Pacific.

Decreasing the meridional circulation in the 3-box model tends to increase the atmospheric CO_2, but this is generally a weak effect, especially at low high-latitude convection rates (Toggweiler and Sarmiento 1985; see their Fig. 6a). These authors suggest that both the high-latitude convection and the meridional circulation were reduced, as this would reduce the atmospheric CO_2 and keep the warm surface ocean $\delta^{13}C$ constant in the face of a −0.5‰ shift due to the addition of dissolved CO_2 from an external organic carbon source. The warm surface ocean productivity decreases with decreasing meridional circulation, so that when this is combined with reduced high-latitude convection (which has little effect on low-latitude biological production), the net result is reduced low-latitude organic carbon productivity. In Toggweiler and Sarmiento's ice age scenario (1985; their Fig. 7), the reduction in low-latitude production is about a factor of 2. Broecker and Peng (1987) have a somewhat similar scenario for the PANDORA model where the low latitude reduction in productivity is reduced by a factor of 3 or 4.

Measurements of the Cd/Ca ratio in Atlantic Ocean benthic foraminifera indicate that the glacial increase in phosphate concentration becomes more pronounced as one proceeds southwards (and deeper) from the northern North Atlantic (Boyle and Keigwin 1987). This suggests that the relative proportion of the North Atlantic Deep Water and Antarctic Bottom Water sources shifted from a predominance of the former in the present ocean to about a 2:1 AABW:NADW source ratio during the glacial. When this change of circulation is combined with an increase by a factor of 2 or 3 in Antarctic biological production in the CYCLOPS model, the effect on the intermediate water is qualitatively the same as in the case of a simple Antarctic productivity increase. Nutrient concentrations in the intermediate water are reduced, although the asymmetry, where the Atlantic has the disproportionately greater reduction, is not as pronounced. Thus, if no other change in model circulation is made, the low-latitude ocean productivity is predicted to decrease on the average by about 35% (Table 1).

Arrhenius (1952) suggested that the ocean upwelling was greater during the ice ages due to a stronger zonal wind velocity field. The sedimentary records along ocean margins appear to be consistent with this idea. As the vertical exchange between the intermediate and warm surface ocean is increased in the CYCLOPS model, warm ocean organic carbon productivity increases. Ocean nutrient distribution and atmospheric CO_2, however, are not strongly affected. Thus the lower warm ocean productivity caused by an increased Antarctic production can be restored or overcome by concurrently increasing the upper ocean exchange. The last entry in Table 1 shows the predicted productivity when the upper ocean exchange is increased to the point where the total upwelling is 12% greater than in the standard present circulation.

A possible constraint on the average ocean productivity during the glacial may be found in the sedimentary record of calcium carbonate dissolution. In the long run, the ocean balances its alkaline input by dissolving the excess biogenic carbonate flux settling out of the surface. Therefore, over time scales of 10^4 yr, lower carbonate production results in better preservation on the seafloor (presupposing the alkaline input has not changed by a similar magnitude). If the warm ocean carbonate production is in fact a function of nutrient limitation, then the observation that the Indian and Pacific Ocean had better preservation during the glacial periods (Berger 1973) would seem to indicate that the average warm ocean production of carbonate, and possibly organic carbon, was somewhat lower in those periods. However, it may be that carbonate production during the glacial was lower for other reasons. Dymond and Lyle (1985) show evidence that the organic carbon to calcium carbonate ratio increases with increasing primary productivity. In the extreme case, this could result from a constant carbonate production over a wide region in which the organic carbon production varied with the flux of upwelling nutrients.

SUMMARY

The various hypotheses as to the cause of the lower atmospheric CO_2 during the ice ages yield differing predictions as to the change in low-latitude organic carbon production. Global ocean nutrient inventory or biomass C/P increases require globally increased production. Increased productivity in the northern North Atlantic causes no low-latitude productivity changes (and not much P_{CO2} change either), and also very little low-latitude production results from a decrease in high-latitude convection in general. An increase in biological production in the Antarctic region results in a significant reduction in low-latitude production due to the depletion of intermediate water nutrients, assuming no change in the ocean circulation.

The nutrient-limited biological production is sensitive to some aspects of the ocean circulation which do not cause very strong changes in the atmospheric CO_2. In models studied to date, this appears to be true for the meridional and the upper ocean vertical circulation. An increase of the vertical exchange due to greater wind velocity fields during glacial periods may superimpose upon other effects on low-latitude production. Thus, the net result of an increase in Antarctic biological production and an increase in upper ocean upwelling may be an increase in production of organic carbon along the upwelling margins, together with little change or a small reduction in productivity away from the margins. The greater preservation of calcium carbonate during the glacial climates suggests that its biogenic production decreased during those times.

Acknowledgements I thank J. Barron, R. Toggweiler, W. Berger and S. Emerson for reviews and comments. This work was supported by Electric Power Research Contract RP986–12 and NSF Grant OCE85–01578.

GLOSSARY

AABW	Antarctic Bottom Water
CYCLOPS	Carbon Cycle and Ocean Productivity Simulation; 13-box ocean model
NADW	North Atlantic Deep Water
PANDORA	11-Box Ocean model of Broecker and Peng (1986)
Productivity	Net particulate transport downward across the ocean pycnocline of biogenic components, such as organic carbon, $CaCO_3$ or silica.

REFERENCES

Arrhenius, G. 1952. Sediment cores from the East Pacific. *Rep. Swed. Deep-Sea Exped. 1947–1948* **5**:1–228.

Barnola, J.; Raynaud, D.; Korotkevich, Y.; and Lorius, C. 1987. Vostok ice core provides 160,000-year record of atmospheric CO_2. *Nature* **329**: 408–414.

Berger, W. 1973. Deep-sea carbonates: Pleistocene dissolution cycles. *J. Foram. Res.* **3**: 187–195.

Boyle, E., and Keigwin, L. 1987. North Atlantic thermohaline circulation during the last 20,000 years: link to high latitude surface temperature. *Nature* **330**: 35–40.

Broecker, W. 1971. Calcite accumulation rates and glacial to interglacial changes in oceanic mixing. In: Late Cenozoic Ages, ed. K. Turekian, pp. 239–265. New Haven: Yale Univ. Press.

Broecker, W. 1982. Glacial to interglacial changes in ocean chemistry. *Prog. Ocean.* **11**: 151–197.

Broecker, W., and Peng, T.-H. 1984. The climate-chemistry connection. In: Climate Processes and Climate Sensitivity, eds. J. Hansen and T. Takahashi. *Geophys. Monog.* **29**: 327–336. Washington, D.C.: Amer. Geophys. Union.

Broecker, W., and Peng, T.-H. 1986. Carbon cycle: 1985 glacial to interglacial changes in the operation of the global carbon cycle. *Radiocarbon* **28**: 309–327.

Broecker, W., and Peng, T.-H. 1987. The role of $CaCO_3$ compensation in the glacial to interglacial atmospheric CO_2 change. *Glob. Biogeochem. Cyc.* **1**: 15–29.

Cooke, D., and Hays, J. 1982. Estimates of Antarctic Ocean seasonal sea-ice cover during glacial intervals. In: Antarctic Geoscience, ed. C. Craddock, pp. 1017–1025. Madison: Univ. Wisconsin Press.

Curry, W., and Lohmann, G. 1985. Carbon deposition rates and deep water residence time in the equatorial Atlantic Ocean throughout the last 160,000 years. In: The Carbon Cycle and Atmospheric CO_2: Natural Variations Archean to Present, eds. E. Sundquist and W. Broecker. *Geophys. Monog.* **32**: 285–301. Washington, D.C.: Amer. Geophys. Union.

Dymond, J., and Lyle, M. 1985. Flux comparisons between sediments and sediment traps in the eastern tropical Pacific: implications for atmosphric CO_2 variations during the Pleistocene. *Limnol. Ocean.* **30**: 699–712.

Emerson, S. 1985. Organic carbon preservation in marine sediments. In: The Carbon Cycle and Atmospheric CO_2: Natural Variations Archean to Present, eds. E. Sundquist and W. Broecker. *Geophys. Monog.* **32**: 78–87. Washington, D.C.: Amer. Geophys. Union.

Ennever, F., and McElroy, M. 1985. Changes in atmospheric CO_2: factors regulating the glacial to interglacial transition. In: The Carbon Cycle and Atmospheric CO_2: Natural Variations Archean to Present, eds. E.Sundquist and W. Broecker. *Geophys. Monog.* **32**: 154–162. Washington, D.C.: Amer. Geophys. Union.

Keir, R. 1988. On the late Pleistocene ocean geochemistry and circulation. *Paleoceanography* **3**: 443–445.

Keir, R., and Berger, W. 1983. Atmospheric CO_2 content in the last 120,000 years: the phosphate extraction model. *J. Geophys. Res.* **88**: 6027–6038.

Knox, F., and McElroy, M. 1984. Changes in atmospheric CO_2: influence of the marine biota at high latitudes. *J. Geophys. Res.* **89**: 4629–4637.

Müller, P., and Suess, E. 1979. Productivity, sedimentation rate and sedimentary organic matter in the oceans. I. Organic carbon preservation. *Deep-Sea Res.* **26**: 1347–1367.

Neftel, A.; Oeschger, H.; Schwander, J.; Stauffer, B.; and Zumbrunn, R. 1982. Ice core sample measurements give atmospheric CO_2 content during past 40,000 years. *Nature* **295**: 220–223.

Pedersen, T. 1983. Increased productivity in the eastern equatorial Pacific during the last glacial maximum (19,000 to 14,000 yr. B.P.). *Geology* **11**: 16–19.

Peng, T.-H.; Broecker, W.; Kipphut, G.; and Shackleton, N. 1977. Benthic mixing in deep sea cores as determined by ^{14}C dating and its implications regarding climate stratigraphy and the fate of fossil fuel CO_2. In: The Fate of Fossil Fuel CO_2 in the Ocean, eds. N. Andersen and A. Malahoff, pp. 355–374. New York: Plenum.

Sarmiento, J., and Toggweiler, R. 1984. A new model for the role of the oceans in determining atmospheric pCO_2. *Nature* **308**: 621–624.

Siegenthaler, U., and Wenk, T. 1984. Rapid atmospheric CO_2 variations and ocean circulation. *Nature* **308**: 624–626.

Toggweiler, R., and Sarmiento, J. 1985. Glacial to Interglacial changes in atmospheric carbon dioxide: the critical role of ocean surface water in high latitudes. In: The Carbon Cycle and Atmospheric CO_2: Natural Variations Archean to Present, eds. E. Sundquist and W. Broecker. *Geophys. Monog.* **32**: 163–184. Washington, D.C.: Amer. Geophys. Union.

Volk, T., and Hoffert, M. 1985. Ocean carbon pumps: analysis of relative strengths and efficiences in ocean-driven atmospheric CO_2 changes. In: The Carbon Cycle and Atmospheric CO_2: Natural Variations Archean to Present, eds. E. Sundquist and W. Broecker. *Geophys. Monog.* **32**: 99–110. Washington, D.C.: Amer. Geophys. Union.

Wenk, T., and Siegenthaler, U. 1985. The high latitude ocean as a control of atmospheric CO_2. In: The Carbon Cycle and Atmospheric CO_2: Natural Variations Archean to Present, eds. E. Sundquist and W. Broecker. *Geophys. Monog.* **32**: 185–194. Washington, D.C.: Amer. Geophys. Union.

Standing left to right:
Hans Schrader, Hans Thierstein, Tim Herbert, John Barron, Rainer Gersonde, Rüdiger Stein, Bernd Mycke

Seated, left to right:
Alan Mix, Bill Curry, Lou Codispoti, Robin Keir

Productivity of the Ocean: Present and Past
eds. W.H. Berger, V.S. Smetacek and G. Wefer, pp. 409–428
John Wiley & Sons Limited

Group Report
Geological Reconstructions of Marine Productivity

T.D. Herbert and W.B. Curry, Rapporteurs
J.A. Barron
L.A. Codispoti
R. Gersonde
R.S. Keir
A.C. Mix
B. Mycke
H. Schrader
R. Stein
H.R. Thierstein

INTRODUCTION

A number of advances in recent years have spurred the study of paleoproductivity by geologists. The evidence from polar ice cores of substantial changes in past atmospheric carbon dioxide has made us more aware of the importance of the carbon cycle in global climate change. Geochronologies have improved greatly on a variety of scales: For example, isotope stratigraphy in the late Pleistocene allows resolution of events in the range of orbital variations, thousands to tens of thousands of years. We are therefore getting close to measuring processes on time scales of the mean ocean circulation and, more readily, of the residence times of nutrients and the carbonate ion in the late Pleistocene ocean. Global correlations over the last 84 Myr with time resolution shorter than 1 Myr should soon be possible, given improvements in biostratigraphy, magnetostratigraphy, and chemostratigraphy. The Deep Sea Drilling Project and its successor, the Ocean Drilling Program, have provided nearly global coverage of ocean sediments into the mid-Cretaceous. Recent drilling has finally allowed good stratigraphic control in the high-latitude southern oceans. With a background of global reconnaissance behind us and a framework of plate tectonics, we now know the first-order history of the oceans into the Cretaceous period and are able to pose focused questions about how the ocean, the Earth's climate, and evolution have interacted to leave the geological record of paleoproductivity.

The challenge to geologists is to exploit the unique features of the record. Although our view is one with lower time and spatial resolution than that of the biologists, we can see changes in productivity much larger than those existing within our lifetimes. Time resolution degrades as we look farther back in the past, but the magnitude of changes increases. For many events which imply changes in ocean productivity we cannot rely on simple statistical relationships within the modern world ("parameterization") because modern analogues do not exist. Ultimately, we must understand more fundamentally the processes that connect both primary and net productivity in the ocean to the preserved record. This meeting has made clear that only a small fraction of the total biological activity in the ocean is preserved in sediments because marine organisms are quite efficient at recycling nutrients and organic compounds (see Williams et al., this volume). Thus, our task as geologists is similar to that of archaeologists reconstructing civilizations from their discarded or abandoned remnants.

The geological perspective of the Earth system is one of constant change, as variables such as mean global temperature, basin connections, weathering inputs, and evolution become free to change on time scales of hundreds of thousands to millions of years. The types of problems geologists address tend to fall into discrete time frames. In the discussion that follows, we outline the geological approach to paleoproductivity in three sections. In the Pleistocene (the last 1.6 Myr), we emphasize the connections between export productivity and the cyclic patterns of glaciation. For the Cenozoic (last 66 Myr), we point to the emerging view of shifting spatial patterns of productivity and the existence of long-term polar cooling. For the Mesozoic interval (66–250 Myr), we illustrate large changes in the physical oceanography and patterns of carbon storage in the ocean, and point to the role of biological evolution. As very little Jurassic and older oceanic crust survives, the record of the marine biota and chemistry becomes more fragmentary (although there are many interesting questions to ask of the pre-Jurassic record of marine productivity, we choose not to address the more ancient record because none of us specialize in this area). We then summarize the strengths and weaknesses of our diverse observational approaches and propose strategies for future research that will advance current understanding of export productivity and the role of the biota in long-term biochemical variations of the oceans and atmosphere.

THE GEOLOGICAL RECORD OF EXPORT PRODUCTIVITY; PLEISTOCENE SCALES (0–1.6 MYR)

During the Pleistocene Epoch, the Earth's climate underwent cyclic fluctuations between two extreme conditions of ice volume. The oxygen isotope record in deep-sea sediments shows that the oscillation has occurred with frequencies characteristic of the Earth's astronomical variation in its configuration with respect to the sun, at periods of about 20, 40, and 100 kyr. We now can infer, from air trapped in ice cores, that the atmospheric CO_2 has been oscillating with similar frequencies (Barnola et al. 1987). Although it is by no means certain, it is possible that CO_2 and climate variation are linked in some way to variation of the marine biological export production of organic carbon and calcium carbonate (the "carbon pump"). The reconstruction of how this biological production may have varied throughout this epoch is a major problem confronting marine geologists and paleoceanographers today.

One of the unique features of Pleistocene paleoceanography is the effective interaction between scientists who gather data and those who model climate. Reconstructions of Ice Age sea surface temperature by the CLIMAP group (1976) help to pinpoint areas where glacial-interglacial productivity might have changed. General circulation climate model experiments have been run with geologically derived Ice Age boundary conditions (Manabe and Broccoli 1985). Immense improvements in the resolution of the Pleistocene marine time scale have been made by the orbital tuning approach of the SPECMAP project to the oxygen isotope record (Imbrie et al. 1984; Martinson et al. 1987). The current SPECMAP emphasis is to dissect the response of such variables as sea surface temperature, carbonate and opal flux, and deep water $\delta^{13}C$ to orbital forcing by looking at the phase relationships of these components relative to the worldwide oxygen isotope chronology.

Cooperation between ocean modelers and marine geologists has been especially close in studies of the carbon cycle. Increasingly complex geochemical box models have been devised to account for the low glacial CO_2 content of the atmosphere, and these have included predictions which are geologically testable (Broecker 1982; Sarmiento and Toggweiler 1984; Siegenthaler and Wenk 1984; Keir, this volume). The newer box models have focused our attention on the high latitudes, where nutrients are currently not completely consumed, as an important control valve on the global export of carbon to the deep ocean. The history of paleoproductivity in the Antarctic and Arctic is not yet clear. Large horizontal translation of the belt of diatom-rich sediments (generally believed to indicate high export fluxes) occurred between glacial and interglacial time in the high-latitude Southern Ocean. An areal integration to quantify total production of silica has not yet been done.

A major stumbling block in the high latitudes is the lack of a high-resolution chronostratigraphy which would allow regional and global correlation of patterns observed in cores and measurement of accumulation rates. Some progress may come from oxygen isotope measurements in opaline silica. This and other potential stratigraphic tools should be considered. Potential productivity indicators, such as those based on species distribution or organic carbon content, should be tried in addition to opal flux.

In the lower latitudes, a high-resolution stratigraphy is provided by the oxygen isotopic record in foraminifera. Here, several methods have been used to estimate both primary and export productivity, including organic carbon burial fluxes (Sarnthein et al. 1987), carbonate burial fluxes (Curry and Lohman 1985), and species distribution of planktonic foraminifera (Mix, this volume). Given the different predictions about warm-ocean productivity in the various carbon cycle models, it is critical that we continue to refine the estimates. Additional tools should also be developed, such as benthic foraminiferal species abundances (Altenbach and Sarnthein, this volume) and organic biomarkers (both for estimating fluxes of relatively well-preserved subsets of organic matter and for quantifying contamination with terrestrial carbon; Prahl, this volume).

Predictions of spatial patterns of productivity made by models of the carbon cycle can be tested by reconstructing paleoproductivity on a global scale. One of the major problems confronting us is to define the link between export productivity, defined as the flux of carbon actually leaving the photic zone, and the geological variables we commonly measure. The efficiency of the biological carbon pump in the ocean can, in principle, be measured using carbon isotopes in benthic and planktonic foraminifera. Shackleton and Pisias (1985) found that changes in the surface deep-ocean gradient in ^{13}C in one core precede changes in the oxygen isotope ice volume indicator. They inferred that the carbon cycle responds first, followed by glaciation. Possible implications are that the carbon system forces most of the glacial-interglacial climate changes through strong climate feedbacks, or that both ice and carbon systems respond independently to the same forcing, but that the carbon system responds faster.

There are several caveats at the present stage of our knowledge. It is not clear to what extent the isotopic indices in a single core reflect local or global effects. For example, local productivity effects could contaminate the ^{13}C record of either the planktic or benthic foraminifera. Even if the ^{13}C record is uncontaminated, there is large spatial variability in ^{13}C signals related to shifting patterns of water mass distributions. For example, the glacial-interglacial changes in Atlantic intermediate waters are of opposite sign to those of deep waters.

CENOZOIC TIMES SCALES (1.6–66 MYR)

Studies of paleoproductivity during the Cenozoic involve considerable problems in both temporal and spatial scaling. We are dealing with vast amounts of time, many orders of magnitude longer than the interval of time studied by most Quaternary paleoceanographers, yet our ability to resolve time in the Cenozoic is at best 100–500 kyr. At the same time, global sampling of discrete time intervals becomes progressively more difficult the farther back we go, especially in the high-latitude and continental shelf areas which we believe to be important in modern export productivity.

Considerable advances have occurred in our ability to tell geologic time during the last decade. Magnetic stratigraphy has been successfully applied to numerous deep-sea cores recovered at all latitudes by the Ocean Drilling Program and other coring programs. High-latitude biostratigraphies using diatoms have been considerably refined for the later Cenozoic (1.6–36 Myr), and these have been well correlated to the magnetic reversal time scale. Strontium isotope stratigraphy, correlated to the paleomagnetic time scale, shows considerable promise in dating carbonate sediments, especially during the last 40 Myr (DePaolo and Ingram 1985). Sea level sequence stratigraphy offers a model for correlating continental margin sediments (Vail et al. 1977). This hypothesis, which assumes globally synchronous changes in sea level, may allow identification of age-equivalent sedimentary facies across the paleoshelf and paleoslope. Consequently, we can begin to map out Cenozoic sediments and their biogenic components on increasingly more refined (and paleoceanographically significant) time scales.

The farther we go back into the Cenozoic, the greater the differences are from the modern world. The early Cenozoic was characterized by a relatively unrestricted equatorial flow of surface currents around the globe that promoted more equable climates between low- and mid-latitude areas (Barron and Baldauf, this volume). The lack of ice on Antarctica and the presence of barriers to circum-Antarctic water flow, including a more southerly position of Australia and closure of the Drake Passage between South America and Antarctica, diminished the importance of the high southern latitudes to early Cenozoic surface and deep water circulation, and also contributed to a reduced latitudinal thermal gradient. There is a need to model thermocline processes under weakened temperature stratification in order to understand how the flux of nutrients to the euphotic zone might have been modified under early Cenozoic conditions.

As the Cenozoic progressed, obstructions to circum-Antarctic surficial flow were removed, while barriers to free equatorial surficial flow were raised (e.g., in the eastern Mediterranean Sea and in Central America). These shifts in oceanic gateways and passages changed both surface and deep water circulation and steepened latitudinal thermal gradients during the Cenozoic (Douglas and Woodruff 1981; Barron and Baldauf, this volume). Ocean productivity and chemistry seem to have been affected. Global changes in the location of biosiliceous sediments occurred, both in the deep sea and along continental margins. Large changes in the calcite compensation depth in different ocean basins (Van Andel 1975) may trace shifts in vertical and horizontal nutrient gradients.

The challenge in studying the Cenozoic geological record is to identify whether these shifts in the distribution of sedimentary facies signal changes in the global input and output of silica and/or other nutrients, or whether they merely reflect shifting foci of deposition because of changes in ocean circulation. In order to address this question it is important to document and to date accurately Cenozoic paleoceanographical changes and to characterize each change with available tools (isotopes, faunal and floral assemblages). Many individual records need to be synthesized into a global picture. Clearly, the first patterns to emerge will be oversimplified; however, they will improve with time as better spatial and temporal coverage is obtained.

The vast expanse of Cenozoic time makes this task imposing. Considerable sentiment exists within our group that intervals of geologic time might be selected for more intensive study along the lines of the CNOP project (see Kennett 1985). One interval proposed for study was the early middle Miocene (15–14 Ma) during which a substantial cooling (+1‰ oxygen isotope shift or about 5°C) of deep waters occurred, presumably due to expansion of the East Antarctic ice cap. Core coverage for this interval is relatively good in most of the ocean basins, and it can readily be identified on the continental margins of the United States and Japan, as well as in the Paratethys region of Central Europe. Diatom biostratigraphy is also refined enough in both high northern and southern latitudes to allow global coverage of middle Miocene time. Such a "time slice" approach might pave the way for detailed work based on "Milankovitch" orbital time control, which would allow for more accurate construction of models of past productivity.

MESOZOIC ANOMALIES

When we look back to the Mesozoic, we see changed continental configurations, higher sea level stands, warmer global climates, and differences related to the evolution of the biosphere (see Thierstein, this volume). Apparently, little polar ice was on the planet during the Cretaceous period. Within this framework, the sedimentary record of the Mesozoic points toward several major global disturbances that are reflected by temporal and spatial shifts in oceanic productivity patterns. Three examples are:

1. The abrupt biological change at the Cretaceous-Tertiary boundary, about 66 Ma ago, during which 80–90% of the fossilized plankton went extinct. This event provides a unique opportunity to study the interaction between a highly disturbed biosphere and cycling of materials in the marine system. Time resolution for such short-term events, as well as

techniques to unscramble the effects of diagenetic alteration, will have to be improved to allow a better understanding of these sudden events.

2. Anomalies of the oceanic carbon cycle on various time scales suggested by the deposition of organic carbon-rich marine sediments ("black shales") in the middle Cretaceous (95–120 Myr). Globally high export productivity could be related to higher delivery of nutrients to the ocean because of accelerated weathering on the continents. However, the time scale of anoxic events was so brief (on the order of a few kyr in many cases) that it seems unlikely that changes in river input of nutrients to the ocean could have been an important mechanism. Alternatively, increased preservation and burial of the organic carbon fixed during photosynthesis may have been linked to lowered deep water overturning. Black shales may record the temporary storage of large amounts of dissolved carbon and nutrients in the deep Cretaceous ocean, which were then returned to the upper ocean during oxygenated, more vigorously circulating times. Uncertainties in the calibration of the Cretaceous time scale and paucity of geographic coverage of available sediment sequences, particularly in the Pacific and high latitudes, are hampering progress in our understanding of this major global disturbance in the carbon cycle.
3. The large scale and seemingly gradual changes in the deposition of biogenous skeletal materials in oceanic environments which occurred in the Mesozoic. Pelagic sediments overlying ophiolite sequences of Jurassic and older age consist almost exclusively of lithified opaline sediments (cherts). From about 140 Ma onward, the calcareous phytoplankton remains have dominated biogenous sedimentation on mid-ocean ridges. Are these changes related to evolutionary adaptations and/or changes in the oceanic nutrient cycling? Have diagenetic processes eliminated a former record of skeleton production as has been documented (Calvert 1974) in the late Neogene? Tracers of the past silica cycle are needed to reconstruct the origin of these cherts.

In studying the Mesozoic ocean, we are faced with the problem that we have few analogues in the modern world (both in the physical and biological regime) which apply to this ancient world. The Mesozoic does present, however, a chance to understand better the fundamental controls on nutrient cycling and marine productivity under changed climatic conditions.

PROXIES

There is no means of measuring directly past ocean circulation and the flux of carbon from the photic zone. Instead, geologists rely on various indirect measures of the processes of interest, loosely termed proxy indicators. At present, most proxies are based on comparison with surface primary productivity rather than export production, since the former has been the only type of data readily available. With the advent of sediment traps and expected progress in understanding the biological controls on export production, we can hope to achieve a more quantitative link of proxy indicators to the ocean carbon pump. Proxy indicators can be broken into two types: those that depend on ratios (isotopes, species composition) and those that depend on fluxes (calcite, opal, and organic carbon accumulation rates). The validity of the latter depend crucially on the accuracy of geological time scales which are used to convert closed sum data such as weight percent carbonate to mass fluxes.

Accumulation of Skeletal Parts

Geologists have long recognized that regional variations in sediment composition parallel surface primary productivity (Lisitzin 1972). Based on these empirical relations, past variations in sediment lithology have been interpreted to reflect changes in the location or intensity of export production.

Opal. Opaline sediments are found today only in areas of high productivity: the equatorial zone, the circumpolar current, and in marginal basins beneath the highly productive eastern boundary currents where these basins are isolated from major input of detrital sediments. It appears that diatom fluxes are a good measure of export production during blooms. Because the ocean is undersaturated everywhere with respect to opal, there is a major problem in relating the amount of opal preserved to its production. In today's ocean, the fraction of opal buried in sediments is only 1–5% of its flux from the surface layer. Local opal preservation depends on at least four factors: the global silica budget of the ocean, pore water dissolved silica concentration, the partitioning of silica between ocean basins, and the thermal history of the sediment column. Until improvements are made in modeling silica diagenesis, we will be far from achieving quantitative reconstructions of opal production on all time scales.

Nevertheless, there do appear to be major trends in the history of opaline sedimentation during the late Phanerozoic (see Barron and Baldauf, this volume). The geological interpretations would be improved by:

1. better constraints on the cycling of silica over geological time scales, including variations in hydrothermal input, volcanic input, and the transport of silica to the oceans from rivers,
2. improved time scales and broader geographic coverage which would allow accurate quantification of opal deposition over time,
3. improved indicators of the extent of opal dissolution in sediments, analogous to indices developed for calcium carbonate dissolution.

Calcium carbonate. The relationship between calcium carbonate accumulation and its production in the overlying water column is more easily understood because undersaturation occurs only in deeper parts of the ocean. Little dissolution of calcite occurs in the water column, so that the carbonate flux to the sediment is nearly equal to its export flux from the upper ocean. With proper site selection, it is therefore possible to obtain representative patterns of carbonate production in most areas. Geologists have generally considered carbonate accumulation as a proxy for export flux of organic carbon because the geographic distribution of carbonate accumulation in deep-sea sediments parallels productivity belts. Recent time series sediment trap data show a good correlation between seasonal variations in the flux of organic carbon and calcium carbonate (Deuser et al. 1981). However, we must consider that carbonate may be partitioned between members of both "regenerative" and "bloom" communities—with planktonic foraminifera falling into the former group, and some coccolithophores into the latter. Such a view implies a more subtle relationship between bulk carbonate flux and export productivity, and suggests that more attention should be given to the relative contributions of planktonic foraminifera and calcareous nannofossils to carbonate content in sediments.

Geologists assume that the relationship between carbonate and organic production has remained constant when they reconstruct past productivity. However, because this rain ratio is under ecological and evolutionary control, it could be modified by climate change and the evolution of the microfossil groups. Planktonic foraminifera, which now contribute 30–40% of the carbonate content of pelagic sediments, generally account for only a few percent of the carbonate fraction in Cretaceous strata. The ratio of opaline to calcite primary producers has probably also increased over the same time as diatoms have become more important members of the photosynthetic community. A decline in coccolith carbonate flux may therefore have compensated for the increased foraminiferal contribution. Reported carbonate accumulation rates for early Cenozoic-Cretaceous sites above the calcite compensation depth fall between 0.5 and 2.5 $g/cm^2/kyr$, in the same range as the modern ocean.

The ecological and evolutionary controls on carbonate production deserve more scrutiny because of the important role carbonate cycling in the ocean plays in the carbon pump. Carbonate production in the surface water partially counteracts the photosynthetic drawdown of CO_2 (Broecker and Peng 1982), and its regeneration at depth modifies the alkalinity of the deep waters which will eventually exchange carbon with the atmosphere.

Organic Carbon

Bulk accumulation measurements. In areas removed from downslope transport of fine-grained material, it has been assumed that the flux of organic carbon reaching the sediment is proportional to its net export from the surface layer. Sediment trap studies indicate a strong depth dependence on the vertical carbon flux because of organic carbon degradation in the water column. We know that only a small fraction of the organic matter which hits the sediment is ultimately buried. Much of what is buried is likely to be refractory material derived from land plants blown or washed into the ocean, and not representative of marine production. It should be emphasized that total organic carbon is not necessarily equivalent to the sum of identifiable terrigenous and marine carbon. Microbial organic carbon plays a significant role in sediments and there is also an inert fraction that cannot be assigned to either of the other fractions due to its high maturity.

The extent to which factors such as oxygenation of deep water and sedimentation rate affect the ratio of carbon burial to flux needs to be assessed (see Jumars et al., this volume). Berner's (1982) survey of organic carbon burial finds that only a few percent of the net carbon storage in sediments occurs today in the vast pelagic realm. The geological record into the mid-Mesozoic indicates that, with a few conspicuous exceptions, deep-sea sediments have not been an important sink for organic carbon removed from the photic zone. The history of organic carbon storage in continental margins and slopes deserves greater attention.

Biomarkers. Organic compounds which are diagnostic of certain organisms present an opportunity to distinguish the relative contributions of terrestrial and marine organic matter in sediments (see Bruland et al. and Jumars et al., both this volume). Biomarkers may allow geologists to expand their knowledge of organisms that lived in the water column beyond those with fossilizable skeletons. Current organic geochemical techniques degrade a portion of the high molecular weight fraction of the sedimentary carbon. Unfortunately, the solvent extractable fraction may not necessarily be representative of the total organic matter. Future research should lead to more efficient degradation of the high molecular weight fraction without loss of information in the degradation products. Promising biomarker research may help determine glacial-interglacial changes in export productivity and the origin of intervals of monospecific fossil content observed in sediment cores.

Species Indices

The use of species distributions has great promise for reconstructing paleoproductivity (e.g., Mix, this volume; Altenbach and Sarnthein, this volume). Many uncertainties remain, however. Our knowledge of the biology of the common fossil groups is limited. This is especially true for foraminifera, radiolarians, and coccoliths, but it is also true for preserved fractions of the diatom flora. It is likely that the planktonic assemblage composition reflect primary productivity rather than export productivity, but the exact relationships are uncertain. Benthic species may reflect the influx of carbon to the seafloor and hence export production if the vertical scaling laws derived from sediment trap data are found to be valid (see Bishop, this volume). For all of the species indices, there may be preservation effects that limit our ability to see true biological relationships. In longer time frames, evolution of species will limit the utility of this type of index. Another challenge will be to isolate the effects of productivity on species distributions from other environmental effects on ecology (e.g., temperature, salinity, etc.). Species-based studies will have to be tried on multiple fossil groups with different production and preservation properties, and refined as we learn more of the biology of the fossilized portion of the biota.

Carbon Isotopes

Carbon isotopic records are among the most important tracers of Ice Age CO_2 changes and Pleistocene oceanic chemistry (Shackleton and Pisias 1985), long-term changes in the burial ratio of carbonate and organic carbon in the ocean (Lasaga et al. 1985), and the partitioning of carbon between terrestrial, atmospheric, marine, and sedimentary reservoirs. Marine plankton preferentially extract ^{12}C relative to the ambient $^{13}C/^{12}C$ seawater concentration during photosynthesis. The remaining dissolved carbon pool in surface waters is therefore shifted to heavier carbon isotopic values (positive shift in $\delta^{13}C$). In turn, respiration of falling organic matter returns lighter organic carbon to the dissolved carbon pool of the deep ocean. Thus the partitioning of carbon isotopes between surface and deep waters depends strongly on the ocean nutrient concentration and its effective utilization. Geologists use the carbon isotopic composition of the carbonate skeletons of planktonic and benthic foraminifera as indicators of the carbon isotopic composition of surface and deep reservoirs. As long as "Redfield" C:N:P ratios apply, $\triangle\delta^{13}C$ is also a proxy for dissolved nutrient content.

Large scale geological changes in organic carbon burial and climate are reflected in geological carbon isotope records. Coupling of the carbon cycle and global climate on a Cenozoic time scale may be recorded by the "Monterey carbon excursion" of Vincent and Berger (1985). A major enrichment ($+0.8^{o}/_{oo}$) of the $\delta^{13}C$ of foraminiferal shells occurred in all the major ocean basins between 17.5 and 14 Ma. This enrichment coincides precisely with the onset of deposition of the biosiliceous (and hydrocarbon-rich) Monterey Formation of California and its many analogs in continental margin basins around the Pacific. Vincent and Berger (1985) invoke vast storage of organic carbon in these marginal basins to explain the $\delta^{13}C$ excursion. They propose a feedback mechanism wherein the enhanced carbon storage lowered atmosphere CO_2 content and triggered the buildup of Antarctic ice. Such a mechanism links productivity and environmental change over a time scale of millions of years, and could explain the middle Miocene $\delta^{18}O$ enrichment of deep waters.

Dean and co-workers (1986) raise the interesting possibility that the marine organic carbon isotopic composition was more negative during the mid-Cretaceous because of substantially higher atmospheric CO_2. Their interpretation relies on prior laboratory experiments which measured the $\delta^{13}C$ of algal organic carbon as a function of artificially maintained CO_2 levels. A global environmental change of another sort appears to be recorded by carbon isotopes across the Cretaceous-Tertiary boundary. Surface to deep carbon gradients fall to near zero in sediments just above the extinction boundary (Hsu and McKenzie 1985). The isotopic evidence suggests that a major drop in export productivity lasted for a few tens of thousands to perhaps 100,000 years after the K/T event.

Reconstruction of paleoproductivity via carbon isotopes becomes less reliable the farther back we go in the geological record. With species now extinct, we can only indirectly verify the faithfulness of the carbon signal in their skeleton to the ambient water $\delta^{13}C$. The effects of evolution are particularly important when trying to evaluate bulk rock $\delta^{13}C$ measurements, as have often been made at the K/T boundary. In this case, the Cretaceous foraminiferal and coccolith species do not continue across the event, so that isotopic measurements are being made on somewhat different recorders. In addition, there is a crucial need for the development of objective methods for assessing the extent of diagenetic modification of carbonate $\delta^{13}C$ from the precipitation of "light" methanogenically derived carbonate and reequilibration of carbonate $\delta^{13}C$ with pore waters during recrystallization.

Nutrient Proxies

At least four nutrients act as controls on export productivity and sedimentation of biogenic particles in the ocean: combined nitrogen, phophorus, iron, and silica. In the present-day ocean, N and P concentrations in the North Atlantic are about half as large as in the North Pacific, but the North Atlantic silica concentrations are an order of magnitude lower than some places in the North Pacific. The inter-ocean nutrient fractionation is controlled by the interaction of horizontal circulation patterns and vertical nutrient gradients, and some calculations suggest that only small changes in circulation are required to cause large changes in horizontal nutrient gradients. For this reason, it is essential to have useful proxies for nutrient concentration that can be examined in the sedimentary record.

With some exceptions (e.g., zones of denitrification), combined nitrogen and biologically available phosphorus (BAP) have similar horizontal and vertical gradients in the sea when the gradients are normalized against their uptake ratio by the biota (the "Redfield" ratio, ca 16:1 by N/P atoms), but the processes by which these elements enter and leave the ocean differ significantly. Nitrogen removal may be affected strongly by processes that enhance denitrification, such as lower oxygen content of ocean waters. Long-term control of the oceanic phosphate concentration is probably set by the inorganic precipitation of phosphorite minerals. In addition, the residence times of BAP and combined nitrogen in the ocean differ by an order of magnitude, being about 10^5 years for BAP and about 10^4 years for combined nitrogen (McElroy 1983; Froelich et al. 1982). Combined nitrogen limitation exerts a greater influence on export productivity in the present-day ocean (Codispoti, this volume), but BAP could have assumed this role at other times.

One nutrient proxy, the Cd/Ca ratios in foraminifera, has already proven useful, and a second method, $^{14}N/^{15}N$, is being tested. Boyle (1986) has shown that the Cd concentration in seawater is proportional to the dissolved phosphate level and that the Cd/Ca ratio in the carbonate of foraminiferal shells accurately reflects the seawater Cd. The distribution of dissolved Cd in the pore waters of the upper few centimeters of the sediment column is not well known; it may be that some infaunal benthic forams do not give reliable indications of ocean water Cd (see Group Report 3).

Denitrification fractionates ^{14}N relative to ^{15}N by about 20 ‰. The short-term variations in this important nitrogen sink could produce large changes in the $\delta^{15}N$ of the ocean's nitrate. If a proxy for $\delta^{15}N$ can be developed, then an important constraint on past nitrate concentration can be achieved. One potential recorder is the $\delta^{15}N$ of the organic matrix of foraminiferal shells (Altabet and Curry, in preparation). Although preliminary results are encouraging, current analytical limitations on the minimum sample size are restrictive. Furthermore, trophic level interactions in the upper water column may alter the $\delta^{15}N$ consumed by foraminifera. This important line of research should be given high priority.

Only a fraction of the biota require silicon, but the oceanic dissolved silica distribution may be an important control on the silica that we find in the sedimentary record. To judge by the occurrence in ancient sediments of diatom and radiolarian tests, in which the original opal has been replaced by calcite or pyrite, it would seem that the ocean has been undersaturated with respect to opal at least into the Mesozoic. Most of the original opal record may therefore be missing. Consequently, the development of a proxy for dissolved silicon (residence time 10^4–10^5 yr) in the ocean will be required to trace the history of the silica cycle. One technique being evaluated is the germanium/silicon ratio in opal tests, where germanium acts analogously to a heavy Si isotope (Froelich and Andreae 1981; Murnane and Stallard 1988).

A resurgence of interest in iron as a limiting nutrient has been spurred by the work of Martin and Fitzwater (1988). The residence time for iron in the ocean is only about 10^2 years, and away from the margins iron is added to the euphotic zone by atmospheric dust. Since geologists are beginning to reconstruct the history of aeolian deposition in the oceans, we may be able to evaluate the effects of iron flux to the ocean on productivity. To date, we know of no geological studies where this has been applied.

FUTURE WORK AND RECOMMENDATIONS

As paleoceanography matures, a number of exciting opportunities will be fulfilled in the study of marine productivity past and present. Particularly productive work will probably come from interaction with scientists in allied disciplines such as physical oceanography and marine biology. Our group has identified a number of research topics which should be important in the next decade.

1. We need to develop a better understanding of the chemical processes which modify sediments at early and late stages of burial. The chemical data on early diagenesis are more extensive than for lithification processes; in particular, we need to know the extent to which carbonate and silica are mobile at later stages (see Jumars et al., this volume, on the effects of early diagenesis on the chemistry of benthic organisms and sedimented organic carbon).
2. Accurate reconstruction of past productivity will require better collaboration of geologists with marine biologists and oceanographers. Geologists need a much better understanding of the biology of microorganisms which leave a fossil record and their relationship to net productivity. In particular, we need to know much more about the ecology of coccolithophores and whether their flux from the surface waters correlates to export production. Additionally, it would be helpful to understand the ecology of certain species which have left nearly monospecific layers in sediment cores (the diatom *Ethmodiscus rex*, the coccolith *Braarudosphaera*, the dinocyst *Thoracosphaera*) and the relationships of these apparent blooms to perturbations in export production.
3. Geologists must collaborate with modelers of past and present biogeochemical cycles in order to test hypotheses of changes in past ocean net productivity. We must think in terms of dynamics and the time constants of various reservoirs which interact to make the sedimentary record. We must also interpret the sedimentary record in ways that will allow us to reconstruct dynamic quantities. This requires transforming closed sum data, such as weight percent carbonate and silica to mass flux per unit area per unit time. We also need theoreticians to include in their models variables that are geologically measurable, i.e., $\delta^{18}O$, $\delta^{13}C$, carbonate accumulation, etc.

4. We need to expand our global coverage of past ocean productivity data. We have little idea whether changes observed in proxy indicators in the geological record are global in extent, or rather involve latitudinal or basin to basin redistribution of high productivity zones. In particular, there is a crucial need for more sediment records in high latitude regions at all times, in continental margin seas, and in intermediate depths in the ocean (0.5–2 km). Some important questions to be addressed are: *High latitudes*. What is the history of siliceous sedimentation in high latitude regions? Can we improve our chronologic resolution in high latitude sedimentary sequences? How can we measure past sea ice extent from the geological record, and what influence did it have on productivity? Recent drilling has added information from the Antarctic region, but we still know almost nothing of the history of the Arctic Ocean, apart from a few shallow cores with abundant, well preserved Cretaceous diatoms (Kitchell and Clark 1982). We do not know whether the present Arctic has a silica and nutrient sink along the Soviet and Canadian margins and certainly have less idea about its past role in the global nutrient cycle.
   Continental margins. Can continental margin export production be included in geochemical productivity box models? How important have continental margins been as sinks for export production? Has productivity along continental margins changed synchronously with open ocean productivity? What are the effects of changing continental configuration and global sea level?
   Intermediate water masses. What is the extent of nutrient redistribution between intermediate and deep water masses in time? Have the changes hypothesized by Boyle and Keigwin (1987) during the last glacial maximum persisted into the more remote past?
5. We need to develop better proxy indicators for the nutrient chemistry of seawater, its dissolved oxygen content, and the silica cycles in the oceans. With rare exceptions, the past ocean has been above the threshold of anoxia. This places a powerful constraint on models of paleofertility and the levels of "preformed" nutrients in the past. We need better indicators of the downward particulate carbon flux, most probably to be found in the various biomarkers which are preserved in sediments (see Bruland et al. and Jumars et al., both this volume), and possibly in barite content (Bishop, this volume).
6. We need to understand better whether the carbon pump-atmospheric CO_2 link has been an effective amplifier of climate change in the past. Redox and calcium carbonate oscillations in pre-Pleistocene sediments suggest that the carbon cycle has been volatile on the same time scales characteristic of the Pleistocene (Fig. 1).
7. The potential of anoxic basins with laminated sediments to give records of changes in export productivity on decadal to millenial time scales

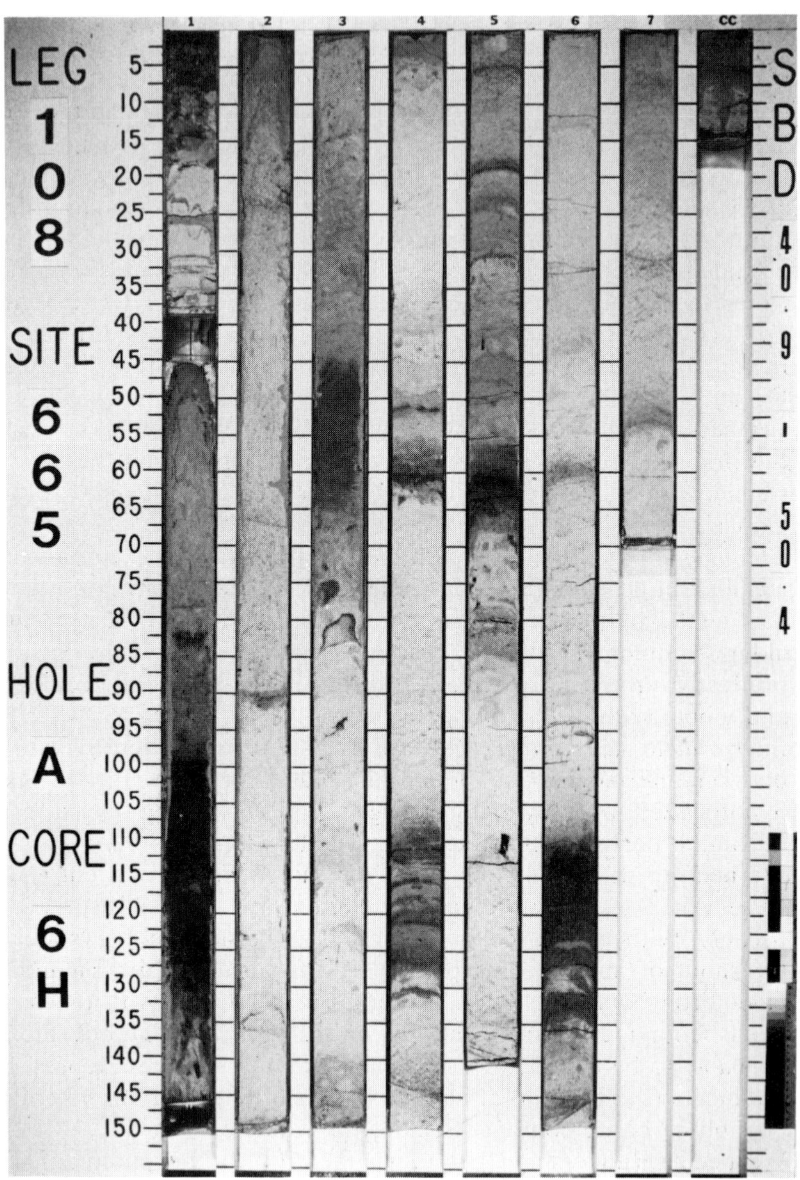

Fig. 1—Variability in the production and accumulation of organic carbon and calcium carbonate in the ocean is not limited to the late Pleistocene. Fig. 1a shows dark sediment bands indicative of oxygen-depleted deep water at ODP site 665A (latest Pliocene), while DSDP site 516F records cyclic oscillations in the production of calcareous organisms 75 million years ago (Fig. 1b).

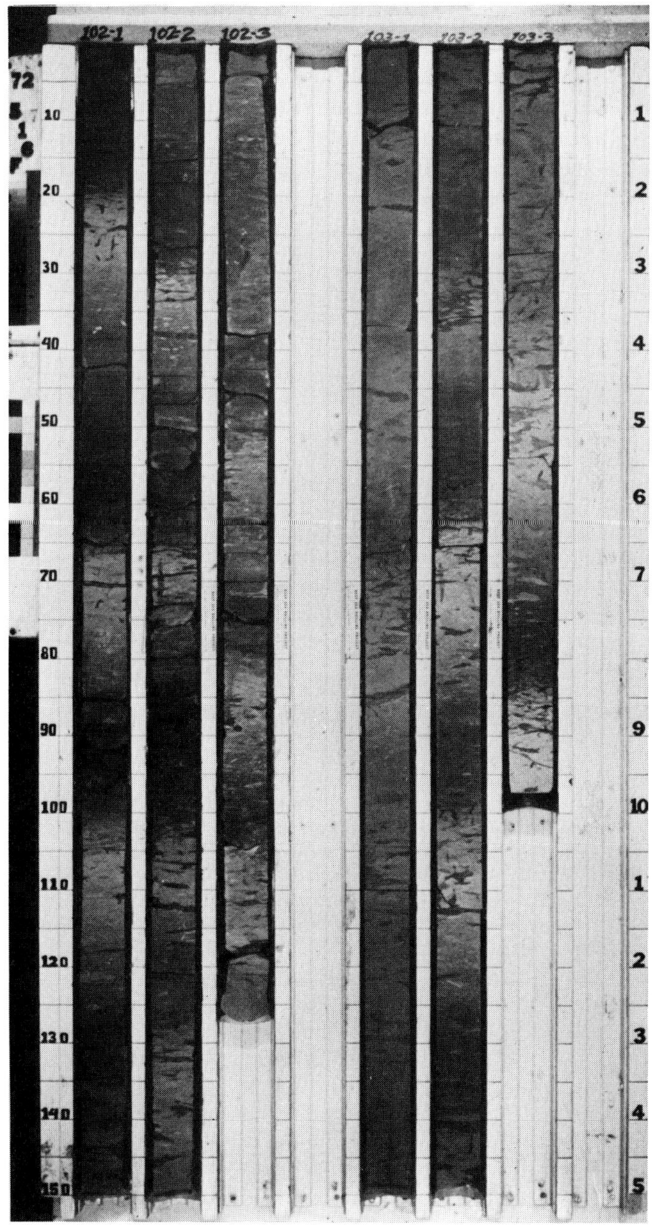
102-1
102-2
102-3
103-1
103-2
103-3
10
20
30
40
50
60
70
80
90
100
110
120
130
140
150
1
2
3
4
5
6
7
9
10
1
2
3
4
5

needs to be explored. Such records might bridge the gap between the frequency range of climate records usually studied by paleoceanographers and the time scales of anthropogenic disturbance of the ocean ecosystem, and give us an idea of "background" productivity variance.

Acknowledgements. We would like to thank Silke Bernhard and the other members of the Dahlem Konferenzen staff for organizing such a stimulating meeting. T.D. Herbert acknowledges support during this work from N.S.F. grant OCE-8711905.

REFERENCES

Barnola, J.M.; Raynaud, D.; Korotkevich, V.S.; and Lorius, C. 1987. Vostok ice core provides 160,000-year record of atmospheric CO_2. *Nature* **329**: 408–414.

Berner, R.A. 1982. Burial of organic carbon and pyrite sulfur in the modern ocean: its geochemical and environmental significance. *Am. J. Sci.* **282**: 451–475.

Boyle, E.A. 1986. Paired carbon isotope and cadmium data from benthic foraminifera: implications for changes in oceanic phosphorus, oceanic circulation, and atmospheric carbon dioxide. *Geochim. Cosmochim. Acta* **50**: 265–276.

Boyle, E.A., and Keigwin, L. 1987. North Atlantic thermohaline circulation during the past 20,000 years linked to high latitude surface temperature. *Nature* **330**: 35–40.

Broecker, W.S. 1982. Ocean chemistry during glacial time. *Geochim. Cosmochim. Acta* **46**: 1689–1705.

Broecker, W.S., and Peng, T.-H. 1982. Tracers in the Sea. Palisades, NY: Eldigio.

Calvert, S.E. 1974. Deposition and diagenesis of silica in marine sediments. *Spec. Publ. Int. Assn. Sediment.* **1**: 273–299.

Curry, W.B., and Lohman, G.P. 1985. Carbon deposition rates and deep water residence time in the equatorial Atlantic throughout the last 160,000 years. In: The Carbon Cycle and Atmospheric CO_2: Natural Variations Archean to Present, eds. E.T. Sundquist and W.S. Broecker. *Geophys. Monog.* **32**: 285–301. Washington, D.C.: Amer. Geophys. Union.

Dean, W.E.; Arthur, M.A.; and Claypool, G.E. 1986. Depletion of ^{13}C in Cretaceous marine organic matter: source, diagenetic, or environmental signal? *Marine Geol.* **70**: 119–157.

De Paolo, D.J., and Ingram, B.L. 1985. High-resolution stratigraphy with strontium isotopes. *Science* **227**: 938–941.

Deuser, W.G.; Ross, E.H.; and Anderson, R.F. 1981. Seasonality in the supply of sediment to the deep Sargasso Sea and implications for the rapid transfer of matter to the deep ocean. *Deep-Sea Res.* **28A**: 495–505.

Douglas, R., and Woodruff, F. 1981. Deep sea benthic foraminifera. In: The Oceanic Lithosphere. The Sea, vol. 7, ed. C. Emiliani, pp. 1233–1327. New York: Wiley.

Froelich, P.N., and Andreae, M.O. 1981. The marine geochemistry of Germanium: Ekasilicon. *Science* **213**: 205–207.

Froelich, P.N.; Bender, M.L.; Luedtke, N.A.; Heath, G.R.; and De Vries, T. 1982. The marine phosphorus cycle. *Am. J. Sci.* **282**: 474–511.

Hsu, K.J., and McKenzie, J.A. 1985. A "Strangelove" ocean in the earliest Tertiary. In: The Carbon Cycle and Atmospheric CO_2: Natural Variations Archean to

Present, eds. E.T. Sundquist and W.S. Broecker. *Geophys. Monog.* **32**: 487–492. Washington, D.C.: Amer. Geophys. Union.

Imbrie, J.; Hays, J.D.; Martinson, D.G.; McIntyre, A.; Mix, A.C.; Morley, J.J.; Pisias, N.G.; and Shackleton, N.J. 1984. The orbital theory of Pleistocene climate: support from a revised chronology of the marine $\delta^{18}O$ record. In: Milankovitch and Climate, eds. A. Berger, J. Imbrie, J. Hays, G. Kukla, and B. Saltzman, pp. 269–305. Hingham, MA: Reidel.

Kennett, J.P., ed. 1985. The Miocene Ocean: Paleoceanography and Biogeography. *Geol. Soc. Amer. Mem.* **163**.

Kitchell, J.A., and Clark, D.L. 1982. Late Cretaceous-Paleocene paleogeography and paleocirculation: evidence of North Polar upwelling. *Paleogeog. Pal. Pal.* **40**: 135–165.

Lasaga, A.C.; Berner, R.A.; and Garrels, R.M. 1985. An improved geochemical model of atmospheric CO_2 fluctuations over the past 100 million years. In: The Carbon Cycle and Atmospheric CO_2: Natural Variations Archean to Present, eds. E.T. Sundquist and W.S. Broecker. *Geophys Monog.* **32**: 397–411. Washington, D.C.: Amer. Geophys. Union.

Lisitzin, A.P. 1972. Sedimentation in the World Ocean. *Soc. Econ. Paleon. Miner. Spec. Publ.* **17**.

Manabe, S., and Broccoli, A.J. 1985. A comparison of climate sensitivity with data from the last glacial maximum. *J. Atmos. Sci.* **42**: 2643–2651.

Martin, J.H., and Fitzwater, S.E. 1988. Iron deficiency limits phytoplankton growth in the north-east Pacific subarctic. *Nature* **331**: 341–343.

Martinson, D.G.; Pisias, N.G.; Hays, J.D.; Imbrie, J.; Moore, T.C.; and Shackleton, N.J. 1987. Age dating and the orbital theory of the ice ages: development of a high-resolution 0-300,000-year chronology. *Quatern. Res.* **27**: 1–29.

McElroy, M.B. 1983. Marine biological controls on atmospheric CO_2 and climate. *Nature* **302**: 328–329.

Murnane, M.J., and Stallard, R.F. 1988. Germanium/Silicon fractionation in biogenic opal. *Paleocean.* **4**: 461–469.

Sarmiento, J.L., and Toggweiler, J.R. 1984. A new model for the role of the oceans in determining atmospheric PCO_2. *Nature* **308**: 621–624.

Sarnthein, M.; Winn, K.; and Zahn, R. 1987. Paleoproductivity of oceanic upwelling on atmospheric CO_2 and climatic change during deglaciation. In: Abrupt Climatic Change, eds. W.H. Berger and L. Labeyrie, pp. 311–337. Dordrecht: Reidel.

Shackleton, N.J., and Pisias, N.G. 1985. Atmospheric carbon dioxide, orbital forcing, and climate. In: The Carbon Cycle and Atmospheric CO_2: Natural Variations Archean to Present, eds. E.T. Sundquist and W.S. Broecker. *Geophys. Monog.* **32**: 303–317. Washington, D.C.: Amer. Geophys. Union.

Siegenthaler, U., and Wenk, T. 1984. Rapid atmospheric CO_2 variations and ocean circulation. *Nature* **308**: 624–626.

Vail, P.R.; Mitchum, R.M., Jr.; Todd, R.G.; Widmier, J.M.; Thompson, S., III; Sangree, J.B.; Bubb, J.N.; and Hatleid, W.G. 1977. Seismic stratigraphy and global changes in sea level. In: Seismic Stratigraphy – Applications to Hydrocarbon Exploration, ed. C.E. Payton, pp. 49–212. *Am. Assn. Petr. Geol. Mem.* **26**.

Van Andel, T.H. 1975. Mesozoic/Cenozoic calcite compensation depth and the global distribution of calcareous sediments. *Earth Planet. Sci. Lett.* **26**: 187–194.

Vincent, E., and Berger, W.H. 1985. Carbon dioxide and polar cooling in the Miocene: the Monterey hypothesis. In: The Carbon Cycle and Atmospheric CO_2: Natural Variations Archean to Present, eds. E.T. Sundquist and W.S. Broecker. *Geophys. Monog.* **32**: 455–468. Washington, D.C.: Amer. Geophys. Union.

SUGGESTED READING

Berger, A.; Imbrie, J.; Hays, J.; Kukla, G.; and Saltzman, B. 1984. Milankovitch and Climate. Hingham, M.A.: Reidel.

Berger, W.H. 1970. Biogenous deep-sea sediments: fractionation by deep-sea circulation. *Geol. Soc. Amer. Bull.* **81**: 1385–1402.

Berner, R.A.; Lasaga, A.C.; and Garrels, R.M. 1983. The carbonate-silicate geochemical cycle and its effect on atmospheric carbon dioxide over the past 100 million years. *Am. J. Sci.* **283**: 641–683.

Brewster, N.A. 1980. Cenozoic biogenic silica sediments in the Antarctic Ocean. *Geol. Soc. Amer. Bull.* **91**: 337–347.

CLIMAP. 1976. The surface of the Ice Age Earth. *Science* **191**: 1131–1137.

Holland, H.D. 1984. The Chemical Evolution of the Atmosphere and Oceans. Princeton, NJ: Princeton Univ. Press.

Leinen, M. 1979. Biogenic silica accumulation in the central equatorial Pacific and its implications for Cenozoic paleoceanography. *Geol. Soc. Amer. Bull.* **90(2)**: 1310–1376.

Redfield, A.C.; Ketchum, B.A.; and Richards, F.A. 1963. The influence of organisms on the composition of sea water. In: The Sea, vol. 2, ed. M.N. Hill, pp. 26–77. New York: Wiley.

Shackleton, N.J., and Summerhayes, C.P. 1986. North Atlantic Paleoceanography. *Geol. Soc. Spec. Publ.* **21**. Oxford: Blackwell Scientific Publications.

Suess, E., and Thiede, J. 1983. Coastal Upwelling – Its Sediment Record. New York: Plenum.

Productivity of the Ocean: Present and Past
eds. W.H. Berger, V.S. Smetacek and G. Wefer, pp. 429–455
John Wiley & Sons Limited

Appendix
Global Maps of Ocean Productivity

W.H. Berger

Scripps Institution of Oceanography
University of California, San Diego
La Jolla, CA 92093, U.S.A.

Abstract. The literature on global productivity dominantly refers to two global maps: the one by Fleming (1957), which relies heavily on the data collected by Steemann Nielsen of the Galathea Expedition, and the one by Koblents-Mishke, Volkovinskyi, and Kabanova (1968), which is based on an extensive but highly heterogeneous data set through the mid-sixties. Both maps attained FAO status and are widely used. Information from the last 20 years has been incorporated in the global map of Berger et al. (1987), which relies on radiocarbon determinations integrated over the photic zone. This latter map agrees well with that of Koblents-Mishke et al. over large areas, especially where coverage is adequate.

Estimates of total annual production based on the maps are as follows: ca. 20 GtC/yr for the Fleming map, 23 GtC/yr for Koblentz-Mishke et al., and 27 GtC/yr for Berger et al. Difficulties in producing a reliable map include poor quality of the data base, seasonal and interannual variation, and problems with the proper definition of productivity. For purposes of paleoceanography and sedimentology, it is sufficient to map an internally consistent productivity index which is closely related to output from the photic zone, such as the Steeman Nielsen productivity. Improvement of such maps will come from incorporation of satellite data into the global patterns and from the development of algorithms relating the productivity of surface waters to physical conditions and to sediments on the seafloor.

INTRODUCTION

Large-scale maps of ocean productivity, showing the carbon fixation rate integrated over the euphotic zone in grams per square meter per unit time, are in great demand for a number of purposes in marine biogeography and ecology, and lately also for studies in geochemistry and paleoceanography. What maps are available to satisfy this demand? How can they be improved? These questions are addressed in the following.

During the Danish Galathea Expedition, a systematic effort was made to establish the global patterns of ocean productivity. Steemann Nielsen and Aabye Jensen (1957) summarized the results on the radiocarbon productivity method, which they introduced and tabulated more than 700 values taken along the path of the globe-circling cruise. For the equatorial and southern Atlantic, they were able to draw semiquantitative maps of primary production (Fig. 1). They used their own data for calibration, and adapted the patterns from previous maps made by Schott, Hentschel, and Wattenberg, regarding color and biomass distributions, and phosphate concentrations (see Fig. 3 of Berger et al., this volume). As we shall see, their estimates of primary production patterns, as reflected in Fig. 1b, are remarkably close to those of later workers, and bear a striking similarity to modern satellite results, as well.

Ideally, one would like to have quantitative maps for several different parameters related to productivity: the average annual fixation rate, the average proportion of export production leaving the photic zone, and the seasonal anomalies in fixation rate both for recycled and for export production. However, for most of the ocean, maps for parameters other than average productivity are not available at present.

What is available suffers from well-known defects, as aptly summarized in Platt and Subba Rao (1975, p. 249): ". . . although an immense amount of effort has been expended in the acquisition of field data, many of the results are of little value for comparative purposes; there is considerable

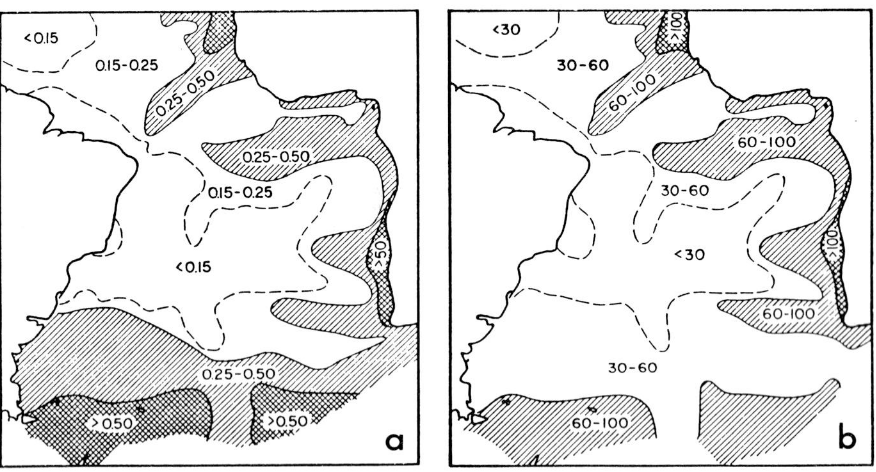

Fig. 1—Primary production in the equatorial and southern Atlantic, as drawn by Steemann Nielsen and Aabye Jensen (1957). (*a*) gross organic productivity in summer ($gCm^{-2}\ d^{-1}$); (*b*) annual net productivity ($gCm^{-2}\ yr^{-1}$).

variation between (and sometimes even within) authors in experimental and analytical methodology; some of the results have been obtained by methods which are now considered unreliable; the data are often incomplete in that they lack one or more critical collateral measurements which would permit the calculation of indices of comparison; there has been little attempt to control variance in the measurements by sound experimental design. It is therefore difficult at best to make valid comparisons between the productivities of different regions, and interpretations are often ambiguous."

The most widely used and reprinted global average productivity maps are based on two sources: Fleming (1957) and Koblentz-Mishke et al. (1970). Large-scale maps for seasonal productivity are available for the Indian Ocean (Kabanova 1968; Zeitzschel 1973; Krey and Babenerd 1976). Here, I summarize this information and compare it with the maps we recently made for paleoceanographic purposes, as given in SIO Reference 87–30 (Berger et al. 1987). Much of what follows is adapted from that report, especially from its appendix (on productivity maps in the literature). The body of that report (presenting the new productivity maps and summarizing the information on export production) appears as a chapter in Agegian (1988). There the reader will find a more extensive reference list than can be given here.

In addition, for the Atlantic I present a "best-guess" average productivity map, which represents an attempt to integrate the map of Koblentz-Mishke et al. with our own, and with satellite-derived information.

FLEMING MAP AND RELATED MAPS

The earliest available quantitative global map of primary production is that of Fleming (1957) (Fig. 2). The map is based on Steemann Nielsen's radiocarbon determinations during the Galathea Expedition, as well as other biological and physical oceanographic data. It is, in essence, an educated guess using a very slim data base. Fleming's discussion nicely summarizes the major factors controlling productivity, including nutrient supply, irradiation, and physical processes such as mixing, upwelling, and divergence. Also, he was aware of the difficulties introduced by seasonality. He writes (p. 103):

"For most ocean areas depletion of plant nutrients in the surface layers is believed to be the chief factor limiting plant productivity, and consequently the processes that will return these essential substances to the sea surface control the total productivity. These processes are chiefly winter mixing, upwelling at coastal boundaries and at current divergences, and the general upward displacement that must result from the formation of deep and intermediate water, and, in coastal water, tidal mixing. Some estimates of production have been made, most recently by Steemann-Nielsen (1954).

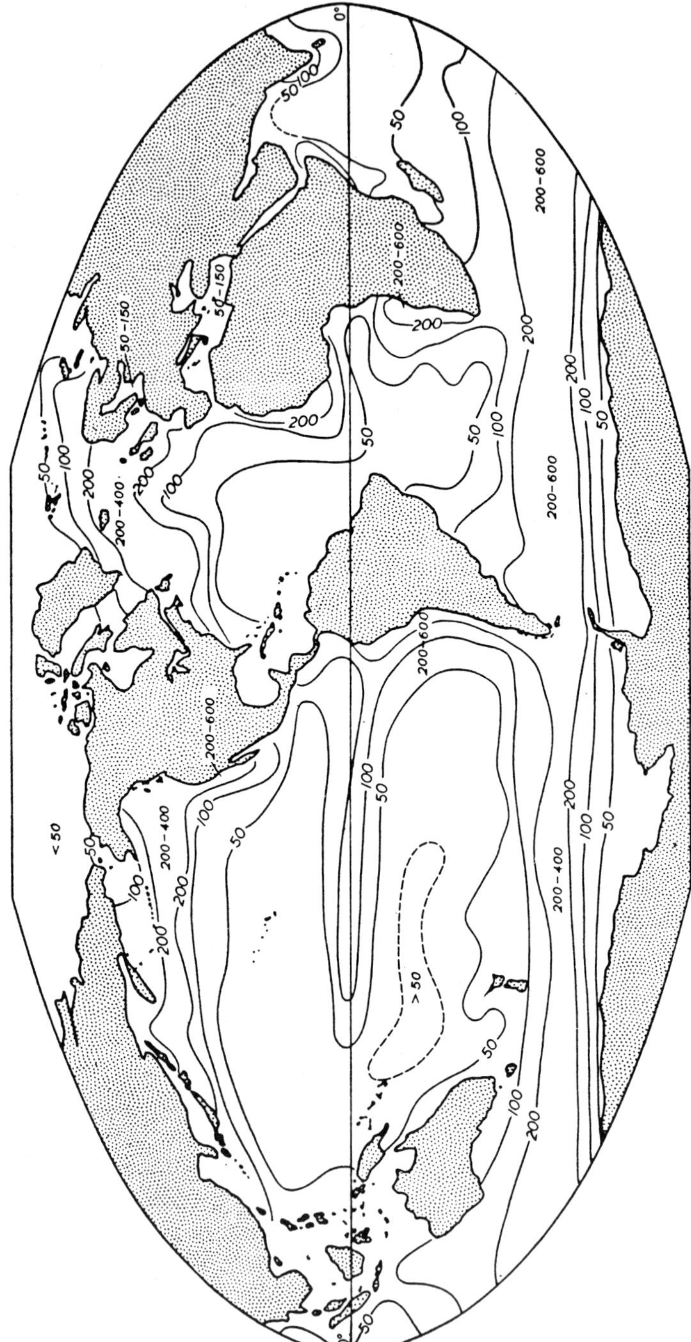

Fig. 2—Estimated annual photosynthetic productivity of the oceans, according to Fleming (1957). Units in $gCm^{-2} yr^{-1}$.

These have been used to construct Figure [2], which shows in a semiquantitative way the major features of the regional productivity (Fleming and Laevastu 1955). The values are given in units of grams of carbon/m^2/year. Although these values are estimates of the total plant production—that is, the gross amount of photosynthesis—they do not give any measure of the standing crops of either plants or animals, although it can be assumed that there will be a qualitative regional agreement. Furthermore, the values represent annual totals and do not indicate the large seasonal variations that are the conspicuous feature of life in the middle and higher latitudes where available light, temperature, and the self-shadowing that occurs with dense phytoplankton populations will greatly limit the rates of photosynthesis at various times of the year. In the Antarctic and Northern Pacific waters depletion of plant nutrients is not the factor that limits the amount of plant growth." (The reference to Fleming and Laevastu (1955) is to an unpublished manuscript.)

A map closely resembling the Fleming map is given in Gessner (1959). It differs from the original in adding a small area of high productivity in the middle of the Norwegian Sea, and in omitting the low productivity zone adjacent to the Antarctic continent. There is a minor modification in the Bering Sea. Also, there is a drafting error involving a large portion of the central South Pacific, where the Fleming map suggests slightly elevated productivity (presumably because of the island effect). In Gessner's map, this region (roughly the size of Australia) is shown as an area of high productivity (100–200 $gCm^{-2}yr^{-1}$).

Gessner states "the map was presented by the FAO at the Symposium on Primary Production, 1957," referring to a meeting by the Conseil International pour l'Exploration de la Mer, in Bergen, September 1957. The Gessner map has been reproduced widely, till about the mid-seventies. For example, the map is reproduced with minor modifications by Lieth in Lieth and Whittaker (1975). In Lieth's map, the South Pacific drafting error is also reproduced. Another version of this map is in Bunt, also in Lieth and Whittaker (1975), again with the South Pacific error, in the form of a computer output credited to a term paper by UNC graduate students. Possibly there are some modifications to the Gessner map: the output is difficult to read. Bunt states that his map yields a total production of 41.6 Gt dry matter/m^2/yr, corresponding to 18.9 GtC/yr, according to his conversion factor of 2.2. The value for total production would appear to depend heavily on the value of productivity assumed for the oligotrophic category (<50 in Fleming).

Fleming's map, with minor modifications (part of which are drafting errors) dominates the (western) literature up to about 1974. Many estimates of total production in the literature are likely to be based in one way or another on this map, and are not necessarily independent from each other. These estimates are near 20 GtC/yr.

MAP OF KOBLENTS-MISHKE, VOLKOVINSKY, AND KABANOVA (1970)

The most widely used productivity map is that of Koblents-Mishke, Volkovinsky, and Kabanova (1970) (referred to in the following as KVK). A very similar (or identical) map appears in Koblents-Mishke et al. (1968).

The KVK map is centered on Asia and the Indian Ocean and has a projection which causes considerable distortion in the Atlantic and the eastern Pacific. The absence of a detailed grid (and inaccuracies in the position of the zero meridian) makes orientation difficult. Nevertheless, over large areas the quality of this map appears to be excellent, that is, it is in good agreement with subsequent data, especially in the Atlantic. It seemed worthwhile, therefore, to redraw the KVK map on a more convenient equal-area projection (Fig. 3). For detail, the better resolved version given

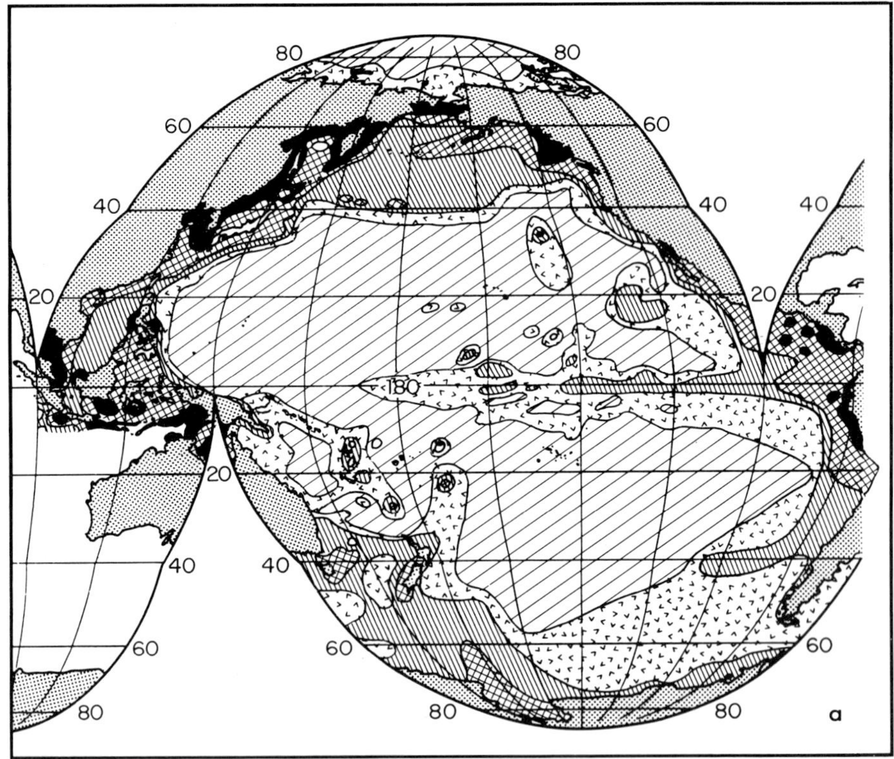

Fig. 3—World ocean primary production according to Koblents-Mishke and co-workers (1970), redrawn on an equal-area projection. Productivity is in $gCm^{-2}\ yr^{-1}$. Categories are, from low to high, <36, 36–54, 54–90, 90–180, >180.

in Romankevich (1984) was used to draw Fig. 3, except for the South Pacific, where Romankevich deviates from the original (Fig. 4).

The Pacific portion of the KVK map is based on surface-production data, recalculated for the entire water column, according to the authors. Scatter plots are given for the relation between primary production in the water column and primary production at the surface. From inspection of the graphs, it appears that the uncertainty in the conversion (standard error) is at least a factor of two.

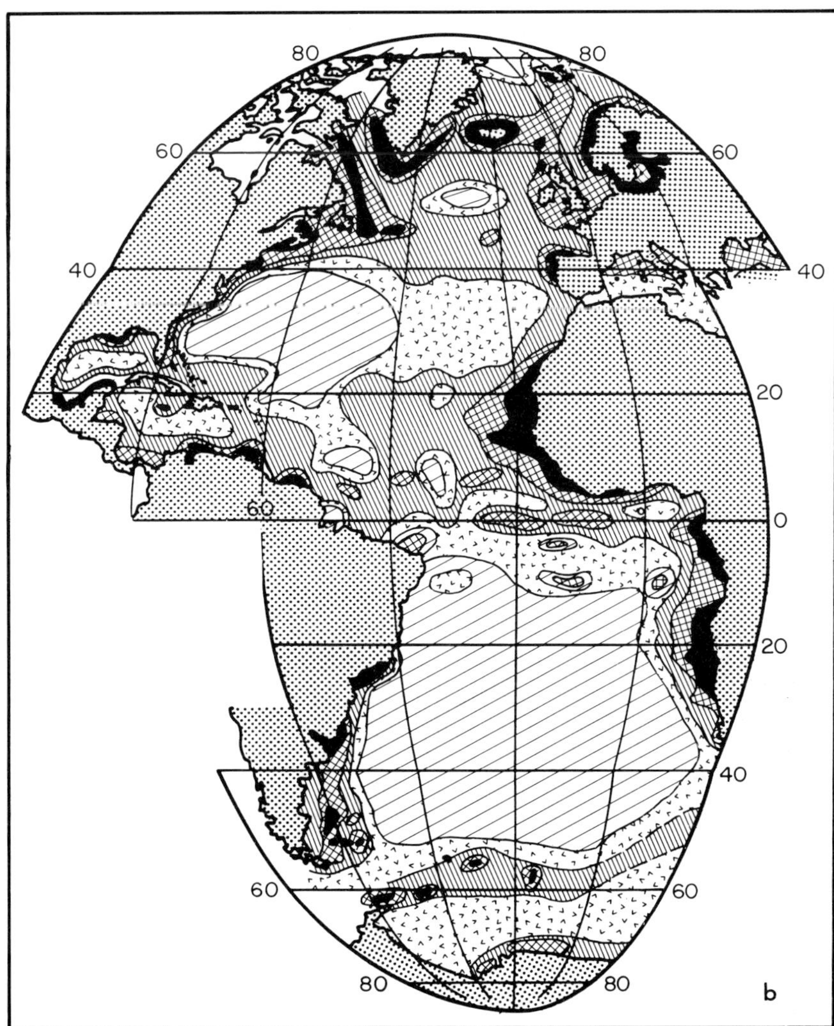

Fig. 3—continued

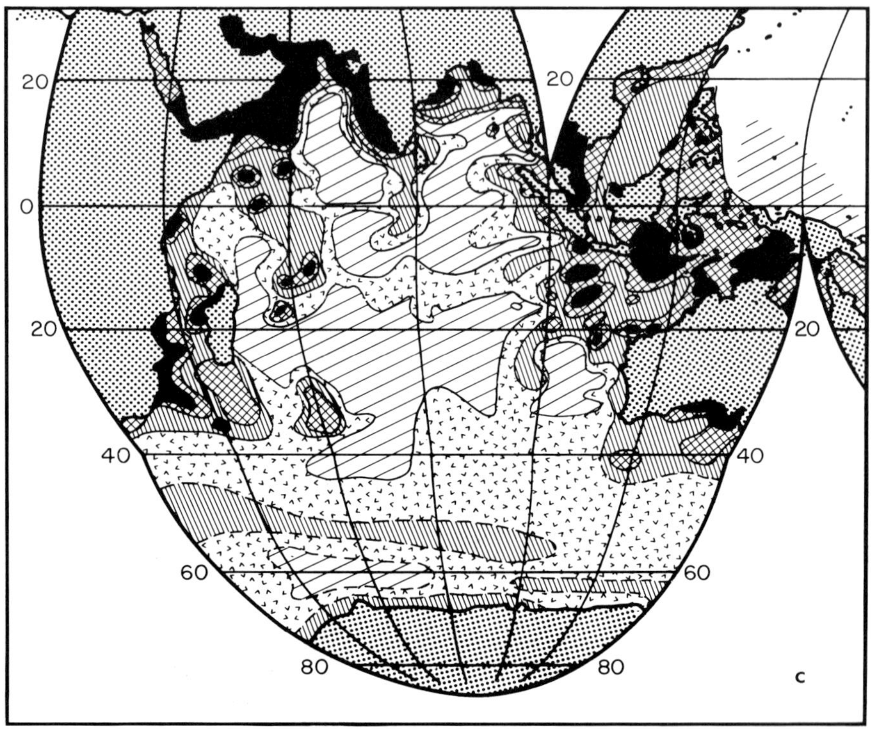

Fig. 3—continued

The basemap for the Pacific, without the portions later filled in and based on indirect indices of productivity ("phytoplankton biomass, hydrogen [=pH], or oxygen saturation," Koblents-Mishke et al. 1970) is given in Koblents-Mishke (1965), together with the density of data coverage. Essentially, there is little or no coverage in the South Pacific east of New Zealand and south of 25°S. This is the region of substantial modification in the Romankevich map, which is in other respects identical to that of Koblents-Mishke et al. (1970).

The basemap for the Indian Ocean is given in Kabanova (1968), showing the coverage of all available expedition results. Radiocarbon determinations were used, integrated over the photic zone. Kabanova, in her article, also offers maps for the two seasons. These show the large variability of productivity in the Indian Ocean, which is expected from the monsoon effect. Cushing (in Zeitzschel 1973) has digitized the seasonal data for 5° squares, showing primary productivity in the Indian Ocean in gCm^{-2} per half year, separately for S.W. monsoon and N.E. monsoon. The ratio of these values (Fig. 5) shows that the S.W. monsoon generates considerably

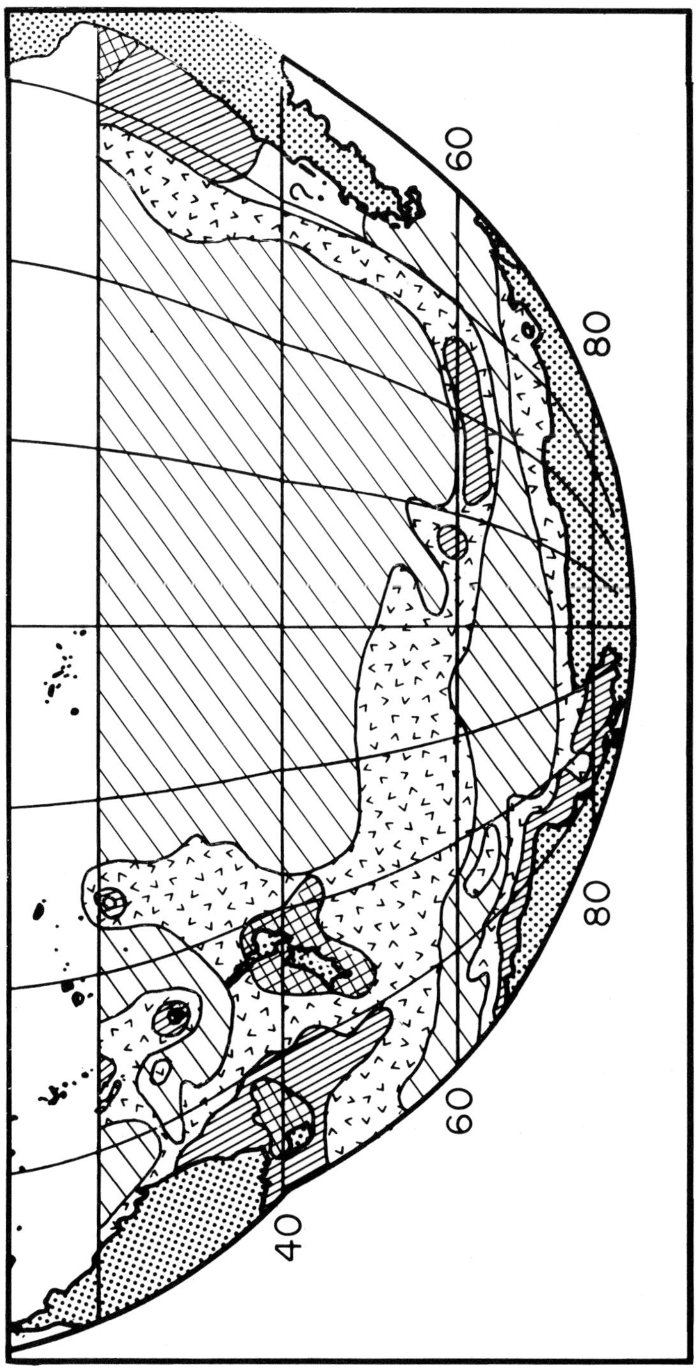

Fig. 4—Portion of the map of Koblents-Mishke et al. (1970) which is modified in Romankevich (1984), redrawn.

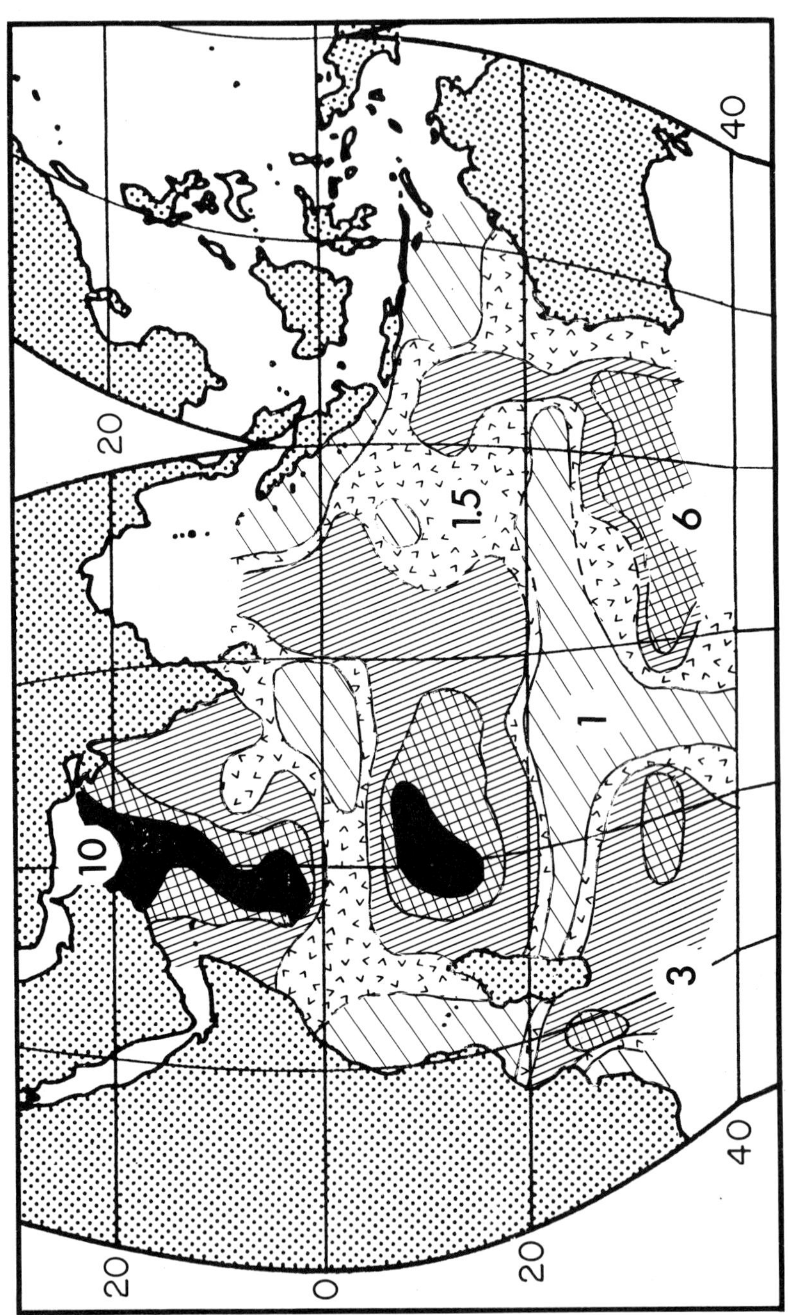

Fig. 5—Seasonal productivity ratios for the Indian Ocean, based on Cushing's (Zeitzschel 1973) digitization of the data compiled by Kabanova (1968). The ratio is the factor by which S.W. monsoon exceeds N.E. monsoon productivity.

higher productivity than the N.E. monsoon over large areas of the Indian Ocean.

Although the Indian Ocean is an extreme case as far as variability within an entire basin, analogous variability in other regions of the ocean interferes substantially with the mapping of meaningful averages. If flux of organic matter toward the seafloor is nonlinearly related to primary productivity (as is likely), the "effective" productivity responsible for most of the export should lie well above the average in regions of great variability.

In the 1970 article, Koblents-Mishke et al. give Doty and Capurro (1961) as the basic source of information for the Pacific and part of the Atlantic, Kabanova (1961) as the source for Indian Ocean information, and a list of about 40 papers as sources for the Atlantic and the inland seas, published between 1946 and 1968, but mainly in the early and middle sixties. The reference to Kabanova apparently is to her 1968 paper. The authors state that they included data from more than 7,000 stations (presumably mostly from the Atlantic). The majority of measurements are based on the Steemann Nielsen method, and some are based on chlorophyll a, on oxygen change, and "algological methods." As mentioned, for the Pacific, the map is based on surface-production data, but for the Atlantic and the Indian Ocean, data on production per square meter were used.

The KVK map appears in a number of treatises on marine ecology, most recently (redrawn) in Parsons et al. (1984). A slightly simplified version of the KVK map, redrawn on a gridded map by Cushing, is in the FAO Atlas of the Living Resources of the Seas (1972). The FAO map is reproduced on the inside of the back cover of Zeitzschel (1973). Simplified versions of this map may be found in various textbooks, commonly credited to FAO 1972.

A clean, redrawn copy of the KVK map, with substantial modifications in the Pacific subantarctic and antarctic sector appears in Romankevich (1984), and was referred to above (Figs. 3 and 4). The Romankevich version also shows divergences and convergences, and ice boundaries, in addition to productivity. The map is peculiar in that it postulates an extensive low productivity zone in the Pacific sector of the Antarctic, where others suggest moderate to high productivity. A recent reappraisal of Antarctic productivity (El-Sayed 1984) does find a trend toward adjustment to lower values in the literature; however, it also emphasizes variability.

ON THE MEANING OF "PRODUCTIVITY"

We have seen that the commonly available sources of information on the global productivity of the ocean draw from two wells: Fleming's guesstimate of 1957, which was largely based on the Galathea work (see Steemann

Nielsen and Aabye Jensen 1957), and the compilation of Koblents-Mishke and co-workers, which includes literature up to 1968. What exactly is being shown by these maps?

Steemann Nielsen has described the radiocarbon method he devised as follows (translated in Bruun et al. 1956, p. 57): "The routine of measuring the production of organic matter below a unit area is to take samples of water from the various depths, pour the water into bottles, add radioactive carbon dioxide, and then lower the bottles to the depths from which the samples were taken, keeping them suspended there for 24 hours. They are then taken up and the algae filtered off. When the radioactivity has been measured, the rest is a matter of straightforward calculation." He adds (ibid, p. 58): "In order to collect a large amount of data in a short space of time we were obliged to make a small alteration in the method. Instead of lowering the samples to their original depth after the addition of ^{14}C, we placed them in a sort of aquarium at the same temperature as in the sea, illuminating them with artificial light of a known intensity."

From this outline the sources of error in the radiocarbon method (that is, variability introduced from the method rather than inherent in the system studied) are evident: (*a*) the bottles are assumed to provide an environment comparable to that of the open ocean, with its turbulence, lack of walls, and chemically pure seawater; (*b*) the effect of handling on the organisms (depressurizing) is thought to be negligible; (*c*) the exact duration of exposure is thought to be of little consequence (4 hours giving twice the yield of 2 hours, for example); (*d*) the artificial illumination is considered a valid replacement for natural illumination, both with regard to spectrum and intensity; (*e*) the interpretation of the results in terms of algal material generated is thought to be straightforward.

Clearly, all these assumptions are tagged with some error, and much discussion has revolved around assessing these errors and correcting for them. Points (*a*) and (*e*) may be the most difficult to treat in this respect: bottle waters do differ from open ocean waters, and organisms other than algae are collected on the filters, in varying proportions. In fact, there is considerable controversy about the meaning of Steemann Nielsen productivity values (see Bender et al. 1987 and refs. therein). It is now commonly assumed that this method distinctly underestimates "actual" productivity, that is, the gross rate of carbon fixation. Factors between 1.4 and 2 have been suggested for this discrepancy (refs. in Sundquist 1985), and even higher ones (2.5 to 4) for polar areas (Rivkin and Putt 1987).

Grazing and mortality during incubation have to be considered. In addition, nano- and pico-plankton may be responsible for a large portion of total photosynthesis (Platt and Li 1986). While such small organisms would be retained on the membrane filters used in ^{14}C studies, the amount caught would be small relative to their productivity, because they are rapidly consumed by microheterotrophic organisms. Such microheterotrophs may be more abundant than hitherto assumed (Smith et al. 1984).

In addition to the problems of measurement, there are problems of integration over depth. Considerable carbon fixation may be taking place at rather low light intensities, below the traditional euphotic zone boundary. Based on the phenomenon of a shallow oxygen maximum in the northern Pacific, Shulenberger and Reid (1981) suggested that "the overall productivity of the open ocean may have been seriously underestimated," a proposition which engendered much discussion (e.g., Platt and Harrison 1986).

These issues are complex and cannot be addressed here. From the viewpoint of sedimentology, that is, the mapping of flux toward the seafloor and distributions on the seafloor, any reasonable definition of productivity will serve as long as it is internally consistent and can be calibrated to the downward flux of organic carbon and other biogenic materials. Thus, "primary production" or "productivity" in the present context means the rate of carbon fixation as measured by traditional methods, that is, the Steemann Nielsen method or an equivalent measure. This measure may be considered a productivity index, which is given in terms of flux, for convenience.

Quantitative mapping of such a productivity index is mainly hindered by the lack of standardization of measurement, and by the sparsity of samples and their variability and inhomogeneity in coverage. Interannual long-term trends in productivity in the open ocean (Venrick et al. 1987) poses another major problem regarding the general validity of productivity maps.

LAI'S COMPILATION

Two decades have passed since the compilation of Koblents-Mishke et al. (1968). Thus, it seemed worthwhile to attempt a new one. Our survey (Berger et al. 1987) was primarily carried out by Chen Lai, doctoral student in marine biology at Scripps Institution of Oceanography, at the time. It includes measurements made from 1944 through 1985, using radiocarbon (more than 90 percent of the data) and oxygen determinations, integrated over the photic zone. Most of the data were taken from reports in English. Also, we used articles in Spanish and French (mainly for the Mediterranean), and we obtained unpublished data from a number of correspondents. Approximately 8000 measurements are represented in the compilation.

Averaging of the data was somewhat arbitrary, because of the great variability of primary production in mid and high latitudes and because of patchiness and seasonality in general. Also, interannual variability made the task difficult. Annual estimates were made for locations where a sufficient number of seasonal values were available. In other cases, map values were estimated from the spot data alone, considering the seasons(s) of measurement. Results are shown as (subjectively drawn) contours of "average primary production" (Fig. 6). In areas of low data density, phosphate concentration at 100 m depth was used to guide interpolation.

In general, our map is quite similar to that of Koblents-Mishke et al. However, the maps differ in some important respects, too. For example, large areas of the ocean in the KVK map are shown as having productivities of less than 36 gCm^{-2} yr^{-1}. In our map, the area with such low values is considerably smaller, especially in the Pacific realm. Also, in the KVK map typical values for the eastern tropical Pacific range from 50 to 100 gCm^{-2} yr^{-1}; in ours the range is typically 60 to 150 gCm^{-2} yr^{-1}. In contrast, in the Atlantic their values for the tropics tend to be somewhat higher than ours (50 to 85 versus 35 to 60, typically). According to W. Gieskes (personal communication, April 1987) the higher values more nearly agree with his measurement series in the region.

There is a great inhomogeneity in coverage for our map, from more than 150 determinations per 10 degree square to zero. The most densely sampled areas are predominantly in the northern Indian Ocean, the northeast Pacific, off the U.S. East Coast, in the northern North Atlantic, and off Northwest Africa. The unsampled regions are in the southern portions of the major oceans, mainly between 30 and 60 degrees South. In fact, some measurements do exist in these regions (e.g., El-Sayed 1984 and refs. therein), but they were only used here if the production was given as integrated over the euphotic zone. Coverage in the Antarctic region is sparse, perhaps enough for educated guesses but probably inadequate considering the importance of the region for the carbon cycle.

THE PHOSPHATE-BASED MAP

Guesswork is unavoidable in producing a global productivity map, given the amount and quality of existing information. Following Fleming (Fig. 2), it is reasonable to fill in the blanks using the distribution of nutrients and light. Care must be taken to weigh these factors properly: Fleming's excessively high estimates for the Southern Ocean indicate a tendency to overvalue the importance of availability of nutrients as compared with that of light (despite his correct assessment in his discussion, as quoted above).

Using the Lai map (Fig. 6) as a target, we employed trial-and-error calculations to find algorithms based on phosphate concentration at 100 meters, latitudinal position (light, seasonality), and distance from land (mixing, eddy formation, upwelling).

Equations are as follows (Berger et al. 1987):

$$PP = \exp\left(P_{100} \cdot (1 - P_{100}/a) + b\right) \qquad (1)$$

where the term a sets the strength of the negative feedback on the exponential growth response to nutrient supply, and b sets the minimum

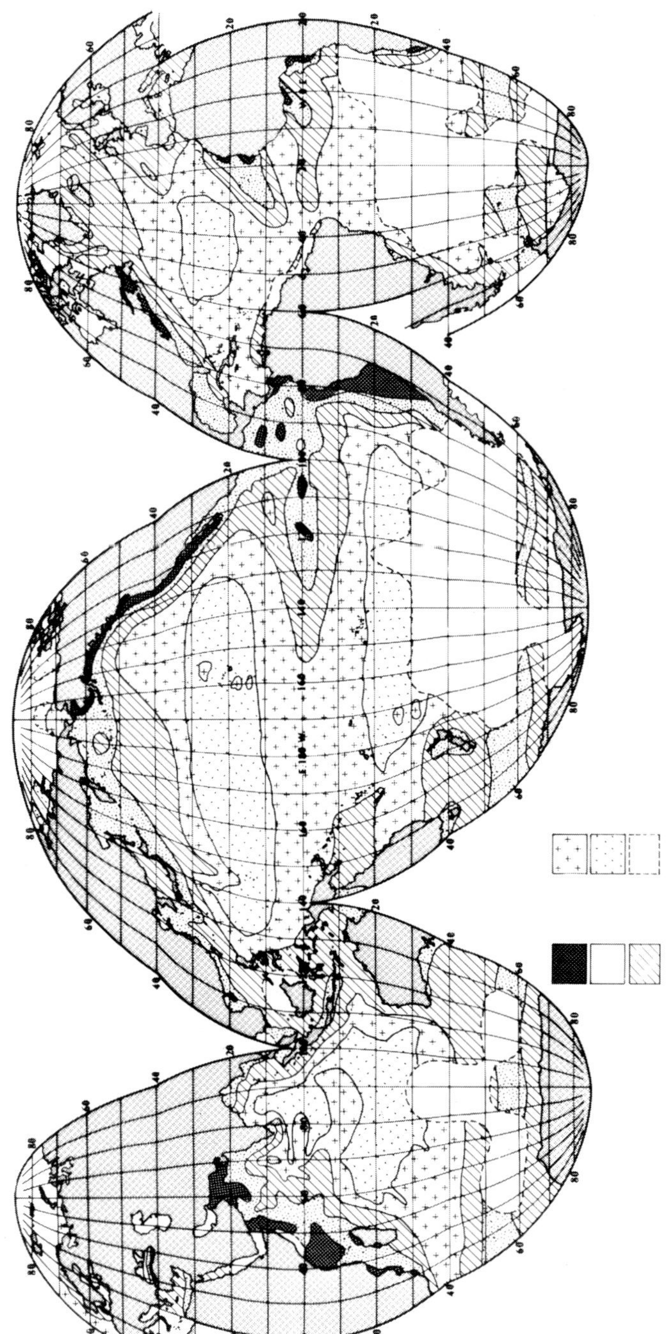

Fig. 6—Average annual primary production of the global ocean based on selected data compiled from the literature (almost entirely radiocarbon determinations). Only vertically integrated productivity measurements were used. In areas with good data coverage syntheses were consulted; in those with spotty coverage (most of the ocean) averages were made considering the season(s) of measurement. From Berger et al. (1987).

productivity, when recycling dominates entirely. Form our data sets, we found that a=10 and b=3.1. With P_{100} measured in mmol per m^3, PP is in $gCm^{-2}\ yr^{-1}$, integrated over the photic zone.

To take account of the importance of light, we define an "effective phosphate concentration" as a function of latitude, as follows:

$$P_{eff} = P_{100} \cdot \cos(L^3 / 79^3 \cdot pi/2) \tag{2}$$

where L is the latitude. The third power on L heavily weights the high latitudes, and the denominator sets the latitude (50°) beyond which a decrease in irradiation first becomes significant. $P_e$ff is entered into Eq. 1 to obtain the estimate for primary production PP.

The rate of supply of nutrients from deeper waters is greatly enhanced along the coast. The reason is that mixing and upwelling are stimulated along the continents through onshore and offshore winds, eddy formation in boundary currents, and tidal dissipation. To account for this increased supply, we formulated a "distance-from-land effect" which simulates the presence of additional phosphate, as follows:

$$P_{apparent} = P_{100} + 10 / \sqrt{D} \tag{3}$$

If the distance from land, D, is less than 100 km it is set to 100 km.

The sequence of synthesizing a primary production map from phosphate at 100 m, then, is Eq. 3 to Eq. 2 to Eq 1. The result of the transform exercise is shown in Fig. 7.

How well did we hit the target provided by the map in Fig. 6?

To answer this question, we averaged the primary production values for 10 degree squares for both maps, and calculated the percent difference. The fit turned out to be quite good: for about one half of the area the transform is within 20% of the target, and over another one third, within 35%. Thus, 85% of the phosphate-based map is well within the error of contouring for the compilation-based map.

Distinctly higher values (>135%) in the transform occur in about 10% of the squares, and lower values (<65%) in 5%. Systematic deviations appear in the monsoon regions (calculated values too low off Africa, too high south of India), in the North Equatorial Current in the Pacific (calculated values much too high), and in the Gulf Stream region (calculated values too low). Reasons presumably have to do with diverging and converging surface waters, and with variability of depth of mixing. In about 15% of the area of the ocean, then, the three factors we used are insufficient to parameterize the physical processes governing primary production. These areas, of course, are some of the more interesting ones, regarding physical and biological dynamics.

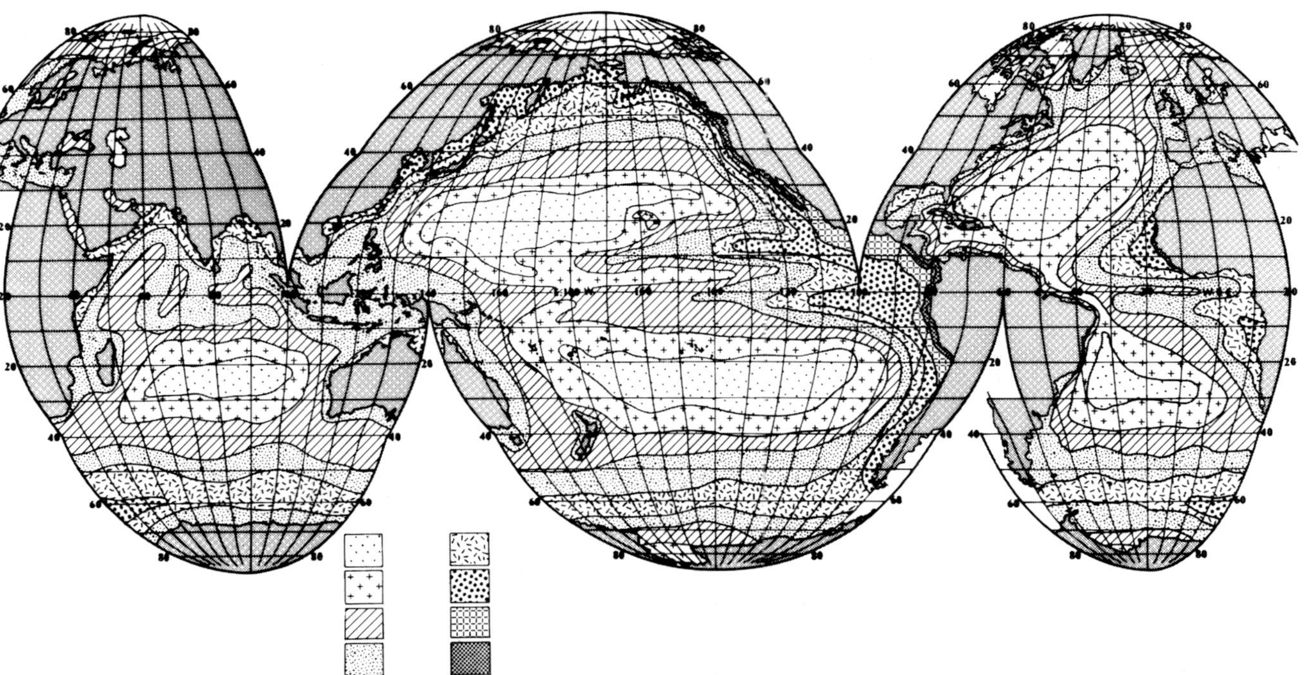

Fig. 7—Map of calculated primary production of the world's ocean. For export production at 100 m use 20% of scale for PP, for 200 m use 10%, and for 500 m 5%. Calculations based on (1) phosphate distributions, (2) latitude, and (3) distance from shore (see text). From Berger et al. (1987).

SATELLITE-DERIVED MAPS

An exciting new global view of ocean productivity is provided by satellite-sensed, ocean-surface color. Algorithm development, relating spectral ratios to chlorophyll abundance and from there to phytoplankton activity, is a field of active research. Esaias et al. (1986) estimate that surface phytoplankton distributions derived from Coastal Zone Color Scanner (CZCS) have an accuracy of from 35% in open oceans to within a factor of 2, generally. Chlorophyll abundance in the upper part of the euphotic zone appears to be quite well correlated with primary production, so that large-scale inferences are possible (Fig. 8).

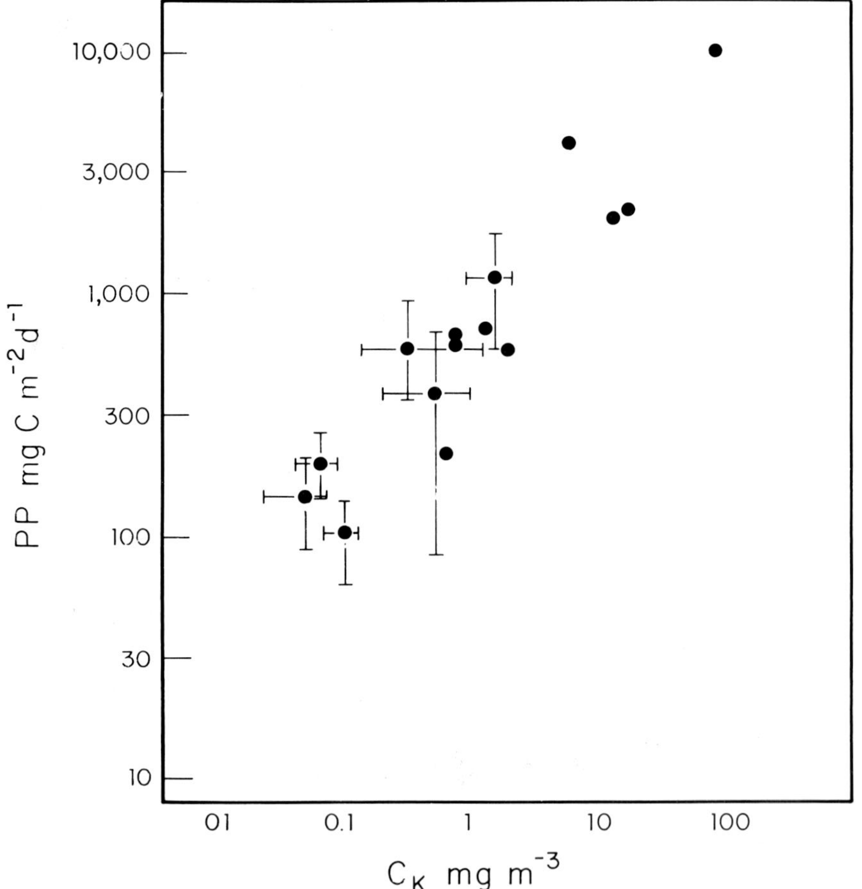

Fig. 8—Correlation between surface chlorophyll and integrated primary productivity, as given by Eppley et al. (1985). Each point represents a data set from a cruise or a number of stations in an area.

Recently, the U.S. Global Flux Study Planning Office at the Woods Hole Oceanographic Institution (1987) has released a global view of "chlorophyll" abundance based on spectral intensities obtained by NASA Nimbus-7, in December 1981. The patterns of high and low abundance of chlorophyll appear to correlate well with those of primary productivity, in low latitudes (Figs. 3, 6, 7) (also see color plates in this book). In both the Pacific and the Atlantic, the equatorial high productivity zone is well expressed, as is the high productivity within the eastern boundary current north of the equator (Fig. 9). Cloud cover, unfortunately, obscures the coastal upwelling regions to the south.

The patterns in the Atlantic have a remarkable resemblance to those mapped more than 30 years ago by Steemann Nielsen and Aabye Jensen (see Fig. 1). In the Pacific, the zone of increased productivity north of the equator suggested by the satellite map has not been observed when mapping productivity measurements, presumably because the pattern is swamped by high variability and patchiness. However, this zone does appear in our phosphate-based map (Fig. 7).

ESTIMATES OF TOTAL OCEAN PRODUCTIVITY

There are a great number of estimates of global ocean productivity in the literature (see tabulations by Platt and Subba Rao 1975; Sundquist 1985), many of which are not independent since they are based in one way or another on the maps of Fleming (1957) and of Koblents-Mishke et al. (1968). Differences in estimates, quite commonly, are based on the application of different "correction factors," rather than on the use of a more reliable data base. The Fleming map yields total estimates near 20 GtC/yr. For example, an evaluation of the "FAO Productivity Map" (Box, in Lieth and Whittaker 1975) yields the following values (gigatons of carbon per year) in different categories of productivity (in $gCm^{-2}\ yr^{-1}$):

0–50, 1.69; 50–100, 4.73; 100–200, 5.18; 200–400, 5.35; >400, 1.62; for a total of 18.56 GtC/yr.

The KVK map yields slightly higher estimates, largely because of assuming a higher productivity in the oligotrophic areas. They obtain a total of 23 GtC/yr. The categories and their values are as follows:

25 (<36), 3.79; 50 (36–54), 4.22; 72 (54–90), 6.31; 122 (90–180), 4.80; 360 (>180), 3.90; for a total of 23.02 GtC/yr.

One notes, in their table, that the "average" productivities are guesses lying somewhere between the lower and upper bound, with an uncertainty of 10 to 20 percent. In one case, the "average" is taken to be close to the upper bound, in two others, in the middle between the limits. From our

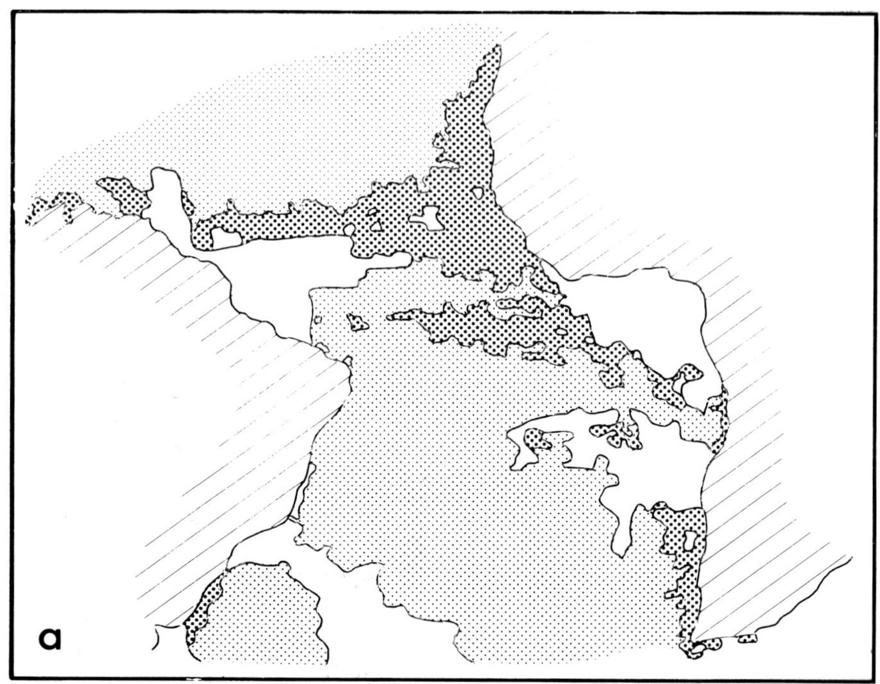

Fig. 9—Relative intensity of "chlorophyll" reflection in the equatorial Atlantic (upper) and in the equatorial Pacific (lower) based on spectral analysis of ocean surface color, NASA Nimbus-7, Dec 1981. From a map distributed by the U.S. Global Flux Study Planning Office, Woods Hole Oceanographic Institute, 1987. White areas: no data (cloud).

experience in contouring productivity values, "averages" between contours are more probably *below* the midpoint.

Our own map yields a total productivity near 27 GtC/yr. Categories are as follows (Berger et al. 1987, Table 1):

15–35, 1.7; 35–60, 5.6; 60–100, 8.3; 100–200, 8.1; 200–500, 3.1; for a total of 26.9 GtC/yr.

As do the other maps, ours underestimates the contribution from the nearshore area. Accounting for this deficit, our guess for the primary production of the ocean, as measured by the radiocarbon method, is 30 GtC/yr. This estimate is the same as that derived by Platt and Subba Rao (1975), who propose a total of 31 GtC/yr, based on an independent survey of available data at that time.

The production per ocean may be split out as follows (Berger et al. 1987; coastal production in parentheses):

Pacific 7.0 (4.1), Indian 2.8 (1.9), Atlantic 4.3 (1.7), Antarctic 5.2. Total: 27 GtC/yr.

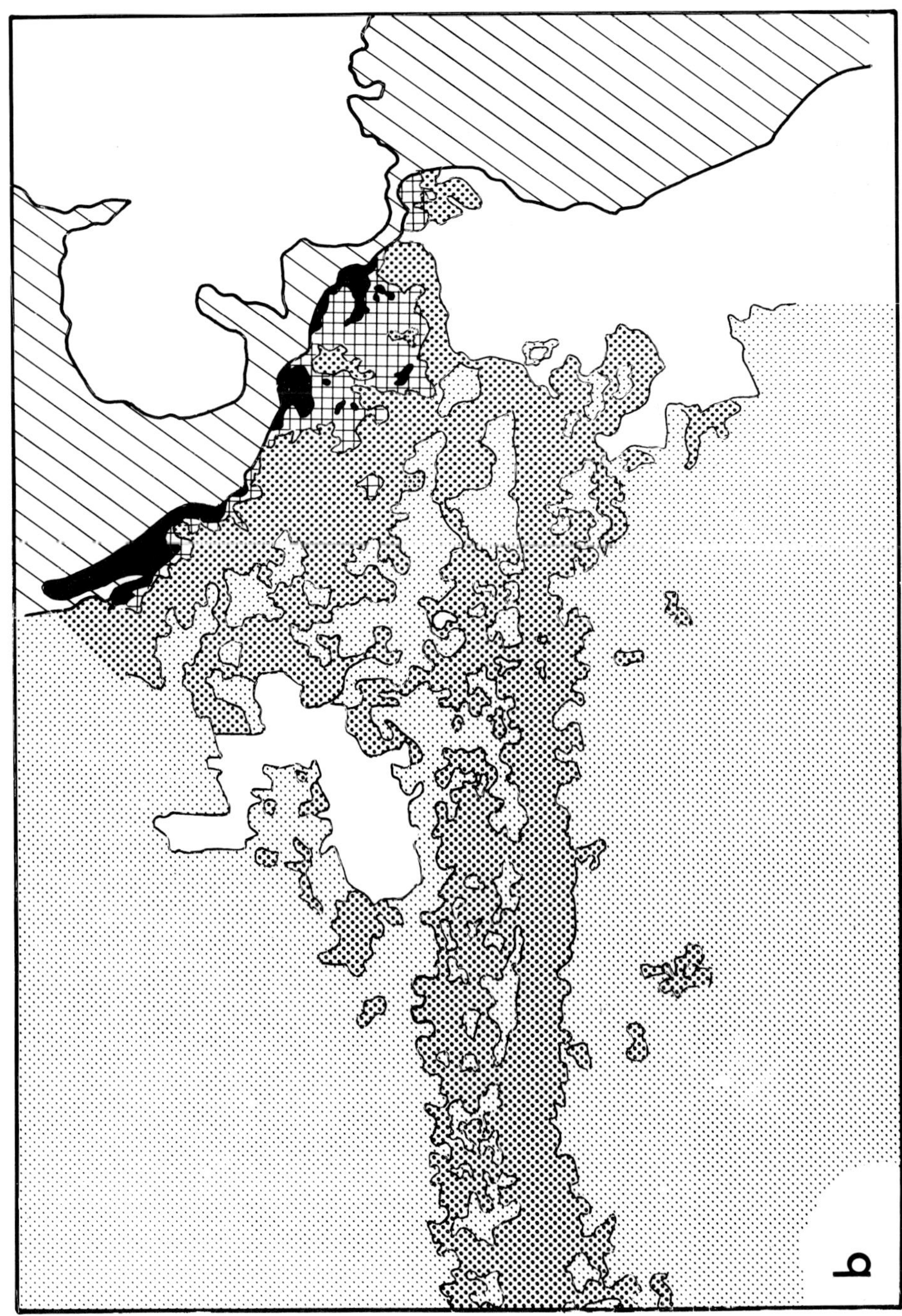
b

For comparison, Platt and Subba Rao (1975) show the following values (shelf production in parentheses):

Pacific 9.4 (2.0), Indian 5.9 (0.7), Atlantic 8.5 (1.3), Antarctic 3.3. Total: 31.1 GtC/yr.

It is apparent that our estimates differ in that "coastal ocean" is a larger category than "shelf" and in that we obtain a distinctly lower total for the Atlantic. Also, we have about twice as much area in "Antarctic" than do Platt and Subba Rao.

Does productivity differ significantly between ocean basins?

Dividing the percentage of the productivities in the regions by the percentages of the areas considered (which sets the average productivity index equal to unity), we get the following values (the ratios based on Platt and Subba Rao are in parentheses):

Pacific 0.93 (0.76); Indian 1.0 (1.05); Atlantic 0.88 (1.28), Antarctic 1.5 (2.2).

Thus, according to our data, the Pacific and Atlantic are equally productive, on the average, and according to Platt and Subba Rao, the Atlantic is the more productive of the two. In our case, the higher deep-water phosphate concentrations in the Pacific are balanced by the grêater proportion of coastal ocean in the Atlantic; in the case of Platt and Subba Rao, the coastal conditions have the greater weight. In both data sets, the Antarctic productivity is significantly above average, as generally assumed until recently.

REFINEMENT FOR ACCURACY

What is the next step in the mapping of global productivity?

Clearly, two avenues are open. First, we can work with the existing data of doubtful quality but reasonable abundance, make the best target maps we can, and improve the physical algorithms which allow us to refine our guesses as to productivity distributions. These can then be compared with, and integrated with, satellite-derived phytoplankton pigment maps. Second, we can disregard much of what has been published or primary productivity, heeding the warning of Platt and Subba Rao (1975), and start over, doing things properly this time. In essence, this means giving up most of the existing data, which were collected over the last 30 years at great expense.

The good agreement between our maps and that of Koblents-Mishke et al. (1970) over large parts of the ocean encourages us to adopt the former strategy: that of step-wise improvement of existing maps. In the following, I outline how such improvement might be achieved, exemplifying the initial steps of the procedure for the Atlantic Ocean.

First, I produced a combination map (LSYNT) from the compilation of Lai (Fig.6) and the phosphate-based algorithm (Fig. 7). I used Fig. 6 for those areas where data density is high, and the algorithm where density is low. The resulting map (Fig. 10) may be compared with the KVK map (Fig. 3). Comparison is somewhat hampered by the different contour categories but shows good agreement in the general patterns.

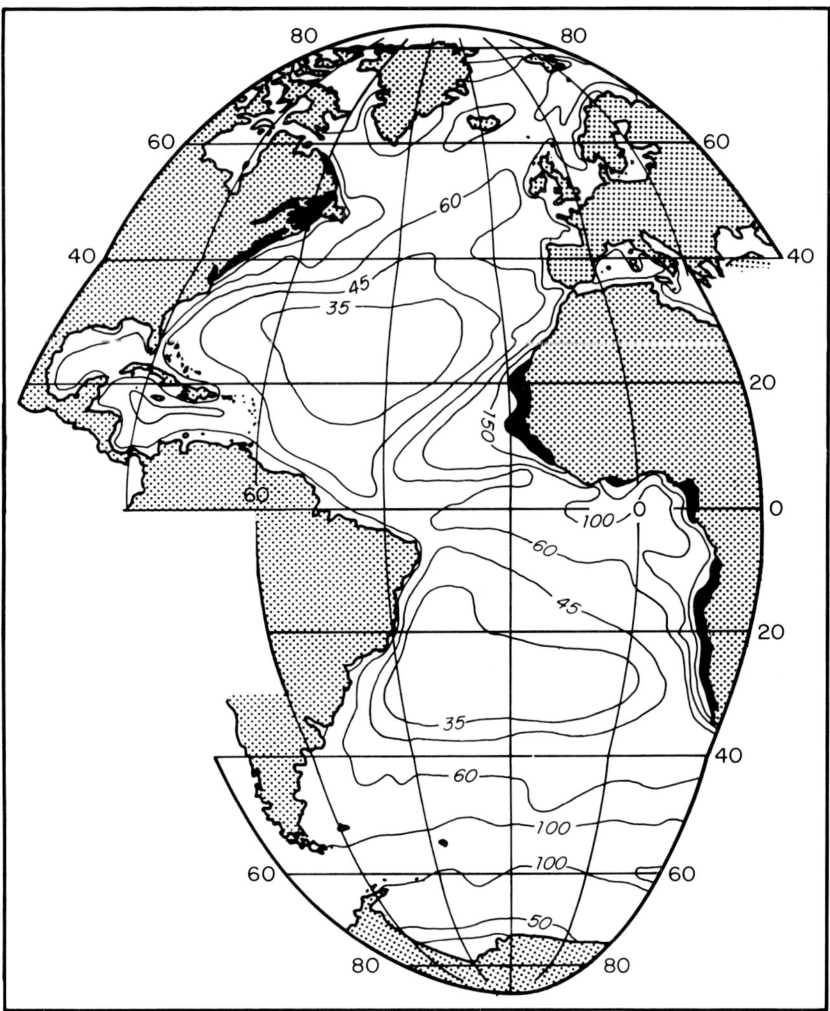

Fig. 10—Productivity map (LSYNT) combining the information from the Lai compilation (Fig. 5) and from the phosphate-based map (Fig. 6). Contours in gCm^{-2} yr^{-1}.

Next, I digitized both the LSYNT map and the KVK map on a 5 degree grid, in preparation of combining them into a best-guess map ("Dahlem" map). Straightforward averaging seemed reasonable in those areas where the patterns agree well. In those where the agreement is poor, I checked the satellite information given in Esaias et al. (1986) and in the Woods Hole map (U.S. Global Flux Study Planning Office 1987) as to whether LSYNT or KVK are more likely to be correct. Accordingly, greater weight was then given to one or the other of the two maps in producing the grid for the Dahlem map. Also, in the final contouring of the grid, the rather detailed information on the satellite phytopigment maps allowed introducing a somewhat higher resolution than expected from combining LSYNT and KVK alone (Fig. 11).

Two independent avenues for further improvement are now open: (*a*) comparison of the Dahlem map with the distribution of planktonic fossils on the seafloor, especially planktonic foraminifera, for finding the transfer functions relating productivity to fossil patterns, with the aim of recalculating the surface productivity patterns from fossils; (*b*) use of the Dahlem map as a target for algorithms based on phosphate distributions, irradiation, and geography, as outlined earlier. Regarding item (*a*), the first step has already been taken by A. Mix (this volume), who showed that transfer equations based on planktonic foraminifera works well with regard to surface productivity. Concerning item (*b*), the simple algorithm used in producing the P-based map (Fig. 7) could presumably be greatly improved if temperature anomalies and horizontal and vertical temperature gradients are employed in addition to phosphate, latitude, and distance from shore. Also, for the shelf, depth of water is a factor which merits consideration.

A number of applications for the Dahlem map (or maps like it) readily come to mind, as follows:

1. Calculation of the fallout of organic matter and other biogenic materials based on Suess-type equations. For organic carbon, we suggest $J(z) = 0.2\ PP/z$ for depths shallower than 1000 m, where z is in units of 100 m, and $J(z) = 0.17\ PP/z + 0.01\ PP$ for depths greater than 1000 m (Berger et al. 1987).
2. Calculation of the supply of organic matter to the seafloor (z = bottom depth), governing mixing intensity of the sediment, and abundance and type of benthic foraminifera. Mixing intensity and supply of carbon together determine carbon burial, which in turn dominates diagenetic reactions within the sediment (and hence all sorts of sediment properties).
3. Calibration of the Dahlem map against seafloor data (fossils, chemistry, other properties) which can be used for reconstruction of the productivity of past oceans, especially in the Pleistocene. This type of information

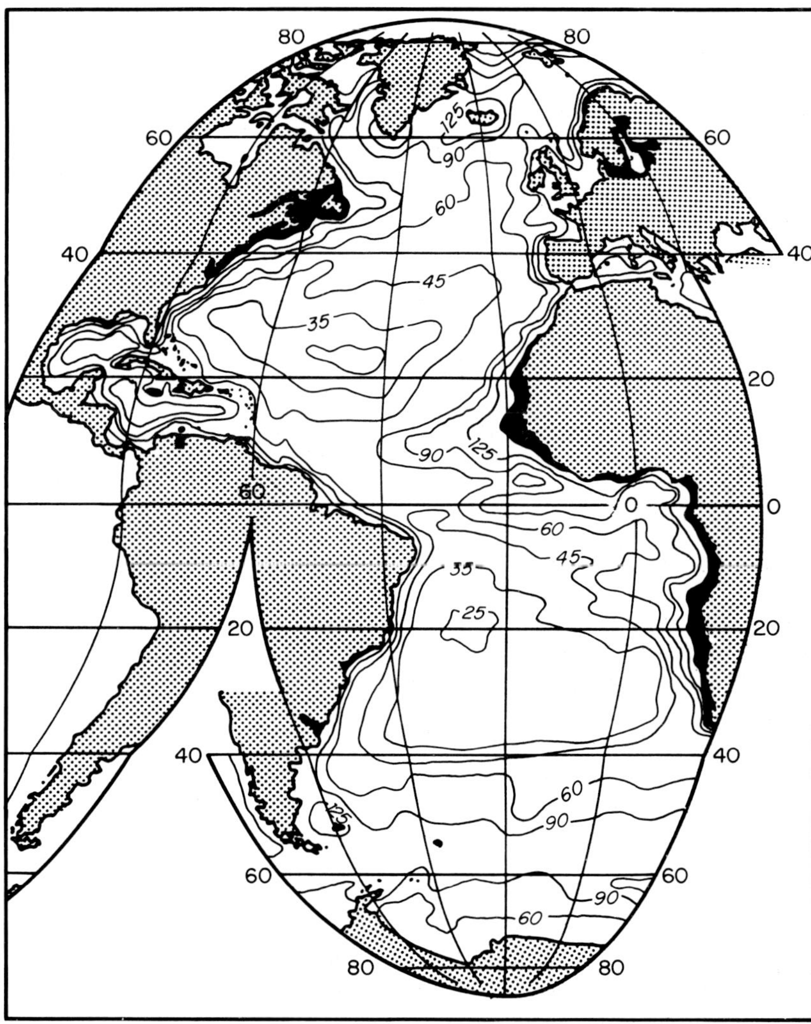

Fig. 11—Productivity map for the Atlantic Ocean ("Dahlem" map), based on LSYNT map (Fig. 9), KVK map (Fig. 2), and satellite information (Esaias et al. 1986; U.S. Global Flux Study Planning Office 1987).

is useful, in turn, for the reconstruction of Pleistocene carbon dioxide levels, for which target values are available in ice cores.

Acknowledgement. A portion of this work was completed while I was in Bremen, and also on board the Forschungsschiff Meteor, at the invitation of Prof. G. Wefer. I profited from discussions with him and with Dr. Richard Eppley, S.I.O. Ms. Evelyn Hegemier (S.I.O.) assisted in preparing the maps for publication.

REFERENCES

Agegian, C., ed. 1988. Biogeochemical Cycling and Fluxes between the Deep Euphotic Zone and other Oceanic Realms. Symp. Series for Undersea Research, NOAA Undersea Res. Prog., vol. 3(2), in press.

Bender, M.; Grande, K.; Johnson, K.; Marra, J.; Williams, P.J. leB.; Sieburth, J.; Pilson, M.; Langdon, C.; Hitchcock, G.; Orchado, J.; Hunt, C.; Donaghay, P.; Heinemann, K. 1987. A comparison of four methods for determining planktonic community production. *Limnol. Ocean.* **32**: 1085–1098.

Berger, W.H.; Fischer, K.; Lai, C.; and Wu, G. 1987. Ocean productivity and organic carbon flux. Part I. Overview and maps of primary production and export production. Univ. of California, San Diego, SIO Reference 87–30.

Bruun, A.F.; Greve, S.; Mielche, H.; and Spaerck, R., eds. 1956. The Galathea Deep Sea Expedition 1950–1952. New York: MacMillan.

Doty, M.S., and Capurro, L.R.A. 1961. Productivity Measurements in the World Ocean. IGY Oceanog. Rep. 4.

El-Sayed, S. 1984. Productivity of the Antarctic waters—a reappraisal. In: Marine Phytoplankton and Productivity, eds. O. Holm-Hansen, L. Bolis, and R. Gilles. Berlin: Springer.

Eppley, R.W.; Stewart, E.; Abbott, M.R.; and Heyman, U. 1985. Estimating ocean primary production from satellite chlorophyll: introduction to regional differences and statistics for the Southern California Bight. *J. Plankton Res.* **7**: 57–70.

Esaias, W.E.; Feldman, G.C.; McClain, C.R.; and Elrod, J.A. 1986. Monthly satellite-derived phytoplankton pigment distribution for the North Atlantic Ocean Basin. *EOS* **67(44)**: 835–837.

FAO Department of Fisheries. 1972. Atlas of the Living Resources of the Seas, 3rd ed. Rome: FAO.

Fleming, R.H. 1957. General features of the ocean. In: Treatise on Marine Ecology and Paleoecology, ed. J.W. Hedgpeth. *Geol. Soc. Amer. Mem.* **67**: 87–107.

Gessner, F. 1959. Hydrobotanik, vol. 2, Stoffhaushalt. Berlin: VEB Deutscher Verlag der Wissenschaften.

Kabanova, Yu.G. 1968. Primary production of the northern part of the Indian Ocean. Translated from Russian. *Oceanol.* **8**: 214–225.

Koblents-Mishke, O.I. 1965. Magnitude of primary production of the Pacific Ocean. In Russian. *Okeanologiia* **5**: 325–337.

Koblents-Mishke, O.I.; Volkovinskiy, V.V.; and Kabanova, Yu.G. 1968. Noviie danniie o velichine pervichnoi produktsii mirovogo okeana. *Doklady Akad. Nauk SSSR* **183**: 1189–1192.

Koblents-Mishke, O.I.; Volkovinsky, V.V.; and Kabanova, Yu.G. 1970. Plankton primary production of the World Ocean. In: Scientific Exploration of the South Pacific, ed. W. Wooster, pp. 183–193. Washington, D.C.: National Academy of Sciences.

Krey, J., and Babenerd, B. 1976. Phytoplankton Production. Atlas of International Indian Ocean Expedition. Kiel: Institut für Meereskunde.

Lieth, H., and Whittaker, R.H. 1975. Primary Productivity of the Biosphere. New York: Springer.

Parsons, T.R.; Takahashi, M.; and Hargrave, B. 1984. Biological Oceanographic Processes, 3rd ed. Oxford: Pergamon.

Platt, T., and Harrison, W.G. 1986. Reconciliation of carbon and oxygen fluxes in the upper ocean. *Deep-Sea Res.* **33**: 273–276.

Platt, T., and Li, W.K.W., eds. 1986. Photosynthetic Picoplankton. *Can. Bull. Fish. Aquat. Sci.* **214**.

Platt, T., and Subba Rao, D.V. 1975. Primary production of marine microphytes. Photosynthesis and productivity in different environments. In: International Biological Programme, vol. 3, pp. 249–279. Cambridge: Cambridge Univ. Press.

Rivkin, R.B., and Putt, M. 1987. Diel periodicity of photosynthesis in polar phytoplankton: influence on primary production. *Science* **238**: 1285–1288.

Romankevich, E.A. 1984. Geochemistry of Organic Matter in the Ocean. Heidelberg: Springer.

Shulenberger, E., and Reid, J.L. 1981. The Pacific shallow oxygen maximum, deep chlorophyll maximum, and primary productivity, reconsidered. *Deep-Sea Res.* **28A**: 901–919.

Smith, R.E.H.; Geider, R.J.; and Platt, T. 1984. Microplankton productivity in the oligotrophic ocean. *Nature* **311**: 252–254.

Steemann Nielsen, E., and Aabye Jensen, E. 1957. Primary oceanic production, the autotrophic production of organic matter in the oceans. *Galathea Rep.* **1**: 49–136.

Sundquist, E.T. 1985. Geological perspectives on carbon dioxide and the carbon cycle. In: The Carbon Cycle and Atmospheric CO_2: Natural Variations Archean to Present, eds. E.T. Sundquist and W.S. Broecker. *Geophys. Monog.* **32**: 5–59. Washington, D.C.: Amer. Geophys. Union.

U.S. Global Flux Study Planning Office. 1987. Map of relative intensity of spectral lines related to chlorophyll concentrations, based on data collected by NASA Nimbus-7, Dec. 1981. Woods Hole, MA: Woods Hole Oceanographic Institution.

Venrick, E.L.; McGowan, J.A.; Cayan, D.R.; and Hayward, T. 1987. Climate and chlorophyll a: long-term trends in the central North Pacific. *Science* **238**: 70–72.

Zeitzschel, B., ed. 1973. The Biology of the Indian Ocean. New York: Springer.

List of Participants with Fields of Research

ALTENBACH, A.V.
Geologisch-Paläontologisches
Institut und Museum
Universität Kiel
Ludewig Meyn Str. 12
2300 Kiel, F.R. Germany
Ecology of benthic foraminifera

BARRON, J.A.
U.S. Geological Survey, MS 915
345 Middlefield Road
Menlo Park, CA 94025, U.S.A.
Diatom biostratigraphy, Cenozoic paleoceanography and time scales

BATHMANN, U.V.
Marine Planktologie
Institut für Meereskunde
Düsternbrooker Weg 20
2300 Kiel 1, F.R. Germany
Pelagic sedimentation: zooplankton contribution to particle production and modification

BERGER, W.H.
Scripps Institution of
Oceanography, A-015
University of California, San Diego
La Jolla, CA 92093-0218, U.S.A.
Ocean productivity and sedimentation

BIENFANG, P.K.
The Oceanic Institute
Makapuu Point
Waimanalo, HI 96795, U.S.A.
Plankton ecology and sedimentation

BISHOP, J.K.B.
Department of Geochemistry
Lamont-Doherty Geological
Observatory
Palisades, NY 10964, U.S.A.
Marine geochemistry: particulate matter composition, distribution, and fluxes

BRULAND, K.W.
Institute of Marine Sciences
University of California
Santa Cruz
Santa Cruz, CA 95064, U.S.A.
Chemical oceanography, marine geochemistry

CODISPOTI, L.A.
Monterey Bay Aquarium Research
Institute
160 Central Avenue, P.O. Box 20
Pacific Grove, CA 93950-0020,
U.S.A.
Chemical oceanography with emphasis on biologically active compounds that affect marine productivity and atmospheric chemistry

CURRY, W.B.
Department of Geology and
Geophysics
Woods Hole Oceanographic
Institute
Woods Hole, MA 02543, U.S.A.
Marine geology, paleoceanography

DE LANGE, G.J.
Department of Geochemistry
Institute of Earth Sciences
Budapestlaan 4
3584 CD Utrecht, Netherlands
Geochemistry of deep-sea sediments and pore waters

EGLINTON, G.
Organic Geochemistry Unit, School of Chemistry
University of Bristol
Cantock's Close
Bristol BS8 1TS, U.K.
Organic geochemistry and marine organic chemistry; fate of biogenic organic compounds in aquatic environments and in sediments; linking of present-day biota and environmental processes with their fossil equivalents and the geological record

EMERSON, S.R.
School of Oceanography, WB-10
University of Washington
Seattle, WA 98195, U.S.A.
Chemical oceanography, marine geochemistry

EPPLEY, R.W.
Scripps Institution of Oceanography
University of California, San Diego
La Jolla, CA 92093-0218, U.S.A.
Biological oceanography

FELDMAN, G.C.
NASA/Goddard Space Flight Center
Code 636
Greenbelt, MD 20771, U.S.A.
Satellite-derived measurements of global primary production

FISCHER, G.
Geowissenschaften, FB5
Universität Bremen
Bibliothekstr.
2800 Bremen 33, F.R. Germany
Particle fluxes in the ocean, stable carbon isotopes in organic matter, C/N ratios

GERSONDE, R.
Alfred-Wegener-Instiut für Polar- und Meeresforschung
Columbusstr.
2850 Bremerhaven, F.R. Germany
Diatom biostratigraphy; diatom particle flux and Neogene paleoceanography in the Southern high-latitude Ocean

HARGRAVE, B.T.
Biological Sciences
Bedford Institute of Oceanography
P.O. Box 1006
Dartmouth, Nova Scotia B2Y 4AZ, Canada
Benthic energy flow: metabolism and transformation of organic matter in sediments

HERBERT, T.D.
Program in Atmospheric and Oceanic Science
P.O. Box 308
Princeton University
Princeton, NJ 08542, U.S.A.
Paleoceanography

ITTEKKOT, V.A.W.
Geologisch-Paläontologisches Institut
Universität Hamburg
Bundesstr. 55
2000 Hamburg 13, F.R. Germany
Organic matter in natural waters: nature and fluxes

JUMARS, P.A.
School of Oceanography, WB-10
University of Washington
Seattle, WA 98195, U.S.A.
Organism-sediment-flow interactions, biology of deposit feeding

KEIR, R.S.
Geological Research Division
Scripps Institution of Oceanography, A-015
University of California, San Diego
La Jolla, CA 92093, U.S.A.
Modeling of the paleocean circulation and carbon cycle, modeling of diagenesis, chemical oceanography

LAMPITT, R.
Institute of Oceanographic Sciences
Deacon Laboratory
Godalming, Surrey GU8 5U8, U.K.
Deep ocean ecology

LEGENDRE, L.
GIROQ
Département de biologie
Université Laval
Québec, Québec G1K 7P4, Canada
Biological oceanography and numerical ecology

MINSTER, J.-F.
UM 39/GRGS
18 av E. Belin
31055 Toulouse Cedex, France
Chemical tracers of the ocean circulation, transport of chemicals in the ocean, satellite oceanography

MIX, A.C.
College of Oceanography
Oregan State University
Cravallis, OR 97331, U.S.A.
Paleoceanography

MÜLLER, P.J.
Fachbereich Geowissenschaften
Universität Bremen
Bibliothekstr., Postfach 330 440
2800 Bremen 33, F.R. Germany
Organic geochemistry: organic matter preservation in sediments, paleoproductivity, amino acid geochronology

MYCKE, B.
Université Louis Pasteur de Strasbourg
Institut de Chimie
1, rue Blaise Pascal, B.P. 296R8
67008 Strasbourg Cedex, France
Organic geochemistry: organic sulfur compounds, degradation of macromolecular organic matter

PRAHL, F.G.
College of Oceanography
Oregan State University
Corvallis, OR 97331, U.S.A.
Organic geochemistry: the use of molecular biomarkers in environmental study

REIMERS, C.E.
Scripps Institution of Oceanography, A-015
University of California, San Diego
La Jolla, CA 92093, U.S.A.
Sediment geochemistry and oceanography

REYNOLDS, C.S.
Freshwater Biological Association
The Ferry House
Ambleside, Cumbria LA22 0LP, U.K.
Freshwater phytoplankton ecology: factors in species selection in space and time, and resultant impacts at community level

SARNTHEIN, M.
Geologisch-Paläontologisches Institut
Universität Kiel
Olshausenstr. 40–60
2300 Kiel, F.R. Germany
Paleoceanography, paleoclimatology

SCHRADER, H.
Geological Institute, Avd. B
University of Bergen
Allegt. 41
7009 Bergen, Norway
Past/present worldwide coastal updwelling

V.S. SMETACEK
Alfred-Wegener-Instiut für Polar- und Meeresforschung
Am Handelshafen 12
2850 Bremerhaven, F.R. Germany
Biological oceanography: phytoplankton ecology, life cycles, magnitude and mechanisms of vertical flux

STEIGER, T.
Institut für Paläontologie und historische Geologie
Richard-Wagner-Str. 10
8000 München 2, F.R. Germany
Jurassic radiolaria, biofacies of Mesozoic sediments (Atlantic-Tetleys)

STEIN, R.
Institut für Geowissenschaften und Lithosphärenforschung
Universität Giessen
Senckenbergstr. 3
6300 Giessen, F.R. Germany
Marine sedimentology/organic geochemistry: accumulation rates of C_{org} *in oxic and anoxic environments, estimates of paleoproductivity*

SUESS, E.
College of Oceanography
Oregan State University
Corvallis, OR 97331, U.S.A.
Mineralization of organic matter in marine sediments, fluid venting along plate subduction zones

THIEDE, J.
GEOMAR Forschungszentrum für Marine Geowissenschaften an der Christian-Albrechts-Universität
Wischofstr. 1–3, Geb. 4
2300 Kiel 14, F.R. Germany
Paleontology, paleoceanography of the Arctic

THIERSTEIN, H.R.
Geologisches Institut
ETH-Zentrum
8092 Zürich, Switzerland
Paleoceanography

TOGGWEILER, J.R.
GFDL/NOAA
Princeton University
P.O. Box 308
Princeton, NJ 08542, U.S.A.
Ocean chemical cycle models

VON BODUNGEN, B.
Marine Planktologie
Institut für Meereskunde
Düsternbrooker Weg 20
2300 Kiel 1, F.R. Germany
Particle flux in polar seas

WALSH, J.J.
Department of Marine Science
University of South Florida
140 Seventh Avenue South
St. Petersburg, FL 33701, U.S.A.
Coastal oceanography, global cycles of carbon and nitrogen, numerical models of coupled physical-biological processes

WEFER, G.
Geowissenschaften
Universität Bremen
Bibliothekstr.
2800 Bremen 33, F.R. Germany
Particle flux in the oceans, paleoceanography

WILLIAMS, P.J. LeB.
School of Ocean Science
Marine Science Laboratory
Menai Bridge
Anglesey, N. Wales, U.K.
Planktonic metabolism of carbon dioxide and oxygen, microbial metabolism and trophodynamics in the sea

Subject Index

Author Index

Dahlem Konferenzen Workshop Reports

Physical, Chemical and Earth Science Research Reports (PC)

PC 6 Resources and World Development: Energy and Minerals, Water and Land
Editors: D.J. McLaren and B.J. Skinner
0 471 91568 8 958pp 1987

PC 7 The Changing Atmosphere
Editors: F.S. Rowland and I.S.A. Isaksen
0 471 92047 9 304pp 1988

PC 8 The Environmental Record in Glaciers and Ice Sheets
Editors: H. Oeschger and C.C. Langway, Jr.
0 471 92185 8 416pp 1989

Prices on application from the Publisher

WILEY

Dahlem Konferenzen Workshop Reports

Life Sciences Research Reports (LS)

LS 37 Mechanisms of Cell Injury: Implications for Human Health
Editor: B.A. Fowler
0 471 91629 3   480pp   1987

LS 38 The Neural and Molecular Bases of Learning
Editors: J.-P. Changeux and M. Konishi
0 471 91569 6   576pp   1987

LS 39 Sexual Selection: Testing the Alternatives
Editors: J.W. Bradbury and M. Andersson
0 471 91683 8   368pp   1987

LS 40 Biological Perspectives of Schizophrenia
Editors: H. Helmchen and F.A. Henn
0 471 91683 8   368pp   1987

LS 41 Humic Substances and Their Role in the Environment
Editors: F.H. Frimmel and R.F. Christman
0 471 91817 2   286pp   1988

LS 42 Neurobiology of Neocortex
Editors: P. Rakic and W. Singer
0 471 91776 1   480pp   1988

LS 43 Etiology of Dementia of Alzheimer's Type
Editors: A.S. Henderson and J.H. Henderson
0 471 92075 4   264pp   1988

LS 45 Complex Organismal Functions: Intergration and Evolution in Vertebrates
Editors: D.B. Wake and G. Roth
0 471 92375 3   approx 400pp   1989   In Press

LS 46 The Structure and Function of Biofilms
Editors: W.G. Characklis and P.A. Wilderer
In Preparation

Prices on application from the Publisher

WILEY